D0083546

New Theory of the Earth

New Theory of the Earth is an interdisciplinary advanced textbook on all aspects of the interior of the Earth and its origin, composition, and evolution: geophysics, geochemistry, dynamics, convection, mineralogy, volcanism, energetics and thermal history. This is the only book on the whole landscape of deep Earth processes that ties together all the strands of the subdisciplines.

This book is a complete update of Anderson's *Theory of the Earth* (1989). It includes dozens of new figures and tables. A novel referencing system using Googlets is introduced that allows immediate access to supplementary material via the internet. There are new sections on tomography, self-organization, and new approaches to plate tectonics. The paradigm/paradox approach to developing new theories is developed, and controversies and contradictions have been brought more center-stage.

As with the *Theory of the Earth*, this new edition will prove to be a stimulating textbook for advanced courses in geophysics, geochemistry, and planetary science, and a supplementary textbook on a wide range of other advanced Earth science courses. It will also be an essential reference and resource for all researchers in the solid Earth sciences.

Don L. Anderson is Professor (Emeritus) of Geophysics in the Division of Geological and Planetary Sciences at the California Institute of Technology (Caltech). He received his B.S. and D.Sc. (Hon) in Geophysics from Rensselaer Polytechnic Institute (RPI), his M.S. and Ph.D. in Mathematics and Geophysics from Caltech, and Doctors Honoris Causa from the Sorbonne, University of Paris. He was Director of the Seismological Laboratory of the California Institute of Technology from 1967–1989. He is a Fellow of the American Academy of Arts and Sciences, the National Academy of Sciences and the American Philosophical Society. He received the Emil Wiechert Medal of the German Geophysical Society, the Arthur L. Day Gold Medal of the Geological Society of America, the Gold Medal of the Royal Astronomical Society, the Bowie Medal of the American Geophysical Union, the Crafoord Prize of the Royal Swedish Academy of Science and the National Medal of Science. He was installed in the RPI Hall of Fame in 2005. He is a Past President of the American Geophysical Union. Professor Anderson's research centers on the origin, evolution, structure and composition of Earth and other planets, and integrates seismological, solid state physics, geochemical and petrological data. He is also interested in the philosophy and logic of science.

From reviews of the previous edition, *Theory of the Earth*:

". . . *Theory of the Earth* is one of the most important books of the decade . . . Anderson is one of a very small group of scientists who have managed to achieve success in both fields [geophysics and geochemistry], providing a dual experience that makes his book an invaluable survey. *Theory of the Earth*, then, is in part an extensive summary of our current state of knowledge of the Earth's interior, . . . drawing on a wide variety of scientific disciplines including not only geophysics and geochemistry but solid-state physics, astronomy, crystallography and thermodynamics. . . . Both as survey and synthesis, Anderson's text, the first in its field, will be of great benefit to students around the world."

Peter J. Smith, *Department of Earth Sciences, Open University*

"Anderson can be congratulated for producing a document that will be a standard taking-off point for many a future graduate seminar."

William S. Fyfe, *Department of Earth Sciences, University of Western Ontario*

". . . much to the envy of the rest of us, there are a few people within the Earth-science community who are, well fairly superhuman. Don Anderson is one of them – as close to being the complete geophysicist/geochemist as anyone is ever likely to be. *Theory of the Earth*, then, is an extensive summary of practically everything 'known' about the physics, chemistry and physicochemical evolution of the Earth's interior. . . . Anderson has produced a remarkable synthesis of our present understanding of the Earth's interior."

Nature

"The appearance of this book is a major event in geoscience literature. It is a comprehensive statement on the Physics and Chemistry of the Earth by one of the great authorities of our time. It will occupy a prominent place on our bookshelves for the rest of our professional lives. When we get into an argument with colleagues or face a fundamental problem that we are unsure about we will reach for it: "Let's see what Anderson says about that". . . . a very valuable book."

Frank Stacey, author of *Physics of the Earth*

". . . as in all good scientific books, there is strong concentration on themes with which Anderson has been closely identified over a number of years. . . . The scope of the book is most impressive: it will be a constantly useful as a source of information that is otherwise extremely time-consuming to track down."

Joe Cann, *Times Higher Education Supplement*

Pre-publication praise of *New Theory of the Earth*

"Anderson's masterful synthesis in *New Theory of the Earth* builds upon his classic 1989 text, weaving an extraordinary breadth of new perspectives and insights into a cogent, provocative and nuanced vision of our planet's history and inner workings. This is a must-read for all scientists seeking to understand the Earth."

Thorne Lay, *Professor of Earth and Planetary Sciences, University of California, Santa Cruz*

"*New Theory of the Earth* can be highly recommended for the book shelf of any serious student of geodynamics. The book contains a wealth of data on a wide variety of subjects in petrology, geochemistry, and geophysics. It is well written and reads smoothly. . . . Many challenging and stimulating views are presented."

Donald L. Turcotte, *Distinguished Professor, Department of Geology, University of California at Davis*

"Don Anderson is the only Earth scientist with the breadth of knowledge and insight necessary to write this book – a fascinating combination of basic data, explanation of concepts, speculation, and philosophy. Now, almost half a century after the realization of plate tectonics, there are rumblings of dissatisfaction over long-held concepts of plumes and mantle convection that are thought to drive plate tectonics, and Don Anderson is leading the charge. This makes *New Theory of the Earth* an especially provocative and exciting reference for all of us scrambling to understand how the Earth works."

Dean C. Presnall, *Department of Geosciences, University of Texas at Dallas and Geophysical Laboratory, Carnegie Institute of Washington*

"This remarkable book by a master geophysicist should be studied by everyone, from junior graduate student to senior researcher, interested in geodynamics, tectonics, petrology, and geochemistry. Here are all the factors omitted from widely accepted models, to their detriment: truly multidisciplinary physics, geophysics, mineral physics, phase petrology, statistics, and much, much more."

Warren B. Hamilton, *Distinguished Senior Scientist, Department of Geophysics, Colorado School Mines*

"An old adage says that there are no true students of the earth because we dig our small holes and sit in them. This book is a striking counter example that synthesizes a broad range of topics dealing with the planet's structure, evolution, and dynamics. Even readers who disagree with some of the arguments will find them insightful and stimulating."

Seth Stein, *William Deering Professor of Geological Sciences, Northwestern University*

New Theory of the Earth

Don L. Anderson

Seismological Laboratory, California Institute of Technology, Pasadena, California

Property of Library
Cape Fear Community College
Wilmington, NC

CAMBRIDGE UNIVERSITY PRESS
Cambridge, New York, Melbourne, Madrid, Cape Town, Singapore, São Paulo

Cambridge University Press
The Edinburgh Building, Cambridge CB2 8RU, UK

Published in the United States of America by Cambridge University Press, New York

www.cambridge.org
Information on this title: www.cambridge.org/9780521849593

© D. L. Anderson 2007

This publication is in copyright. Subject to statutory exception
and to the provisions of relevant collective licensing agreements,
no reproduction of any part may take place without
the written permission of Cambridge University Press.

First published 2007

Printed in the United Kingdom at the University Press, Cambridge

A catalog record for this publication is available from the British Library

Library of Congress Cataloging in Publication data
Anderson, Don L.
The new theory of the Earth / Don L. Anderson.
p. cm.
Includes bibliographical references and index.
ISBN 0-521-84959-4 (hardback)
1. Earth sciences–Textbooks. I. Anderson, Don L. Theory of the Earth. II. Title.
QE26.3.A53 2006
551.1′1–dc22 2006014726

ISBN-13 978-0-521-84959-3 hardback

Cambridge University Press has no responsibility for the persistence or accuracy of
URLs for external or third-party internet websites referred to in this publication,
and does not guarantee that any content on such websites is, or will remain,
accurate or appropriate.

It was a long time before man came to understand that any true theory of the earth must rest upon evidence furnished by the globe itself and that no such theory could properly be framed until a large body of evidence had been gathered together.

Sir Archibald Geike, 1905

We now know that science cannot grow out of empiricism alone, that in the constructions of science we need to use free invention which only a posteriori can be confronted with experience as to its usefulness . . . the more primitive the status of science is, the more readily can the scientist live under the illusion that he is a pure empiricist.

Albert Einstein

Contents

Preface and Philosophy

A mind is a fire to be kindled, not a vessel to be filled.

Plutarch

Go not where the path leads; go where there is no path and leave a trail.

Ralph Waldo Emerson

Science progresses by interchanging the roles of prejudice, paradox and paradigm. Yesterday's prejudice leads to today's paradox and tomorrow's 'truth.' An accumulation of paradoxes, enigmas and coincidences means that it is time to step back and start anew. Plate tectonics, mantle convection, isotope geochemistry and seismic tomography are now mature sciences, but they share an uncomfortable coexistence. They are all part of what may be described as the not-yet-unified standard model of mantle dynamics. Evidence for this disunification is the number of times that the words *paradox, enigma, surprise, unexpected, counter-intuitive* and *inconsistent* appear in the current literature of mantle geochemistry and tomography, and the number of meetings dedicated to solving 'long standing paradoxes' between geophysics and geochemistry. In the jargon of the day, present models of geodynamics are not *robust*.

The maturing of the Earth sciences has led to a fragmentation into subdisciplines that speak imperfectly to one another. Some of these subdisciplines are field geology, petrology, mineralogy, geochemistry, geodesy and seismology, and these in turn are split into even finer units. The science has also expanded to include the planets and even the cosmos. The practitioners in each of these fields tend to view Earth in completely different ways. Discoveries in one field diffuse only slowly into the consciousness of a specialist in another. In spite of the fact that there is only one Earth, there are more Theories of the Earth than there are of astronomy, particle physics or cell biology where there are uncountable samples of each object. Even where there is cross-talk among disciplines, it is usually in code and mixed with white noise. Too often, one discipline's unproven assumptions or dogmas are treated as firm boundary conditions for a theoretician in a slightly overlapping area. The data of each subdiscipline are usually consistent with a range of hypotheses. More often, the data are completely consistent with none of the standard models. The possibilities can be narrowed considerably as more and more diverse data and ways of thinking are brought to bear on a particular problem. The questions of origin, composition and evolution of the Earth require input from astronomy, cosmochemistry, meteoritics, planetology, geology, petrology, mineralogy, crystallography, fluid dynamics, materials science and seismology, at a minimum. To a student of the Earth, these are artificial divisions, however necessary they are to make progress on a given front. New ways of looking at things, new sciences, keep things lively. Advances in materials science, statistics, chaos theory, far-from-equilibrium thermodynamics, geochemistry and tomography make this an appropriate time to update our theory of the Earth.

The timing is also appropriate in that there is a widespread feeling of crisis and frustration amongst workers in mantle dynamics and geochemistry.

The paradigm of layered mantle convection was established nearly 20 years ago, mostly based on geochemical mass balance and heat budget arguments. It is now stumbling over the difficulty imposed by convection models to maintain a sharp interface in the mantle at mid-depth and by overwhelming tomographic evidence that at least some of the subducting lithospheric plates are currently reaching the core-mantle boundary. The present situation, however, remains frustrating because the reasons why the layered convection model was defended in the first place are still there and do not find a proper answer with the model of homogeneous mantle convection.

(www.theconference.com/JConfAbs/6/Albarede.html)

Recent discoveries in a variety of fields are converging on a simple model of geodynamics and geochemistry that is inconsistent with current widely held views. These developments include noble-gas measurements, mantle tomography, convection simulations, statistics, quantum-mechanical equations of state, age dating, paleomagnetism, petrology and techniques to infer temperatures and small-scale heterogeneity of the mantle. Recognition that density variations as small as 1%, which are unavoidable in the accretion and differentiation of the Earth, can irreversibly stratify the mantle is one such development.

Multidisciplinarity is more essential than ever. But we must also honor the venerable rules of logic and scientific inference. Fallacies and paradoxes are waiting to surprise and annoy us, but they tell us that we are making bad assumptions or that we are living in the wrong paradigm.

A seismologist struggling with the meaning of seismic velocity anomalies beneath various tectonic provinces, or in the vicinity of a deeply subducting slab, is apt to interpret seismic results in terms of temperature variations in a homogeneous, isotropic half-space or relative to a standard model. However, the petrological aspects – variations in mineralogy, crystal orientation or partial melt content – are much more important than temperature. These, in turn, require knowledge of phase equilibria, mineralogy, anisotropy and material properties.

An isotope geochemist, upon finding evidence for several components in the rocks and being generally aware of the geophysical evidence for a crust and a 650 km discontinuity, will tend to interpret the chemical data in terms of ancient isolated reservoirs, a 'normal' mantle source and a lower mantle source. The 'standard' petrological model is a homogeneous peridotite mantle containing about 20% basalt, available as needed, to fuel the midocean ridges with uniform magmas. Exotic basalts are assumed to be from the core–mantle boundary. The crust and shallow mantle may be inhomogeneous, but the rest of the mantle is viewed as well homogenized by convection. Numerous paradoxes occur in the standard 'box' models of mantle geochemistry.

The convection theoretician, for 'simplicity', treats the mantle as a homogeneous fluid or as a two-layered system, with constant physical properties, driven by temperature-induced buoyancy, ignoring melting and phase changes and even pressure. Thermodynamic self-consistency and realistic boundary conditions – such as the inclusion of continents – can completely change the outcome of a convection simulation.

In *New Theory of the Earth* I attempt to assemble the bits and pieces from a variety of disciplines, including new disciplines, which are relevant to an understanding of the Earth. Rocks and magmas are our most direct source of information about the interior, but they are biased toward the properties of the crust and shallow mantle. Seismology is our best source of information about the deep interior; however, the interpretation of seismic data for purposes other than purely structural requires input from solid-state physics and experimental petrology. One cannot look at a few selected color cross-sections of the mantle, dramatic as they are, and infer temperature, or composition or the style of mantle convection. There is not a simple scaling between seismic velocity and temperature.

The new theory of the Earth developed here differs in many respects from conventional views. Petrologist's models for the Earth's interior usually focus on the composition of mantle samples contained in basalts and kimberlites from the shallow mantle. The 'simplest' hypothesis based on these samples is that the observed basalts and peridotites bear a complementary relation to one another, that peridotites are the source of basalts or the residue after their removal, and that the whole mantle is identical in composition to the inferred chemistry of the upper mantle and the basalt source region. The mantle is therefore homogeneous in composition, and thus all parts of the mantle eventually rise to the surface to provide basalts. Subducted slabs experience no barrier in falling through the mantle to the core–mantle boundary.

Geochemists have defined a variety of distinct reservoirs, or source regions, based on imperfect understanding of seismic results and of statistics, particularly of the central limit theorem. Midocean ridge basalts are viewed as a unique component of the mantle, rather than as an average composition of a heterogeneous population. In some models the mantle is still grossly homogeneous but contains blobs of isotopically distinct materials so that it resembles a marble cake. Most of the mantle is generally considered to be accessible, undegassed and nearly primordial in composition.

Seismologists recognize large lateral heterogeneity in the upper mantle and several major seismic discontinuities. The discontinuities represent equilibrium phase changes rather than reservoir boundaries and major changes in mantle chemistry. High-resolution seismic techniques have identified about 10 other discontinuities and numerous small-scale scatterers in the mantle. These could be due to changes in chemistry or rock type. The oceanic and continental lithospheres represent material that is colder, stronger *and* chemically different from the underlying mantle. Recent discoveries include megastructures in the lower mantle – unfortunately called 'megaplumes' – which are not due to temperature variations. Simple physical scaling arguments suggest that these are ancient features; they are not buoyant plumes.

Current Earth paradigms are full of paradoxes and logical fallacies. It is widely believed that the results of seismology and geochemistry for mantle structure are discordant, with the former favoring whole-mantle convection and the later favoring layered convection. However, a different view arises from recognizing effects usually ignored in the construction of these models. Self-compression and expansion affect material properties that are important in all aspects of mantle geochemistry and dynamics, including the interpretation of tomographic images. Pressure compresses a solid and changes physical properties that depend on volume and does so in a highly non-linear way. Intrinsic, anelastic, compositional and crystal structure effects also affect seismic velocities; temperature is not the only parameter. Deep-mantle features may be convectively isolated from upper-mantle processes. Major chemical boundaries may occur near 1000 and 2000 km depths. In contrast to standard geochemical models the deeper layers may not be accessible to surface volcanoes.

Tomographic images are often interpreted in terms of an assumed velocity–density–temperature correlation, e.g. high shear velocities (blue regions) are attributed to cold dense slabs, and low shear velocity (red regions) are interpreted as hot rising blobs. There are many factors controlling shear velocity and some do not involve temperature or density. Cold, dense regions of the mantle, such as eclogite sinkers, can have low shear velocities. Likewise, large igneous provinces and hotspots are usually viewed as results of particularly hot mantle. But the locations and magnitudes of melting anomalies depend on fertility of the mantle and the stress state of the lithosphere, perhaps more so than on temperature.

From their inception, the standard models of petrology and geochemistry, involving a uniform pyrolite mantle, or a layered primordial mantle, have had paradoxes; lead isotopes in general and the lead paradoxes in particular, the helium paradoxes and various heat-flow paradoxes. Paradoxes have, in fact, multiplied since the first edition of *Theory of the Earth* (TOE). New isotopic systems have been brought on line – Os, Hf, W and other short-lived isotopes – and they show that the Earth accreted and differentiated and formed a core in the first tens of millions of years of its existence; the cold undegassed geochemical model does not make sense. Paradoxes are a result of paradigms and assumptions; sometimes we can make progress by dropping assumptions, even cherished ones, and abandoning the paradigm. Sometimes new embellishments and complications to the standard model are made simply to overcome problems, or paradoxes, created by the original unphysical assumptions.

A theme running throughout the first edition of TOE was that the energy of accretion of the Earth was so great, and the melting temperatures and densities of the products so different, that early and extensive – and irreversible – chemical stratification of the Earth is the logical outcome. Basalts and the incompatible elements – including K, U and Th – are expected to be concentrated toward the surface, and dense refractory – and depleted – crystals are expected to settle toward the interior. Although mantle homogeneity may be the simplest hypothesis for mantle geochemists and convection modelers, any scenario that results in a cold origin, primordial reservoirs, or a homogeneous mantle is incredibly complex and contrived. In addition to geology, chemistry and physics, one must understand Occam's Razor and the difference between cause and effect. Another theme was that seismic velocities depend on many things; tomographic images are not temperature maps. The framework was established for interpreting seismic velocities. A heterogeneous mantle, involving eclogite and isotopically enriched domains, was another theme.

The title of this book was not picked casually. The year of the first edition was the two-hundredth anniversary of the publication of *Theory of the Earth; or an Investigation of the Laws Observable in the Composition, Dissolution, and Restoration of Land Upon the Globe* by James Hutton, the founder of modern geology. It was not until much progress had been made in all the physical and natural sciences that geology could possess any solid foundations or real scientific status. Hutton's knowledge of chemistry and mineralogy was considerable, and his powers of observation and generalization were remarkable, but the infancy of the other basic sciences made his *Theory of the Earth* understandably incomplete. In the last century the incorporation of physics, chemistry and biology into geology and the application of new tools of geophysics and geochemistry has made geology a science that would be unrecognizable to the Founder, although the goals are the same. Hutton's uniformitarian principle demanded an enormous time period for the processes he described to shape the surface of the Earth, and Hutton could see that the different kinds of rocks had been formed by diverse processes. These are still valid concepts, although we now recognize catastrophic and extraterrestrial events as well. Hutton's views prevailed over the *precipitation theory*, which held that all rocks were formed by mineral deposits from the oceans. Ironically, a currently emerging view is that crystallization of rocks from a gigantic magma ocean was an important process in times that predate the visible geological record. Uniformitarianism, as an idea, can be carried too far. Episodic and non-steady-state processes, and *evolving self-organized systems*, are the keys to understanding mantle evolution and the onset of plate tectonics.

The new sciences of chaos, far-from-equilibrium thermodynamics, self-organization and *ab initio* equations of state have been applied to deep-Earth problems. Sampling theory and other branches of statistics are starting to threaten some of the cherished dogmas about reservoirs, mantle homogeneity, convection and volcanism. These are new topics in this edition.

The word *theory* is used in two ways. A theory is the collection of facts, principles and assumptions that guide workers in a given field. Well-established theories from physics, chemistry, biology and astrophysics, as well as from geology, are woven into the Earth sciences. Students of the Earth must understand solid-state physics, crystallography, thermodynamics, quantum mechanics, Hooke's Law, optics and, above all, the principles of logical inference. Yet these collections of theories do not provide a theory of the Earth. They provide the tools for unraveling the secrets of Earth and for providing the basic facts which in turn are only clues to how the Earth operates. By assembling these clues we hope to gain a better understanding of the origin, structure, composition and evolution of our planet. This better understanding is all that we can hope for in developing a new theory of the Earth.

NOTE ON REFERENCES

The Web has completely changed the way researchers and students do research, teach and learn. Search engines can be used to supplement textbooks and monographs. Conventional references are included in this book, but occasionally a `Googlet` is inserted with key search words for a given topic. These `Googlets` when used with a search engine can find pertinent recent references, color pictures, movies and further background on the subject of interest. For example, if one wants to investigate the relationship between the Deccan traps and Reunion one can insert [`Reunion Deccan mantleplumes`] into Google. If one wants to know more about shear-wave splitting or the Love Rayleigh discrepancy one just types into Google [shear-wave splitting] or [Love Rayleigh discrepancy]. Often the author and a keyword can replace a list of references e.g. [Anderson tomography]. These convenient Googlets will be sprinkled throughout the text. The use of these is optional and the book can be used without interrogating the Web. But if this resource is available, if used, it can cut down the time required to find references and supplementary material. For the ordinary reader, these Googlets should be no more distracting than *italics* or **boldface** and much less distracting that the usual form of referencing and footnoting. They can be treated as keywords, useful but not essential. The key phrases have been designed so that, when used in a search, the top hits will contain relevant information. There may be, of course, some un-useful and redundant hits in the top five. Supplementary and current material can be found with keywords Don L Anderson and mantleplumes.

NOTE ON THIS EDITION

At the time of writing of the First Edition, there were some assumptions that were holding up progress in the study of the Earth's interior. Seismologists were mainly assuming that the Earth was isotropic and that seismic waves did not depend on frequency or direction. Tomography was a brand-new science. Seismic velocities were assumed to depend mainly on temperature. In Theory of the Earth (TOE) (this is the first Googlet in the book) there were therefore extensive chapters on anisotropy, anelasticity, anharmonicity and asphericity. These are now mainstream sciences and there are monographs on each, so these chapters have been trimmed back. Mantle convection is a branch of thermodynamics but there are textbooks on this venerable science so the chapters on thermo have been trimmed. We are still awaiting a fully thermodynamic self-consistent treatment of mantle convection but a recent mantle convection monograph on mantle convection fills the need for a background on this. One can even find mantle convection movies on the Web. But new topics have moved in to take their place. Scaling relations, top-down convection, self-organization, pressure effects on convection, the eclogite engine, lower crustal delamination, seismic scattering, chemical stratification and variably fertile mantle are issues that are receiving more attention. The perception of mantle plumes and hotspots is currently undergoing a dramatic paradigm shift. *Plate tectonics itself is a more powerful concept than generally believed.* Sampling theory and the roles of reservoirs versus components, and sampling vs. stirring are receiving more attention. The various noble gas paradoxes have forced a rethinking of geochemical models and the assumptions they are based on. These topics are almost completely ignored in current texts, monographs and reviews and therefore receive more emphasis in *New Theory of the Earth*.

I thank my colleagues and students for stimulating discussions over the years and for their numerous contributions to the ideas and materials in this book. I especially appreciate the wisdom of Hiroo Kanamori, Don Helmberger and Adam Dziewonski, but the names of those who contributed in one way or another to my general world view are too numerous to list. Most recently, I have received considerable support and wise council from Jim Natland, Gillian Foulger, Anders Meibom, Jerry Winterer, Seth Stein, Bruce Julian and Dean Presnall. The more direct products of these collaborations can be seen on www.mantleplumes.org, and in Plates, Plumes and Paradigms, which provide much supplementary material to this book. I again acknowledge my debts to Nancy and my family for their patience and understanding.

Abbreviations and acronyms (also see Appendix)

ABM	absorption-band model
AOB	alkali olivine basalts
BAB	backarc basins
BABB	BAB basalts
BSE	bulk silicate Earth
C	common component
C	a Bullen region; the TZ; 400- to 1000-km depth
CC	continental crust
CFB	continental flood basalts
CMB	core–mantle boundary
D	a Bullen region; the lower mantle, starting at 1000-km depth
D′	a Bullen subregion; between 1000-km depth and D″; mesosphere
D″	a Bullen subregion; the lowermost mantle
DM	depleted mantle
DMORB	depleted MORB
DUM	depleted upper mantle; BSE-CC
DUPAL	a geochemical component, possible delaminated lower CC
EH	high iron enstatite chondrites
EM	enriched mantle; EM+DM+CC=BSE
EM1	a geochemical component in basalts; possibly sediments
EM2	a geochemical component; possibly continental in origin
EMORB	enriched MORB
FOZO	FOcal ZOne; a common endmember component of basalts; possibly melted peridotite (see C, PHEM, UMR)
G	gravitational constant
GA	billion years ago
HFSE	high field strength elements
HIMU	a geochemical component of basalts based on Pb isotopes
HREE	heavy rare-earth elements
IAB	island-arc basalts
IAV	island-arc volcanics
IC	inner core
IDP	interplanetary dust particles
KIMB	kimberlite; sometimes the Q component
KREEP	K, REE, and P-rich lunar material
LIL	large ion lithophile
LIP	large igneous province
LM	lower mantle; Bullen's region D; between 1000-km depth and the CMB
LONU	low 3He/(U,Th) ratio; yields low $^3He/^4He$ ratio basalts
LREE	light rare-earth elements
LVZ	low-velocity zone
m	mass
Ma	million years ago
MORB	midocean-ridge basalts
MREE	middle rare-earth elements
NMORB	normal MORB
OC	oceanic crust
OIB	ocean-island basalts
OPB	oceanic-plateau basalts
P	primary seismic wave; 'compressional' wave
PHEM	primary helium mantle component
PKJKP	a seismic wave that traverses the IC as a shear wave
PLUME	primary layer of upper mantle enrichment
PM	primitive mantle
PMORB	plume-type MORB
PN	the P wave that refracts along the top of the mantle
PREM	preliminary reference Earth model
PREMA	prevalent mantle component (see C, FOZO); probably a peridotite
PUM	primitive upper mantle, prior to differentiation
P′P′	a seismic wave that goes through the core and reflects off of the opposite side of the Earth
Q	seismic quality factor; also quintessence or the fifth essential component, and heat flow
QCT	qualitative chromotomography; visual or intuitive interpretations
REE	rare-earth elements
SCLM	subcontinental lithospheric mantle

SH	shear wave, horizontal polarization
SOFFE	self-organized, far-from-equilibrium
SV	shear wave, vertical polarization
SUMA	statistical upper-mantle assemblage
SUMA	sampling upon melting and averaging
TMORB	transitional MORB
TPW	true polar wander
TZ	transition zone; 410- to 650-km
TR	transition region; 410- to 1000-km depth
UDS	undepleted source
UM	upper mantle
UMR	ultramafic rock; a geochemical component

Part I

Planetary perspective

I want to know how God created this world. I am not interested in this or that phenomenon, in the spectrum of this or that element. I want to know his thoughts, the rest are details.

Albert Einstein

Overview

Earth is part of the solar system and it cannot be completely understood in isolation. The chemistry of meteorites and the Sun provide constraints on the composition of the planets. The properties of the planets provide ideas for and tests of theories of planetary formation and evolution. The Earth is often assumed to have been formed by the slow accumulation of planetesimals – small cold bodies present in early solar system history. In particular, types of stony meteorite called *chondrites* have been adopted as the probable primary material accreted by the Earth. This material, however, has to be extensively processed before it is suitable.

Study of the Moon, Mars and meterorites demonstrates that melting and basaltic volcanism is ubiquitous, even on very small bodies. Planets form hot, or become hot, and begin to differentiate at a very early stage in their evolution, probably during accretion. Although primitive objects have survived in space for the age of the solar system, there is no evidence for the survival of primitive material once it has been in a planet. One would hardly expect large portions of the Earth to have escaped this planetary differentiation, and to be 'primordial' and undegassed. The present internal structure of the Earth was mainly established 4.57 billion years ago. This is not a central dogma of current geochemical models but the use of high-precision short-lived isotope data promises to change this.

A large amount of gravitational energy is released as particles fall onto an accreting Earth, enough to raise the temperature by tens of thousands of degrees and to evaporate the Earth back into space as fast as it forms. Melting and vaporization are likely once the proto-Earth has achieved a given size. The mechanism of accretion and its time scale determine the fraction of the heat that is retained, and therefore the temperature and heat content of the growing Earth. The 'initial' temperature of the Earth was high. A rapidly growing planet retains more of the gravitational energy of accretion, particularly if there are large impacts.

The magma-ocean concept was developed to explain the petrology and geochemistry of the Moon. It proved fruitful to apply this to the Earth, taking into account the petrological differences required by the higher pressures on the Earth.

We now know that plate tectonics, at least the recycling kind, is unique to Earth, perhaps because of its size or water content. The thickness and average temperature of the lithosphere and the role of phase changes in basalt are important. *Any theory of plate tectonics must explain why the other terrestrial planets do not behave like Earth.* (Reminder: key words are embedded in the text, with the type face of the preceding sentence. These words and phrases can be entered into search engines to obtain background material, definitions and references.)

Chapter 1

Origin and early history

Earth is the namesake of the terrestrial planets, also known as the inner or rocky planets. The chemistry of meteorites and the Sun provide constraints on the composition of the bulk of these planets and they provide tests of theories of planetary formation and evolution. In trying to understand the origin and structure of the Earth, one can take the geocentric approach or the *ab initio* approach. In the former, one describes the Earth and attempts to work backward in time. For the latter, one attempts to track the evolution of the solar nebula through collapse, cooling, condensation and accretion, hoping that one ends up with something resembling the Earth and other planets. Planets started hot and had a pre-history that cannot be ignored. The large-scale chemical stratification of the Earth reflects accretionary processes.

Condensation of the nebula

The equilibrium assemblage of solid compounds that exists in a system of solar composition depends on temperature and pressure and, therefore, location and time. The condensation behavior of the elements is given in Figures 1.1 and 1.2.

At a nominal nebular pressure of 10^{-1} atm, the material would be a vapor at temperatures greater than about 1900 K. The first solids to condense at lower temperature or higher pressure are the refractory metals (such as W, Re, Ir and Os). Below about 1750 K refractory oxides of aluminum, calcium, magnesium and titanium condense, and metallic iron condenses near 1470 K (Table 1.1 and Figure 1.2). Below about 1000 K, sodium and potassium condense as feldspars, and a portion of the iron is stable as fayalite and ferrosilite with the proportion increasing with a further decrease in temperature. FeS condenses below about 750 K. Hydrated silicates condense below about 300 K.

Differences in planetary composition may depend on the location of the planet, the location and width of its feeding zone and the effects of other planets in sweeping up material or perturbing the orbits of planetesimals. In general, one would expect planets closer to the Sun and the median plane of the nebula to be more refractory rich than the outer planets. On the other hand, if the final stages of accretion involve coalescence of large objects of different eccentricities, then there may be little correspondence between bulk chemistry and the present position of the terrestrial planets (Table 1.2).

There is evidence that the most refractory elements condensed from the solar nebula as a group, unfractionated from one another, at temperatures above the condensation temperature of the Mg-silicates. Hence, the *lithophile refractory elements* (Al, Ca, Ti, Be, Sc, V, Sr, Y, Zr, Nb, Ba, rare-earth elements, Hf, Ta, Th and U and, to some extent, W and Mo) can be treated together. From the observed abundance in samples from the Moon, Earth and achondrites, there is strong support for the idea that these elements are present in the same ratios as in Cl chondrites. The abundance of the refractory elements in a given planet can be weakly constrained from

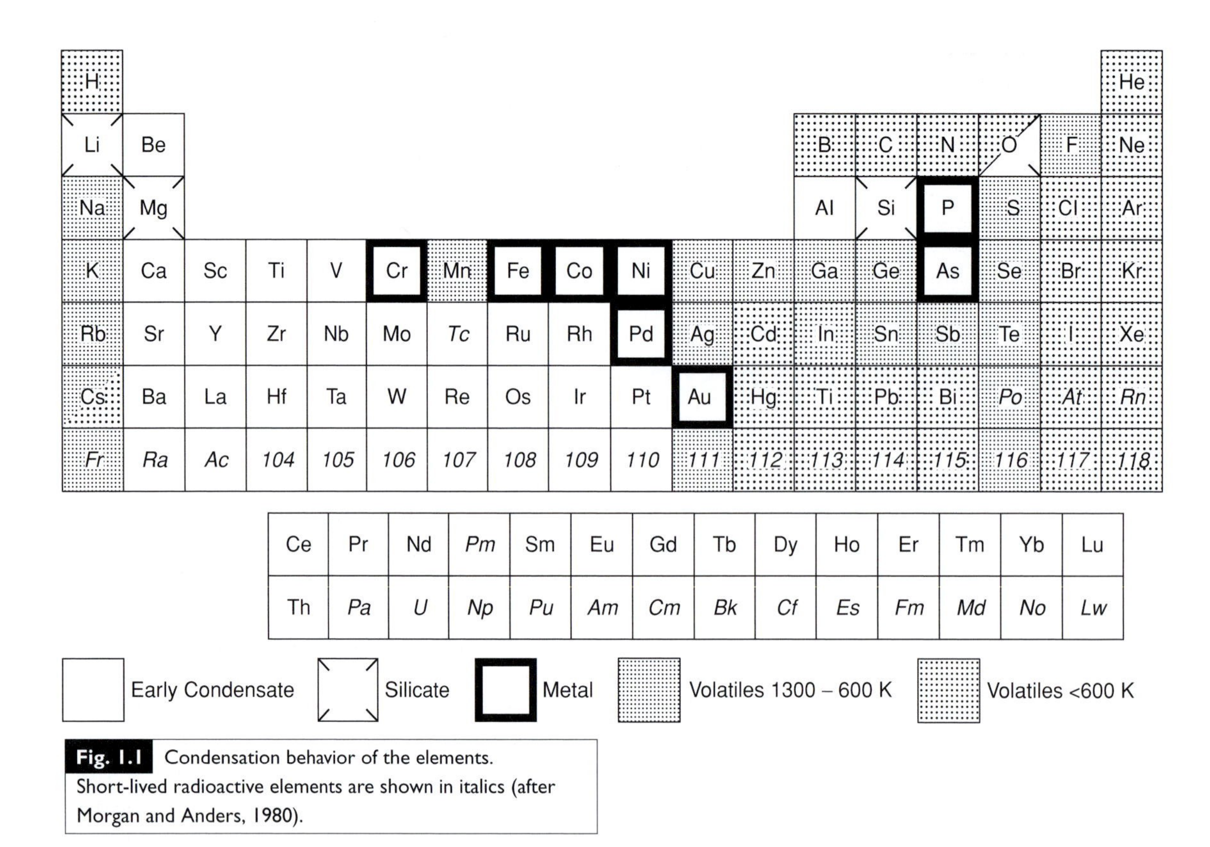

Fig. 1.1 Condensation behavior of the elements. Short-lived radioactive elements are shown in italics (after Morgan and Anders, 1980).

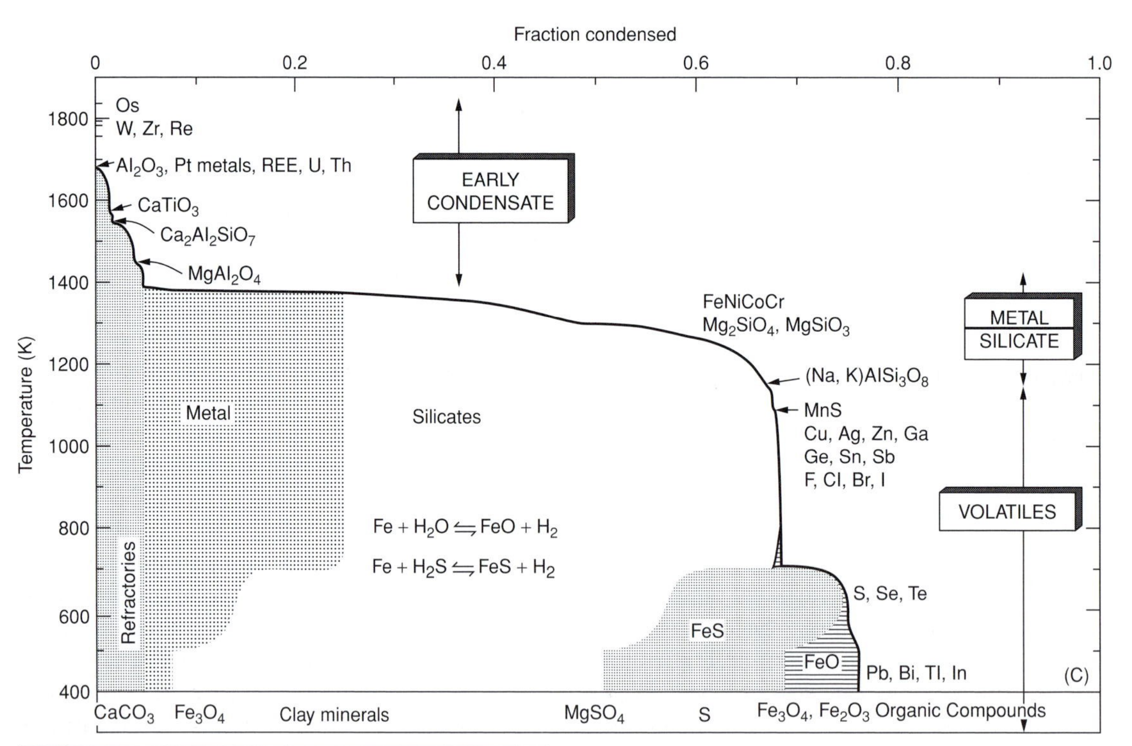

Fig. 1.2 Condensation of a solar gas at 10^{-4} atm (after Morgan and Anders, 1980)

Table 1.1 Approximate sequence of condensation of phases and elements from a gas of solar composition at 10^{-3} atm total pressure

Phase	Formula	Temperature
Hibonite	$CaAl_{12}O_{19}$	1770 K
Corundum	Al_2O_3	1758 K
Platinum	Pt, W, Mo, Ta	
metals	Zr, REE, U, Th	
	Sc, Ir	
Perovskite	$CaTiO_3$	1647 K
Melilite	$Ca_2Al_2SiO_7$–	
	$Ca_2Mg_2Si_2O_7$	1625 K
	Co	
Spinel	$MgAl_2O_4$	1513 K
	Al_2SiO_5	
Metallic iron	Fe, Ni	1473 K
Diopside	$CaMgSi_2O_6$	1450 K
Forsterite	Mg_2SiO_4	1444 K
Anorthite	$CaAl_2Si_2O_8$	1362 K
	Ca_2SiO_4	
	$CaSiO_3$	
Enstatite	$MgSiO_3$	1349 K
	Cr_2O_3	
	P, Au, Li	
	$MnSiO_3$	
	MnS, Ag	
	As, Cu, Ge	
Feldspar	$(Na,K)AlSi_3O_8$	
	Ag, Sb, F, Ge	
	Sn, Zn, Se, Te, Cd	
Reaction	$(Mg,Fe)_2SiO_4$	1000 K
products	$(Mg,Fe)SiO_3$	
Troilite,	FeS, (Fe, Ni)S	700 K
pentlandite	Pb, Bi, In, Tl	
Magnetite	Fe_3O_4	405 K
Hydrous	$Mg_3Si_2O_7 2H_2O$, etc.	
minerals		
Calcite	$CaCO_3$	<400 K
Ices	H_2O, NH_3, CH_4	<200 K

Anders (1968), Grossman (1972), Fuchs and others (1973), Grossman and Larimer (1974).

the inferred abundance of their heat-producing members, uranium and thorium, and the global heat flux. But the present surface heat flow does not accurately represent the current rate of heat production. A large fraction of the present heat

Table 1.2 Properties of the terrestrial planets

	GM 10^{18} cm^3/s^2	R km	ρ g/cm^3	I/MR^2	D* km
Earth	398.60	6371	5.514	0.3308	14
Moon	4.903	1737	3.344	0.393	75
Mars	42.83	3390	3.934	0.365	>28
Venus	324.86	6051	5.24	?	?
Mercury	22.0	2440	5.435	?	?

*Estimated crustal thickness.

flow is due to cooling of the Earth, which means that only an upper bound can be placed on the uranium and thorium content. Nevertheless, this is a useful constraint particularly when combined with the lower bound on potassium provided by argon-40 and estimates of K/U and Th/U provided by magmas and the crust. There is little justification for assuming that the volatile elements joined the planets in constant proportions. In this context the volatiles include the alkali metals, sulfur and so forth in addition to the gaseous species.

Theories of planetary formation

The nature and evolution of the solar nebula and the formation of the planets are complex subjects. The fact that terrestrial planets did in fact form is a sufficient motivation to keep a few widely dispersed scientists working on these problems. There are several possible mechanisms of planetary growth. Either the planets were assembled from smaller bodies (planetesimals), a piece at a time, or diffuse collections of these bodies, clouds, became gravitationally unstable and collapsed to form planetary-sized objects. The planets, or protoplanetary nuclei, could have formed in a gas-free environment or in the presence of a large amount of gas that was subsequently dissipated. Some hypotheses speculate that large amounts of primordial helium dissolved in an early molten Earth. Others assume that the bulk of the Earth assembled gas-free and volatiles were brought in later. The intermediate stages of planetary assembly involved impacts of large objects. The final stages involved sweeping up the debris and collecting an outer veneer of exotic materials from the Sun and the outer solar system.

The planets originated in a slowly rotating disk-shaped 'solar nebula' of gas and dust with solar composition. The temperature and pressure in the hydrogen-rich disk decreased radially from its center and outward from its plane. The disk cooled by radiation, mostly in the direction normal to the plane, and part of the incandescent gas condensed to solid 'dust' particles. As the particles grew, they settled to the median plane by collisions with particles in other orbits, by viscous gas drag and gravitational attraction by the disk. The total gas pressure in the vicinity of Earth's orbit may have been of the order of 10^{-1} to 10^{-4} of the present atmospheric pressure. The particles in the plane formed rings and gaps. The sedimentation time is rapid, but the processes and time scales involved in the collection of small objects into planetary-sized objects are not clear. Comets, some meteorites and some small satellites may be left over from the early stages of accretion.

The accretion-during-condensation, or inhomogeneous-accretion, hypothesis leads to radially zoned planets with refractory and iron-rich cores, and a compositional zoning away from the Sun; the outer planets are more volatile-rich because they form in a colder part of the nebula. Superimposed on this effect is a size effect: the larger planets, having a larger gravitational cross section, collect more of the later condensing (volatile) material but they also involve more gravitational heating.

In the widely used `Safronov cosmogonical theory` (1972) it is assumed that the Sun initially possessed a uniform gas–dust nebula. The nebula evolves into a torus and then into a disk. Particles with different eccentricities and inclinations collide and settle to the median plane within a few orbits. As the disk gets denser, it breaks up into many dense accumulations where the self-gravitation exceeds the disrupting tidal force of the Sun. As dust is removed from the bulk of the nebula, the transparency of the nebula increases, and a large temperature gradient is established.

If the relative velocity between planetesimals is high, fragmentation rather than accumulation will dominate and planets will not grow. If relative velocities are low, the planetesimals will be in nearly concentric orbits and the collisions required for growth will not take place. For plausible assumptions regarding dissipation of energy in collisions and size distribution of the bodies, mutual gravitation causes the mean relative velocities to be only somewhat less than the escape velocities of the larger bodies. Thus, throughout the entire course of planetary growth, the system regenerates itself such that the larger bodies would always grow. The formation of the giant planets, however, may have disrupted planetary accretion in the inner solar system and the asteroid belt.

The initial stage in the formation of a planet is the condensation in the cooling nebula. The first solids appear in the range 1750–1600 K and are oxides, silicates and titanates of calcium and aluminum and refractory metals such as the platinum group. These minerals (such as corundum, perovskite, melilite) and elements are found in white inclusions (chondrules) of certain meteorites, most notably in Type III carbonaceous chondrites. These are probably the oldest surviving objects in the solar system. Metallic iron condenses at relatively high temperature followed shortly by the bulk of the silicate material as forsterite and enstatite. FeS and hydrous minerals appear at very low temperature, less than 700 K. Volatile-rich carbonaceous chondrites have formation temperatures in the range 300–400 K, and at least part of the Earth must have accreted from material that condensed at these low temperatures. The presence of He, CO_2 and H_2O in the Earth has led some to propose that the Earth is made up almost entirely of cold carbonaceous chondritic material – the *cold-accretion hypothesis*. Even in some current geochemical models, the lower mantle is assumed to be gas-rich, and is speculated to contain as much helium as the carbonaceous chondrites. This is unlikely. The volatile-rich material may have come in as a late veneer – the inhomogenous accretion hypothesis. Even if the Earth accreted slowly, compared to cooling and condensation times, the later stages of accretion could involve material that condensed further out in the nebula and was later perturbed into the inner solar system. A drawn-out accretion time does not imply a cold initial condition. Large impacts reset the thermometer.

The early history of planets was a very violent one; collisions, radioactive heat and core formation provided enough energy to melt the planet. Cooling and crystallization of the planet over timescales of millions of years resulted in its chemical differentiation – segregation of material according to density. This differentiation left most of the Earth's mantle different in composition from that part of the mantle from which volcanic rocks are derived. There must be material that is complementary in composition to the materials sampled by volcanoes.

The Earth and the Moon are deficient in the very volatile elements that make up the bulk of the Sun and the outer planets, and also the moderately volatile elements such as sodium, potassium, rubidium and lead. Mantle rocks contain some `primordial noble gas isotopes`. (*Reminder*: `primordial noble gas isotopes` *is a Googlet. If it is typed into a search engine it will return useful information on the topic, including definitions and references. These Googlets will be sprinkled throughout the text to provide supplementary information.*) The noble gases and other very volatile elements were most likely brought in after the bulk of the Earth accreted and cooled. The ^{40}Ar content of the atmosphere demonstrates that the Earth is an extensively degassed body; the atmosphere contains about 70% of the ^{40}Ar produced by the decay of ^{40}K over the whole age of the Earth. This may imply that most of the K and other incompatible elements are in the crust and shallow mantle.

Magma ocean

A large amount of gravitational energy is released as particles fall onto an accreting Earth, enough to evaporate the Earth back into space as fast as it forms. Even small objects can melt if they collide at high velocity. The mechanism of accretion and its time scale determine the fraction of the heat that is retained, and therefore the temperature and heat content of the growing Earth. The 'initial' temperature of the Earth was likely to have been high even if it formed from cold planetesimals. A rapidly growing Earth retains more of the gravitational energy of accretion, particularly if there are large impacts that can bury a large fraction of their gravitational energy. Evidence for early and widespread melting on such small objects as the Moon and various meteorite parent bodies attests to the importance of high initial temperatures, and the energy of accretion of the Earth is more than 15 times greater than that for the Moon. The intensely cratered surfaces of the solid planets provide abundant testimony of the importance of high-energy impacts in the later stages of accretion.

During accretion there is a balance between the gravitational energy of accretion, the energy radiated into space and the thermal energy produced by heating of the body. Latent heats associated with melting and vaporization are also involved when the surface temperature gets high enough. The ability of the growing body to radiate away part of the heat of accretion depends on how much of the incoming material remains near the surface and how rapidly it is covered or buried. Devolatization and heating associated with impact generate a hot, dense atmosphere that serves to keep the surface temperature hot and to trap solar radiation. One expects the early stages of accretion to be slow, because of the small gravitational cross section and absence of atmosphere, and the terminal stages to be slow, because the particles are being used up. The temperature profile resulting from this growth law gives a planet with a cold interior, a temperature peak at intermediate depth, and a cold outer layer. Superimposed on this is the temperature increase with depth due to self-compression and possibly higher temperatures of the early accreting particles. However, large late impacts, even though infrequent, can heat and melt the upper mantle. Formation of 99% of the mass of Earth probably took place in a few tens of millions of

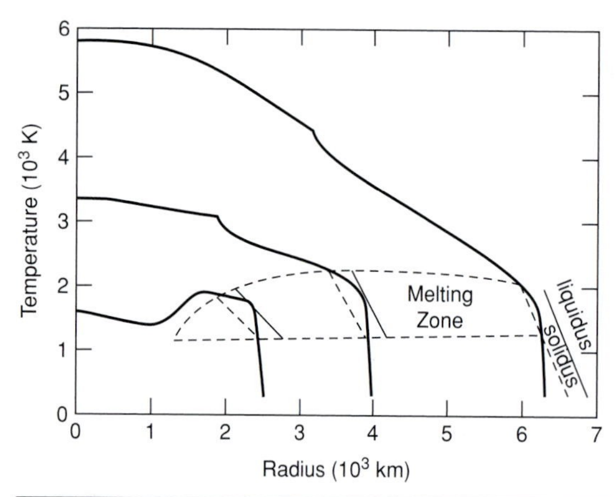

Fig. 1.3 Schematic temperatures as a function of radius at three stages in the accretion of a planet (heavy lines). Temperatures in the interior are initially low because of the low energy of accretion. The solidi and liquidi and the melting zone in the upper mantle are also shown. Upper-mantle melting and melt-solid separation is likely during most of the accretion process. Silicate melts, enriched in incompatible elements, will be concentrated toward the surface throughout accretion. The Earth, and perhaps the mantle, will be stratified by intrinsic density, during and after accretion. The Melting Zone in the upper mantle or a near-surface magma ocean processes accreting material. Temperature estimates provided by D. Stevenson.

years, around 4.55 billion years ago. The core was forming during accretion and was already in place by its end. There was likely not a `core-forming event`.

Accretional calculations, taking into account the energy partitioning during impact, have upper-mantle temperatures in excess of the melting temperature during most of the accretion time (Figure 1.3). If melting gets too extensive, the melt moves toward the surface, and some fraction reaches the surface and radiates away its heat. A hot atmosphere, a thermal boundary layer and the presence of chemically buoyant material at the Earth's surface, however, insulates most of the interior, and cooling is slow. Extensive cooling of the interior can only occur if cold surface material is subducted into the mantle. This requires a very cold, thick thermal boundary layer that is denser than the underlying mantle. This *plate tectonic mode* of mantle convection – with subduction and recycling – may only extend back into Earth history about 1 Ga (10^9 years ago). An extensive accumulation of basalt or olivine near the Earth's surface during accretion forms a buoyant layer that resists subduction. An extensively molten, slowly cooling, upper mantle, and a very slowly cooling deeper mantle are predicted.

A magma ocean freezes from the bottom but a thin chill layer may form at the surface. As various crystals freeze out of the ocean they will float or sink, depending on their density. On the Moon, plagioclase crystals float when they freeze and this is one explanation of the anorthositic highlands. On the much larger Earth, the aluminum enters dense garnet crystals and a deep eclogite-rich layer is the result. Although a magma ocean may be convecting violently when it is hot, or being stirred by impacts, at some point it cools through the crystallization temperatures of its components and the subsequent gravitational layering depends on the relative cooling rate and sinking rates of the crystals. Meanwhile, new material is being added from space and is processed in the magma ocean. A chemically stratified Earth is the end result. Accretional and convective stirring is unlikely to dominate over gravitational settling.

Magma is one of the most buoyant products of mantle differentiation and will tend to stay near the surface. A hot, differentiated planet cools by the `heat-pipe cooling mechanism of mantle convection`; pipes or sheets of magma remove material from the base of the proto-crust and place it on top of the basaltic pile, which gets pushed back into the mantle, cooling the interior. As a thick basalt crust cools, the lower portions eventually convert to dense eclogite – instead of melting – and delaminate. This also cools off the interior. As the surface layer cools further, the olivine-rich and eclogitic parts of the outer layer become denser than the interior and subduction initiates. At this point portions of the upper mantle are rapidly cooled and the thermal evolution of the Earth switches over to the plate-tectonic era. Plate tectonics is a late-stage method for cooling off the interior, but it is restricted to those parts of the interior that are less dense than slabs. A dense primitive atmosphere and buoyant outer layers are effective insulators and serve to keep the crust and upper mantle from cooling and crystallizing as rapidly as a homogenous fluid, radiating to outer space.

On a large body, such as the Earth, the dense mineral garnet forms and sinks into the interior; on a small body plagioclase forms and rises to the surface. In both cases an aluminum-poor, residual mantle forms, composed of olivine and pyroxene. The chemical stratification that forms during accretion and magma ocean crystallization may be permanent features of the planet. The importance of these processes during the earliest history of the planet cannot be over-emphasized. No part of the interior is likely to have escaped extensive heating, melting and degassing. What happened at high temperature and relatively low pressure is unlikely to be reversed.

The 'initial' state of a planet

Partial differential equations require boundary conditions and initial conditions; so do geodynamic and evolutionary models. The present surface boundary condition of the Earth is a continuously evolving system of oceanic and continental plates. The initial condition usually adopted employs one edge of Occam's razor; *the mantle started out cold and homogenous and remains homogenous today*. The more probable initial condition is based on the other edge of Occam's razor. Although a homogenous mantle with constant properties is the simplest imaginable assumption about the *outcome*, it is not consistent with a simple *process*. No one has simply explained how the mantle may have arrived at such a state, except by slow, cold, homogenous accretion. This is an unstated assumption in the standard models of mantle geochemistry. The accretion of Earth was more likely to have been a violent high temperature process that involved repeated melting and vaporization and the probable end result was a hot, gravitationally differentiated body.

That the Earth itself is efficiently differentiated there can be no doubt. Most crustal elements are in the crust, possibly all the ^{40}Ar – depending on the uncertain potassium content – is in the atmosphere and most of the siderophile elements – such as Os, Ir – are in the core. Given these circumstances, it is probable that the mantle is also zoned by chemistry and density. Large-degree melts from primitive mantle can have relatively unfractionated ratios of such elements as Sm, Nd, Lu and Hf, giving 'chondritic' isotope ratios. This has confused the issue regarding the possible presence of *primordial unfractionated reservoirs*.

The assumed starting composition for the Earth is usually based on *cosmic* or *meteoritic abundances*. The refractory parts of carbonaceous, ordinary or enstatite chondrites are the usual choices. These compositions predict that the lower mantle has more silicon than the olivine-rich buoyant shallow mantle and that only a small fraction of the mantle, or even the upper mantle, can be basaltic. The volatile components that are still in the Earth were most likely added to Earth as a late veneer after most of the mass had already been added and the planet had cooled to the point where it could retain volatiles.

A process of RAdial ZOne Refining (RAZOR) during accretion may remove incompatible and volatile elements and cause purified dense materials to sink. Crystallizing magma oceans at the surface are part of this process. The formation of a deep reservoir by perovskite fractionation in a magma ocean is not necessary. The magma ocean may always have been shallower than the perovskite-phase boundary – roughly 650 km depth – but as the Earth accretes, the deeper layers will convert to high-pressure phases. There is no need for material in the upper mantle to have been in equilibrium with the dense phases that now exist at depth.

Prior to the era of plate tectonics, the Earth was probably surfaced with thick crustal layers, which only later became dense enough to sink into the mantle. But because of the large stability field of garnet, there is a subduction barrier, currently near 600 km. The great buoyancy of young and thick oceanic crust, particularly oceanic plateaus, dehydration of recycled material, the low melting temperature of eclogite, and the subduction barrier to eclogite (and harzburgite) probably prevents formation of deep fertile and radioactive layers, even after the onset of plate tectonics.

The RAZOR process sets the initial stage for mantle evolution, including the distribution of radioactive elements. This step is often overlooked in geochemical and geodynamic models; it is usually assumed that most of the radioactive elements are still in the deep mantle. The initial temperatures may have been forgotten but the stratification of major and radioactive elements may be permanent.

Evolution of a planet

Isotopic studies indicate that distinct geochemical components formed in the mantle early in its history. *Zone refining during accretion* and crystallization of a deep magma ocean are possible ways of establishing a chemically zoned planet (Figure 1.4). At low pressures basaltic melts are less dense than the residual refractory crystals, and they rise to the surface, taking with them many of the trace elements. The refractory crystals themselves are also less dense than undifferentiated mantle and tend to concentrate in the shallow mantle.

As the Earth accretes and grows, the crustal elements are continuously concentrated into the melts and rise to the surface. When these melts freeze, they form the crustal minerals that are rich in silicon, calcium, aluminum, potassium and the large-ion lithophile (LIL) elements. Melts generally are also rich in FeO compared to primitive material. This plus the high compressibility of melts means that the densities of melts and residual crystals converge, or even cross, as the pressure increases. They cross again as phase changes increase the density of the solids. Melt separation is therefore difficult at depth, and melts may even drain downward at very high pressure, until the silicate matrix undergoes a phase change. During accretion the majority of the melt-crystal separation occurs at low pressure. All of the material in the deep interior has passed through this low-pressure melting stage in a sort of continuous zone refining. The magnesium-rich minerals, Mg_2SiO_4 and $MgSiO_3$, have high melting temperatures and are fed through the melting zone into the interior. Even if the accreting material is completely melted during assembly of the Earth, these minerals will be the first to freeze, and they will still separate from the remaining melt. The downward separation of iron-rich melts, along with nickel, cobalt, sulfur and the trace siderophile elements, strips these elements out of the crust and mantle.

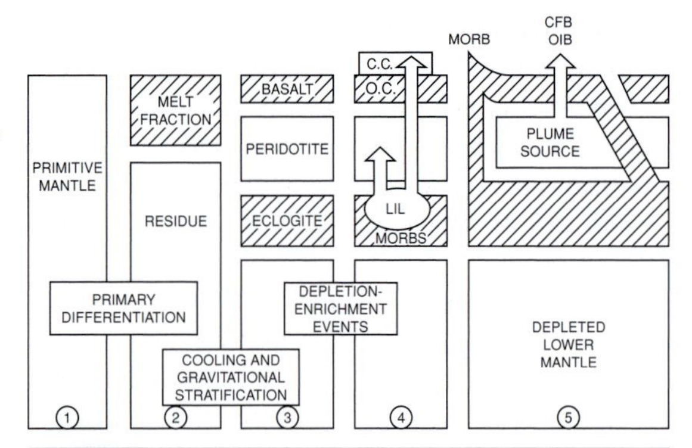

Fig. 1.4 A model for the early evolution of the mantle. Primitive mantle (1) is partially molten either during accretion or by subsequent whole-mantle convection, which brings the entire mantle across the solidus at shallow depths. Large-ion lithophile (LIL) elements are concentrated in the melt. The deep magma ocean (2) fractionates into a thin plagioclase-rich surface layer and deeper olivine-rich and garnet-rich cumulate layers (3). Late-stage melts in the eclogite-rich cumulate are removed (4) to form the continental crust (C.C.), enrich the shallow peridotite layer and deplete MORBs, the source region of oceanic crust (O.C.) and lower oceanic lithosphere. Partial melting of PLUME – or Primary Layer of Upper Mantle Enrichment – in the upper mantle (5) generates continental flood basalts (CFB), ocean-island basalts (IOB) and other enriched magmas, leaving a depleted residue (harzburgite) layer – perisphere – that stays in the upper mantle because of its buoyancy. Enriched or hot-spot magmas (EMORB, OIB, CFB) may be from a shallow part of the mantle and may represent delaminated C.C. Most of the mantle has been processed through the melting zone and is depleted in the heat-producing elements such as U and Th, which are now in the crust and upper mantle.

The aluminum, calcium, titanium and sodium contents in chondritic and solar material, restrict the amount of basalt that can be formed, but are adequate to form a crust some 200 km thick. The absence of such a massive crust on the Earth might suggest that the Earth has not experienced a very efficient differentiation. On the other hand, the size of the core and the extreme concentration of the large-ion, magmaphile elements into the crust suggest that *differentiation has been extremely efficient*. The lack of a thick plagioclase-rich crust on Earth has been used as an argument that early Earth, in contrast to the early Moon, did not have a magma ocean.

There is an easy solution to this apparent paradox. At pressures corresponding to depths of the order of 50 km, the low-density minerals of the crust convert to a mineral assemblage denser than olivine and orthopyroxene. Most of the original crust therefore is unstable and sinks into the mantle. Any magma below 200–400 km may suffer the same fate. Between 50 and 500 km the Al_2O_3–CaO–Na_2O-rich materials crystallize as clinopyroxene and garnet, a dense eclogite assemblage that is denser than peridotite. Eclogite transforms to a garnet solid solution, which is still denser and which is stable between about 500 and 800 km, depending on temperature. Peridotite also undergoes a series of phase changes that prevent most eclogites from sinking deeper than about 650 km. There is likely to be a midmantle garnet-rich layer in the mantle. This would explain both the absence of a thick crust, and the presence of an olivine-rich shallow mantle. Mars may also have a perched garnet-rich layer. However, when the planets are much smaller than Mars-size, eclogite could sink to the core–mantle boundary.

Subsequent cooling and crystallization of the Earth introduces additional complications. A chemically stratified mantle cools more slowly than a homogenous Earth. Phase change boundaries are both temperature and pressure dependent, and these migrate as the Earth cools. The initial crust of the Earth, or at least its deeper portions, can become unstable and plunge into the mantle. This is an effective way to cool the mantle and to displace lighter and hotter material to the shallow mantle where it can melt by pressure release – `adiabatic decompression melting` – providing a continuous mechanism for bringing melts to the surface. This mechanism can also, on the smaller planets, cool the core and, perhaps, start a dynamo.

The separation of melts and crystals is a process of differentiation. Convection is often thought of as a homogenization process, tantamount to stirring. Differentiation, however, can be irreversible. Melts that are separated from the mantle when the Earth was smaller, or from the present upper mantle, crystallize to assemblages that have different phase relations than the residual crystals or original mantle material. If these rocks are returned to the mantle, they will not in general have neutral buoyancy, nor are they necessarily denser than "normal" mantle at all depths. Eclogite, for example, is denser than peridotite when the latter is in the olivine, β-spinel and γ-spinel fields but is less dense than the deeper mantle.

Removal of crystals from a crystallizing magma ocean and drainage of melt from a cooling crystal mush (also, technically a magma) are very much faster processes than cooling and crystallization times. Therefore, an expected result of early planetary differentiation is a stratified composition. Because of the combined effects of temperature and pressure on physical properties, shallow stratification may be reversible – leading to plate tectonics – while deep dense layers may be trapped at depth.

Chapter 2

Comparative planetology

The sun, with all those planets revolving around it and dependent on it, can still ripen a bunch of grapes as if it had nothing else in the universe to do.

Galileo Galilei

Before the advent of space exploration, Earth scientists had a handicap almost unique in science: they had only one object to study. Compare this with the number of objects available to astronomers, particle physicists, biologists and sociologists. Earth theories had to be based almost entirely on evidence from Earth itself. Although each object in the solar system is unique, we have learned some lessons that can be applied to Earth.

(1) Study of the Moon, Mars and the basaltic achondrites demonstrated that early melting is ubiquitous.
(2) Although primitive objects, such as the carbonaceous chondrites, have survived for the age of the solar system, there is no evidence for the survival of primitive material once it has been in a planet.
(3) The magma-ocean concept proved useful when applied to the Earth, taking into account the differences required by the higher pressures on the Earth.
(4) The importance of great impacts in the early history of the planets is now clear.
(5) Material was still being added to the Earth and Moon after the major accretion stage and the giant impacts, and is still being added, including material much richer in the noble metals and noble gases than occur in the crust or mantle.
(6) The difference in composition of the atmospheres of the terrestrial planets shows that the original volatile compositions, the extent of outgassing – or the subsequent processes of atmospheric escape – have been quite different.
(7) We now know that plate tectonics, at least the recycling kind, is unique to Earth. The thickness and average temperature of the lithosphere and the role of phase changes in basalt seem to be important. Any theory of plate tectonics must explain why the other terrestrial planets do not behave like Earth.

Although the Earth is a unique body, and is the largest of the terrestrial planets, we can apply lessons learned from the other objects in the solar system to the composition and evolution of the Earth. The Earth is also an average terrestrial planet; if we take one part Mercury, one part each of Venus and Mars, and throw in the Moon, we have a pretty good Earth, right size and density, and about the right size core. The inner solar system has the equivalent of two Earths.

Planetary crusts

The total crustal volume on the Earth is anomalously small, compared with other planets, and compared with its crust-forming potential, but it nevertheless contains a large fraction of the terrestrial inventory of incompatible elements. The thin crust on Earth can be explained by crustal recycling and the shallowness of the basalt–eclogite boundary in the Earth. Most of Earth's 'crust' probably resides in the transition region of the mantle. Estimates of bulk Earth chemistry can yield a basaltic layer of about 10% of the mass of the mantle.

The crust of the Earth is enriched in Ca, Al, K and Na in comparison to the mantle, and ionic-radii considerations and experimental petrology suggest that the crust of any planet will be enriched in these constituents. A maximum average crustal thickness for a fully differentiated chondritic planet can be obtained by removing all of the CaO, with the available Al_2O_3, as anorthite to the surface. This operation gives a crustal thickness of about 100 km for Mars. Incomplete differentiation and retention of CaO and Al_2O_3 in the mantle will reduce this value, which is likely to be the absolute upper bound (Earth's crust is much thinner due to crustal recycling, delamination and the basalt–eclogite phase change). In the case of the Earth, up to 60–70% of some large-ion elements are in the crust, implying that about 30–40% of the crustal elements are in the mantle. *This does not require that 30–40% of the mantle is still in a* `primordial undegassed state` as some geochemists believe.

The average thickness of the crust of the Earth is only 15 km, which amounts to 0.4% of the mass of the Earth. The crustal thickness is 5–10 km under oceans and 30–50 km under older continental shields. The thickest crust on Earth – about 80 km – is under young actively converging mountain belts. The parts deeper than about 50 km may eventually convert to eclogite, and fall off. The situation on the Earth is complicated, since new crust is constantly being created at midoceanic ridges and consumed at island arcs. The continental crust loses mass by erosion and by delamination of the lower eclogitic portions. Continental crust is recycled but its total volume is roughly constant with time. Both the Moon and Mars have crustal thicknesses greater than that of the Earth in spite of their much smaller sizes, and probable less efficient differentiation.

Mercury – first rock from the Sun

Mercury is 5.5% of the mass of the Earth, but it has a very similar density, 5.43 g/cm^3. Its radius is 2444 km. Any plausible bulk composition is about 60% iron and this iron must be largely differentiated into a core. Mercury has a perceptible magnetic field, appreciably more than either Venus or Mars, probably implying that the core is molten. *Mercury's surface is predominantly silicate, but apparently not basaltic.* A further inference is that the iron core existed early in its history; a late core-formation event would have resulted in a significant expansion of Mercury.

Mercury's shape may have significantly changed over the history of the planet. Tidal despinning results in a less oblate planet and compressional tectonics in the equatorial regions. Cooling and formation of a core cause a change in the mean density and radius. A widespread system of arcuate scarps on Mercury, which appear to be thrust faults, provides evidence for compressional stresses in the crust. The absence of normal faults suggests that Mercury has contracted. This is evidence for cooling of the interior.

One factor affecting the bulk composition of Mercury is the probable high temperature in its zone of the solar nebula; it may have formed from predominantly high-temperature condensates. If the temperature was held around 1300 K until most of the uncondensed material was blown away, then a composition satisfying Mercury's mean density can be obtained, since most of the iron will be condensed, but only a minor part of the magnesian silicates. Since the band of temperatures at which this condition prevails is quite narrow, other factors must be considered. Two of these are (1) dynamical interaction among the material in the terrestrial planet zones, leading to compositional mixing, and (2) collisional differentiation. A large impact after core formation may have blasted away much of the silicate crust and mantle. Our Moon may have been the result of such an impact on proto-Earth. On Mars, the crust is locally thinner under the large impact basins.

Terrestrial bodies were subjected to a high flux of impacting objects in early planetary history. The high-flux period can be dated from lunar studies at about 3.8 billion years ago. The large basins on the surface of Mercury formed during this period of high bombardment. Later cooling and contraction apparently were responsible for global compression of the outer surface and may have shut off volcanism. On the Earth, volcanism is apparently restricted to the extending regions.

Venus

Venus is 320 km smaller in radius than the Earth and is about 4.9% less dense. Most of the difference in density is due to the lower pressure, giving a smaller amount of self-compression and deeper phase changes. Venus is a smoother planet than the Earth but has a measurable triaxiality of figure and a 0.34 km offset of the center of the figure from the center of mass. This offset is much smaller than those of the Moon (2 km), Mars (2.5 km) and Earth (2.1 km).

In contrast to the bimodal distribution of Earth's topography, representing continent–ocean differences, Venus has a narrow unimodal height distribution with 60% of the surface lying within 500 m of the mean elevation. This difference is probably related to erosion and isostatic differences caused by the presence of an ocean on Earth. For both Earth and Venus the topography is dominated by long-wavelength features. Most of the surface of Venus is gently rolling terrain. The gravity and topography are positively correlated at all wavelengths. On Earth most of the long-wavelength geoid is uncorrelated with surface topography and is due to deep-mantle dynamics or density variations.

The other respects in which Venus differs markedly from the Earth are its slow rotation rate, the absence of a satellite, the virtual absence of a magnetic field, the low abundance of water, the abundance of primordial argon, the high surface temperature and the lack of obvious signs of subduction. From crater counts it appears that the age of the surface of Venus is 300–500 million years old, much less than parts of the Earth's surface. The oceanic crust on the Earth is renewed every 200 million years but the continents survive much longer.

If Venus had an identical bulk composition and structure to the Earth, then its mean density would be about 5.34 g/cm^3. By 'identical structure' I mean that (1) most of the iron is in the core, (2) the crust is about 0.4% of the total mass and (3) the deep temperature gradient is adiabatic (an assumption). The high surface temperature of Venus, about 740 K, would have several effects; it would reduce the depth at which the convectively controlled gradient is attained, it would deepen temperature-sensitive phase changes and it may prevent mantle cooling by subduction.

The density of Venus is 1.2–1.9% less than that of the Earth after correcting for the difference in pressure. This may be due to differences in iron content, sulfur content, oxidization state and deepening of the basalt–eclogite phase change. Most of the original basaltic crust of the Earth subducted or delaminated when the upper-mantle temperatures cooled into the eclogite stability field. The density difference between basalt and eclogite is about 15%. Because of the high surface temperature on Venus, the upper-mantle temperatures are likely to be 200–400 K hotter in the outer 300 km or so than at equivalent depths on Earth, or melting is more extensive. This has interesting implications for the phase relations in the upper mantle and the evolution of the planet. In particular, partial melting in the upper mantle would be much more extensive than is the case for the Earth except for the fact that Venus is probably deficient in the volatile and low-molecular-weight elements that also serve to decrease the melting point and viscosity. Crust can be much thicker because of the deepening of the basalt–eclogite phase boundary.

Schematic geotherms are shown in Figure 2.1 for surface temperatures appropriate for Earth and Venus. With the phase diagram shown, the high-temperature geotherm crosses the solidus at about 85 km. With other plausible phase relations the eclogite field is entered at a depth of about 138 km. For Venus, the lower gravity and outer-layer densities increase these depths by about 20%; thus, we expect a surface layer

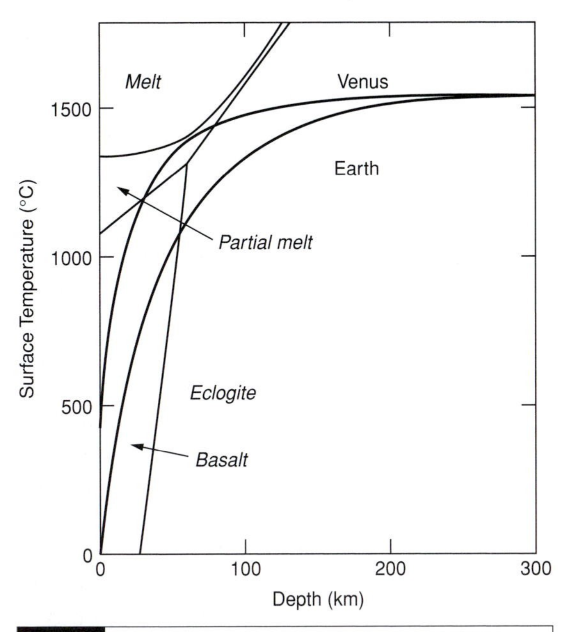

Fig. 2.1 Schematic geotherms for the Earth with different surface temperatures. Note that the eclogite stability field is deeper for the higher geotherms and that a partial melt field intervenes between the basaltic crust and the rest of the upper mantle. Basaltic material in the eclogite field will sink through much of the upper mantle and will be replaced by peridotite. Shallow subduction of basaltic crust leads to remelting in the case of Venus and the early Earth. Conversion to eclogite leads to lower crustal delamination or deep subduction for the present Earth. This figure explains why the low-density crust can be no thicker than about 50 km on Earth. The depth scale is for an Earth-size planet with the colder geotherm and present crust and upper-mantle densities. For Venus, with smaller gravity, higher temperatures and low-density crust replacing part of the upper mantle, the depths are increased by about 20%.

of 100–170 km thickness on Venus composed of basalt and partial melt. On the present Earth, the eclogite stability field is entered at a depth of 40–60 km. If the interior of Venus is dry it will be stronger at a given temperature and will have a higher solidus temperature.

A large amount of basalt has been produced by the Earth's mantle, but only a thin veneer is at the surface at any given time. There must therefore be a substantial amount of eclogite in the mantle, the equivalent of about 200 km in thickness. If this were still at the surface as basalt, the Earth would be several percent less dense. Correcting for the difference in temperature, surface gravity and mass and assuming that Venus is as well differentiated as Earth, only a fraction of the basalt in Venus would have converted to eclogite. This would make the uncompressed density of Venus about 1.5% less than Earth's without invoking any differences in composition or oxidation state. Thus, Venus may be close to Earth in composition. It is possible that the present tectonic style on Venus is similar to that of Earth in the Archean, when temperatures and temperature gradients were higher. If the Moon, Mercury and Mars have molten iron core, it is probable that Venus does as well.

The youthful age of the surface of Venus has been attributed to a `global resurfacing event`. The cratering record indicates that the global resurfacing event, about 300 my ago, was followed by a reduction of volcanism and tectonism. Delamination of thick basaltic crust, foundering of a cold thermal boundary layer and a massive reorganization of mantle convection are candidates for the resurfacing event. Mantle convection itself is strongly controlled by surface processes and changes in these processes.

Mars

Mars is about one-tenth of the mass of Earth. The uncompressed density is substantially lower than that of Earth or Venus and is very similar to the inferred density of a fully oxidized (minus C and H_2O) chondritic meteorite. The moment of inertia, however, requires an increase in density with depth over and above that due to self-compression and phase changes, indicating the presence of a dense core. This in turn indicates that Mars is a differentiated planet.

The tenuous atmosphere of Mars suggests that it either is more depleted in volatiles or has experienced less outgassing than Earth or Venus. It could also have lost much of its early atmosphere by large impacts. Geological evidence for running water on the surface of Mars suggests that a large amount of water is tied up in permafrost and ground water and in the polar caps. Whether Mars had standing water – oceans and lakes – for long periods of time is currently being debated. The high $^{40}Ar/^{36}Ar$ ratio on Mars, ten times the terrestrial value, suggests either a high potassium-40 content plus efficient outgassing, or a net depletion of argon-36 and, possibly, other volatiles. If Mars is volatile-rich, compared to Earth, it should have more K and hence more argon-40. Early outgassed argon-36 could also have been removed from the planet.

SNC meteorites have trapped rare-gas and nitrogen contents that differ from other meteorites but closely match those in the martian atmosphere. If SNC meteorites come from Mars, then a relatively volatile-rich planet is implied, and the atmospheric evidence for a low volatile content for Mars would have to be rationalized by the loss of the early accretional atmosphere. Mars is more susceptible to atmospheric escape than Venus or Earth owing to its low gravity. The surface of Mars appears to be weathered basalt. The dark materials at the surface contain basaltic minerals and hematite and sulfur-rich material, and there is evidence for the past action of liquid water. The large volcanoes on Mars are similar in form to shield volcanoes on Earth. Andesite – a possible indicator of plate tectonics – has been proposed as a component of martian soil but this is controversial; weathered basalt can explain the available data.

The topography and gravity field of Mars indicate that parts of Mars are grossly out of hydrostatic equilibrium and that the crust is highly variable in thickness. If variations in the gravity field are attributed to variations in crustal thickness, reasonable values of the density contrast imply that the average crustal thickness is at least 45 km, and the maximum crustal thickness may reach 100 km. Giant impacts may have removed most of the crust beneath the basins, replacing crustal material by uplifted mantle. If so, the crust was in place in early martian history, consistent with other evidence throughout the solar system for rapid early planetary differentiation. On Earth, delamination of lower crust produces a thinning but the whole crust is not involved.

The only direct evidence concerning the internal structure of Mars is the mean density, moment of inertia, topography and gravity field. The mean density of Mars, corrected for pressure, is less than that of Earth, Venus and Mercury

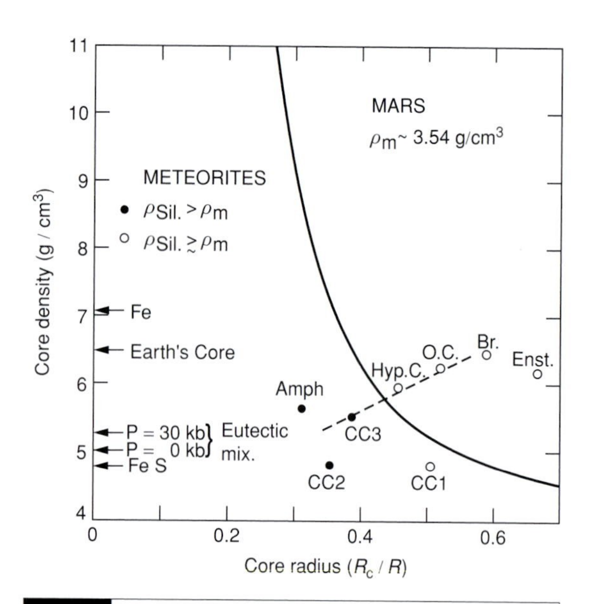

Fig. 2.2 Radius of the core versus density of core for Mars models. The points are for meteorites with all of the FeS and free iron and nickel differentiated into the core. The dashed line shows how core density is related to core size in the Fe–FeS system.

but greater than that of the Moon. This implies either that Mars has a small total Fe-Ni content or that the FeO/Fe ratio varies among the planets. Plausible models for Mars can be constructed that have solar or chondritic values for iron, if most or all of it is taken to be oxidized. With such broad chemical constraints, mean density and moment of inertia and under the assumption of a differentiated planet, it is possible to trade off the size and density of the core and density of the mantle.

The mantle of Mars is presumably composed mainly of silicates, which can be expected to undergo one or two major phase changes, each involving a 10% increase in density. To a good approximation, these phase changes will occur at one-third and two-thirds of the radius of Mars. The deeper phase change will not occur if the radius of the core exceeds one-third of the radius of the planet.

The curve in Figure 2.2 is the locus of possible Mars models. Clearly, the data can accommodate a small dense core or a large light core. The upper limit to the density of the core is probably close to the density of iron, in which case the core would be 0.36 of Mars' radius, or about 8% of its mass. To determine a lower limit to the density, one must consider possible major components of the core. Of the potential core-forming materials, iron, sulfur, oxygen and nickel are by far the most abundant elements. The assumption of a chondritic composition for Mars leads to values of the relative radius and mass of the core: $R_c/R = 0.50$ and $M_c/M = 0.21$. The inferred density of the mantle is less than the density of the silicate phase of most ordinary chondrites.

Three kinds of chondrites, HL (high iron, low metal), LL (low iron, low metal) and L (low iron or hypersthene–olivine) chondrites, all have lower amounts of potentially core-forming material than is implied for Mars, although HL and LL have about the right silicate density (3.38 g/cm^3). If completely differentiated, H (high iron) chondrites have too much core and too low a silicate density (3.26–3.29 g/cm^3). We can match the properties of H chondrites with Mars if we assume that the planet is incompletely differentiated. If the composition of the core-forming material is on the Fe side of the Fe–FeS eutectic, and temperatures in the mantle are above the eutectic composition, but below the liquidus, then the core will be more sulfur-rich and therefore less dense than the potential core-forming material.

Carbonaceous chondrites are extremely rich in low-temperature condensates, as well as carbon and 'organic matter.' If we ignore the water and carbon components, a fully differentiated planet of this composition would have a core of 15% by mass, composed mainly of FeS (13.6% FeS, 1.4% Ni), and a mantle with a density of about 3.5 g/cm^3. However, these meteorites also contain about 19% water, most of which must have escaped if Mars is to be made up primarily of this material. Otherwise, the mantle would not be dense enough. But there is abundant evidence for water near the surface of Mars in the past.

In ordinary high-iron chondrites, the free iron content averages 17.2% by weight. The FeS content is approximately 5.4% (3.4% Fe, 2.0% S) and the nickel content is 1.6%. A planet assembled from such material, if completely differentiated, would yield a core of 24% of the mass of the planet, with Fe: S: Ni in the approximate proportions of 21:2:2 by weight. Low-iron chondrites would yield a core of 15% of the mass of the planet, with proportions of 12:1:2. Enstatite chondrites contain between 15–25 weight% free iron. The oxidation state and the oxygen isotopes of enstatite chondrites make them an attractive major component for forming the Earth. These are not trivial considerations since oxygen is the major elemnt, by volume, in a terrestrial planet.

Carbonaceous chondrites have little or no free iron but contain 7–25% by weight FeS and about 1.5% nickel. The average core size for a planet made of carbonaceous chondrites would be 15% by mass, Fe: S: Ni being in the proportions 18:10:3. An absolute minimum core density can probably be taken as 4.8 g/cm^3, corresponding to a pure FeS core with a fractional core radius of 0.6 and a fractional mass of 26%. On these grounds, the mass of the martian core can be considered to lie between 8 and 24% of the mass of the planet.

A third possibility would be to assemble Mars from a mixture of meteorites that fall above and below the curve. The meteorites below the curve are relatively rare, although Earth may not be collecting a representative sample.

The size of the core and its density can be traded off. By using the density of pure iron and the density of pure troilite (FeS) as reasonable upper and lower bounds for the density of the core, its radius can be considered to lie between 0.36 and 0.60 of the radius of the planet. It is probably that there is sulfur in the core. Therefore, a core about half the radius of the planet, with a low melting point, is a distinct possibility.

The zero-pressure density of the mantle implies an FeO content of 21–24 wt.% unless some free iron has been retained by the mantle. The presence of CO_2 and H_2O, rather than CO and H_2, in the martian atmosphere suggests that free iron is not present in the mantle. Chondrites may therefore be an appropriate guide to the major-element composition of Mars. The core of Mars is smaller and less dense than the core of Earth, and the mantle of Mars is denser than that of Earth. Mars contains 25–28% iron, independent of assumptions about the overall composition or distribution of the iron. Earth is clearly enriched in iron or less oxidized than Mars and most classes of chondritic meteorites.

The lithosphere on Mars is capable of supporting large surface loads. The evidence includes the roughness of the gravity field, the heights of the shield volcanoes, the lack of appreciable seismicity and thermal history modeling. There is some evidence that the lithosphere has thickened with time. *Olympus Mons* is a volcanic construct with a diameter of 700 km and at least 20 km of relief, making it the largest known volcano in the solar system. It is nearly completely encircled by a prominent scarp several kilometers in height and it coincides with the largest gravity anomaly on Mars. The origin of the `Olympus Mons scarp` is controversial; it may be, in part, due to spreading of the volcano, and, in part, to erosion.

The surface of Mars is more complex than that of the Moon and Mercury. There is abundant evidence for volcanic modification of large areas after the period of heavy bombardment, subsequent to 3.8 Ga. Mars has a number of gigantic shield volcanoes and major fault structures. In contrast to Mercury there are no large thrust or reverse faults indicative of global contraction; all of the large tectonic features are extensional. The absence of terrestrial-style plate tectonics is probably the result of a thick cold lithosphere. In any event, the absence of plate tectonics on other planets provides clues to the dynamics of a planet that are unavailable from the Earth. Apparently, a planet can be too small or too dry or too old or too young, for it to have plate tectonics. Liquid water, magnetic fields, plate tectonics and life are all unique to Earth and there may be a reason for this.

Moon

Strange all this difference should be
'Twixt tweedle-dum and tweedle-dee.

John Byrom

The Moon is deficient in iron compared with the Earth and the other terrestrial planets. It is also apparently deficient in all elements and compounds more volatile than iron. The density of the Moon is considerably less than that of the other terrestrial planets, even when allowance is made for pressure. Venus, Earth and Mars contain about 30% iron, which is consistent with the composition of stony meteorites and the non-volatile components of the Sun. They therefore fit into any scheme that has them evolve from solar material. Mercury is overendowed with iron, which has led to the suggestion that part of the mantle was blasted away by impacts. Because iron is the major dense element occurring in the Sun, and in the preplanetary solar nebula, the Moon is clearly depleted in iron, and in a number of other elements as well. A common characteristic of many of these elements and their compounds is volatility. Calcium, aluminum and titanium are the major elements involved in high-temperature condensation processes; minor refractory elements include barium, strontium, uranium, thorium and the rare-earth elements. The Moon is enriched in all these elements, and we are now sure that more than iron–silicate separation must be involved in lunar origin.

The surface samples of the Moon are remarkably depleted in such volatile elements as sodium, potassium, rubidium and other substances that, from terrestrial and laboratory experience, we would expect to find concentrated in the crust, such as water and sulfur. The refractory trace elements – such as barium, uranium and the rare-earth elements – are concentrated in lunar surface material to an extent several orders of magnitude over that expected on the basis of cosmic or terrestrial abundances. Some of these elements, such as uranium, thorium, strontium and barium, are large-ion elements, and one would expect them to be concentrated in melts that would be intruded or extruded near the surface. However, other volatile large-ion elements such as sodium and rubidium are clearly deficient, in most cases, by at least several orders of magnitude from that expected from cosmic abundances. The enrichment of refractory elements in the surface rocks is so pronounced that several geochemists proposed that refractory compounds were brought to the Moon's surface in great quantity in the later stages of accretion. The reason behind these suggestions was the belief that the Moon, overall, must resemble terrestrial, meteoritic or solar material and that it was unlikely that the whole Moon could be enriched in refractories. In these theories the volatile-rich materials must be concentrated toward the interior. In a cooling-gas model of planetary formation, the refractories condense before the volatiles, and it was therefore implied that the Moon was made inside out! The standard geochemical model of terrestrial evolution also invokes a volatile-rich interior, one that is rich in ^{3}He, but there is no evidence for this. The strange chemistry of the Moon is consistent with condensation from a gas–dust cloud caused by a giant impact on proto-Earth.

Large-ion refractory elements are concentrated in the lunar-mare basalts by several orders of magnitude over the highland plagioclase-rich material, with the notable exception of europium, which is retained by plagioclase. Compared to the other rare-earth elements, europium is depleted in basalts and enriched in anorthosites. The "europium anomaly" was one of the early mysteries of the lunar sample-return program and implied that plagioclase was abundant somewhere on the Moon. The predicted material was later found in the highlands. Similarly, in terrestrial samples, there are missing elements that imply eclogitic and kimberlitic material at depth.

The maria on the Moon are remarkably smooth and level; slopes of less than one-tenth of a degree persist for hundreds of kilometers, and topographic excursions from the mean are generally less than 150 m. By contrast, elevation differences in the highlands are commonly greater than 3 km. The mean altitude of the terrace, or highlands, above maria is also about 3 km. The center of mass is displaced toward the Earth and slightly toward the east by about 2 km.

Seismic activity of the Moon is much lower than on Earth, both in numbers of quakes and their size, or magnitude. Their times of occurrence appear to correlate with tidal stresses caused by the varying distance between the Moon and the Earth. Compared with the Earth, they occur at great depth, about half the lunar radius. The Moon today is a relatively inactive body. This conclusion is consistent with the absence of obvious tectonic activity and with the low level of stresses in the lunar interior implied by gravity and moment-of-inertia data.

The lunar crust

The thickness and composition of the lunar highland crust indicate that the Moon is both a refractory-rich body and an extremely well differentiated body. The amount of aluminum in the highland crust may represent about 40% of the total lunar budget. This is in marked contrast to the Earth, where the amount of such major elements as aluminum and calcium in the crust is a trivial fraction of the total in the planet. On the other hand the amount of the very incompatible elements such as rubidium, uranium and thorium in the Earth's crust is a large fraction of the terrestrial inventory. This dichotomy between the behavior of major elements and incompatible trace elements can be understood by considering the effect of pressure on the crystallization behavior of calcium- and aluminum-rich phases. At low pressures these elements enter low-density phases such as plagioclase, which are then concentrated toward the surface. At higher pressures these elements enter denser phases such as clinopyroxene and garnet. At still higher pressures, equivalent to depths greater than about 300 km in the Earth, these phases react to form a dense garnet-like solid solution that is denser than such upper-mantle phases as olivine and pyroxene. Therefore, in the case of the Earth, much of the calcium and aluminum is buried at depth. The very incompatible elements, however, do not readily enter any of these phases, and they are concentrated in light melts. The higher pressures in the Earth's magma ocean and the slower cooling rates of the larger body account for the differences in the early histories of the Earth and Moon.

In the case of the Moon, the anorthositic component is due to the flotation of plagioclase aggregates during crystallization of the ocean. Later basalts are derived from cumulates or cumulus liquids trapped at depth, and the KREEP (K, REE, P rich) component represents the final residual melt. The isotopic data (Pb, Nd, Sr) require large-scale early differentiation and uniformity of the KREEP component. About 50% of the europium and potassium contents of the Moon now reside in the highland crust, which is less than 9% of the mass of the Moon. Estimates of the thickness of the magma ocean are generally in excess of 200 km, and mass-balance calculations require that most or all of the Moon has experienced partial melting and melt extraction. Evidence in support of the magma ocean concept, or at least widespread and extensive melting, include: (1) the complementary highland and mare basalt trace-element patterns, particularly the europium anomaly; (2) the enrichment of incompatible elements in the crust and KREEP; (3) the isotopic uniformity of KREEP; and (4) the isotopic evidence for early differentiation of the mare basalt source region, which was complete by about 4.4 Ga.

The separation of crustal material and fractionation of trace elements is so extreme that the concept of a deep magma ocean plays a central role in theories of lunar evolution. The cooling and crystallization of such an ocean permits efficient separation of various density crystals and magmas and the trace elements that accompany these products of cooling. This concept does not require a continuous globally connected ocean that extends to the surface nor one that is even completely molten. Part of the evidence for a magma ocean on the Moon is the thick anorthositic highland crust and the widespread occurrence of KREEP, an incompatible-element-rich material best interpreted as the final liquid dregs of a Moon-wide melt zone. The absence of an extensive early terrestrial anorthositic crust and the presumed absence of a counterpart to KREEP have kept the magma ocean concept from being adopted as a central principle in theories of the early evolution of the Earth. However, a magma ocean is also quite likely for the Earth and probably the other terrestrial planets as well.

Tables 2.1 and 2.2 gives comparisons between the crusts of the Moon and the Earth. In spite of the differences in size, the bulk composition and magmatic history of these two bodies, the products of differentiation are remarkably similar. The lunar crust is less silicon rich and poorer in volatiles, probably reflecting the overall depletion of the Moon in volatiles. The lunar-highland crust and the mare basalts are both more similar to the terrestrial oceanic crust than to the continental crust. Depletion of the Moon in siderophiles and similarity of the Earth and Moon in oxygen isotopes are consistent with the Moon forming from the Earth's mantle, after separation of the core.

Table 2.1 | Crustal compositions in the Moon and Earth

	Lunar Highland Crust	Lunar Mare Basalt	Terrestrial Continental Crust	Terrestrial Oceanic Crust
	Major Elements (percent)			
Mg	4.10	3.91	2.11	4.64
Al	13.0	5.7	9.5	8.47
Si	21.0	21.6	27.1	23.1
Ca	11.3	8.4	5.36	8.08
Na	0.33	0.21	2.60	2.08
Fe	5.1	15.5	5.83	8.16
Ti	0.34	2.39	0.48	0.90
	Refractory Elements (ppm)			
Sr	120	135	400	130
Y	13.4	41	22	32
Zr	63	115	100	80
Nb	4.5	7	11	2.2
Ba	66	70	350	25
La	5.3	6.8	19	3.7
Yb	1.4	4.6	2.2	5.1
Hf	1.4	3.9	3.0	2.5
Th	0.9	0.8	4.8	0.22
U	0.24	0.22	1.25	0.10
	Volatile Elements (ppm)			
K	600	580	12,500	1250
Rb	1.7	1.1	42	2.2
Cs	0.07	0.04	1.7	0.03

Taylor (1982), Taylor and McLennan (1985).

The origin of the Moon

Prior to the Apollo landings in 1969 there were three different theories of lunar origin. The fission theory, proposed by G. H. Darwin, Charles Darwin's son, supposed that the moon was spun out of Earth's mantle during an early era of rapid Earth rotation. The capture theory supposed that the moon formed somewhere else in the solar system and was later captured in orbit about the Earth. The co-accretion or 'double planet' theory supposed that the Earth and moon grew together out of a primordial swarm of small 'planetesimals.' All three theories made predictions at

Table 2.2 | Composition of the continental crust

Species	A	B	C
SiO_2 (percent)	58.0	63.7	57.3
TiO_2	0.8	0.5	0.9
Al_2O_3	18.0	15.8	15.9
FeO	7.5	4.7	9.1
MnO	0.14	0.07	—
MgO	3.5	2.7	5.3
CaO	7.5	4.5	7.4
Na_2O	3.5	4.3	3.1
K_2O	1.5	2.0	1.1
P_2O_5	—	0.17	—
Rb (ppm)	42	55	32
Sr	400	498	260
Th	4.8	5.1	3.5
U	1.25	1.3	0.91
Pb	10	15	8

A: Andesite model (Taylor and McLennan, 1985).
B: Amphibolite–granulite lower-crustal model (Weaver and Tarncy, 1984).
C: Theoretical model (Taylor and McLennan, 1985).

variance with the observations that the moon has no substantial metallic iron core, and that its rocks are similar in composition to the Earth's mantle (its oxygen isotopic ratios are identical to the Earth's), but are strongly depleted in volatiles. It was then realized that accreting matter would form embryonic planets with a large range of sizes.

The final stages of planetary formation would involve giant impacts in which bodies of comparable size collided at high speed. A giant impact produces rock-vapor that preferentially retains the refractory elements as it condensed. Shortly after the formation of the Earth, a large object is inferred to have hit the Earth at an oblique angle, destroying the impactor and ejecting most of that body along with a significant amount of the Earth's silicate portions. Some of this material then coalesced into the Moon. This event also melted a large fraction of the Earth. From the angular momentum of the present Earth–Moon system the projectile is inferred to have had a mass comparable to Mars and the Earth was smaller than it is today. After the collision the Earth is a very hot body indeed. The idea of a cold primordial undegassed Earth can no longer be entertained. The present lower mantle is more likely to be refractory and gas-poor than to be primordial and gas-rich. The `giant impact theory` is the now dominant `theory for the formation of the Moon`. The theory was proposed in 1975 by Hartman and Davis (see Hartman *et al.*, 1986).

Chapter 3

The building blocks of planets

For as the sun draws into himself the parts of which he has been composed, so earth receives the stone as belonging to her, and draws it toward herself . . .

Plutarch

The Earth is part of the solar system and the composition of the Sun, meteorites, comets, interplanetary dust particles and other planets provide information that may be useful in deducing the overall composition of our planet, most of which is inaccessible to direct observation. `Carbonaceous chondrites` (CI) appear to be the most primitive and low-temperature extraterrestrial (ET) objects available to us. Even if these were the only building blocks of planets the final planet would differ in composition from them because of vaporization during accretion, and loss of low-molecular-weight material. But the ratios of refractory elements in ET materials may provide a useful constraint. Since most of the volume of a terrestrial planet is oxygen, the oxygen isotopes of candidate materials play a key role in deciding how to assemble a planet. Oxygen isotopes require that the Earth either be made of *enstatite meteorites* or a mixture of meteorites that bracket the isotopic composition of the Earth or enstatite meteorites. The bulk oxygen-isotopic composition of the Earth precludes more than a few percent of carbonaceous chondritic material accreting to the Earth. Mars has a different oxygen-isotopic composition from the Earth, suggesting that distinct oxygen reservoirs were available in the early solar system over relatively small annuli of heliocentric distance.

Most workers assume that the Earth accreted from some sort of *primitive material* delivered to the Earth as meteorites and probably originating in the asteroid belt. But it is understood that when the mantle is referred to as having the composition of `CI chondrites`, or *chondritic*, it is usually only the refractory parts that are meant. In the cosmological context *refractory* and *volatile* refer to the condensation temperature in a cooling nebula of solar composition. The Earth is clearly deficient in elements more volatile than about Si, and this includes Na and K. In some models constructed by noble gas geochemists, however, the lower mantle is actually assumed to be *primordial* or undegassed and to approach CI in overall composition. The actual material forming the Earth is unlikely to be represented by a single meteorite class. It may be a mixture of various kinds of meteorites and the composition may have changed with time. The oxidation state of accreting material may also have changed with time.

Meteorites

Using terrestrial samples, we cannot see very far back in time or very deep into a planet's interior. Meteorites offer us the opportunity to extend both of these dimensions. Some meteorites, the chondrites, are chemically *primitive*, having compositions – volatile elements excluded – very

Table 3.1 | Compositions of chondrites (wt.%)

		Ordinary		Carbonaceous			
	Enstatite	H	L	CI	CM	CO	CV
Si	16.47–20.48	17.08	18.67	10.40	12.96	15.75	15.46
Ti	0.03–0.04	0.06	0.07	0.04	0.06	0.10	0.09
Al	0.77–1.06	1.22	1.27	0.84	1.17	1.41	1.44
Cr	0.24–0.23	0.29	0.31	0.23	0.29	0.36	0.35
Fe	33.15–22.17	27.81	21.64	18.67	21.56	25.82	24.28
Mn	0.19–0.12	0.26	0.27	0.17	0.16	0.16	0.16
Mg	10.40–13.84	14.10	15.01	9.60	11.72	14.52	14.13
Ca	1.19–0.96	1.26	1.36	1.01	1.32	1.57	1.57
Na	0.75–0.67	0.64	0.70	0.55	0.42	0.46	0.38
K	0.09–0.05	0.08	0.09	0.05	0.06	0.10	0.03
P	0.30–0.15	0.15	0.15	0.14	0.13	0.11	0.13
Ni	1.83–1.29	1.64	1.10	1.03	1.25	1.41	1.33
Co	0.08–0.09	0.09	0.06	0.05	0.06	0.08	0.08
S	5.78–3.19	1.91	2.19	5.92	3.38	2.01	2.14
H	0.13	—	—	2.08	1.42	0.09	0.38
C	0.43–0.84	—	—	3.61	2.30	0.31	1.08
Fe^0/Fe_{tot}	0.70–0.75	0.60	0.29	0.00	0.00	0.09	0.11

Mason (1962).

similar to that of the sun. The volatile-rich carbonaceous chondrites are samples of slightly altered, ancient planetesimal material that condensed at moderate to low temperatures in the solar nebula. The nonchondritic meteorites are *differentiated* materials of nonsolar composition that have undergone chemical processing like that which has affected all known terrestrial and lunar rocks.

Meteorites are assigned to three main categories. Irons (or siderites) consist primarily of metal; stones (or aerolites) consist of silicates with little metal; stony irons (or siderolites) contain abundant metal and silicates.

Carbonaceous chondrites

Carbonaceous chondrites contain high abundances of volatile components such as water and organic compounds, have low densities, and contain the heavier elements in nearly solar proportions. They also contain carbon and magnetite. These characteristics show that they have not been strongly heated, compressed or altered since their formation; that is, they have not been buried deep inside planetary objects.

The CI or Cl meteorites are the most extreme in their primordial characteristics and are used to supplement solar values in the estimation of cosmic composition. The other categories of carbonaceous chondrites, CII (CM) and CIII (CO and CV), are less volatile-rich.

Some carbonaceous chondrites contain calcium–aluminum-rich inclusions (CAI), which appear to be high-temperature condensates from the solar nebula. Theoretical calculations show that compounds rich in Ca, Al and Ti are among the first to condense in a cooling solar nebula. Highly refractory elements are strongly enriched in the CAI compared to Cl meteorites, but they occur in Cl, or cosmic, ratios.

Cl 'chondrites' are fine grained, do not contain chondrules and are chemically similar, to the true chondrites (see Table 3.1).

As the name suggests, *ordinary chondrites* are more abundant, at least in Earth-crossing orbits, than all other types of meteorites. They are chemically similar but differ in their contents of iron

Table 3.2 | Normative mineralogy of ordinary chondrites (Mason, 1962)

Species	High Iron	Low Iron
Olivine	36.2	47.0
Hypersthene	24.5	22.7
Diopside	4.0	4.6
Feldspar	10.0	10.7
Apatite	0.6	0.6
Chromite	0.6	0.6
Ilmenite	0.2	0.2
Troilite	5.3	6.1
Ni–Fe	18.6	7.5

Mason (1962).

Table 3.3 | Element ratios (by weight) in four subtypes of chondritic meteorites

Ratio	CI	H	L	E6
Al/Si	0.080	0.063	0.063	0.044
Mg/Si	0.91	0.80	0.79	0.71
Ca/Al	1.10	1.11	1.08	1.06
Cr/Mg	0.025	0.025	0.026	0.024

and other siderophiles, and in the ratio of oxidized to metallic iron. As the amount of oxidized iron decreases, the amount of reduced iron increases. Olivine is the most abundant mineral in chondrites, followed by hypersthene, feldspar, nickel-iron, troilite and diopside with minor apatite, chromite and ilmenite (Table 3.2). The composition of the olivine varies widely, from 0 to 30 mole% Fe_2SiO_4 (Fa). Enstatite chondrites are distinguished from ordinary chondrites by lower Mg/Si ratios (Table 3.3), giving rise to a mineralogy dominated by $MgSiO_3$ and having little or no olivine. They formed in a uniquely reducing environment and contain silicon-bearing metal and very low FeO silicates. They contain several minerals not found elsewhere (CaS, TiN, Si_2N_2O). In spite of these unusual properties, enstatite chondrites are within 20% of solar composition for most elements. They are extremely old and have not been involved in major planetary processing. They have been suggested as possible constituents of the Earth because of their high free-iron content, their oxidation state and oxygen isotopic ratios. If Earth is to be made out of a single meterorite class, the enstatite chondrites are the closest match.

Achondrites

The achondrites are meteorites of igneous origin that are thought to have been dislodged by impact from small bodies in the solar system. Some of these may have come from the asteroid belt, others are almost certainly from the Moon, and one subclass (the SNC group) have apparently come from Mars. Many of the achondrites crystallized between 4.4 and 4.6 billion years ago. They range from almost monomineralic olivine and pyroxene rocks to objects that resemble lunar and terrestrial basalts. Two important subgroups, classified as basaltic achondrites, are the *eucrites* and the *shergottites.* Two groups of meteoritic breccias, the *howardites* and the *mesosiderites,* also contain basaltic material. The eucrites, howardites, mesosiderites and diogenites appear to be related and may come from different depths of a common parent body. They comprise the *eucritic association.* The shergottites, nakhlites and chassignites form another association and are collectively called the SNC meteorites.

Eucrites are plagioclase–pyroxene rocks similar to basalts and have textures similar to basalts. However, terrestrial basalts have higher abundances of sodium, potassium, rubidium and other volatile elements and have more calcium-rich pyroxenes. Eucrite plagioclase is richer in calcium and poorer in sodium than terrestrial basaltic feldspar. The presence of free iron in eucrites demonstrates that they are more reduced than terrestrial basalts.

Studies of basalts from the Moon and the eucrite parent body have several important implications for the early history of the Earth and the other terrestrial planets. They show that even very small bodies can melt and differentiate. The energy source must be due to impact, rapid accretion, short-lived radioactive isotopes or formation in a hot nebula. The widespread occurrence of chondrules in chondritic meteorites also is evidence for high temperatures and melting in the early solar system.

The depletion of volatiles in eucrites and lunar material suggests that small planet and the early planetesimal stage of planet formation, may be characterized by volatile loss. These extraterrestrial basalts also contain evidence that free iron was removed from their source region. Alternatively, these objects are fragments of giant impacts that caused melting and silicate/iron separation. Nevertheless, the process of core formation must start very early and is probably contemporaneous with accretion.

Shergottites are remarkably similar to terrestrial basalts. They are unusual, among meteorites, for having very low crystallization ages, about 10^9 years, and, among basalts, for having abundant shocked plagioclase. The shergottites are so similar to terrestrial basalts that their source regions must be similar to the upper mantle of the Earth. The similarities extend to the trace elements, be they refractory, volatile or siderophile, suggesting a similar evolution for both bodies. The young crystallization ages imply that the shergottites are from a large body, one that could maintain igneous processes for 3 billion years. Cosmic-ray-exposure ages show that they were in space for several million years after ejection from their parent body.

Shergottites are slightly richer in iron and manganese than terrestrial basalts, and, in this respect, they are similar to the eucrites. They contain no water and have different oxygen isotopic compositions than terrestrial basalts. The major-element chemistry is similar to that inferred for the martian soil. The rare-gas contents of shergottites are similar to the martian atmosphere, giving strong circumstantial support to the idea that these meteorites may have come from the surface of Mars. In any case, these meteorites provide evidence that other objects in the solar system have similar chemistries and undergo similar processes as the Earth's upper mantle.

The growing Earth probably always had basalt at the surface and, consequently, was continuously zone-refining the incompatible elements toward the surface. The corollary is that the deep interior of a planet is refractory and depleted in volatile and incompatible elements. The main difference between the Earth and the other terrestrial planets, including any meteorite parent body, is that the Earth can recycle material back into the interior. Present-day basalts on Earth may be recycled basaltic material that formed during accretion and in early Earth history rather than initial melts from a previously unprocessed peridotitic parent. Indeed, no terrestrial basalt shows evidence, if all the isotopic and geochemical properties are taken into account, of being from a primitive, undifferentiated reservoir.

Cosmic abundances

The Sun and planets probably formed more or less contemporaneously from a common mass of interstellar dust and gas. There is a close similarity in the relative abundances of the *condensable elements* in the atmosphere of the Sun, in chondritic meteorites and in the Earth. To a first approximation one can assume that the planets incorporated the condensable elements in the proportions observed in the Sun and the chondrites. On the other hand, the differences in the mean densities of the planets, corrected for differences in pressure, show that they cannot all be composed of materials having exactly the same composition. Variations in iron content and oxidation state of iron can cause large density variations among the terrestrial planets. The giant, or Jovian planets, must contain much larger proportions of low-atomic-weight elements than Mercury, Venus, Earth, Moon and Mars.

With the exception of a few elements such as Li, Be and B, the composition of the solar atmosphere is essentially equal to the composition of the material out of which the solar system formed. The planets are assumed to accrete from material that condensed from a cooling primitive solar nebula. Various attempts have been made to compile tables of 'cosmic' abundances. The Sun contains most of the mass of the solar system; therefore, when we speak of the elemental abundances in the solar system, we really refer to those in the Sun. The spectroscopic analyses of elemental abundances in the solar photosphere do not have as great an accuracy as chemical analyses of solid materials. Carbonaceous chondrite meteorites, which appear to be the most representative samples of the relatively nonvolatile constituents of the solar system, are used for compilations of the abundances of most of the elements (Tables 3.4 to 3.6). For the very

Table 3.4 | Cosmic abundances of the elements (Atoms/10^6 Si)

1	H	2.72×10^{10}	24	Cr	1.34×10^4	48	Cd	1.69	72	Hf	0.176
2	He	2.18×10^9	25	Mn	9510	49	In	0.184	73	Ta	0.0226
3	Li	59.7	26	Fe	9.00×10^5	50	Sn	3.82	74	W	0.137
4	Be	0.78	27	Co	2250	51	Sb	0.352	75	Re	0.0507
5	B	24	28	Ni	4.93×10^4	52	Te	4.91	76	Os	0.717
6	C	1.21×10^7	29	Cu	514	53	I	0.90	77	Ir	0.660
7	N	2.48×10^6	30	Zn	1260	54	Xe	4.35	78	Pt	1.37
8	O	2.01×10^7	31	Ga	37.8	55	Cs	0.372	79	Au	0.186
9	F	843	32	Ge	118	56	Ba	4.36	80	HG	0.52
10	Ne	3.76×10^6	33	As	6.79	57	La	0.448	81	Tl	0.184
11	Na	5.70×10^4	34	Se	62.1	58	Ce	1.16	82	Pb	3.15
12	Mg	1.075×10^6	35	Br	11.8	59	Pr	0.174	83	Bi	0.144
13	Al	8.49×10^4	36	Kr	45.3	60	Nd	0.836	90	Th	0.0335
14	Si	1.00×10^6	37	Rb	7.09	62	Sm	0.261	92	U	0.0090
15	P	1.04×10^4	38	Sr	23.8	63	Eu	0.0972			
16	S	5.15×10^5	39	Y	4.64	64	Gd	0.331			
17	Cl	5240	40	Zr	10.7	65	Tb	0.0589			
18	Ar	1.04×10^5	41	Nb	0.71	66	Dy	0.398			
19	K	3770	42	Mo	2.52	67	Ho	0.0875			
20	Ca	6.11×10^4	44	Ru	1.86	68	Er	0.253			
21	Sc	33.8	45	Rh	0.344	69	Tm	0.0386			
22	Ti	2400	46	Pd	1.39	70	Yb	0.243			
23	V	295	47	AG	0.529	71	Lu	0.0369			

Anders and Ebihara (1982).

abundant volatile elements, solar abundance values are used.

The very light and volatile elements (H, He, C, N) are extremely depleted in the Earth relative to the Sun or carbonaceous chondrites. Moderately volatile elements (such as K, Na, Rb, Cs and S) are moderately depleted in the Earth. Refractory elements (such as Ca, Al, Sr, Ti, Ba, U and Th) are generally assumed to be retained by the planets in their cosmic ratios. It is also likely that magnesium and silicon occur in a planet in chondritic or cosmic ratios with the more refractory elements. The Mg/Si ratio, however, varies somewhat among meteorite classes. Sometimes it is assumed that magnesium, iron and silicon may be fractionated by accretional or pre-accretional processes, but these effects, if they exist, are slight.

The upper mantle of the Earth is olivine-rich and has a high Mg/Si ratio compared with the cosmic ratio (Figures 3.1 to 3.3). If the Earth is chondritic in major-element chemistry, then the deeper mantle must be rich in pyroxene and garnet and their high-pressure phases. Figure 3.3 is a schematic illustration of how the original accreting silicate material of a planet (*primitive mantle*) may fractionate into a melt (*magma ocean*) and dense refractory crystals. Crystallization of the magma ocean creates the materials that we sample from the upper mantle. In a large planet, the original differentiation may be irreversible because of the effects of pressure on material properties, such as the thermal expansion coefficient.

Composition of the terrestrial planets

The mean uncompressed densities of the terrestrial planets decreases in the order Mercury, Earth, Venus, Mars, Moon (Figure 3.4). Some

Table 3.5 | Solar and cosmic abundances in atoms/1000 Si atoms

Z	Element	Corona (1)	Photosphere (1)	'Cosmic' (2)
6	C	2350	6490	12 100
7	N	700	2775	2480
8	O	5680	22 900	20 100
9	F	0.28	1.1	0.843
10	Ne	783	3140	3760
11	Na	67.0	67.0	57.0
12	Mg	1089	1089	1075
13	Al	83.7	83.7	84.9
14	Si	1000	1000	1000
15	P	4.89	9.24	10.4
16	S	242	460	515
17	Cl	2.38	9.6	5.24
18	Ar	24.1	102	104
19	K	3.9	3.9	3.77
20	Ca	82	82	61.1
21	Sc	0.31	0.31	0.034
22	Ti	4.9	4.9	2.4
23	V	0.48	0.48	0.295
24	Cr	18.3	18.3	13.4
25	Mn	6.8	6.8	9.51
26	Fe	1270	1270	900
27	Co	<18.1	<18.1	2.25
28	Ni	46.5	46.5	49.3
29	Cu	0.57	0.57	0.514
30	Zn	1.61	1.61	1.26

(1) Breneman and Stone (1985).
(2) Anders and Ebihara (1982).

Table 3.6 | Short table of cosmic abundances (Atoms/Si)

Element	Cameron (1982)	Anders and Ebihara (1982)
O	18.4	20.1
Na	0.06	0.057
Mg	1.06	1.07
Al	0.085	0.0849
Si	1.00	1.00
K	0.0035	0.003 77
Ca	0.0625	0.0611
Ti	0.0024	0.0024
Fe	0.90	0.90
Ni	0.0478	0.0493

typical compositions of possible components of the terrestrial planets are listed in Table 3.7. There are some constraints on the amounts or ratios of a number of key elements in a planet. For example, the mean density of a planet, or the size of the core, constrains the iron content. Using cosmic ratios of elements of similar geochemical properties (say Co, Ni, refractory siderophiles), a whole group of elements can be constrained. The uranium and thorium content are constrained by the heat flow and thermal history calculations. The K/U ratio, roughly constant in terrestrial magmas, is a common constraint in this kind of modeling. The Pb/U ratio can be estimated from lead isotope data. The amount of argon-40 in the atmosphere provides a lower bound on the amount of potassium in the crust and mantle. Most of these are very weak constraints, but they do allow rough estimates to be made of the refractory, siderophile, volatile and other contents of the Earth and terrestrial planets. The elements that are correlated in magmatic processes have very similar patterns of geochemical behavior, even though they may be strongly fractionated during nebular condensation. Thus, some abundance patterns established during condensation tend not to be disturbed by subsequent planetary melting and igneous fractionation. On the other hand, some elements are so strongly fractionated from one another by magmatic and core formation processes that discovering a 'cosmic' or 'chondritic' pattern can constrain the nature of these processes.

The outer planets and satellites are much more volatile-rich than the inner planets. Meteorites also vary substantially in composition and volatile content. The above considerations suggest that there may be an element of inhomogeneity in the accretion of the planets, perhaps caused by temperature and pressure gradients in the early solar nebula. Early forming planetesimals would have been refractory- and iron-rich and the later forming planetesimals more volatile-rich. If planetary accretion was occurring simultaneously with cooling and condensation,

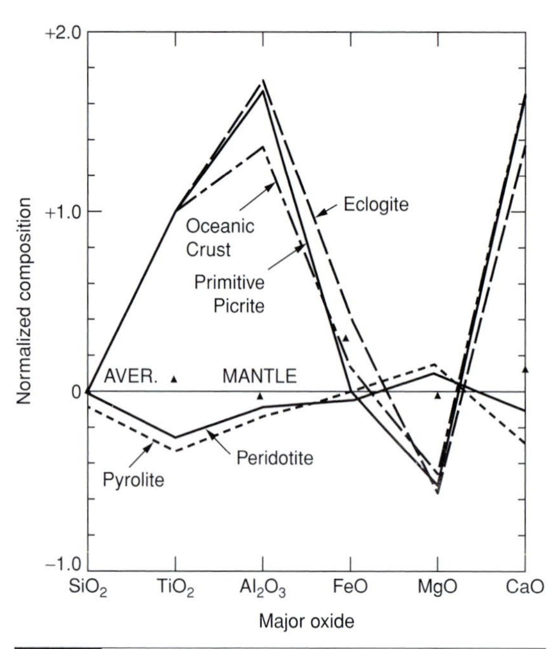

Fig. 3.1 Bulk chemistry of ultramafic rocks (peridotite) and basic, or basaltic, rocks (oceanic crust, picrite, eclogite) normalized to average mantle composition based on cosmochemical considerations and an assumption about the FeO content of the mantle. Pyrolite is a hypothetical upper rock but it has been proposed to be representative of the whole mantle. If so, the Mg/Si ratio of the mantle will not be chondritic. A composition equivalent to 80% peridotite and 20% eclogite (or basalt), shown by triangles, is a mix that reconciles petrological and cosmochemical major-element data. Allowance for trace-element data and a possible $MgSiO_3$-rich lower mantle reduces the allowable basaltic component to 15 weight% or less.

then the planets would have formed inhomogeneously. As a planet grows, the gravitational energy of accretion increases, and impact vaporization becomes more important for the larger planets and for the later stages of accretion. The assumption that Earth has cosmic abundances of the elements is therefore only a first approximation but is likely to be fairly accurate for the involatile elements. There is little dispersion of the refractory elements among the various stony meteorite classes, suggesting that these elements are not appreciably fractionated by preaccretional processes. Fortunately, the bulk of a terrestrial planet is iron, magnesium, silicon, calcium, and aluminum and their oxides. The bulk composition of a terrestrial planet can therefore be discussed with some confidence.

The composition of the Sun, meteorites, comets, interplanetary dust particles and other planets provide information that may be useful in deducing the overall composition of our planet, most of which is inaccessible to direct observation. Compilations of solar and cosmic abundances agree fairly closely for the more significant rock-forming elements. From these data simple models can be made (Tables 3.8, 3.9) of the Earth's bulk chemical composition and mineralogy, assuming that the mantle is completely oxidized. The composition in the first column of Table 3.9 is based on cosmic abundances. This is converted to weight fractions via the molecular weight and renormalization.

The Fe_2O requires some comment. Based on cosmic abundances, it is plausible that the Earth's core is mainly iron; however, from seismic data and from the total mass and moment of inertia of the Earth, there must be a light alloying element in the core. Of the candidates that have been proposed (O, S, Si, N, H, He and C), only oxygen and silicon are likely to be brought into a planet in refractory solid particles – the others are very volatile elements and will tend to be concentrated near the surface or in the atmosphere or lost to space. The hypothetical high-pressure phase Fe_2O has about the right density to match core values. If most of the iron is in the core, in Fe_2O proportions, then the mass of the core will be 30–34 weight% of the planet. The actual mass of the core is 33%. There may also be some sulfur, carbon, and so on in the core, but little or none seems necessary.

Since most of the volume of a terrestrial planet is oxygen, the oxygen isotopes of candidate materials play a key role in deciding what to assemble a planet from. Oxygen isotopes imply that the Earth is made of enstatite meteorites or a mixture of meteorites that bracket the isotopic composition of the Earth and these meteorites. The bulk oxygen-isotopic composition of the Earth precludes more than a few percent of carbonaceous chondritic material accreting to the Earth.

When the mantle is referred to as having the composition of CI chondrites, or 'chondritic,' it is

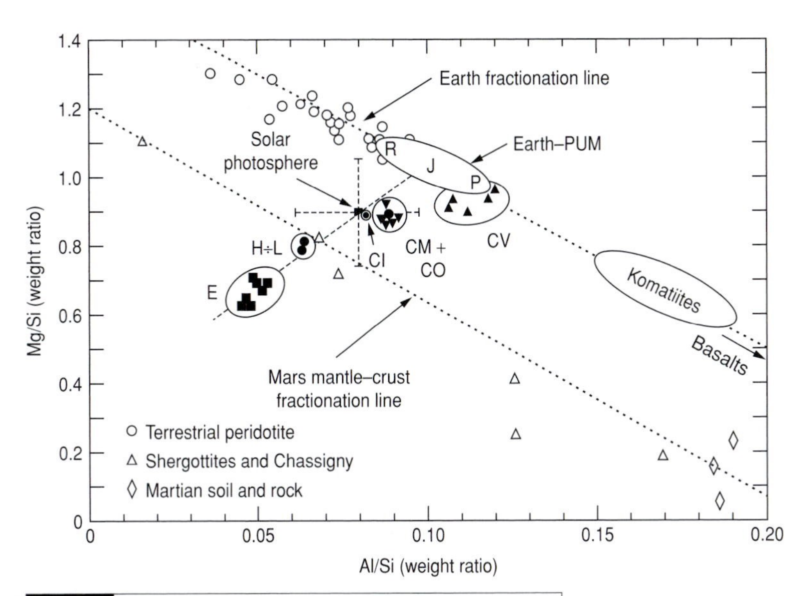

Fig. 3.2 The major-element composition of material in the inner Solar System is not of uniform composition, but exhibits trends in Mg/Si versus Al/Si ratios in chondritic, terrestrial and martian materials. If the average composition of the Earth's mantle is similar to chondrites then the lower mantle is perovskite-rich and komatiites and basalts represent only a small fraction of the mantle. The Earth fractionation trend then only represents the final stages of mantle differentiation. Abbreviations: enstatite (E), ordinary (H, L) and carbonaceous (CI, CM, CO and CV) chondrites. Circles are terrestrial peridotites. R, J and P are estimates of the bulk silicate Earth composition. The martian trend is defined by Chassigny, shergottites and martian samples. (Drake & Righter, 2002).

usually only the refractory parts that are meant. The exception is the so-called `standard model of noble gas geochemistry`; in this model the high $^3He/^4He$ ratios of some ocean island basalts is assumed to imply a He-rich reservoir, deep in the mantle, with abundances of helium similar to undegassed carbonaceous chondrites.

Javoy (1995) presented a model in which the Earth is built from essentially pure EH (high-iron enstatite achondrites) material. This is justified by the fact that most elements in these meteorites exist in very refractory phases, more so than in most other types of meteorites. Such a model implies a chemically stratified mantle, layered convection, and limitations on the chemical interchanges between lower and upper mantle. Formation of Earth from EH material involves a

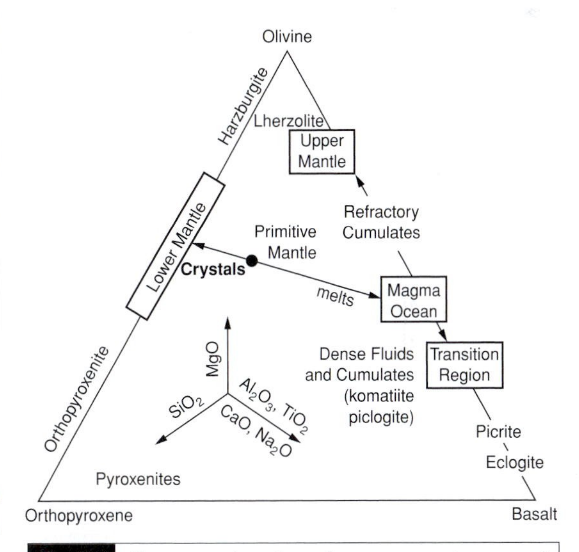

Fig. 3.3 Representation of mantle components in terms of olivine and orthopyroxene (the high-melting-point minerals) and basalt (the most easily fusible component). Primitive mantle is based on cosmic abundances. Melting of chondritic material at high temperature gives an MgO-rich melt (basalt + olivine) and a dense refractory residual (olivine + orthopyroxene). Crystallization of a magma ocean separates clinopyroxene and garnet from olivine. Melting during accretion tends to separate components according to density and melting temperature, giving a chemically zoned planet. If melting and melt-crystal separation occur primarily at low pressure, the upper mantle will be enriched in basalt, olivine and the incompatible elements relative to the lower mantle and relative to the chondritic starting material.

Table 3.7 | Compositions of possible components of the terrestrial planets (% or ppm)

Species	CI	EC	HTC
SiO_2	30.9	39.1	20.2
TiO_2	0.11	0.06	1.9
Al_2O_3	2.4	1.9	36.5
Cr_2O_3	0.38	0.35	—
MgO	20.8	21.3	7.1
FeO	32.5	1.7	
MnO	0.25	0.14	—
CaO	2.0	1.6	34.1
Na_2O	1.0	1.0	
K (ppm)	800	920	—
U (ppm)	0.013	0.009	0.19
Th (ppm)	0.059	0.034	0.90
Fe	0	26.7	—
N	1.3	1.7	—
S	8.3	4.5	—

Cl: Average Cl carbonaceous chondrite, on a C-, H_2O-free basis (Wood, 1962).
EC: Average enstatite chondrite (Wood, 1962).
HTC: High-temperature condensate (Grossman, 1972).

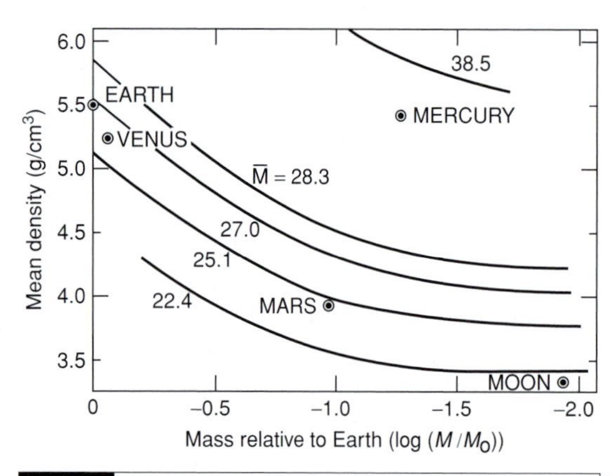

Fig. 3.4 Mean density versus mass, relative to Earth, of planets having the same structure as the Earth and various metal/silicate ratios, expressed as $\overline{M}$, mean atomic weight. Earth and Venus have similar bulk chemistries while Mars and Moon are clearly deficient in iron. Mercury is enriched in iron. Earth and Venus might be considered *average terrestrial planets*.

two-step redox process, and a very late accretion of 0.8% of CI type material. Formation of the core and mantle from fully oxidized chondritic material is even more complex since little free iron is present.

The accessible and sampled parts of Earth are clearly deficient in elements more volatile than Li and Si, and this includes Rb, Na and K. However, in geochemical models based on noble gases, these missing elements are assumed to be sequestered in the lower mantle or core rather than being deficient in the Earth as a whole.

Although there is no inconsistency – from a major element and refractory element point of view – in building the planets from known meteorite classes, or from estimates of cosmic abundances, part of the Earth may have accreted from hydrous materials that are not represented in meteorite collections; that is, they are no longer extant in the inner Solar System, at least as meteorite-size objects. This is not surprising. Comet dust and interplanetary dust particles (IDP) may be important for certain elements or isotopes. The isotopic compositions of at least the lighter rare gases are similar to those in IDPs falling on the Earth today.

Minerals in mantles

Several cosmochemical estimates of mineralogy of Earth's mantle are given in Table 3.8. These tend to be less rich in olivine than estimates of the composition of the upper mantle (column 4) and some estimates of the whole mantle, which in turn are based on the assumption of a homogenous mantle. The mineralogy changes with depth because of solid–solid phase and compositional changes.

Although Al_2O_3, CaO and Na_2O are minor constituents of the average mantle, their presence changes the mineralogy, and this in turn affects the physical properties. The effect on density can result in chemical stratification of the mantle and concentration of these, and related, elements into certain layers. They also influence the melting point and tend to be concentrated in melts.

Olivine is an essential component in most groups of meteorites except the irons. Pallasites

Table 3.8 | Mineralogy of Earth's mantle

Species	Whole-Mantle Models (1)	(2)	(3)	Upper Mantle (4)
Olivine	47.2	36.5	37.8	51.4
Orthopyroxene	28.3	33.7	33.2	25.6
Clinopyroxene	12.7	14.6	11.8	11.0
Jadeite	9.8	2.2	1.8	0.65
Ilmenite	0.2	0.5	0.24	0.57
Garnet	1.53	11.6	14.2	9.6
Chromite	0.0	1.6	0.94	0.44

(1) Equilibrium condensation (BVP, 1980).
(2) Cosmochemical model (Ganapathy and Anders 1974).
(3) Cosmochemical model (Morgan and Anders, 1980).
(4) Pyrolite (Ringwood, 1977).

Table 3.9 | Simple Earth model based on cosmic abundances

Oxides	Molecules	Molecular Weight	Grams	Weight Fraction
MgO	1.06	40	42.4	0.250
SiO_2	1.00	60	60.0	0.354
Al_2O_3	0.0425	102	4.35	0.026
CaO	0.0625	56	3.5	0.021
Na_2O	0.03	62	1.84	0.011
Fe_2O	0.45	128	57.6	0.339
Total			169.7	1.001

are composed of nickel–iron and olivine. Olivine is a major constituent in all the chondrites except the enstatite chondrites and some of the carbonaceous chondrites, and it is present in some achondrites. For these reasons – and ignoring the enstatite chondrites – olivine is usually considered to be the major constituent of the mantle. The olivines in meteorites, however, are generally much richer in FeO than mantle olivines. The olivine compositions in chondrites generally lie in the range 19 to 24 mole% fayalite. Removal of iron from meteoritic olivine, either as Fe or FeO, would decrease the olivine content of the mantle.

Earth models based on cosmic abundances, with Mg/Si approximately 1 (molar), give relatively low total olivine contents. Since the upper mantle appears to be olivine-rich, this results in an even more olivine-poor transition region and lower mantle. It is usually assumed, however, that the basaltic fraction of the Earth is still mostly dispersed throughout the mantle. This is the assumption behind the pyrolite model and most geochemical models of the mantle. Basalts were probably liberated during accretion of the Earth and concentrated in the upper mantle. Efficient remixing or rehomogenization can be ruled out.

Samples of the Earth's upper mantle are distinct in composition from that of any kind of extant meteorite. A hypothetical so-called primitive upper mantle (PUM) – a mixture of continental crust and ultramafic upper mantle rocks – also differs from meteorites in significant ways. Pyrolite – a hypothetical rock constructed from basalts and peridotites from the upper mantle – also does not have the same ratios of major elements as do meteorites. Several suggestions have been made to explain the elevated Mg/Si and Al/Si ratios in the Earth's shallowest mantle relative to undifferentiated meteorites. These include sequestering Si in the core of the Earth, raising the Mg/Si and Al/Si ratios in the silicate mantle, or appealing to the possibility that the mantle is chemically stratified with the lower mantle – which is the bulk of the Earth – having a different composition from the upper mantle. The latter is the most plausible suggestion.

There is no particular reason why the Earth should have accreted olivine in preference to pyroxene. There is no compelling evidence that Si is or can be extracted into the core and the idea is also inconsistent with the upper mantle abundances of V, Cr and Mn; they should be depleted far below observed abundances if conditions were sufficiently reducing to allow Si to dissolve in molten iron.

Estimates of the Earth's upper mantle composition are distinct from any known type of meteorite, but there are numerous petrological reasons why the Earth's mantle should not have survived in a homogenous state. Some workers argue that it is more reasonable that the Earth accreted from material with major-element compositions that were distinct from any meteorite type than to accept chemical stratification and petrological differentiation of the mantle.

Late veneer

Most of the volatiles in the Earth may have been brought in as a late veneer, or late veneers. The late veneer material is mixed into the upper mantle, and the term does not imply a thin surface coating. The heterogenous accretion hypothesis makes dynamical sense in that the *feeding zone* of the Earth must have extended further out from the Sun as the growth of planets increased the relative velocities and the eccentricities of the accreting material. Earth probably developed magma oceans throughout the accretion process, including late in its accretion, effectively processing and fractionating incoming material. Magma oceans and extensive melting are sometimes viewed as homogenizers rather than fractionators or filters.

Free water in the Earth and the trace siderophiles in the upper mantle, and possibly the lighter rare gases, may all have been delivered to the Earth's surface long after the bulk of the Earth had formed and the core was in place, and the bulk of the mantle had cooled, differentiated and solidified. Comets may have delivered some fraction of the water. Solar wind and IDP have helium and neon isotopic characteristics suitable to be an end member in some basalts. The similarity of Os-isotope ratios in the mantle and ordinary chondrites and oxygen isotopic composition with the enstatite meteorites is evidence for the existence of Earth-building materials sharing some properties with various meteorites.

No meteorite or other ET material has all of the properties of Earth material, at least without processing by accretional or petrological processes. This is not unexpected since they cannot avoid processing by accretional or petrological processes. Both the bulk of the Earth, and the late veneer probably have multiple sources and it is too much to expect that all the chemical characteristics of the Earth must be found in one kind of extraterrestrial object or its fragments which themselves have a history of processing. The alternatives are that all the distinctly Earth-forming material is either contained in the Earth and Moon or was ejected from the inner Solar System, during giant impacts, making the Earth–Moon system unique in the Solar System.

There are two views about the source of volatiles on Earth. One view holds that temperatures were too high at 1 AU for hydrous and volatile-rich phases to exist in the accretion disk, so that the bulk of the Earth accreted 'dry.' Water, and noble gases, were delivered from sources such as comets or meteorites after the bulk of the Earth had formed. An alternative view holds that the Earth accreted 'wet' and gas-rich, with anhydrous and hydrous silicate phases among the material accreted to the growing planet throughout the accretionary history. In this view, Earth's water, and other volatiles such as the noble gases, have an indigenous origin. This implies a cold origin and early history and reaction of water with Fe. The late veneer of the Earth may have come from more than one source. The noble metals, noble gases and water may have been brought in by different components and at different times. The Os-isotopic compositions of the crust and upper mantle of the Earth, are inconsistent with water-bearing carbonaceous chondrites, but are consistent with anhydrous ordinary chondrites. Various components have also been recognized in the noble gas inventory of the Earth. These may have been in large accreting objects, or in dust particles affected by Solar radiation. They may, in part, be due to cosmic radiation at the surface of early Earth.

Part II

Earth: the dynamic planet

This is the fourth time that I have taken part in a public discussion of this theory. In each previous one a distinguished biologist or geologist has presented the case for drift, and has been followed by equally distinguished ones who have pointed out facts that it would render more difficult to explain . . . The present impasse suggests that some important factor has been overlooked.

Sir Harold Jeffreys, 1951

Overview

Plate tectonics on Earth, at present, consists of about a dozen large semi-coherent entities – called plates – of irregular shape and size that move over the surface, separated by boundaries that meet at triple junctions. There are also many broad zones of deformation.

Plate tectonics is often regarded as `simply the surface, or the most important, manifestation of thermal convection in the mantle` [this phrase, and phrases in the same typeface, is a Googlet; see Preface or type it into a search engine]. In this view the plates are driven by thermal and density variations in the mantle. Cooling plates and sinking slabs can also be regarded as driving themselves, *and* driving convection in the underlying mantle; they create chemical, thermal and density anomalies in the mantle.

Plate tectonics qualifies as a branch of *complexity theory*. Plate tectonics may be a `far-from-equilibrium self-organized system` powered by heat and gravity from the mantle and organized by dissipation in and between the plates. Mantle convection, below the plates, may not drive or organize the plates; it may be the other way around. Plate buoyancy and dissipation control plate motions, stresses, and locations of plate boundaries, intraplate extensional zones and volcanic chains. The cold stiff outer shell of Earth is the active element and the template; the underlying *convective mantle* is passive.

The outer shell of the Earth is not just a thermal boundary layer or a cold strong layer. It is, in part, the accumulated buoyant residue of mantle differentiation including the on-going process of seafloor spreading and building of island arcs. It is composed of fertile melts, dikes, sills and cumulates, and infertile refractory residues. It is, in part, isolated from the low-viscosity fertile interior. Earthquakes and volcanoes not only mark plate boundaries but they anticipate new plate boundaries, changes in boundary conditions and dying plate boundaries. If the shallow mantle is close to the melting point or partially molten, volcanoes have a simple cause, stress.

The mode of convection in the Earth depends on the distribution of radioactive elements and physical properties and how these properties depend on temperature and pressure and melting point. An Earth with most of the radioactive elements in the crust and upper mantle, and with strongly pressure-dependent thermal properties will not behave as a uniform fluid being heated on a stove. These effects, plus continents and sphericity, break the symmetry between the top and bottom thermal boundary layers.

The main plate tectonic cycle is the ridge-trench-slab system, primarily playing out in the ocean basins. There is a secondary cycle involving underplating, freezing at depth, delamination and asthenospheric upwelling. A mafic lower crust, if it thickens and cools sufficiently, will convert to a high density mineral assemblage, leading to a gravitationally unstable configuration in which the lower crust can sink into the underlying lower-density mantle, cooling it and fertilizing it.

The solid Earth can rotate rapidly underneath its spin axis through a process known as *true polar wander* (*TPW*). The spinning Earth continuously aligns its maximum moment of inertia with the spin axis. Melting ice caps, plate motions, continental uplift and drift and ridge-trench annihilations can all cause TPW. The magnetic and rotational poles are good terrestrial reference systems; the hotspot frame is not. But fertile patches in the asthenosphere move more slowly than the plates and plate boundaries and therefore melting anomalies appear to be relatively fixed.

Chapter 4

The outer shells of Earth

> When I use a word, it means just what I choose it to mean – neither more nor less.
>
> The question is, whether you CAN make words mean so many different things.
>
> The question is, which is to be master – that's all.
>
> *Humpty Dumpty and Alice*

Plate tectonics involves the concepts of plates, lithospheres, cratonic keels and thermal boundary layers; these are not equivalent concepts. *Lithosphere* means *rocky shell* or *strong layer*. It can support significant geologic loads, such as mountains, and can bend, as at trenches, without significant time-dependent deformation; it behaves elastically. *Plates* are not necessarily strong or elastic. *Thermal boundary layers* conduct heat out of the underlying regions; they are defined by a particular thermal gradient.

Plates

A *plate* is a region of the Earth's surface that translates coherently. The word *plate* as in *plate tectonics* implies strength, brittleness and permanence and often the adjectives *rigid* and *elastic* are appended to it. But plates are collages, held together by stress and adjacent portions rather than by intrinsic strength. Plates break at suture zones (former plate boundaries), fracture zones and subplate boundaries, often generating volcanic chains in the process. A plate can be *rigid* in the sense that relative plate motions can be described by rotations about Euler poles on a sphere but can still have meter-wide cracks, which is all that is needed to create volcanic chains from the hot underlying mantle and its low-viscosity magmas. What is meant by *rigidity* is relative coherence in motions, not absolute strength.

Because rocks are weak under tension, the conditions for the existence of a plate probably involve the existence of lateral compressive forces. Plates have been described as rigid but this implies long-term and long-range strength. They are better described as coherent entities organized by stress fields and rheology. The corollary is that volcanic chains and plate boundaries are regions of extension. Plates possibly organize themselves so as to minimize dissipation.

The term *plate* itself has no agreed-upon formal definition. If *plate* is defined operationally as that part of the outer shell that moves coherently then several interpretations are possible.

(1) Plates are strong and rigid (the conventional interpretation).
(2) Plates are those regions defined by lateral compression since plate boundaries are formed by lateral extension.
(3) Plates move coherently because the parts experience similar forces or constraints.

With the first definition, if there is to be spreading and volcanism, local heating or stretching must overcome the local strength; this reasoning spawned the *plume hypothesis*. With the second definition the global stress field, dictated by plate boundary and subplate conditions, and cooling plates, controls the locations of stress conditions appropriate for the formation of dikes and volcanic chains, and incipient plate boundaries; the underlying mantle is already at or near the melting point. This is the *plate hypothesis*.

Plates are not permanent; they are temporary alliances of subplates. Global plate reorganization processes episodically change the orientations of spreading centers, the directions and speeds of plates, and redefine the plates. Plates annex and lose territory to adjacent plates and they break up or coalesce. New plate boundaries do not form all at once but evolve as age-progressive chains of volcanoes. Volcanic chains can also be extinguished if lateral compression takes over from local extension. Volcanism can be turned on and off by changing stress but it is not so easy to turn off plume volcanism, or to suddenly reduce the temperature of the mantle.

Important aspects of plate tectonics are the necessity for ridges and trenches to migrate, for triple junctions and boundary conditions to evolve, and for plates to interact and to reconfigure when boundary conditions change. Second order features of plates and plate boundaries (e.g. fracture zones, accreted terranes, transform faults, broad diffuse zones, swells, sutures, lithospheric architecture and microplates) and boundary reorganizations are actually intrinsic and provide the key for a more general view of plate tectonics than contained in the *rigid plate-fixed hotspot scheme*.

The many lithospheres

The lithosphere is that part of the cold outer shell of the Earth that can support stresses elastically. The lithosphere is defined by its rheological behavior. There are other elements of the outer shell that involve lateral motions, buoyancy, chemistry, mineralogy or conductivity and these may or may not be part of the lithosphere. *Lithosphere* is not the same as *thermal boundary layer* or *plate*. Since mantle silicates flow readily at high temperatures and flow more rapidly at high stress, the lithosphere appears to be thicker at low stress levels and short times than it does for high stress levels and long times. Thus, the elastic lithosphere is thick when measured by seismic or postglacial-rebound techniques. At longer times the lower part of the instantaneous elastic lithosphere relaxes and the effective elastic thickness decreases. Thus, the elastic lithosphere is relatively thin for long-lived loads such as seamounts and topography. Estimates of the flexural thickness of the lithosphere range from 10–35 km for loads having durations of millions of years. A more complete definition of the lithosphere is *that part of the crust and upper mantle that deforms elastically for the load and time scale in question*.

The viscosity and strength of the mantle depend on composition – including water content – mineralogy and crystal orientation as well as on temperature and stress. If the upper mantle is compositionally layered, then the lithosphere–asthenosphere boundary may be controlled by factors other than temperature. For example, if the subcrustal layer is dry olivine-rich harzburgite, it may be stronger at a given temperature than a damp peridotite, or a clinopyroxene–garnet-rich layer. If the latter is weak enough, the lithosphere–asthenosphere boundary may represent a chemical boundary rather than an isotherm. Likewise, a change in the preferred orientation of the dominant crystalline species may also markedly affect the creep resistance. The boundary may represent a dehydration boundary – wet minerals are weak. The effective elastic thickness of the lithosphere depends on many parameters but these do not necessarily include the parameters that define plates, thermal boundary layers and cratonic keels.

The layer that translates coherently, the *plate* of *plate tectonics*, is often taken to be identical with the *elastic lithosphere*. This is probably a valid approximation if the stresses and time scales of the experiment that is used to define the flexural thickness are similar to the stresses and time scales of plate tectonics. It must be kept in mind, however, that mantle silicates are anisotropic in their flow and thermal characteristics, and that the stresses involved in plate tectonics may have different orientations and magnitudes than the stresses involved in surface loading experiments. We do not know the thickness of the plate or how well it is coupled to the underlying mantle. We do not even know the sign of the basal drag force.

In a convecting or cooling mantle there is a surface thermal boundary layer (TBL) through which heat must pass by conduction. The thickness of the thermal boundary layer is controlled by such parameters as conductivity and heat flow and is not related in a simple way to the thickness of the elastic layer or the plate. Since temperature increases rapidly with depth in the conduction layer, and viscosity decreases rapidly with temperature, the lower part of the boundary layer probably lies below the elastic lithosphere; that is, only the upper part of the thermal boundary layer can support large and long-lived elastic stresses. Unfortunately, the conduction layer too is often referred to as the lithosphere. In a chemically layered Earth there can be TBLs between internal layers. These TBLs act as thermal bottle-necks and slow down the cooling of an otherwise convective mantle.

Continental cratons have high seismic velocity roots, or keels, extending to 200–300 km depth. These are referred to as *archons*. They persist because of low density and high viscosity, and because they are protected from high stress. They are more than just part of the strong outer shell. These keels can last for billions of years.

Most models of the Earth's mantle have an *upper-mantle low-velocity zone*, LVZ, overlain by a layer of higher velocities, referred to as the LID. The LID is also often referred to as the lithosphere. Seismic stresses and periods are much smaller than stresses and periods of geological interest. If seismic waves measure the relaxed modulus in the LVZ and the high-frequency or unrelaxed modulus in the LID, then, in a chemically homogenous mantle, the LID should be much thicker than the elastic lithosphere. If the LID is chemically distinct from the LVZ, then one might also expect a change in the long-term rheological behavior at the interface. If the LID and the elastic lithosphere turn out to have the same thickness, then this would be an argument for chemical, water or crystallographic control, rather than thermal control, of the mechanical properties of the upper mantle.

In summary, the following 'lithospheres' appear in the geodynamic literature (*'When I make a word do a lot of work like that,' said Humpty Dumpty, 'I always pay it extra.'*).

(1) The elastic, flexural or *rheological lithosphere*. This is the closest to the classical definition of a rocky, or strong, outer shell. It can be defined as that part of the crust and upper mantle that supports elastic stresses of a given size for a given period of time. The thickness of this lithosphere depends on stress and load duration.
(2) The *plate*. This is that part of the crust and upper mantle that translates coherently in the course of plate tectonics. The thickness of the plate may be controlled by chemical or buoyancy considerations or by stress, as well as by temperature, but there is no known way to measure its thickness. Plates are ephemeral.
(3) The *chemical* or *compositional lithosphere*. The density and mechanical properties of the lithosphere are controlled by chemical composition and crystal structure as well as temperature. If chemistry and mineralogy dominate, then the elastic lithosphere and LID may be identical. If the lithosphere, below the crust, is mainly depleted peridotite or harzburgite, it may be buoyant relative to the underlying mantle. A cratonic root, or *archon*, is often called the *continental lithosphere* or *subcontinental lithospheric mantle* (*SCLM*) and has been proposed as a geochemical reservoir.
(4) The *thermal boundary layer* or *conduction layer* should not be referred to as the lithosphere, which is a mechanical concept, but if the lithospheric thickness is thermally controlled, the thickness of the lithosphere should be proportional to the thickness of the thermal boundary layer. If TBLs get too thick, they can sink, or delaminate.
(5) *The seismic LID* is a region of high seismic velocity that overlies the low-velocity zone. At high temperatures the seismic moduli measured by seismic waves may be relaxed,

in which case they can be of the order of 10% less than the high-frequency or *unrelaxed* moduli. High-temperature dislocation relaxation and partial melting are two mechanisms that decrease seismic velocities. The boundary between the LID and the LVZ would be diffuse and frequency dependent if thermal relaxation is the mechanism. A sharp interface would be evidence for a chemical or mineralogical boundary.

The best evidence for cooling of the oceanic plate, or thickening of the thermal boundary layer, comes from the deepening of bathymetry as a function of time. The simple square-root-of-time relationship for bathymetry fails after 70–80 million years, indicating that the thermal boundary layer has reached an equilibrium thickness or that a thermal event prior to ~80 Ma affected the parts of the oceanic lithosphere that have been used to calculate the bathymetry–age relation. Dikes and sills intruded into the plate may reset the thermal age. There is some evidence that the seismic LID continues to follow the root-*t* dependency to the oldest ages. This would mean that density and seismic velocity are not correlated.

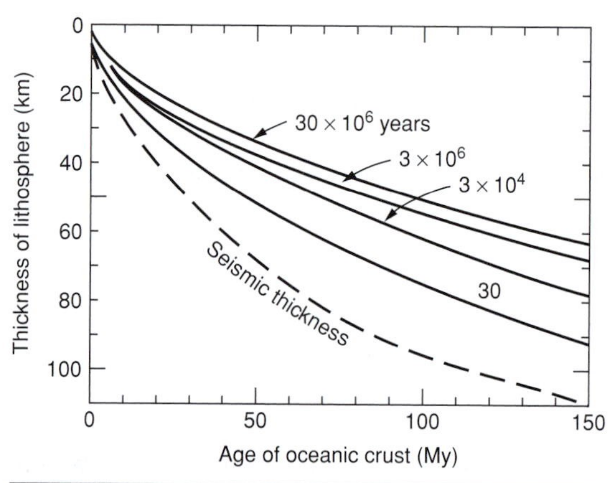

Fig. 4.1 Relationship between effective thickness of *elastic lithosphere* (also called the *rheological* or *flexural lithosphere*), age of oceanic crust and duration of loading for 1 kbar stress. For a given load the effective elastic thickness decreases with time. The elastic thickness generally follows the 600–700 degree C isotherm, and is roughly half the thickness of the thermal boundary layer (TBL). The latter should not be confused with the lithosphere and it should not be called the *lithosphere* or even the *thermal lithosphere;* these are unnecessary and misleading terms. The *plate*, the *tectonosphere* and the *tectosphere* are different concepts. The dashed line is the upper bound of seismic determinations of LID thicknesses from surface waves and the assumption of isotropy. The LID is actually anisotropic and when this is allowed for the thickness and shear velocity are less than in isotropic inversions. Because of relaxation effects the elastic model of the plate, over geological time scales, may be less than the seismic moduli.

Effective elastic thickness

The `thickness of the lithosphere` is a more complicated concept than is generally appreciated. Figure 4.1 gives estimates of the thickness of the oceanic rheological lithosphere as a function of age of crust and duration of load, using oceanic geotherms and a stress of 1 kilobar. The depth to the rheological asthenosphere is defined as the depth having a characteristic time equal to the duration of the load. For example, a 30 My load imposed on 80-My crust would yield a thickness of 45 km if the lithosphere did not subsequently cool, or if the cooling time is longer than the relaxation time. A 30-My load on currently 50-My crust would give a theological lithosphere 34 km thick with the above qualifications. The thickness of the high seismic velocity layer overlying the low-velocity zone (the seismological LID) is also shown in Figure 4.1. For load durations of millions of years, the rheological lithosphere is about one-half the thickness of some of the older estimates of the thickness of the seismological lithosphere, assuming that it is isotropic. Note that the rheological thickness decreases only gradually for old loads. On the other hand, the lithosphere appears very thick for young loads, and the apparent thickness decreases rapidly. The elastic or flexural thickness is roughly half the thickness of the thermal boundary layer (TBL).

Asthenosphere

The physical properties of the mantle depend on stress, mineralogy, crystal orientation, temperature and pressure. Small amounts of water and CO_2 can have significant effects. In the outer 100–200 km the increasing temperature with depth is more important than the increasing pressure. The melting point increases with pressure, as does viscosity, density and thermal conductivity. For a homogenous mantle, there should be a minimum in viscosity, density, seismic velocities and thermal conductivity near 100–200 km depth. This region is known as the *asthenosphere*. Below this depth, the temperature of the mantle diverges from the melting point. The coefficient of thermal expansion also goes through a minimum, and this plus the effects of partial melting and other phase changes in the shallow mantle means that small changes in temperature cause large changes in buoyancy. The minimum in thermal conductivity means that this region will have a steep thermal gradient until it starts to convect. The minimum in viscosity means that it will flow easily. Pressure stiffens the mantle, raises the melting point and makes it easier to conduct heat. All of these effects serve to concentrate deformation in the upper mantle, and serve to at least partially decouple plate motions from the interior. They also explain why the *seismic low-velocity zone* is roughly equivalent to the *asthenosphere*.

Thermal boundary layers

The thermal boundary layer (TBL) thickness of a fluid cooled from above or heated from below grows as

$$h \sim (\kappa t)^{1/2}$$

where κ is thermal diffusivity, and t is time. The TBL becomes unstable, and detaches when the *local or sublayer Rayleigh* number

$$\mathrm{Ra_c} = \alpha g(\delta T)h^3/\kappa\nu$$

exceeds about 1000 (g is acceleration due to gravity; ν is kinematic viscosity and δT is the temperature increase across the TBL). The combination $\alpha/\kappa\nu$ decreases with V, thereby lowering $\mathrm{Ra_c}$ at high P or low T.

The local Rayleigh number is based on TBL thickness. At the surface of the Earth the issue is complicated because of water and because the crust and refractory peridotite part of the outer shell are not formed by conductive cooling (they are intrinsically buoyant) and because the viscosity, conductivity and $\dot{\alpha}$ at low P are strongly T-dependent. For parameters appropriate for the top of the mantle, treated as a constant viscosity fluid, the surface TBL becomes unstable at a thickness of about 100 km. The time-scale is about 10^8 years, approximately the lifetime of surface oceanic plates. The top boundary is very viscous, stiff and partly buoyant, and the instability (called subduction or delamination) is controlled, in part, by faulting; a viscous instability calculation is not entirely appropriate. For the bottom boundary layer the deformation is more likely to be purely viscous but there may also be intrinsic stabilizing density effects. Realistic convection simulations must include all of the above effects plus self-consistent thermodynamics.

The critical thickness of the lower TBL, ignoring radiative transfer, is about ten times larger than at the surface, or about 1000 km. If there is an appreciable radiative component to thermal conductivity, or if there is a chemical component to the density, then the scale-lengths at the base of the mantle can be greater than this. Tomographic anomalies in the lower third of the mantle are very large, much larger than upper mantle slabs (cold plumes), consistent with scaling theory. The lifetime of the lower TBL, scaled from the upper mantle value, is about 3 billion years. The surface TBL cools rapidly and becomes unstable quickly because of the magnitude of the thermal properties. The same theory, scaled for the density increase across the mantle, predicts large scale and long-lived features in the deep mantle.

Slabs

When a plate starts to sink it is called a *slab*. In fluid dynamics, both cold sinking features and hot rising features are called *plumes*. Slabs are cold when they enter the mantle but they immediately start to warm up toward ambient temperature; they get warmed up from both sides. The cold portions of slabs may have high seismic velocities, compared with mantle of the same composition, and can be denser than normal mantle. Slabs are composed of oceanic crust, depleted harzburgite and serpentinized peridotite; CO_2 and water occur in the upper part. Some of these lower the melting point and seismic velocities of slabs compared with dry refractory peridotite. After the conversion of basalt to eclogite the melting point is still low. Even small amounts of fluid or melt can drastically lower the seismic velocity, even if the density remains high. Cold eclogite itself has low shear velocities and high density compared with warm peridotite.

The time scale for heating the slab to above the dehydration and melting points is a small fraction of the age of the plate upon subduction since the basalt and volatiles are near the top of the slab. Recycled slabs can be fertile and can have low seismic velocities; these are the characteristics associated with plumes. As far as the slab is concerned the surrounding mantle is an infinite heat source. Low-velocity material in the mantle can be low-temperature. Seismic velocity is controlled by composition, mineralogy and volatile content as well as by temperature. If the melts and volatiles completely leave the slab then it can become a high seismic velocity anomaly. Delaminated continental crust also sinks into the mantle but it differs in composition, density and temperature from subducted oceanic crust. Fertile patches in the mantle can have low seismic velocities and can cause melting anomalies, with no need for high-temperature plumes.

The fates of slabs

The material entering subduction zones have a large range of ages, crustal thicknesses and buoyancies (Figures 4.2 and 4.4). They therefore equilibrate different depths (Figure 4.3). A variety of evidence supports the notion that some slabs sink to 650 km depth. There is some evidence that slabs may pile up at 650 km and at ~1000 km. Some slabs may reach a position of neutral buoyancy at much shallower depths. About 15% of the surface area of the ocean floor is composed of young (<20 My) – possibly buoyant – lithosphere; roughly 0.2 km^3/year of such material is currently entering trenches. The basalt part of this is entering the mantle at about the same rate as the rate of 'midplate' magmatism.

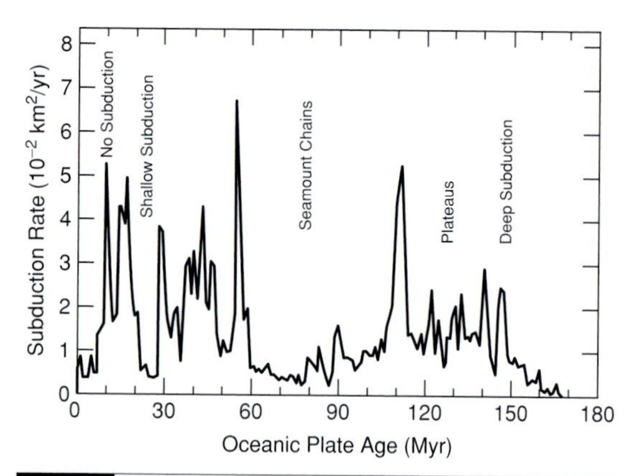

Fig. 4.2 The age distribution of oceanic plates about to enter subduction zones derived from a study of the rate of plate creation and destruction (Rowley, 2002). The younger plates do not sink far and will thermally equilibrate rapidly. The same is true for delaminated lower continental crust, which is warmer than subducted oceanic crust. Older plates will sink deeper, and will take longer to equilibrate.

A similar area (plateaus, aseismic ridges) has thick crust. Young and thick-crust material may never subduct to great depths. Over-thickened continental crust delaminates at a rate equivalent to about 10% of the rate at which oceanic crust subducts. These subducted and delaminated crustal materials contribute to the chemical and lithologic heterogeneity of the mantle and may be sampled again by partial melting at leaky transform faults, extensional regions of the lithosphere and by migrating ridges. In contrast, thicker slabs of older oceanic crust and colder lithosphere are more likely to sink deeper into the mantle. The evidence from obduction, ophiolites, flat subduction and exhumed 200 km deep slab fragments confirms the shallow, and temporary, nature of some recycling. The inference that some thick slabs, representing old oceanic plates, break through the 650 km mantle discontinuity does not require that they all do; slabs have a variety of fates. It is often assumed that subducted oceanic crust is recycled through the mantle and reappears at oceanic islands and hotspots. But the rate of subduction and the storage capacity of the mantle are such that no such recycling is required by any mass balance calculation. The evidence for recycling must come from elsewhere.

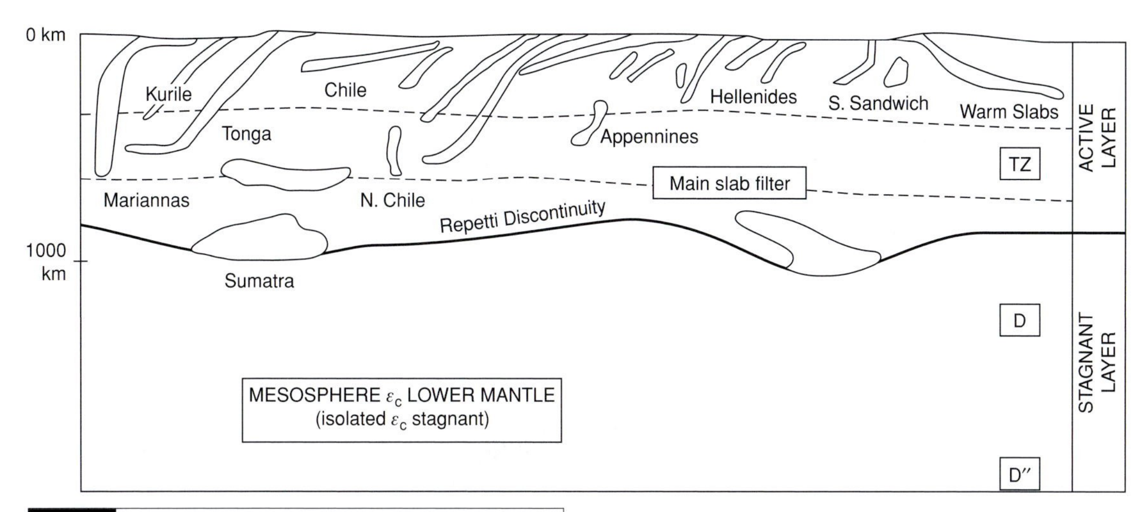

Fig. 4.3 Slabs have a variety of fates in the mantle. They sink to various depths and warm up at different rates. Some underplate continents, some have shallow dips, some are nearly vertical. A very few sink into the transition region (400–1000 km depth).

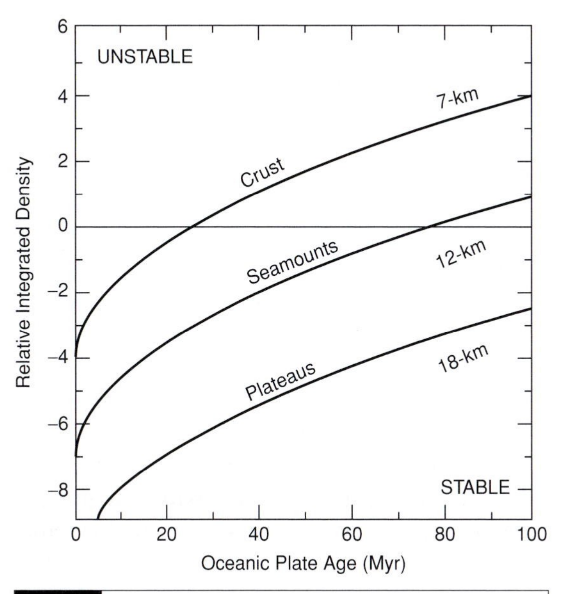

Fig. 4.4 Buoyancy of plates as a function of age and crustal thickness. Young plates and plates with thick crust (seamount chains, plateaus) may stay in the shallow mantle and be responsible for melting, or fertility, anomalies (after Van Hunen *et al.*, 2002).

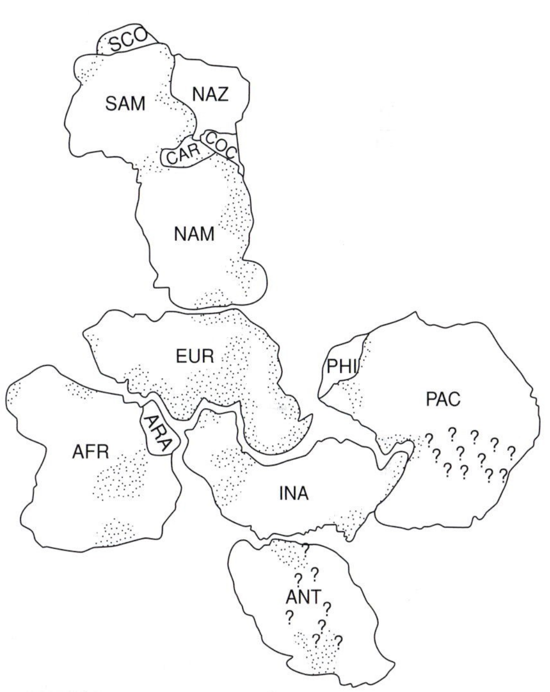

Fig. 4.5 A disjointed Lambert equal-area projection centered on each plate. Note the band of medium-sized plates encircling the large African and Pacific plates. Note the similarity in shape of AFR and PAC, and of EUR and INA.

The topology of plates

Plate tectonics on Earth, at present, consists of about a dozen large plates, of irregular shapes and sizes that move over the surface. They are separated by boundaries that meet at triple junctions. There are also many broad zones of deformation (see Figure 4.5). The seven major plates account for 94% of the surface area of the Earth (Table 4.1). Some authors recognize 20 plates and an equal number of broad zones of

Table 4.1 | Plate parameters

Plate		Area (millions of km^2)	Growth rate (km^2/100 years)
Pacific	PAC	108	−52
Africa	AFR	79	30
Eurasia	EUR	69	−6
Indo-Australia	INA	60	−35
N. America	NAM	60	9
Antarctic	ANT	59	55
S. America	SAM	41	13
Nazca	NAZ	15	−7
Arabia	ARA	4.9	−2
Caribbean	CAR	3.8	0
Cocos	COC	2.9	−4
Philippine	PHI	5	−1
Somalia	SOM		
Juan de Fuca	JdF		
Gorda	GOR		
Scotia	SCO		
SE Asia	SEA		
Total		507.6	0

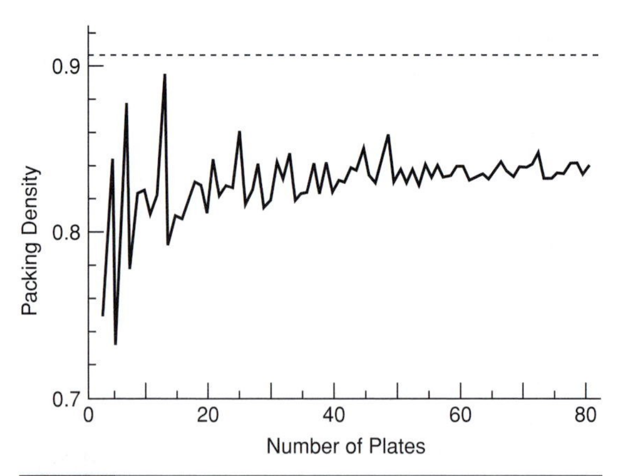

Fig. 4.6 When tiles of a given size and shape are packed on a sphere, without overlap, the packing density depends on the number *n*. The optimal packing density is achieved for $n = 12$. This figure is for spherical caps but other shapes give similar results. The voids between the tiles have dimensions of order 10–15% of the tile dimensions so only small tiles or plates can be accommodated in the interstices (after Clare and Kepert, 1991).

deformation. Eurasia (EUR) and North America (NAM) are collages of `accreted terranes`; the large Pacific plate (PAC) grew by annexing neighboring plates but it is now shrinking. Africa (AFR) and PAC are the largest plates; AFR is growing, by the migration of ridges, and PAC is shrinking by the encroachment of continents and slab rollback. Table 4.1 gives the parameters of some of the current plates.

Three of the larger plates have large fractions of their areas occupied by diffuse deformation zones and do not qualify in their entirety as *rigid* plates. In fact, at least 15% of the Earth's surface violates the rules of rigid plates and localized boundaries. It is interesting that packing of similar sized polygons or circles on a sphere leaves about 15% void space (except when *n* is 6 or 12) (see Figure 4.6). The minor plates, in aggregate, are smaller than the smallest major plate. In a close-packed or random assemblage of discs or spherical caps on a sphere the number of contacting neighbors decreases as the size disparity of the discs increases. At some point the system loses rigidity, typically at 15% void space.

The Indo-Australian (IAU) plate can be divided into an Indian (IND), Australian (AUS) and `Capricorn plate` (CAP) and diffuse zones of compression. There are few earthquakes or volcanoes to mark these boundaries. The Pacific plate has several bands of earthquakes and volcanoes that could be cited as possible diffuse or incipient plate boundaries. One of these zones extends from Samoa and Polynesia to the East Pacific Rise (EPR) – and then to Chile – and includes most of the intraplate earthquakes and active volcanoes in the Pacific. There is little evidence for relative motion between the north and south Pacific plates but current motions between EUR, Antarctica (ANT), and AFR are also very slow. Therefore, there are somewhere between 8 and 20 plates with 12 being a frequently quoted number. Plate reconstructions in the past also recognize about 12 persistent plates. Each plate, on average, has about five nearest neighbors and five next nearest neighbors.

Ridges and trenches account about equally for 80% of the plate boundaries, the rest being transform faults. Most plates are not attached to a slab or bounded by a transform. Therefore, in the real world, plates are not equal, in either size or configuration. Most plates are not attached to slabs and their freedom to move is constrained by the surrounding plates.

Plate boundaries, such as ridges and trenches, are not static and in fact, must move. The whole global mosaic of plates periodically reorganizes itself and new plate boundaries form while others close up. In some sense, plates behave as floating bubbles, or foams, constantly coalescing and changing shapes and sizes. Islands of bubbles may move coherently for a time, even though they are far from rigid. Bubbles may even support small loads and deform locally as elastic sheets.

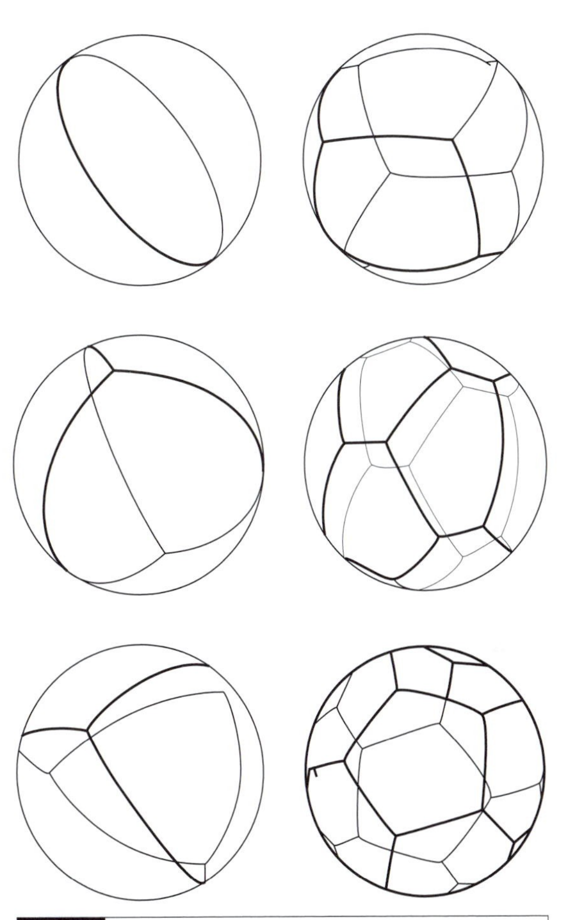

Fig. 4.7 There are ten possible configurations of great-circle arcs on a sphere that meet three at a time with angles of 120°. Six of these are shown here. Only five have identical faces. These represent possible optimal shapes of plates and plate boundaries. These figures have 2, 3, 4, 6, 10 and 12 faces or plates.

Plate theory without physics

Nature tends to organize herself with little attention to the details of physics. Large interacting systems tend to self-organize. Thus, mudcracks, honeycombs, frozen ground, convection, basalt columns, foams, fracture patterns in ceramic glazes and other natural features exhibit similar hexagonal patterns. On curved surfaces there are mixtures of pentagons and hexagons. In many cases there is a *minimum* or *economy* principle at work; minimum energy, area, stress, perimeter, work and so on. Dynamic systems, including plate tectonics, may involve dynamic minimization principles such as least dissipation. Natural processes spontaneously seek a minimum of some kind. These minima may be local; systems may jam before achieving a global minimum. Large complex interacting systems can settle down into apparently simple patterns and behavior.

On a homogenous sphere one expects that `surface tessellations` due to physical processes will define a small number of identical or similar domains (e.g. an American soccer ball or a European football). This will be true whatever the organizing mechanism unless symmetry breaking and bimodal domains are essential for the operation of plate tectonics. There are only a few ways to tessellate a sphere with regular spherical polygons that fill space and are bounded by geodesics and triple junctions. Whether the surface is subdivided into 'faces' by deep mantle convection forces or by superficial (Platonic) forces, we expect the surface of a homogenous sphere to exhibit a semi-regular pattern, although this pattern will not be stable if the triple junctions are not. Figure 4.7 illustrates some of the ways to tessellate a sphere with great circles and triple junctions.

In contrast to the beehive and other isoperimetric problems, and the bubble problems, we do not know what is minimized, if anything, in plate tectonics. It could be boundary or surface energy, toroidal energy or dissipation. Nevertheless, it may be fruitful to approach plate geometry initially as a tiling, packing or isoperimetric exercise. The problems of tessellations of spheres and global tectonic patterns are venerable ones. No current theory addresses these issues. The planform of a freely convecting spherical shell may have little to do with the sizes, shapes and number of plates. The plates may self-organize and serve as the template that organizes mantle flow. This would definitely be true if the outer shell contains most of the buoyancy of the system and most of the dissipation.

In the Earth sciences, perceived polyhedra forms on the surface have historically been used to support global contraction, expansion and drift theories [e.g. mantleplumes]. But, identical geometric forms can often be generated by very different forces; it is impossible to deduce from observed patterns alone which forces are acting. The corollary is that pattern formation may be understood, or predicted, at some level without a complete understanding of the physical details.

The classic problem of convection goes back to experiments by Benard in 1900. The hexagonal pattern he observed was attributed to thermal convection for many decades but is now known to be due to variable surface tension at the top of the fluid [Marangoni convection]. Plan views of foams also exhibit this pattern. In all of these cases space must be filled and principles of economy are at work.

The optimal arrangement of tiles on a sphere has a correspondence with many distinct physical problems (carbon clusters, clathrates, boron hydrides, quasicrystals, distribution of atoms about a central atom and bubbles in foam). It is often found that straight lines (great circles) and equant cells (squares, pentagons or hexagons rather than rectangles) of identical size serve to minimize such quantities as perimeter, surface area, energy and so on. There is an energy cost for creating boundaries and larger entities often grow at the expense of smaller ones. In phenomena controlled by surface tension, surface energy, stress, elasticity and convection one often finds tripartite boundaries (called triple junctions or valence-3 vertices) and hexagonal patterns.

Bubbles and minimal surfaces

In plate tectonic animations one sees that the plates are constantly changing shape and size much like a froth on a beer. Foams are collections of surfaces that minimize area under volume constraints. Confined bubbles have 12 pentagonal faces and consist of a basic pentagonal dodecahedron with two, three or four extra hexagonal faces. These structures have only pentagonal and hexagonal faces, with no adjacent hexagons. Clathrates and zeolites also adopt these structures. Carbon cages involve the basic pentagonal dodecahedron unit supplemented by hexagonal faces with an isolated pentagon rule for the bigger molecules. These structures are constrained minimal energy surfaces and may serve as models for the tessellation of the Earth. All vertices of these structures are triple-junctions, meeting at 120°; all faces are pentagons or hexagons. Sheared bubbles in foams can adopt complex shapes, such as boomerangs, but they are still minimal surfaces. In close-packed arrays of bubbles the dominant hexagonal coordination is interrupted by linear defects characterized by five-coordination. The midpoints between objects packed on a sphere often define a pentagonal network.

Jamming

Bubble rafts, or 2D foams, are classic minimum energy systems and show many similarities to plate tectonics. They are examples of *soft matter*. They readily deform and *recrystallize* (coarsen). Foams are equilibrium structures held together by surface *tension*. A variety of systems, including granular media and colloidal suspensions are *fragile*; they exhibit non-equilibrium transitions from a fluid-like to a solid-like state characterized by jamming of the constituent particles. A jammed solid can be refluidized by heat, vibration, or by an applied stress. Although plates are often treated as rigid objects, in the long term they act more like fragile or soft matter, with ephemeral shapes and boundaries.

Granular material and colloids tend to self-organize so as to be compatible with the load on them. They are held together by *compression*. They are rigid or elastic along compressional *stress chains* but they collapse and reorganize in response to other stresses, until they jam again in a pattern compatible with the new stresses. The system is weak to incompatible loads. Changes in *porosity*, *temperature* and *stress* are equivalent and can trigger reorganization and apparent changes in rigidity. The jamming of these materials prevents them from exploring phase space so their ability to self-organize is restricted, but is dramatic when it occurs.

In the plate tectonic context it is compression – or lack of lateral extension – that keeps plates together. When the stress changes, and before new compatible stress circuits are established, plates may experience extension and collapse. New `compatible plate boundaries` must form. Widespread volcanism is to be expected in these un-jamming and reorganization events if the mantle is close to the melting point. These events are accompanied by changes in stress and in the locations and nature of plate boundaries and plates, rather than by abrupt changes in plate motions. Volcanic chains, which may be thought of as chains of tensile stress, will reorient, even if plate motions do not. Jamming theory may be relevant to plate sizes, shapes and interactions. The present plate mosaic is presumably consistent with the stress field that formed it but a different mosaic forms if the stresses change.

In the ideal world

Suppose that the outlines of plates, or their 'rigid' cores, are approximated by circles. The fraction of a flat plane occupied by close-packed circles is 0.9069. Packing efficiency of circlular caps on a sphere depends on the number of circles (spherical caps). The area covered ranges from 0.73 to 0.89 for $n \leq 12$ and then oscillates about 0.82 (Figure 4.6). Packing of more than six regular tiles on a sphere is inefficient except for 12 equal spherical pentagons, which can tile a sphere with no gaps. The efficiency of packing, the sizes of the voids and their aggregate area depends little on the size distribution or shape, within limits. The voids between regular tiles on a sphere, when close packed, are typically 10% of the radius of the discs. Typically, equal-sized circular or polygonal non-overlapping caps can cover only about 70–85% of the surface of a sphere. About 15% void space occurs even for optimal packing of large numbers of circular or pentagonal caps. However, one can efficiently arrange 12 caps onto a sphere, with only 0–10% void space. This is much more efficient than for, say, 10, 13, 14 or 24 caps. Furthermore, the difference between the sizes of caps which pack most efficiently (least void space), and cover the whole surface most economically (least overlap), is relatively small for 12 caps (the difference is zero for regular spherical pentagons).

The ideal plate-tectonic world may have n identical faces (plates, tiles) bounded by great circle arcs which meet three at a time at 120°. This simple conjecture dramatically limits the number of possibilities for tessellation of a sphere and possibly, for the ground state of plate tectonics. In soap bubbles and plate tectonics, junctions of four or more faces are unstable and are excluded. There are ten such possible networks of great circles on a sphere, some of which are shown in Figure 4.7. If plate boundaries approximate great circles meeting three at a time at 120° then there can be a maximum of 12 plates. The ideal plate may be bounded by five or six edges and ideal plate boundaries may approximate great circles which terminate at triple junctions dominated by 120° angles.

The study of convective planforms and pattern selection is a very rich and fundamental field in thermal convection and complexity theory. Even in complex convection geometries regular polyhedral patterns are common. Pattern selection in the plate tectonics system, however, may have little to do with an imposed pattern from mantle convection although it is commonly assumed to do so.

Close-packed networks of objects are jammed or rigid. However, even open networks can jam by the creation of load-bearing stress chains which freeze the assemblage so it cannot minimize the open space. These networks can be mobilized by changing the stress. Since plates are held together by networks of compressional forces it is important that they pack efficiently. On the other hand, they must be mobile, and cannot be a permanently jammed system. Materials with this rigid-fluid dichotomy – `fragile matter` – may be better analogues for the outer layers of Earth than implied by the terms *shells* or *plates*.

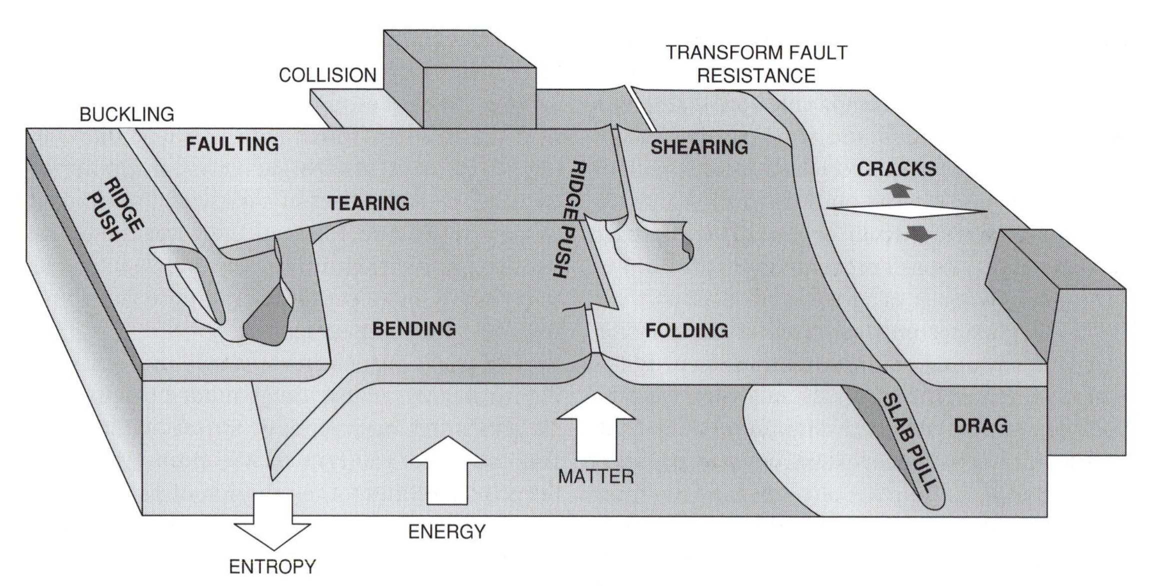

Fig. 4.8 Much of the energy and (negative) buoyancy associated with plate tectonics and mantle convection occurs in the surface layer. The mantle provides energy and matter but the plates and slabs organize and dissipate this. Far-from-equilibrium self-organized systems (SOFFE) require an outside source of energy and matter but their structure and dynamics is self-controlled.

Dynamics

How might the forces acting on the surface of a planet tend to subdivide it? The forces involve thermal contraction, slab pull, ridge push, tearing (changes in dips of bounding slabs), stretching (changes in strike of boundaries), bucking, flexure, convection, jamming, force chains and so on (Figure 4.8). In far-from-equilibrium systems the minimizing principle, if any, is not always evident. The first step in developing a theory is recognition of any patterns and identification of the rules. To date, most of the emphasis on developing a theory for plate tectonics has involved thermal convection in ideal fluids driven from within and below. *If plates drive and organize themselves, and organize mantle convection, or if plates are semi-rigid and can jam, then a different strategy is needed.*

The definition of a plate and a plate boundary is subjective. Things are constantly changing. It is improbable that there is a steady-state or equilibrium configuration of plates since only a few of the 16 possible kinds of triple junctions are stable. Nevertheless, there seems to be a pattern in the plate mosaic and there are similarities with so-called minimal surfaces. From strictly static and geometric considerations the ideal (Platonic) world may consist of 12 tiles with five NN and five NNN coordinations. The real dynamic world contains a hint of this ideal structure. A regular pentagonal dodecahedron with rigid faces would be a jammed structure (Plato argued that true knowledge was forever fixed in the ideal realm of Platonic forms). Fewer, or more, plates, deformable plates and a bimodal distribution may be required to mobilize the surface mosaic. Vibrations (earthquakes) temporarily mobilize sections of plate boundaries.

Far-from-equilibrium self-organized systems

In *complexity theory*, a system with multiple interactions and feedbacks can spontaneously organize itself into a whole series of complex structures, in apparent contradiction with the second law of thermodynamics. A *far-from-equilibrium system* can act as a whole, just as does a flock of birds, in spite of the short-range character of the interactions. Plate tectonics can be considered as a non-equilibrium process in the sense that a small perturbation in the system can cause a complete reorganization in the configuration and sizes of plates and the planform of mantle convection. Such perturbations include stress changes in a plate, and ridge–trench and continent–continent collision and separation. Changes in boundary conditions are transmitted essentially instantaneously throughout the system and new plates and new boundaries form. These topside changes cause mantle convection to reorganize or reverse. The lithosphere is not *rigid*, as generally assumed, but is fractured and is weak in extension. It reassembles itself into new plates, held together by lateral compression. Regions of extension become volcanic chains and plate boundaries. The sign of the horizontal stress field determines what becomes a plate and what becomes a plate boundary. Plate tectonics is, therefore, an example of *self-organization*. If the convecting mantle was bounded by isothermal free-slip surfaces it would be free to self-organize and to organize the surface motions, but it is not and cannot.

Near-equilibrium systems respond to a fluctuation by returning to equilibrium. Systems far from equilibrium may respond by evolving to new states. The driving forces of plate tectonics are thermal and gravitational and therefore change slowly; mass distributions, mantle temperature and viscosity are slow to change. Dissipation forces, such as friction and continent–continent collision, however, change rapidly as do normal stresses across subplate boundaries and tensile stresses in plate interiors. These stress-related fluctuations can be large and rapid and can convert the slow steady thermal and gravitational stresses into episodic 'catastrophes'. This is the essence of *self-organization*. This is the reverse of the view that it is mantle convection that causes breakup and reorganization of plates and giant igneous events, i.e. a rapidly changing and localized event in the convecting mantle or at the core–mantle boundary.

In plate tectonics and mantle convection, as in other slow viscous flow problems, there is a balance between buoyancy forces and dissipation. Basic questions are: where is most of the buoyancy – or negative buoyancy – that drives the motion, and where is most of this motion resisted? In normal Rayleigh–Bénard convection the buoyancy is generated in thermal boundary layers and is due to thermal expansion. The dissipation – resistance to vertical flow – is the viscosity of the mantle. In a homogenous layer of fluid heated from below and cooled from above the upper and lower TBLs contribute equally to the buoyancy and there is a basic symmetry to the problem. In a spherical container with mostly internal heating, the upper TBL removes more heat than the lower TBL and the heated regions move around so all parts of the interior can deliver their heat to the surface. If viscosity and strength are temperature dependent then the cold surface layers will resist motions and deformation, more so than the hotter interior. The main resistance to plate motions and mantle convection may therefore be at the surface. In one limit, the surface does not participate in the motions and acts simply as a rigid shell. In another limit, the plate acts as a low-viscosity fluid and has little effect on interior motions.

Plates and slabs are driven by gravity and resisted by mantle viscosity, plate bending and friction between plate subunits. The driving forces change slowly but resisting and dissipating forces in and between plates can change rapidly, and are communicated essentially instantaneously throughout the system. Episodic global plate reorganizations are inevitable. Even slow steady changes can reverse the normal stresses across fracture zones and can start or shut off volcanic chains. Construction and erosion of volcanoes change the local stress field and can generate self-perpetuating volcanic chains; the load of one triggers the next. All of these phenomena are controlled by the lithosphere itself, not by a hot convective template from the underlying mantle or introduction of core–mantle plumes into the shallow mantle.

Self-organized systems evolve via dissipation when a large external source of energy is available, and the systems are far from equilibrium. This self-ordering is in apparent violation of the second law of thermodynamics. This process is called `dissipation controlled self-organization`. A fluid heated uniformly from below is an example of far-from-equilibrium self-organization. The static fluid spontaneously organizes itself into convection cells but one cannot predict when, or where the cells will be, or their sense of motion. In contrast, if the heating or cooling is not uniform, the boundary conditions control the planform of convection. The fluid is no longer free to self-organize. In most mantle convection simulations the boundary conditions are uniform and various patterns and styles evolve that do not replicate conditions inside the Earth or at the surface. In many published simulations the effect of pressure on material properties is ignored. The few simulations that have been done with more realistic surface boundary conditions and pressure-dependent properties are more Earth-like, but plate-like behavior and realistic plate tectonics do not evolve naturally from the equations of fluid mechanics. Two-dimensional simulations, in planar or cylindrical configurations, cannot hope to capture the complexity of 3D mantle convection.

Far-from-equilibrium systems do strange things. They become inordinately sensitive to external or internal influences. Small changes can yield huge, startling effects up to and including reorganization of the entire system. We expect self-organization in slowly driven interaction-dominated systems. The resulting patterns do not involve templates or tuning. The dynamics in complex systems is dominated by mutual interactions, not by individual degrees of freedom. Periods of gradual change or calm quiescence are interrupted by periods of hectic activity. Such changes in the geologic record may be due to plate interactions and plate reorganization, rather than events triggered by mantle convection (mantle overturns, mantle avalanches).

Dissipation

Although plate-driving forces are now well understood, the resisting or dissipative forces are the source of self-organization. The usual thought is that mantle viscosity is what resists plate and convective motions. But bending forces in plates, for example, may control the cooling rate of the mantle. Sources of dissipation include friction along faults, internal plate deformation, continent–continent collision, and so on. These forces also generate local sources of heating. Dissipative forces and local stress conditions can change rapidly.

An important attribute of plate tectonics is the large amount of energy associated with toroidal (strike-slip and transform fault) motion. This does not arise directly from buoyancy forces involved in normal convection. In a convecting system the buoyancy potential energy is balanced by viscous dissipation in the fluid. In plate tectonics both the buoyancy and the dissipation are generated by the plates. The slab provides most of the driving buoyancy and this is balanced almost entirely by slab bending and by transform fault resistance as characterized by the toroidal/poloidal energy partitioning of plate motions. This further confirms the passive role of the mantle.

In SOFFE systems, such as thermal convection, it is not always clear what it is, if anything, that is being minimized or maximized. At one time it was thought that the pattern of convection in a heated fluid was self-selected to maximize heat flow but this did not turn out to be the case. The plate system may reorganize itself to minimize dissipation, but this is just one conjecture out of many possibilities. It can do this most effectively by changing the lengths, directions and normal stresses across transform faults, reducing the toroidal component of plate motions, localizing deformational heating, increasing the sizes of plates, changing trench-rollback and ridge migration rates and so on. If plate dynamics is a far-from-equilibrium system, sensitive to initial conditions or fluctuations, then we expect each planet to have its own style of behavior, and the Earth to have behaved differently in the past and, perhaps, during different supercontinent cycles. One does not necessarily expect a simple thermal history or a balance between heat production and heat loss.

Dissipation is also involved in the style of small-scale convection. Tabular or linear upwelling evolve as they rise to 3D plume-like upwellings. Longitudinal and transverse rolls develop under moving plates, possibly to minimize dissipation. Two-dimensional convection simulations do not capture these 3D effects. These are some of the reasons why an independent style of convection, and source of heat, is often invoked to explain 'hotspots.' Plate tectonics on a sphere has additional out-of-plane sources of lithospheric stress that may account for linear volcanic chains.

Summary

Complexity theory is a recent phrase but an ancient concept. It is usefully applied to plate tectonics. Plate tectonics is a slowly driven, interaction-dominated system where many degrees of freedom are interacting. Stresses in plate interiors can change much more rapidly than plate motions since the latter are controlled by the integral of slowly varying buoyancy forces. Apparent changes in orientation and activity of volcanic chains are therefore more likely to be due to changes in stress than to changes in plate motions. The *stress-crack hypothesis* seems to apply to most volcanic chains while the *core-plume hypothesis* accounts for only a fraction of intraplate magmatism. Dramatic changes in plate configurations in part of the world (e.g. Pangea breakup) are likely to be accompanied by a global reorganization (e.g. formation of new plate boundaries and triple junctions in the Pacific and Farallon plates) and creation of new plates. Minimization of dissipation appears to be a useful concept in global tectonics. Obviously, formation of a new crack, ridge or suture may reduce TF or collisional resistance, and may change or reverse mantle convection.

Plates, like other dynamical and complex systems, organize themselves into structures that interact. They are provided with energy from the mantle and they evolve under the influence of driving forces and interaction forces. There is no single characteristic plate size or characteristic time scale. Such systems organize themselves without any significant tuning from the outside. The fundamental question in mantle dynamics and thermal history is whether the mantle is primarily a *top-down* (plate tectonics), *inside-out* (normal mantle convection with simple boundary conditions) or *bottoms-up* (pot-on-a-stove, plume) system. In these kinds of problems, the outcome can be entirely the result of the assumptions, even if a large amount of computation intervenes. One gains intuition in complexity problems by exploring large numbers of cases; a single experiment or calculation is usually worth less than none.

Hotspots

Volcanic chains and midplate volcanism motivated an amendment to the plate tectonic hypothesis. Wilson (1973) suggested that time-progressive volcanism along the Hawaiian chain could be explained by the lithosphere moving across a 'jetstream of lava' in the mantle under the island of Hawaii. Volcanic islands appeared to be thermal anomalies so the term *hotspot* was coined to explain them, the assumption being that high absolute temperature was the controlling parameter; magma volume was used as a proxy for mantle temperature. Mantle fertility, magma focusing, passive heterogeneities, cracks and lithospheric stress can also localize melting, but the term *hotspot* is well established. Hotspots are also called *midplate volcanism, melting anomalies* and *plumes*. These are all somewhat misleading terms since *anomaly* implies background homogeneity and constancy of stress, mantle composition and temperature that is inconsistent with plate tectonics. Hotspots, in common with other volcanic features, occur in extensional regions of the lithosphere, either plate boundaries or intraplate. Many 'midplate' volcanoes started on ridges which have since moved relative to the anomaly or been abandoned. Some may be expressions of incipient or future plate boundaries.

Plates, Plumes and Paradigms provide catalogs and maps of those volcanic, tectonic and geochemical features that have become known as *hotspots*, including those that may have a shallow plate-tectonic or asthenospheric origin [see also mantleplumes, global mantle hotspot maps, scoring hotspots]. Many proposed hotspots, including isolated structures and the active portions, or inferred ends, of seamount chains, do not have significant swells, substantial magmatic output or tomographic anomalies [hotspots tomography]. A *melting anomaly*, or *hotspot*, may be due to localized high absolute mantle temperature, or a localized fertile, or fusible patch of the asthenosphere. Some have been called *wetspots* and some have been called *hotlines*. The localization may be due to lithospheric stress or architecture.

What is a Hotspot?

Morgan (1971) identified some 20 volcanic features that he proposed were underlain by deep mantle plumes equivalent in strength to the hypothetical Hawaiian plume. Most were islands or on land and most are now known to be on major fracture zones. Modern seafloor maps and global tectonic maps have many more features, including fracture zones and fields of seamounts, and there are many more active or recently active volcanoes than were recognized by Wilson and Morgan when they developed their ideas about hotspots and island chains. Most proposed plumes/hotspots are in tectonic locations that suggest lithospheric or stress control, e.g., Yellowstone, Samoa, Afar, Easter island, Louisville, Iceland, Azores and Tristan da Cuhna. There is little evidence from tomography, heat flow or magma temperatures that the mantle under hotspots is particularly hot [mantleplumes, 'global hotspot maps'].

Wilson described his concept of hotspots as follows; 'At a level of 400 to 700 km, the mantle becomes opaque, so that heat slowly accumulates until, due to local irregularities, cylindrical plumes start to rise like diapirs in the upper part of the mantle. These plumes reach the surface, which they uplift, while their excess heat gives rise to volcanism. The lavas at these uplifts are partly generated from material rising from depths of several hundred kilometers, which is thus chemically distinct from that generated at shallow levels. The plumes are considered to remain steady in the mantle for millions of years'. This hypothesis is very different from Morgan's, and both are very different from modern concepts of relatively weak, deep mantle plumes, which are easily 'blown in the mantle wind,' and that provide only a small fraction of the Earth's heat flow and magma. The common theme of hotspot hypotheses is that they are caused by hot upwelling mantle from a deep thermal boundary layer (TBL) well below the upper mantle; the upwellings are active, that is, they are driven by their own thermal buoyancy rather than passively responding to plate tectonics and subduction. A TBL instability gives rise to narrow buoyant upwellings. The concepts of 'plumes' and 'hotspots' have been coupled since these early papers, even though volcanoes, volcanic chains, time-progressive volcanism and swells can exist without plumes. Passive upwellings such as those triggered by spreading or by subduction are not considered to be plumes, and neither are intrusions such as dikes triggered by magma buoyancy, despite the fact that all these phenomena are plumes in the fluid dynamics sense.

The terms *hotspot* and *plume* refer to different concepts but they are generally used interchangeably; they do not have well-defined meanings that are agreed upon in the Earth science community [see global hotspot maps]. For some, *hotspot* is simply a region of magmatism that is unusual – and sometimes not so unusual – in location, volume or chemistry, compared with some segments of the ocean ridge system. Geochemists use the term *plume* to refer to any feature that has 'anomalous' geochemistry but *anomalous* is not used in any statistical or consistent sense. *Geochemical plumes* are simply regions that provide basalts that differ in some detail from what has been defined as normal MORB; the phrase *fertile blob* could be substituted for *plume* in isotope geochemistry papers since there is no constraint on temperature, melting temperature or depth of origin. In fluid dynamics a *plume* is a thermal upwelling or downwelling. The word *plume* is now applied to tomographic, geochemical and other 'anomalies' that have little or no surface expression, thereby decoupling the concepts of hotspots and melting anomalies from plumes. Some seismologists use *plume* to refer to any region of the mantle having lower than average seismic velocity. One assumption behind this usage is that any low-velocity structure is hot and buoyant. Early tomographic maps used red and other warm colors for low seismic velocities and this was interpreted by later workers as indicating that low-velocity regions must be hot. A low-velocity region may, however, be due to volatiles or a low melting point or a different chemistry, and need not be buoyant, or even hot compared with the surrounding mantle. Dense eclogite sinkers from delaminated continental crust, for example, can show up as low shear wave velocity features.

Table 4.2 is a compilation of features that have been consistently labeled as 'hotspots,' along with their locations [scoring hotspots]. Some of these features have little or no surface expression. Some are the extrapolated ends of volcanic chains, or were invented to extend assumed monotonic age progressions to the present day, assuming a fixed hotspot or plume origin. Some volcanic regions have multiple centers that are attributed to independent hotspots. Individual hotspots, volcanic chains, mechanisms and interpretations are discussed in Foulger *et al.* (2005) and www.mantleplumes.org. Very few hotspots have associated upper mantle or lower mantle tomographic anomalies [mantle hotspots] and most of these do not have substantial swells or linear volcanic chains. Most hotspots, including many with large swells and noble-gas anomalies lack evidence for a deep origin. The most impressive and geochemically distinct hotspots, including Hawaii, Iceland, Yellowstone, the Galapagos, Afar and Reunion are not underlain by lower-mantle P-wave seismic anomalies.

Mechanisms for forming hotspots

Linear volcanic chains are the defining characteristics of plate tectonics and of active plate boundaries. They are also expected in the waxing and waning stages of plate boundaries and plate creation and destruction. The common characteristics of features designated as *hotspots* or *midplate volcanism* also suggest an underlying common cause. High temperature, compared to midocean ridges, or a common chemistry do not seem to be requirements [mantleplumes]. Mechanisms involving lithospheric architecture and stress, and variable mantle fertility may satisfy the observations for most hotspots. It is generally agreed that *most* melting anomalies along ridges and midplate volcanism are *not* due to deep mantle plumes and that *hotspots* and *plumes* are not the same thing. Nevertheless, a plume is the default explanation for a hotspot or a geochemical anomaly, particularly in the field of geochemistry where one sees reference to 'the' Shona plume, 'the' Hollister plume and so on, even for minor volcanic features. What is meant is that some isotopic signature in a sample does not conform to what is traditionally associated with midocean-ridge basalts.

The lithosphere is cracked and variable in thickness. The underlying mantle is close to the melting point from the base of the plate to about 200 km depth. Low-melting-point material is recycled into the shallow mantle by subduction and lower continental crust delamination or foundering. Plate boundaries are continually forming and reforming, and being reactivated. Under these conditions volcanism is expected to be widespread, simply as a result of near-surface conditions, but neither uniform nor random. A small change in local stress can cause dikes, extrusions and volcanoes where formerly there were sills, intrusion and underplating that did not reach the surface. Nothing except stress need change. A change in water content may also change the eruptive style.

Plate tectonics explains most volcanic features. Mechanisms for *melting anomalies* along plate boundaries or for features not obviously associated with active plate boundaries include crack propagation, self-perpetuating volcanic chains, reactivated plate boundaries, incipient plate boundaries, membrane and extensional stresses, gravitational anchors, reheated slabs, delamination, buoyant decompression melting of mantle heterogeneities, dike propagation, leaky transform faults and rifting unrelated to uplift. The lithosphere and asthenosphere control the locations of many, if not all, midplate volcanoes and volcanic chains (Foulger *et al.*, 2005).

Plume and anti-plume tectonics

Mort de ma vie! all is confounded, all! Reproach and everlasting shame Sits mocking in our plumes.

Shakespeare

Melting anomalies at the surface of the Earth, such as volcanic chains, or midocean ridges, can result from hot or fertile regions of the shallow mantle – hotspots or fertile spots – or from hot upwelling jets – plumes. They can be the result of variations in thickness or stress of the plates. Focusing, edge effects, ponding and interactions of surface features with a partially molten asthenosphere can create melting

Table 4.2 Hotspot locations

Long.	Lat.	Most recent activity (Ma)	Hotspot		Location	Age of crust (Ma)	Tectonics	
42	12		Afar, Ethiopia		Africa	Ridge	TJ	
77	−37	0.0	Amsterdam/St. Paul		Indian Ocean	Ridge	Major FZ	
−14	−8	2.0	Ascension		Equatorial Atlantic	7	FZ	
−140	−29		Australs/Cook		Aitutaki, Macdonald			
3	46		Auvergne	Chaine de Puys	France			
−28	38	0.0	Azores		North Atlantic	32	Major FZ	TJ
−113	27		Baja/Guadalupe		Mexico	Ridge		
165	−67		Balleny		Antarctic plate			
−60	30	30.0	Bermuda		N. Atlantic	117	Edge?	
3	−54	0.0	Bouvet		S. Atlantic	Ridge	RRR	
−135	53	0.0	Bowie Seamount		NE pacific	18	Leaky TF	
−5	−16		Cameroon line		C. Atlantic	41	FZ	
−17	28	1.6	Canary Islands		N. Atlantic	155	FZ	
−24	15	0.0	Cape Verde		off NW Africa	140	Edge?	
166	5	1.1	Caroline Islands		W. Equatorial pacific	153		
80	−35		Christmas		S. Indian Ocean	Ridge		
−9	−8		Circe/Ascension		Equatorial Atlantic	Ridge	FZ	
−130	46	0.1	Cobb	Juan de Fuca	NE Pacific	4		
44	−12	0.0	Comores Islands		N. of Madagascar	128		
−52	37	73.0	Corner		N. Atlantic	98		
45	−46		Crozet/Pr. Edward		SW Indian Ocean	80		
−109	−27	0.1	Easter		SE Pacific	Ridge(4.5 Ma)	Easter FZ	
7	50		Eifel		Europe			
167	−78		Erebus/Ross Sea		Antarctica		Rift	
−32	−4		Fernando De Noronha		Off NE S. America		Edge	3 FZ
−115	−26	0.1	Foundation Smts.		S. Central Pacific	Ridge		
−86	1	0.0	Galapagos		E. Equatorial Pacific	Near Ridge	Fragmentation	

−10	−40	2.6	Gough		S. Atlantic	Tristan Chain	
28	−30		Great Meteor		N. Atlantic	80	
−118	29		Guadalupe		Mexico	Ridge	Abandoned Ridge
−155	19	0.0	Hawaii		C. Pacific	93	
6	23		Hoggar Mountains		Africa		
−20	64		Iceland		N. Atlantic	Ridge	
−8	71		Jan Mayen		Arctic Ocean	Ridge	Microcontinent
231	45		Juan De Fuca		Pacific NW	Dying Ridge	
−79	−34		Juan Fernandez		SE Pacific		FZ, Tear?
63	−49		Kerguelen		S. Indian Ocean	118	Microcontinent, Shear
159	−31		Lord Howe		Off E. Australia		
−138	−50	0.5	Louisville Hotspot	Inferred	SW Pacific	85	Eltanin FZ System
−140	−29	0.0	Macdonald		Australs/Cook	42	
−17	33	1.0	Madeira		N. Atlantic	129	Edge
38	−47		Marion	Prince Edward	S. of Madagascar	Ridge	
−139	−8	0.5	Marquesas Islands		S. Central Pacific	63	
−154	−21		Marshall-Gilbert Islands		SW Pacific		
57	−20		Mauritius		Indian Ocean	59	Old Ridge/TF
1	−52		Meteor		S. Atlantic		
167	−78	0.0	Mount Erebus		Antarctica	Rift	Rift
28	−32	85.0	New England		NW Atlantic	95	FZ
−130	−24	1.0	Pitcairn (Unnamed seamount)		SW Pacific	23	FZ
−159	−22	1.1	Rarotonga (Cook)		S. Pacific	>90	
−104	36		Raton	New Mexico	New Mexico		Shear Zone
−111	19		Revilla Gigedo	Socorro	Off Baja		

(cont.)

Table 4.2 | *(cont.)*

Long.	Lat.	Most recent activity (Ma)	Hotspot		Location	Age of crust	Tectonics
56	−21	0.0	Reunion (Fournaise)		Indian Ocean	64	Abandoned Ridge-TF
64	−20	1.5	Rodriguez		Indian Ocean	10.6	
−170	−14		Samoa		SW Pacific	100	
−80	−26		San Felix		SE Pacific		Tear
−118	17	0.0	Shimada Smt.		N. Central Pacific	20	Isolated
0	−52		Shona		S. Atlantic	Ridge	Ridge Jump
−148	−18	0.0	Societies		Mehetia		
−111	19		Socorro		E. Pacific		Abandoned Ridge
−10	−17	13.0	St. Helena		C. Atlantic	41	Leaky FZ
78	−39	0.0	St. Paul Island	Amsterdam	Indian Ocean	4	FZ
−148	−18		Tahiti	Mehetia	Societies		
17	21		Tibesti, Chad		Africa		
−29	−21	1.5	Trinidade/Marten		Off E. S. America	90	Edge
−12	−37	0.1	Tristan da Cunha	Walvis Ridge	S. Atlantic	Near Ridge	
−111	44		Yellowstone		NW USA	Edge Craton	Shallow

TJ = Triple junction.
TF = Transform fault.
FZ = Fracture zone.
RRR = Ridge TJ.

anomalies at the surface. Adiabatic decompression melting can be caused by passive upwellings – such as material displaced upwards by sinking slabs – changes in thickness of the lithosphere, or by recycling of basaltic material with a low melting temperature. The usual explanation for melting anomalies is that they result from active hot upwellings from a deep thermal boundary layer. In the laboratory, upwellings thought to be analogous to these are often created by the injection of hot fluids, not by the free circulation of a fluid.

In the fluid dynamics literature upwelling and downwelling features in a fluid that are maintained by thermal buoyancy are called *plumes*. *Thermal plumes* form at *thermal boundary layers*, and rise from boundaries heated from below. As they rise, they migrate horizontally along the boundary generating both small-scale and large-scale flow, including rotation of the whole fluid layer, a sort of convective wind. The term *plume* in the Earth sciences is not always consistently used or precisely defined. What a geophysicist means by a plume is not always understood to be the case by a geochemist or a geologist, or a fluid dynamicist.

The mantle is not the ideal homogenous fluid heated entirely from below – or cooled from above – that is usually envisaged in textbooks on fluid dynamics and mantle dynamics. Normal convection in a fluid with the properties of the mantle occurs on a very large scale, comparable to the lateral scales of plates and the thicknesses of mantle layers. In geophysics, *plumes* are a special form of *small-scale convection* originating in a thin, lower, thermal boundary layer (TBL) heated from below; in this sense not all upwellings, even those driven by their own buoyancy, are plumes. Narrow downwellings in the Earth are fluid dynamic plumes but they are called *slabs*. The dimension of a plume is controlled by the thickness of the boundary layer. There is likely to be a thermal boundary layer at the core–mantle boundary (CMB), and there is one at the Earth's surface. There is no reason to believe that these are the only ones, however. In the Earth, boundary layers tend to collect the buoyant products of mantle differentiation at the surface one (continents, crust, harzburgite) and the dense dregs at the lower one. They are, therefore, not strictly thermal; they are *thermo-chemical boundary layers*.

The presence of deep TBLs does not require that they form narrow upwelling instabilities that rise to the Earth's surface. A TBL is a necessary condition for the formation of a plume – as it is understood in the geophysics and geochemistry literature – but is not a sufficient condition. Likewise, the formation of a melting anomaly at the surface, or a buoyant upwelling, does not require a deep TBL.

Internal and lower thermal boundary layers in the mantle need not have the same dimensions and time constants as the upper one. Plate tectonics and mantle convection can be maintained by cooling of plates, sinking of slabs and secular cooling, without any need for a lower thermal boundary layer – particularly one with the same time constants as the upper one. Buoyant decompression melting, caused by upwelling, can also be generated without a lower thermal boundary layer. Because of internal heating and the effects of pressure, the upper and lower thermal boundary layers are neither symmetric nor equivalent. The Earth's mantle does not have the same kind of symmetry regarding convection that is exhibited by a pot of water on a stove that is heated from below and cooled from above.

Upwellings in the mantle can be triggered by spreading, by hydration, by melting, by phase changes and by displacement by sinking materials (passive upwelling). In a heterogenous mantle with blobs of different melting points and thermal expansion coefficients, upwellings can form without thermal boundary layers. Convection in the mantle need not involve *active* upwellings. The active, or driving, elements may be dense downwellings with complementary passive upwellings. Upper mantle convection is primarily passive; the plates and the slabs are the active elements.

Cooling of the surface boundary layer creates dense slabs. Extension of the lithosphere allows the intrusion of dikes. These are all plumes in the strict fluid dynamic sense but in geophysics the term is restricted to narrow hot upwellings rooted in a deep thermal boundary layer, and having a much smaller scale than normal mantle convection and the lateral dimensions of plates. Sometimes geophysical plumes are considered to be the way the core gets rid of its heat. The corollary is that plate tectonics is the way that the mantle gets rid of its heat. Actually, the mantle and the surface and boundaries layers are coupled.

If much of the upper mantle is at or near the melting point, which seems unavoidable, then there is a simple explanation for island chains, extensive diking and large igneous provinces. The criterion for diking is that the least compressive axis in the lithosphere is horizontal and that melt buoyancy can overcome the strength of the plate. Magma can break the lithosphere at lower deviatoric stresses than it would otherwise take. Dikes propagate both vertically and along strike, often driven by the hydrostatic head of an elevated region. Dikes, like cracks, can propagate laterally. Typical crack propagation velocities are 2 to 10 cm/year, comparable to plate velocities. These velocities are known from the study of *ridge propagators* and the penetration of ridges into continents. These propagating cracks are volcanic so they are probably examples of propagating magma-filled cracks or dikes [giant radiating dikes].

Geophysical plumes

One finds many kinds of plumes in the Earth science literature; diapiric, cavity, starting, incubating, plume heads, plume tails, zoned plumes, plumelets, megaplumes and so on. When the whole lower TBL goes unstable we have a diapiric plume. Scaling relations show that these will be huge, slow to develop and long-lived. When a thin low-viscosity layer near the core feeds a plume head, we have a cavity plume; the physics is similar to a hot-air balloon. Temperature dependence of viscosity is essential for the formation of cavity plumes with large bulbous plume heads and narrow plume tails. Temperature dependence and internal heating, however, reduce the temperature drop across the lower thermal boundary layer and the viscosity contrasts essential for this kind of plume. Pressure broadens the dimensions of diapiric plumes and cavity plume heads. It is this 'pressure-broadening' that makes intuitive concepts about plumes, based on unscaled laboratory simulations, implausible for the mantle.

The critical dimension of lower-mantle thermal instabilities is predicted to be about 10 times larger than at the surface, or about 1000 km. The timescale of deep thermal instabilities scaled from the upper mantle value is $\sim 3 \times 10^9$ years. Radiative transfer and other effects may increase thermal diffusivity, further increasing timescales. The surface TBL cools rapidly and becomes unstable quickly. The same theory, scaled for the density and pressure increase across the mantle, predicts large and long-lived features above the core.

Geophysical plume theory is motivated by experiments that inject narrow streams of hot fluid into a stationary tank of low viscosity fluid. These injected streams are not the same as the natural instabilities that form at the base of a fluid heated from below. Computer simulations often mimic the laboratory experiments by inserting a hot sphere at the base of the fluid and watching it evolve with time during the period that it rises to the surface. These are not cyclical or steady-state convection experiments, and they ignore the effects of pressure on physical properties. Because of the inordinate sensitivity of mantle convection and lithospheric stress to surface conditions it is necessary to exhaust the various plate-tectonic explanations for episodicity, plate reorganization, volcanic chains, periods of massive magmatism, continental breakup etc. before one invokes unstable deep thermal layers and narrow active upwellings.

Boussinesq approximation

The Boussinesq approximation is widely used in fluid dynamics. It simply means that all physical properties are assumed to be independent of depth and pressure and, except for density, independent of temperature. This is not a good approximation for the deep mantle. Some workers try to fix up the limitations of this approximation by assigning depth or temperature dependence to some of the physical properties, but if this is not done in a thermodynamically self-consistent way, it can make things worse. High Prandtl number essentially means high viscosity and low thermal conductivity, a characteristic of the mantle but not of laboratory fluids.

Anti-plumes: when the crust gets too thick

Plumes are Rayleigh–Taylor instabilities caused by temperature. A chemical layer can also get denser than the underlying layers or less dense than the overlying layers. A salt dome is an example of the latter. Many of the phenomena that have been attributed to thermal instabilities from a thermal boundary layer at the core–mantle boundary can also be explained by lithologic instabilities of the upper boundary layer. Plumes, as used in the geophysical literature, are thermal instabilities of the lower thermal boundary layer, heated from below by the core. Anti-plumes are chemical instabilities triggered by phase changes in the upper boundary layer, which is cooled from above. They are 'anti-' in all these respects. These are generally due to delamination of the lower continental crust.

If continental crust gets too thick, the dense eclogitic (or garnet pyroxenitic) bottom detaches, causing uplift, asthenospheric upwelling and pressure-release melting. Delamination of lower crust explains a number of geologic observations, such as uplift, uncompensated high elevations, high heat flow and magmatism. These are the predicted consequences of passive upwelling of hot asthenospheric mantle in response to the removal of the lower crust. Delamination may be important in creating compositional heterogeneity in the mantle. It may also be important in creating magmatic regions that have been labeled as 'hotspots.' Once a sufficiently thick eclogite layer forms, it will detach and founder because of its high density. This results in uplift, extension and magmatism. Delamination of a 10-km-thick eclogite layer can lead to 2 km of uplift and massive melt production within about 10–20 Myr. Lower-crust density exceeds the mantle density by 3–10% when it converts to eclogite. Crustal delamination is a very effective and non-thermal way of thinning the lithosphere, extending the melting column and creating massive melting and uplift. In contrast to thermal models, uplift occurs during and after the volcanism and the crustal thinning is rapid. Delamination may also be involved in the formation of oceanic plateaus such as Ontong Java plateau.

There are several ways to generate massive melting; one is to bring hot material adiabatically up from depth until it melts; the other is to insert low-melting-point fertile material – delaminated lower arc crust, for example – into the mantle from above and allow the mantle to heat it up. Both mechanisms may be involved in the formation of large igneous provinces – LIPs. The time-scale for heating and recycling of lower-crust material is much less than for subducted oceanic crust because the former starts out much hotter and does not sink as deep.

The removal of dense mafic roots from over-thickened continental crust may not be a Rayleigh–Taylor instability; roots may be sheared, faulted or peeled off, or entrained, but the results regarding the fertilization of the mantle and local lowering of the melting point are the same.

Chapter 5

The eclogite engine

The World's great age begins anew,
The golden years return,
The Earth doth like a snake renew
Her winter weeds outgrown

Shelley

The water cycle drives geological processes at the surface. The fact that water coexists as fluid, vapor and solid is crucial in shaping the Earth's surface. The fact that conditions in the upper mantle can readily convert eclogite to magma to basalt, and back, with enormous density changes, is crucial in global magmatism and tectonics. Phase changes in the mafic components of the upper mantle are larger than thermal expansion effects and they drive the *eclogite engine.*

There are several ways to generate massive melting in the mantle; one is to bring hot material adiabatically up from depth until it melts; the other is to insert low-melting point fertile material – delaminated lower arc-crust, for example – into the mantle from above and allow the mantle to heat it up. Both mechanisms may be involved in the formation of large igneous provinces – LIPs. The timescale for heating and recycling of lower-crust material is much less than for subducted oceanic crust because the former starts out much hotter and does not sink as deep.

The standard petrological models for magma genesis involve a homogenous pyrolite mantle, augmented at times by small-scale pyroxenite veins or recycled oceanic crust. It is increasingly being recognized that large blocks of eclogite in the mantle may be an important fertility source. Delaminated continental crust differs in many important respects from recycled MORB. It does not go through subduction zone and seafloor processing, it starts out hotter than MORB, it may occur in bigger blobs, and it is not accompanied by the same amount, if any, of buoyant infertile harzburgite. Figure 5.1 illustrates the lower crust delamination cycle. The crust thickens by tectonic or igneous processes, eventually forming dense eclogite that detaches and sinks into the mantle. It reaches a level of neutral buoyancy and starts to warm up. Eventually it rises and forms a warm fertile patch in the mantle. If the overlying continents have moved off, a midplate magmatic province is the result.

Thermal expansion is the main source of buoyancy in thermal convection of simple fluids. Phase changes can be more important in the mantle. When basalt converts to eclogite, or when it melts, there are changes in density that far exceed those associated with thermal expansion. Removal of the dense lower continental crust is an important element of plate tectonics that complements normal subduction zone and ridge processes. It causes uplift and magmatism and introduces distinctive materials into the mantle that are dense, fertile and have low melting points. After delamination, eclogitic lower crust sinks into the mantle where it has relatively low seismic velocities and melting point (Figures 5.2 and 5.4) compared with normal mantle peridotite. These fertile mafic blobs sink to various depths where they warm up, melt and

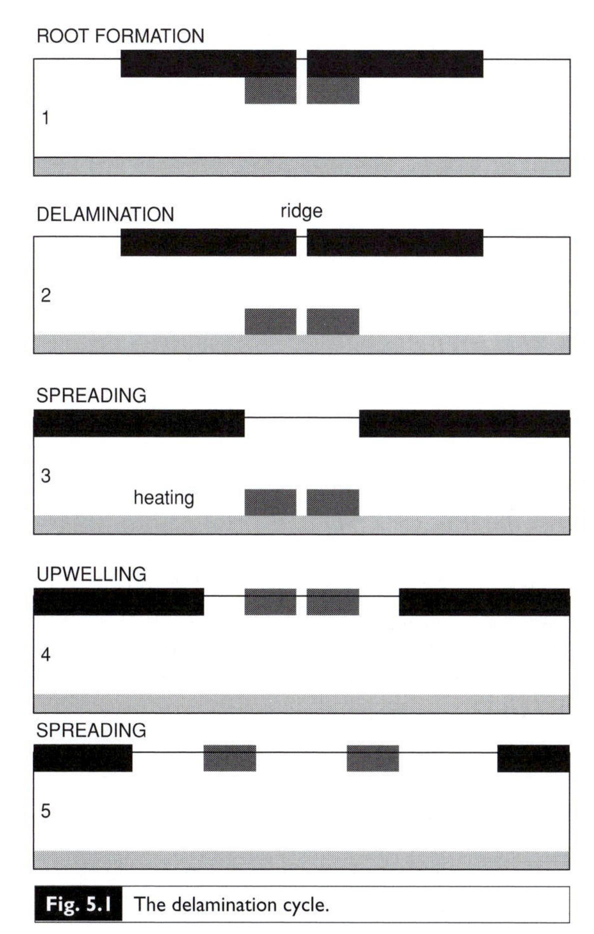

Fig. 5.1 The delamination cycle.

return to the surface as melting anomalies, often at ridges. Large melting anomalies that form on or near ridges and triple junctions may be due to the resurfacing of fertile blobs, including delaminated continental crust. This is expected to be especially prevalent around the 'passive' margins of former supercontinents: Bouvet, Kerguelen, Broken Ridge, Crozet, Mozambique ridge, Marion, Bermuda, Jan Mayen, Rio Grande rise and Walvis ridge may be examples; the isotopic signatures of these plateaus are expected to reflect lower crustal components.

The eclogite cycle

Cold oceanic crust and warm lower continental crust are continuously introduced into the mantle by plate tectonic and delamination processes. These materials melt at temperatures higher than their starting temperatures, but lower than normal mantle temperatures. They heat up by conduction from the surrounding mantle (Figure 5.2); they are fertile blobs in the mantle, in spite of being cold.

Trenches and continents move about on the Earth's surface, refertilizing the underlying mantle. Ridges also migrate across the surface, sampling and entraining whatever is in the mantle beneath them. Sometimes a migrating ridge will override a fertile spot in the asthenosphere; a melting anomaly ensues, an interval of increased magmatic output. Ridges and trenches also die and reform elsewhere. What is described is a form of convection, but it is quite different from the kind of convection that is usually treated by geodynamicists. It is more akin to fertilizing and mowing a lawn. The mantle is not just convecting from a trench to a ridge; the trenches and ridges are visiting the various regions of the mantle. The fertile dense blobs sink, more-or-less vertically into the mantle and come to rest where they are neutrally buoyant, where they warm up and eventually melt (Figures 5.3 and 5.4). They will represent more-or-less fixed points relative to the overlying plates and plate boundaries. They are chemically distinct from 'normal' mantle. Eclogites are not a uniform rock type; they come in a large variety of flavors and densities and end up at various depths in the mantle.

Mantle stratigraphy

The densities and shear-velocities of crustal and mantle minerals and rocks are arranged in order of increasing density (Figure 5.2). Given enough time, this is the stratification toward which the mantle will evolve - the neutral-density profile of the mantle. Such density stratification is already evident in the Earth as a whole, and in the crust and continental mantle. This stratification, at least of the upper mantle, is temporary, however, even if it is achieved. Cold eclogite is below the melting point but eclogite melts at much lower temperatures than the surrounding peridotite. As eclogite warms up, by conduction of heat from the surrounding mantle, it will melt and become buoyant, creating a form of yo-yo tectonics.

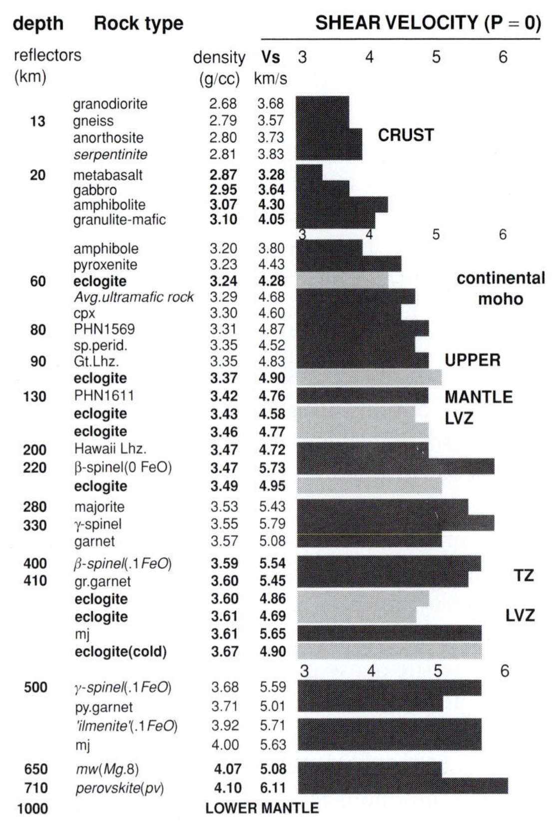

Fig. 5.2 Density and shear-velocity of crustal and mantle minerals and rocks at STP, from standard compilations. The ordering approximates the situation in an ideally chemically stratified mantle. The materials are arranged in order of increasing density. The STP densities of peridotites vary from 3.3 to 3.47 g/cm^3; eclogite densities range from 3.45 to 3.75 g/cm^3. The lower density eclogites (high-MgO, low-SiO2) have densities less than the mantle below 410 km and will therefore be trapped at that boundary, even when cold, creating a LVZ. Eclogites come in a large variety of compositions, densities and seismic velocities. They have much lower melting points than peridotites and will eventually heat up and rise, or be entrained. If the mantle is close to its normal (peridotitic) solidus, then eclogitic blobs will eventually heat up and melt. Cold oceanic crust can contain perovskite phases and can be denser than shown here. Regions of over-thickened crust can transform to eclogite and become denser than the underlying mantle. The upper mantle may contain peridotites, lherzolites and some of the less garnet-rich, high-MgO, eclogites.

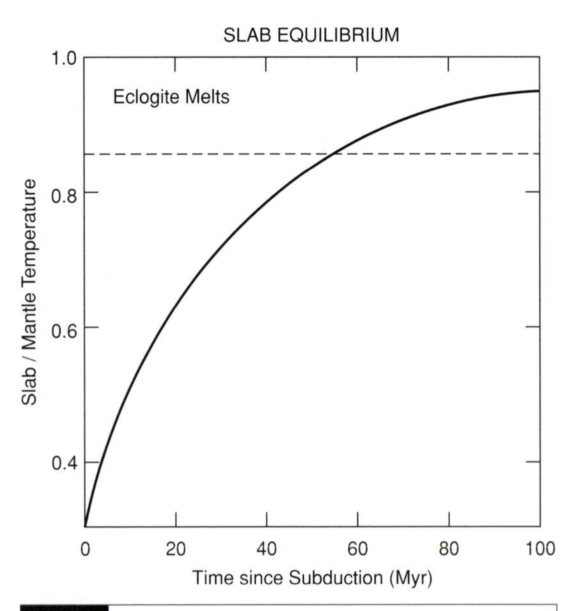

Fig. 5.3 Heating rates of subducted slabs due to conduction of heat from ambient mantle. Delaminated continental crust starts hot and will melt quickly. The total recycle time, including reheating, may take 30–75 Myr. If delamination occurs at the edge of a continent, say along a suture belt, and the continent moves off at 3.3 cm/year, the average opening velocity of the Atlantic ocean, it will have moved 1000 km to 2500 km away from a vertically sinking root. One predicts paired igneous events, one on land and one offshore (see Figure 5.5).

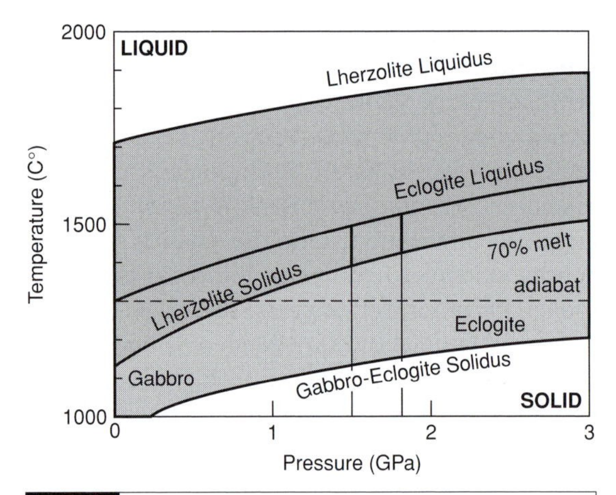

Fig. 5.4 Melting relations in dry lherzolite and eclogite based on laboratory experiments. The dashed line is the 1300-degee mantle adiabat, showing that eclogite will melt as it sinks into normal temperature mantle, and upwellings from the shallow mantle will extensively melt gabbro and eclogite. Eclogite will be about 70% molten before dry lherzolite starts to melt [compiled by J. Natland, personal communication].

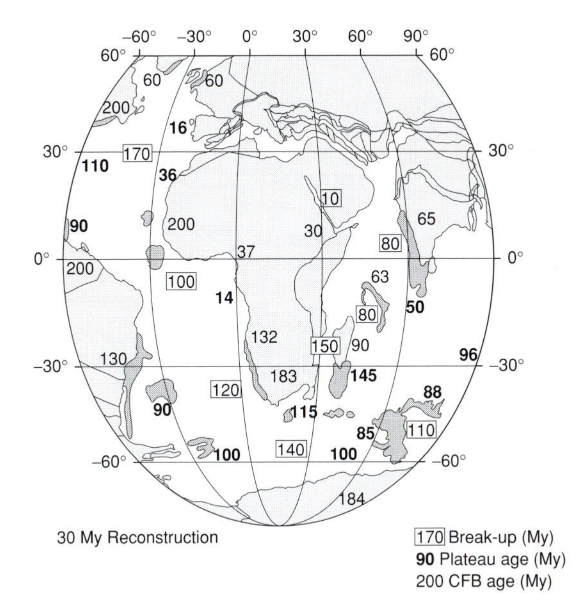

Fig. 5.5 Distributions and ages of LIPs in the Gondwana hemisphere. The ages of continental breakup and the ages of volcanism or uplift are shown. Delamination of lower crust may be responsible for the LIPs in the continents, which usually occur along mobile belts, island arcs and accreted terranes. If the continents move away from the delamination sites, it may be possible to see the re-emergence of fertile delaminated material in the newly formed ocean basins, particularly where spreading ridges cause asthenospheric upwellings.

As it warms up by conduction it will rise, adiabatically decompress and melt more. It will either erupt, causing a melting anomaly (Figure 5.5), or underplate the lithosphere, depending on the stress state of the plate. Eclogites have a wide variety of properties, depending on the amount of garnet, and the Fe and Na contents. At shallow depths eclogites have higher seismic velocities than peridotites. At greater depth, eclogite can create a LVZ. Cold dense sinking eclogite can create a low-velocity feature that could be mistaken for a hot rising plume. As eclogite melts it will have even lower shear velocities.

There are two main sources of eclogite, subducted oceanic crust, and delaminated thick crust at batholiths and island arcs (arclogites). The density differences between basalts and gabbros, eclogites and magmas are enormous compared with density variations caused by thermal expansion and normal compositional effects. The dynamics of the outer Earth will be strongly influenced, if not controlled, by this 'eclogite engine.'

Do we need to recycle oceanic crust?

If current rates of oceanic crust recycling operated for 1 Gyr, the total oceanic crust subducted would account for 2% of the mantle and it could be stored in a layer only 70 km thick. The surprising result is that most subducted oceanic crust need not be recycled or sink into the lower mantle in order to satisfy any mass balance constraints.

The recycling rate of lower crustal cumulates (arclogites) implies that about half of the continental crust is recycled every 0.6 to 2.5 billion years. In contrast to oceanic crust one can make a case that eroded and delaminated continental material is not stored permanently or long-term or very deep in the mantle; it is re-used and must play an important role in global magmatism and shallow mantle heterogeneity. Sediments, altered oceanic crust, oceanic lithosphere and delaminated continental crust probably all play some role in the source regions of various mantle magmas, including midocean-ridge basalts, ocean-island basalts and continental flood basalt. These recycled components, however, are unlikely to be stirred efficiently into the mantle sources of these basalts. The melting and eruption process, however, is a good homogenizer.

Chapter 6

The shape of the Earth

When Galileo let his balls run down an inclined plane with a gravity which he had chosen himself . . . then a light dawned upon all natural philosophers.

I. Kant

Terrestrial planets are almost spherical because of gravity and the weakness of rock in large masses. The largest departures from sphericity are due to rotation and variations in buoyancy of the surface and interior shells. Otherwise, the overall shape of the Earth and its heat flow are manifestations of convection in the interior and conductive cooling of the outer layers. The style of convection is uncertain. There are various hypotheses in this field that parallel those in petrology and geochemistry. The end-members are whole-mantle convection in a chemically uniform mantle, layered convection with interchange and overturns, and irreversible chemical stratification with little or no interchange of material between layers. Layered schemes have several variants involving a primitive lower mantle or a depleted (in U and Th) lower mantle. In a convecting Earth we lose all of our reference systems. The mantle is heated from within, cooled from above and experiences secular cooling. Global topography and gravity provide constraints on mantle dynamics.

Topography

Although the Earth is not flat or egg-shaped, as previously believed at various times, neither is it precisely a sphere or even an ellipsoid of revolution. Although mountains, ocean basins and variations in crustal thickness contribute to the observed irregular shape and gravity field of the Earth, they cannot explain the long-wavelength departures from a hydrostatic figure.

The distribution of elevations on the Earth is distinctly bimodal, with a peak near +0.1 km representing the mean elevation of continents and a peak near –4.7 km corresponding to the mean depth of the oceans [see Google `Images hypsometry`]. This bimodal character contrasts with that of the other terrestrial planets. The spherical harmonic spectrum of the Earth's topography shows a strong peak for $l = 1$, corresponding to the distribution of continents in one hemisphere, and a regular decrease with increasing n. The topography spectrum is similar to that of the other terrestrial planets. There are small peaks in the spectrum at $l = 3$ and $l = 9 - 10$, the latter corresponding to the distribution of subduction zones and large oceanic swells.

The wavelength, in kilometers, is related to the spherical harmonic degree l and the circumference of the Earth (in km) approximately by

$$\text{Wavelength} = 40\,040/(l + 0.5)$$

Thus, a wavelength of 10 degrees or 1100 km corresponds to a spherical harmonic degree of about 40.

Active orogenic belts such as the Alpine and Himalayan are associated with thick crust, and high relief, up to 5 km. Older orogenic belts such as the Appalachian and Caledonian, because of erosion and lower crustal delamination, are associated with low relief, less than 1 km, and thinner crusts. Regional changes in the topography of the continents are generally accompanied by changes in mean crustal thickness. Continents stand high because of thick, low-density crust, compared with oceans. There is a sharp cut-off in crustal thickness at about 50 km, probably due to delamination of over-thickened crust at the gabbro–eclogite phase change boundary. As the dense root grows, the surface subsides, forming sedimentary basins. Upon delamination, the surface pops up, forming a swell, often accompanied by magmatism. Many continental flood basalt provinces (CFB) erupt on top of sedimentary basins and the underlying crust is thinner than average for the continents.

The long-wavelength topography of the ocean floor exhibits a simple relationship to crustal age, after averaging and smoothing. The systematic increase in the depth of the ocean floor away from the midocean ridges can be explained by simple cooling models for the evolution of the oceanic lithosphere. The mean depth of ocean ridges is 2.5 km below sealevel although regional variations off 1 km around the mean are observed. Thermal subsidence of the seafloor is well approximated by an empirical relationship of the form

$$d(t) = d_o + At^{1/2}$$

where d is seafloor depth referred to sea-level and positive downward, d_o is mean depth of mid-ocean ridges and t is crustal age. The value of A is around 350 m/(my)$^{1/2}$ if d and d_o are expressed in meters and t in my. *Depth anomalies* or *residual depth anomalies* refer to oceanfloor topography minus the expected thermal subsidence. Although there is a large literature on the interpretation of positive depth anomalies – swells – it should be kept in mind that in a convecting Earth, with normal variations in temperature and composition, the depth of the seafloor is not expected to be a simple function of time or age. Geophysical anomalies, both positive and negative, are well outside the normal expected variations for a uniform isothermal mantle.

Data from the western North Atlantic and central Pacific Oceans, for seafloor ages from 0 to 70 Myr, topography are described by

$$d(t) = 2500 + 350t^{1/2}$$

where t is crustal age in Myr and $d(t)$ is the depth in meters. Older seafloor does not follow this simple relationship, being shallower than predicted, and there is much scatter at all ages. Slightly different relations hold if the seafloor is subdivided into tectonic corridors. There are large portions of the ocean floor where depth cannot be explained by simple thermal models; these include oceanic islands, swells, aseismic ridges and oceanic plateaus as well as other areas where the effects of surface tectonics and crustal structure are not readily apparent. Simple cooling models assume that the underlying mantle is uniform and isothermal and that all of the variation in bathymetry is due to cooling of a thermal boundary layer (TBL). The North Atlantic is generally too shallow for its age, and the Indian Ocean between Australia and Antarctica is too deep. Continental insulation, a chemically heterogenous mantle and accumulated slabs at depth may explain these anomalies. There is no evidence that shallow regions are caused by particularly hot mantle. In fact, there is evidence for moderate mantle temperature anomalies associated with hotspot volcanism.

Residual depth anomalies, the departure of the depth of the ocean from the value expected for its age, in the ocean basins have dimensions of order 2000 km and amplitudes greater than 1 km. Part of the residual anomalies are due to regional changes in crustal thickness. This cannot explain all of the anomalies. Positive (shallow) depth anomalies – or swells – are often associated with volcanic regions such as Bermuda, Hawaii, the Azores and the Cape Verde Islands. These might be due to thinning of the plate, chemically buoyant material in the shallow mantle, or the presence of abnormally hot upper mantle. Patches of eclogite in the mantle are dense when they are colder than ambient mantle, but they melt at temperatures some 200 °C colder than peridotite and can therefore be responsible for elevation and melting anomalies.

Shallow areas often exceed 1200 m in height above the expected depth and occupy almost the entire North Atlantic and most of the western

Pacific. Almost every volcanic island, seamount or seamount chain surmounts a broad topographic swell. The swells generally occur directly beneath the volcanic centers and extend along fracture zones. Small regions of anomalously shallow depth occur in the northwestern Indian Ocean south of Pakistan, in the western North Atlantic near the Caribbean, in the Labrador Sea and in the southernmost South Pacific. They are not associated with volcanism but are slow regions of the upper mantle as determined from seismic tomography.

Shallow regions probably associated with plate flexure border the Kurile Trench, the Aleutian Trench and the Chile Trench. Major volcanic lineaments without swells include the northern end of the Emperor Seamount chain, the Cobb Scamounts off the west coast of North America and the Easter Island trace on the East Pacific Rise. Bermuda and Vema, in the southeast Atlantic, are isolated swells with no associated volcanic trace. For most of the swells explanations based on sediment or crustal thickness and plate flexure can be ruled out. They seem instead to be due to variations in lithospheric composition or thickness, or abnormal upper mantle. Dike and sill intrusion, underplating of the lithosphere by basalt or depleted peridotite, serpentinization of the lithosphere, delamination, or reheating and thinning the lithosphere are mechanisms that can decrease the density or thickness of the lithosphere and cause uplift of the seafloor. A higher temperature asthenosphere, greater amounts of partial melt, chemical inhomogeneity of the asthenosphere and upwelling of the asthenosphere are possible sublithospheric mechanisms.

A few places are markedly deep, notably the seafloor between Australia and Antarctica – the Australian–Antarctic Discordance or AAD – and the Argentine Basin of the South Atlantic. Other deep regions occur in the central Atlantic and the eastern Pacific and others, most notably south of India, are not so obvious because of deep sedimentary fill. Most of the negative areas are less than 400 m below the expected depth, and they comprise a relatively small fraction of the seafloor area. They represent cold mantle, lower melt contents, dense lower crust or an underlying and sinking piece of subducted slab or delaminated lower crust.

Dynamic topography

The long-wavelength topography is a dynamic effect of a convecting mantle. It is difficult to determine because of other effects such as crustal thickness. Density and thermal variations in a convecting mantle deform the surface, and this is known as the `mantle dynamic topography`. The `long-wavelength geoid` of the Earth is controlled by density variations in the deep mantle and has been explained by circulation models involving whole mantle flow. However, the relationship of long-wavelength topography to mantle circulation has been a puzzling problem in geodynamics. Dynamic topography is mainly due to density variations in the upper mantle. Layered mantle convection, with a shallow origin for surface dynamic topography, is consistent with the spectrum, small amplitude and pattern of the topography. `Layered mantle convection, with a barrier near 1000 km` *depth provides a self-consistent geodynamic model for the amplitude and pattern of both the long-wavelength geoid and surface topography.*

The geoid

The centrifugal effect of the Earth's rotation causes an equatorial bulge, the principal departure of the Earth's surface from a spherical shape. If the Earth were covered by oceans then, apart from winds and internal currents, the surface would reflect the forces due to rotation and the gravitational attraction of external bodies, such as the Sun and the Moon, and effects arising from the interior. When tidal effects are removed, the shape of the surface is due to density variations in the interior. Mean sea level is an equipotential surface called the geoid or figure of the Earth. Crustal features, continents, mountain ranges and midoceanic ridges represent departures of the actual surface from the geoid, but mass compensation at depth, isostasy, minimizes the influence of surface features on the geoid. To first

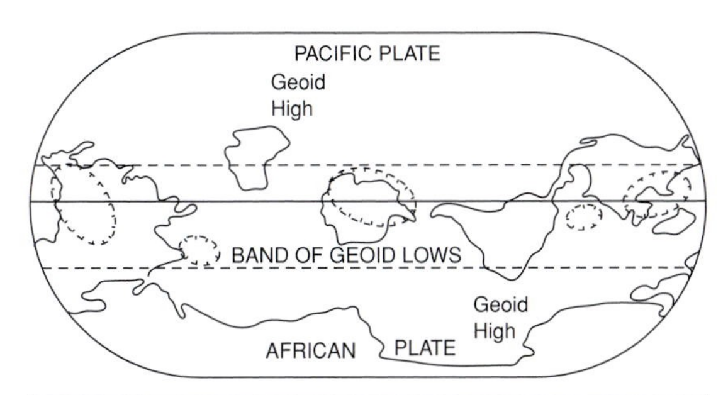

Fig. 6.1 Geoid lows are concentrated in a narrow polar band passing through Antarctica, the Canadian Shield and Siberia. Most of the continents and smaller tectonic plates are in this band. Long-wavelength geoid highs and the larger plates (Africa, Pacific) are antipodal and are centered on the equator. The geoid highs control the location of the axis of rotation. Large-scale mass anomalies in the deep mantle control the long-wavelength geoid. These in turn can affect the stress in the surface plates.

order, near-surface mass anomalies that are compensated at shallow depth have no effect on the geoid.

The shape of the geoid is now known fairly well, particularly in oceanic regions, because of the contributions from satellite geodesy [see geoid images]. Apart from the geoid highs associated with subduction zones, there is little correlation of the long-wavelength geoid with such features as continents and midocean ridges. The geoid reflects temperature and density variations in the interior, but these are not simply related to the surface expressions of plate tectonics.

The largest departures of the geoid from a radially symmetric rotating spheroid are the equatorial and antipodal geoid highs centered on the central Pacific and Africa (Figures 6.1 and 6.2). The complementary pattern of geoid lows lie in a polar band that contains most of the large shield regions of the world. The largest geoid highs of intermediate scale are associated with subduction zones. The most notable geoid high is centered on the subduction zones of the southwest Pacific near New Guinea, again near the equator. The equatorial location of geoid highs is not accidental; mass anomalies in the mantle control the moments of inertia of the Earth and, therefore, the location of the spin axis and the equator. The largest intermediate-wavelength geoid lows are found south of India, near Antarctica (south of New Zealand) and south of Australia. The locations of the mass anomalies responsible for these lows are probably in the lower mantle. Many shield areas are in or near geoid lows, some of which are the result of deglaciation and incomplete rebound. The thick continental crust would, by itself, raise the center of gravity of continents relative to oceans and cause slight geoid highs. The thick lithosphere ($\sim$150 km) under continental shields is cold, but the seismic velocities and xenoliths from kimberlite pipes suggest that it is olivine-rich and garnet-poor; the temperature and petrology have compensating effects on density. The longterm stability of shields indicates that, on average, the crust plus its underlying lithosphere is buoyant. Midocean ridges show mild intermediate-wavelength geoid highs, but they occur on the edges of long-wavelength highs. Hotspots, too, are associated with geoid highs. The long-wavelength features of the geoid are probably due to density variations in the lower mantle and the resulting deformations of the core–mantle boundary and other boundaries in the mantle (Richards and Hager, 1984).

Geoid anomalies are expressed as the difference in elevation between the measured geoid and some reference shape. The reference shape is usually either a spheroid with the observed flattening or the theoretical hydrostatic flattening associated with the Earth's rotation, the equilibrium form of a rotating Earth. The latter, used in Figure 6.2, is the appropriate geoid for geophysical purposes and is known as the nonhydrostatic geoid. The geometric flattening of the Earth is 1/298.26. The hydrostatic flattening is 1/299.64.

The maximum geoid anomalies are of the order of 100 m. This can be compared with the 21 km difference between the equatorial and polar radii. To a good approximation the net mass of all columns of the crust and mantle are equal when averaged over dimensions of a few hundred kilometers. This is one definition of isostasy. Smaller-scale anomalies can be supported by the strength of the crust and lithosphere. The geoid anomaly is nonzero in such cases and depends on the distribution of mass. A negative $\Delta\rho$, caused for example by thermal

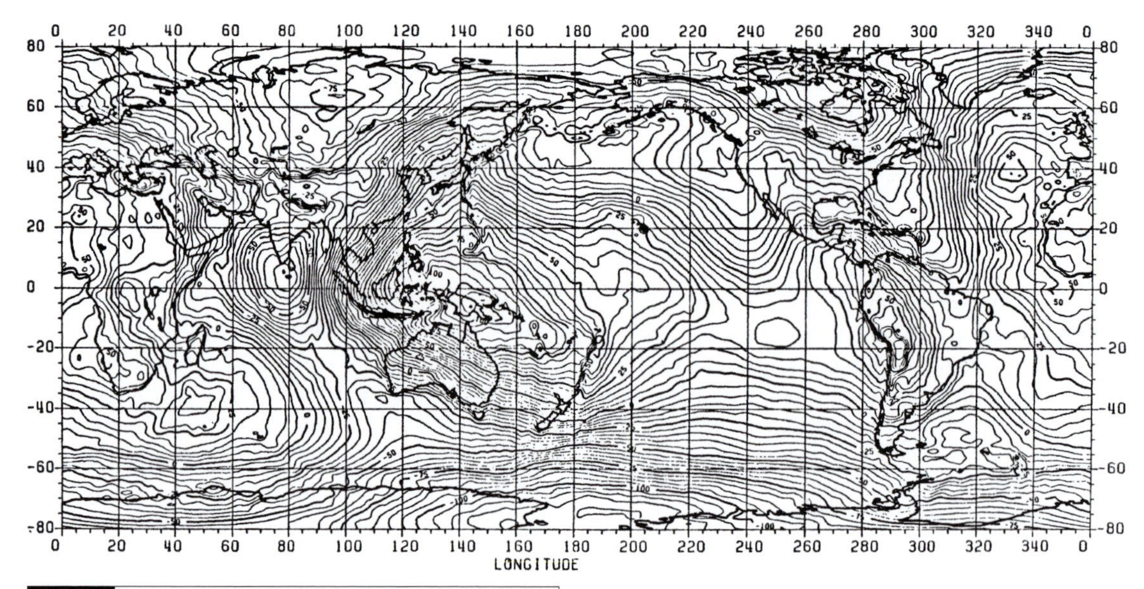

Fig. 6.2 Geoid undulations (to degree 180) referred to a hydrostatic shape, flattening of 1/299.638 [called the non-hydrostatic flattening of the geoid]. Contour interval is 5 m (after Rapp, 1981).

expansion, will cause the elevation of the surface to increase ($\Delta\rho$ = positive) and gives a positive geoid anomaly because the center of mass is closer to the Earth's surface. The mass deficiency of the anomalous material is more than canceled out by the excess elevation.

All major subduction zones are characterized either by geoid highs (Tonga and Java through Japan, Central and South America) or by local maxima (Kuriles through Aleutians). The long-wavelength part of the geoid is about that expected for the excess mass of the cold slab. The shorter wavelength geoid anomalies, however, are less, indicating that the excess mass is not simply rigidly supported. There is an excellent correlation between the geoid and slabs; this can be explained if the viscosity of the mantle increases with depth by about a factor of 30. The high viscosity of the mantle at the lower end of the slab partially supports the excess load. Phase boundaries and chemical boundaries may also be involved. The deep trenches represent a mass deficiency, and this effect alone would give a geoid low. The ocean floor in back-arc basins is often deeper than equivalent-age normal ocean, suggesting that the mass excess associated with the slab is pulling down the surface. A thinner-than-average crust or a colder or denser shallow mantle could also depress the seafloor.

Cooling and thermal contraction of the oceanic lithosphere cause a depression of the seafloor with age and a decrease in the geoid height. Cooling of the lithosphere causes the geoid height to decrease uniformly with increasing age, symmetrically away from the ridge crest. The change is typically 5–10 m over distances of 1000–2000 km. The elevation and geoid offset across fracture zones is due to the age differences of the crust and lithosphere. The long-wavelength topographic highs in the oceans generally correlate with positive geoid anomalies, giving about 6–9 meters of geoid per kilometer of relief.

There is a good correlation between intermediate-wavelength geoid anomalies and seismic velocities in the upper mantle; slow regions are geoid highs and vice versa. Subduction zones are slow in the shallow mantle, presumably due to the hot, partially molten mantle wedge under back-arc basins.

In subduction regions the total geoid anomaly is the sum of the positive effect of the dense sinker and the negative effects caused by boundary deformations. For a layer of uniform viscosity, the net dynamic geoid anomaly caused by a dense sinker is negative; the effects from the deformed boundaries overwhelm the effect from the sinker itself. For an increase in viscosity with depth, the deformation of the upper boundary is less and the net geoid anomaly is positive.

Shorter wavelength features

There is a broad range of dominant wavelengths – or peaks in the spectrum – in the geoid and bathymetry, ranging in wavelength from 160 km to 1400 km. Although these have been interpreted as the scales of convection and thermal variations they could also be caused by density variations due to chemistry and, perhaps, partial melt content, in the upper mantle. Several of the spectral peaks are similar in wavelength to chemical variations along the ridges. The shorter wavelengths may be related to thermal contraction and bending of the lithosphere. The longer wavelengths probably correspond to lithologic (major element) variations in the asthenosphere and, possibly, fertility and melting point variations. Intermediate-wavelength (400–600 km) geoid undulations are continuous across fracture zones and some have linear volcanic seamount chains at their crests.

Profiles of gravity and topography along the zero age contour of oceanic crust are perhaps the best indicators of mantle heterogeneity. These show some very long wavelength variations, ~5000 km and ~1000 km, and also abrupt changes. Ridges are not uniform in depth, gravity or chemical properties. Complex ridge-plume interactions have been proposed, the assumption being that normal ridges should have uniform properties. The basalts along midocean ridges are fairly uniform in composition but nevertheless show variations in major oxide and isotopic compositions. *Major and minor element chemistry shows spectral peaks with wavelengths of 225 and 575 km.* In general, one cannot pick out the ridge-centered and near-ridge hotspots from profiles of gravity, geoid, chemistry and seismic velocity. This suggests that short-wavelength elevation anomalies, e.g. *hotspots*, do not have deep roots or deep causes. Some hotspots have low seismic velocities at shallow depths, shallower than 200 km, consistent with low-melting-point constituents in the asthenosphere.

Basalt chemistry exhibits lateral variations on length scales of 150 and 400 km that may be related to intrinsic heterogeneity of the mantle. Large variations in magma output along volcanic chains occur over distances of hundreds to thousands of km; most chains – often called *hotspot tracks* – are less than 1000 km long. These dimensions may be the characteristic scales of mantle chemical and fertility variations. This provides a straightforward explanation of the order of magnitude variations in volcanic output along long volcanic chains and along spreading ridges.

Interpreting the geoid

Quantitative interpretations of the geoid are often based on relations such as wavelength vs. spherical harmonic degree; the geoid bears little relation to global tectonic maps or to present tectonic features of the Earth other than trenches. The Earth's largest positive geoid anomalies have no simple relationship to continents and ridges. The Mesozoic supercontinent of Pangea, however, apparently occupied a central position in the Atlantic–African geoid high. This and the equatorial Pacific geoid high contain most of the world's hotspots although there is little evidence that the mantle in these regions is particularly hot. The plateaus and rises in the western Pacific formed in the Pacific geoid high, and this may have been the early Mesozoic position of a subduction complex, the fragments of which are now the Pacific rim portions of the continents. Geoid highs that are unrelated to present subduction zones may be the former sites of continental aggregations, the centers of large long-lived plates – which cause mantle insulation and, therefore, hotter than normal mantle. The pent-up heat causes uplift, magmatism, fragmentation, and the subsequent formation of plateaus, aseismic ridges and seamount chains. However, the effect must be deep in order to also affect the long wavelength geoid.

When the subduction-related geoid highs are removed from the observed field, the residual geoid shows broad highs over the central Pacific and the eastern Atlantic–African regions. Like the total geoid, the residual geoid does not reflect the distribution of continents and oceans and shows little trace of the ocean-ridge system. Residual geoid highs, however, correlate with regions of anomalously shallow ocean floor and sites of extensive Cretaceous volcanism.

The lack of correlation of the large geoid anomalies to present-day plate boundaries and tectonics requires that the anomalies reflect a deep-mantle structure that is unrelated to plate tectonics, or, perhaps, to an ancient configuration of plates. The correspondence of the Atlantic–African anomaly with the Mesozoic continental assemblage and of the antipodal central Pacific anomaly with extensive Cretaceous volcanism in the Pacific is suggestive, but may be coincidental. Surface-wave tomography shows a good correlation of intermediate-wavelength geoid highs and slow regions of the upper mantle. However, the very-long-wavelength components of the geoid correlate best with tomography of the lower mantle. Most of the present continents, except Africa, and most of the present subduction zones (except Tonga–Fiji) are in long-wavelength geoid lows and therefore probably overlie denser than average lower mantle.

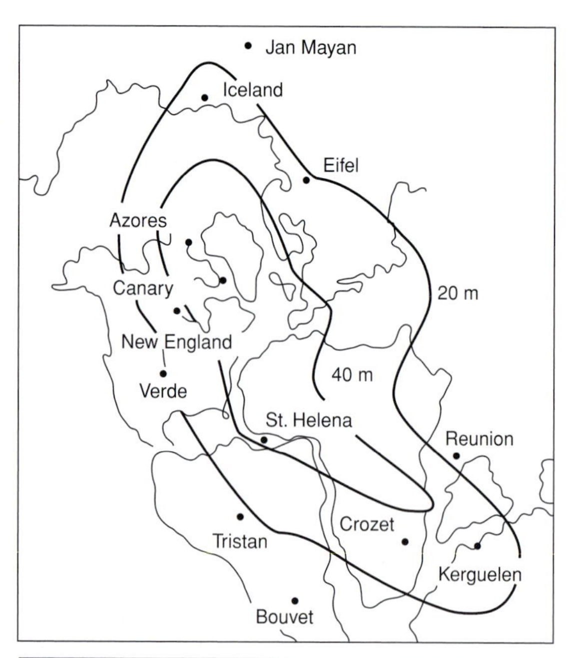

Fig. 6.3 The Pangea and Atlantic–Africa geoid high plotted relative to the 200 Ma (200 million years ago) positions of the continents and hotspots.

The Atlantic-African geoid high extends from Iceland through the north Atlantic and Africa to the Kerguelen plateau and from the middle of the Atlantic to the Arabian Peninsula and western Europe (Figure 6.3). Most of the Atlantic, Indian Ocean, African and European hotspots are inside this anomaly, but so are spreading ridges. The hotspots Iceland, Trinidade, Tristan, Kerguelen, Reunion, Afar, Eiffel and Jan Mayen form the 20-m boundary of the anomaly and appear to control its shape. The Azores, Canaries, New England seamounts, St. Helena, Crozet and the African hotspots are interior to the anomaly.

Although the geoid high cuts across present-day ridges and continents, there is a remarkable correspondence of the pre-drift assemblage of continents with both the geoid anomaly and hotspots. Reconstruction of the mid-Mesozoic configuration of the continents reveals, in addition, that virtually all of the large shield areas of the world are contained inside the geoid high (Figures 6.1 and 6.3). These include the shield areas of Canada, Greenland, Fennoscandia, India, Africa, Antarctica and Brazil. Most of the Phanerozoic platforms are also in this area. In contrast, today's shields and platforms are concentrated near geoid lows. They may have drifted into, and come to rest over, these geoid lows. The area inside the geoid high is also characterized by higher-than-normal elevations, for example in Africa, the North Atlantic and the Indian Ocean southeast of Africa. This holds true also for the axial depth of oceanic ridges.

Large plates insulate the mantle and allow radioactive heat to build up. When a supercontinent – or super oceanic plate – breaks up, we expect active volcanism in the wake, both at newly opening ridges and intra-plate settings. This is the result of extensional stress as well as high fertility at former sutures and higher temperatures. This may also trigger delamination and foundering of the deeper portions of over-thickened continental crust, and uplift. These phenomena at new plate boundaries and the edges of continents are often attributed to plumes, but they are a natural part of plate tectonics.

Most of the continental areas were above sea level from the Carboniferous and Permian through the Triassic, at which time there was subsidence in eastern North and South America, central and southern Africa, Europe and Arabia. The widespread uplift, magmatism, breakup and initial dispersal of the Pangean landmass apparently occurred while the continents were centrally located with respect to the present geoid anomaly, assuming that it is long-lived. The subsequent motions of the plates, by and large, were and are directed away from the anomaly. This suggests that the residual geoid high, hotspots, the distribution of continents during the late Paleozoic and early Mesozoic, and their uplift and subsequent dispersal and subsidence are all related. The shields have abnormally thick, cold – but buoyant – keels. The high viscosity and low thermal expansion of the lower mantle, and the relatively small amount of radioactive and core heating, means that the features responsible for the long-wavelength geoid are probably very long-lived.

At 100 Ma Europe, North America and Africa were relatively high-standing continents. This was after breakup commenced in the North Atlantic but before significant dispersal from the pre-breakup position. North America suffered widespread submergence during the Late Cretaceous while Africa remained high. Europe started to subside at about 100 Ma. This is consistent with North America and Europe drifting away from the center of the geoid high while Africa remained near its center, as it does today.

Horizontal temperature gradients can drive continental drift. The velocities decrease as the distance increases away from the heat source and as the thermal anomaly decays. Thick continental lithosphere then insulates a new part of the mantle, and the cycle repeats. Periods of rapid polar motion and continental drift follow periods of continental stability and mantle insulation.

The relationship between surface tectonic features and the geoid changes with time. Supercontinents periodically form and insulate the underlying mantle and also control the locations of mantle cooling (subduction zones). When the continent breaks up, the individual fragments move away from the hot part of the mantle and the geoid high, and come to rest over cold mantle, in geoid lows. Large long-lived oceanic plates can also insulate the mantle, generating broad topographic swells.

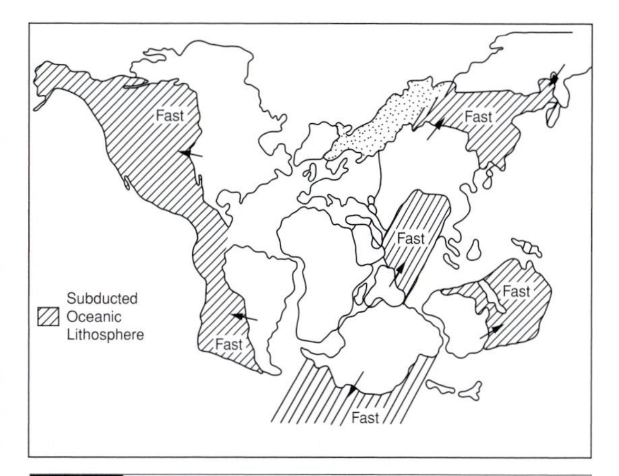

Fig. 6.4 Reconstruction of the continents and their motion vectors at about 110 Ma. The hatched areas represent former oceanic lithosphere. These regions, in general, have high seismic velocities in the transition region, consistent with the presence of cold subducted lithosphere. They are also, in general, geoid lows. Dots represent possible convergence areas.

The locations of hot and cold regions in the upper mantle may also be influenced by thermal anomalies in the lower mantle. The lower mantle contribution to the geoid is probably long-lived. Empirically, subduction zones and continents are primarily in long-wavelength geoid lows and over long-wavelength fast seismic regions of the lower mantle. This can be understood if continents come to rest in geoid lows and if subduction zones, on average, are controlled by the advancing edges of continents. Midocean ridges tend to fall between the long-wavelength highs and lows. By long wavelength, we mean features having dimensions of thousands of kilometers.

Figure 6.4 shows the approximate locations of the continents just after breakup of Pangea commenced. The hatched regions show oceanic lithosphere that has been overridden by the

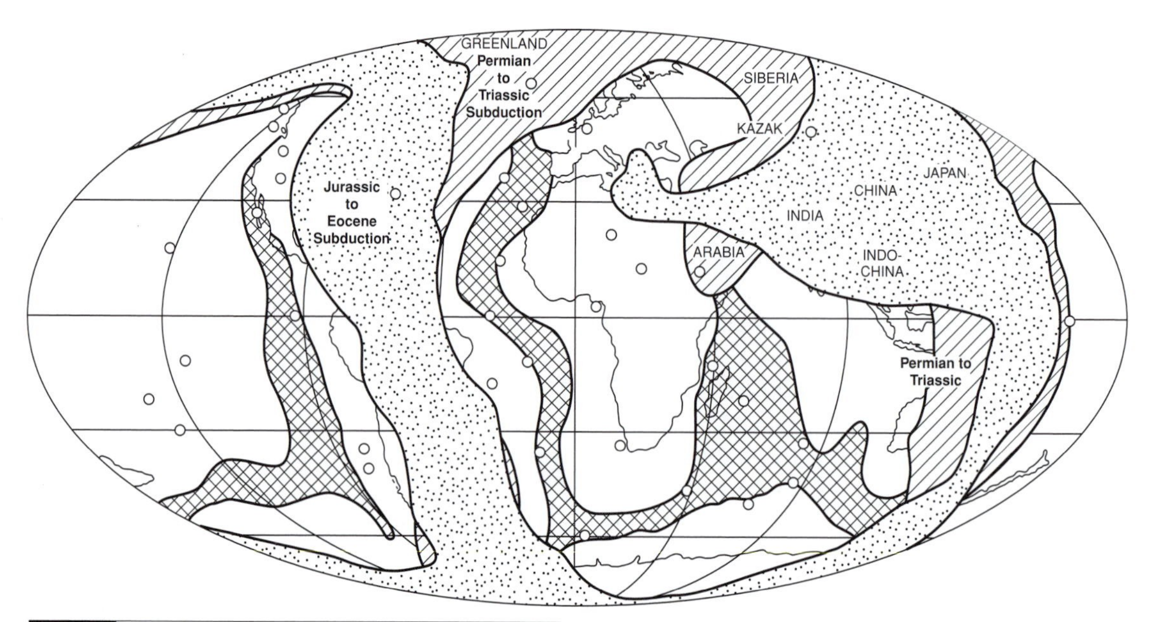

Fig. 6.5 The approximate locations of subducted slabs or subduction zones during the last two supercontinent cycles. Note the similarity with the long wavelength geoid. This pattern also matches that of seismic tomography at a depth of about 800–1000 km depth, suggesting that slabs bottom out at ~1000 km depth.

advancing continents. These are labeled 'fast' because these are seismically fast regions of the transition region, where cold lithosphere may have cooled off the mantle. The arrows represent the motions of the continents over the past 110 Myr. Most of the hatched regions are also geoid lows. Figure 6.5 shows the inferred locations of subduction zones over the past two supercontinent cycles. Slabs and delaminated lower crust have entered the mantle in these regions, both cooling it and fertilizing it.

Involvement of the lower mantle

Tomographic techniques can be applied to the problem of lateral heterogeneity of the lower mantle [`tomography geoid lower mantle, Hager O'Connell`]. Long-wavelength velocity anomalies in the lower mantle correlate well with the $l \sim 2, 3$ geoid. Phenomena such as tides, Chandler wobble, polar wander and the orientation of the Earth's spin axis depend on the $l = 2$ component of the geoid; the lower mantle is important for these problems. By contrast the upper mantle has only a weak correlation with the $l = 2$ and 3 geoid. The effects of pressure on viscosity and thermal expansion are such that we expect the lower mantle to convect very sluggishly. The large features in the lower mantle are probably ancient, and not caused by recent plate tectonic cycles. The implication is that the geoid and the rotation axis are relatively stable. The plates and the upper 1000 km of the mantle are the active layers.

Polar wander

Because the Earth is a dynamic body, it is impossible to define a permanent internal reference frame. There are three reference frames in common use: the rotation axis, the geomagnetic reference frame and the hotspot reference frame. The rotational frame is controlled by the size of the mass anomalies and their distance from the axis of rotation. Upper-mantle effects are important because lateral heterogeneity is greater than lower-mantle or core heterogeneity and because they are far from the center of the Earth. The lower mantle is important because of its large volume, but a given mass anomaly has a greater effect in the upper mantle. The location of the magnetic pole is controlled by convection in the core, which in turn is influenced by the rotation of the Earth and the temperatures at the base of the mantle. On average the rotational pole and the magnetic pole are close together, but the instantaneous poles can be quite far apart. The hotspot frame is no longer considered valid. The fixed-hotspot hypothesis led to the view that hotspots are anchored deep in the mantle and may reflect a different kind of convection than that which is responsible for large-scale convection in the mantle. The apparent motions of hotspots has been used to argue for `true polar wander` (TPW) but this is not a well-founded argument.

Density inhomogeneities in the mantle grow and subside, depending on the locations of continents and subduction zones. The resulting geoid highs reorient the mantle relative to the spin axis. Whenever there was a major continental assemblage in the polar region surrounded by subduction, as was the case during the Devonian through the Carboniferous, the stage was set for a major episode of true polar wandering.

The outer layers of the mantle, including the brittle lithosphere, do not fit properly on a reoriented Earth. Membrane stresses generated as plates move around the surface, or as the rotational bulge shifts, may be partly responsible for the breakup and dispersal of Pangea. In this scenario, true polar wandering and continental drift are intimately related. A long period of continental stability allows thermal and geoid anomalies to develop. A shift of the axis of rotation can cause plates to split. Horizontal temperature gradients, along with slab pull, force continental fragments to drift away from the thermal anomalies that they caused. The continents drift toward cold parts of the mantle and, in fact, make the mantle cold as they override oceanic lithosphere.

Polar wandering can occur on two distinct time scales. In a slowly evolving mantle the rotation axis continuously adjusts to changes in the moments of inertia. This will continue to be the case as long as the major axis of inertia remains close to the rotation axis. If one of the other axes becomes larger, the rotation vector swings quickly to the new major axis (Figure 6.6). This is called `inertial interchange true polar wander` (IITPW). The generation and decay of thermal perturbations in the mantle

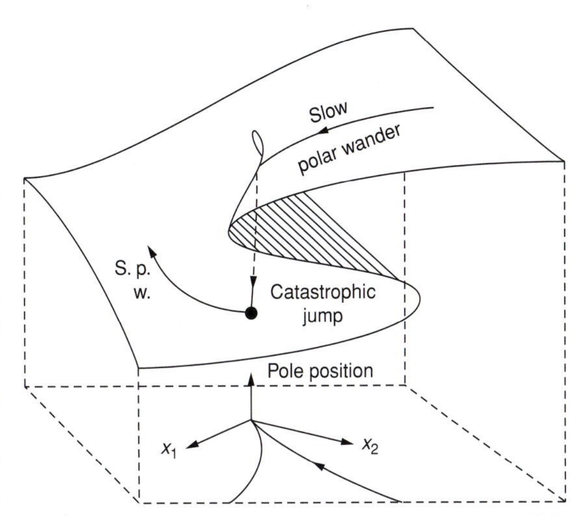

Fig. 6.6 The principal moments of inertia shown on a cusp catastrophe diagram. As the moments of inertia vary, due to convective processes in the interior, the pole will slowly wander unless the ratios of the moments x_1, and x_2, pass through unity, at which point a catastrophe will occur, leading to a rapid change in the rotation axis.

are relatively gradual, and continuous small-scale polar wandering can be expected. The interchange of moments of inertia, however, occurs more quickly, and a large-scale 90-degree shift can occur on a timescale limited only by the relaxation time of the rotational bulge. The rate of polar wandering at present is much greater than the average rate of relative plate motion, and it would have been faster still during an interchange event. The relative stability of the rotation axis for the past 200 million years suggests that the geoid highs related to hotspots have existed for at least this long. On the other hand, the rapid polar wandering that started 500 Ma may indicate that the Atlantic–African geoid high was forming under Gondwana at the time and had become the principal axis of inertia. With this mechanism a polar continental assemblage can be physically rotated to the equator as the Earth tumbles.

The southern continents all underwent a large northward displacement beginning sometime in the Permian or Carboniferous (280 Ma) and continuing to the Triassic (190 Ma). During this time the southern periphery of Gondwana was a convergence zone, and a spreading center is inferred along the northern boundary. One would expect that this configuration would be consistent with a stationary or a southern migration of Gondwana, unless a geoid high centered on or near Africa was rotating the whole assemblage toward the equator. The areas of very low upper-mantle velocities in northeast Africa and the western Indian Ocean may be the former site of the center of Gondwana.

Thus, expanding the paradigm of continental drift and plate tectonics to include continental insulation and true-polar wandering may explain the paradoxes of synchronous global tectonic and magmatic activity, rapid breakup and dispersal of continents following long periods of continental stability, periods of static pole positions separated by periods of rapid polar wandering, sudden changes in the paths of the wandering poles, the migration of rifting and subduction, initiation of melting, the symmetry of ridges and fracture zones with respect to the rotation axis, and correlation of tectonic activity and polar wandering with magnetic reversals. Tumbling of the mantle presumably affects convection in the core and orientation of the inner core and offers a link between tectonic and magnetic field variations. Global plate reorganizations are a necessary part of plate tectonics on a sphere. New plate boundaries are often accompanied by extensive volcanism and enriched magmas, presumably from the shallow mantle.

The largest known positive gravity anomaly on any planet is associated with the Tharsis volcanic province on Mars. Both geologic and gravity data suggest that the positive mass anomaly associated with the Tharsis volcanoes reoriented the planet with respect to the spin axis, placing the Tharsis region on the equator. There is also evidence that magmatism associated with large impacts reoriented the Moon. The largest mass anomaly on Earth is centered over New Guinea, and it is also almost precisely on the equator. The long-wavelength part of the geoid correlates well with subduction zones, and these appear to control the orientation of the mantle relative to the spin axis. Thus, we have the possibility of a feedback relation between geologic processes and the rotational dynamics of a planet. Volcanism and continental collisions cause mass excesses to be placed near the surface. These reorient the planet, causing large stresses that initiate rifting and faulting, which in turn affect volcanism and subduction. Curiously, Earth scientists have been more reluctant to accept the inevitability of true-polar wandering than to accept continental drift, even though the physics of the former is better understood.

Chapter 7

Convection and complexity

. . . if your theory is found to be against the second law of thermodynamics, I can give you no hope; there is nothing for it but to collapse in deepest humiliation.

Eddington

Contrary to current textbooks . . . the observed world does not proceed from lower to higher "degrees of disorder", since when all gravitationally-induced phenomena are taken into account the emerging result indicates a net decrease in the "degrees of disorder", a greater "degree of structuring" . . . classical equilibrium thermodynamics . . . has to be completed by a theory of 'creation of gravitationally-induced structures' . . .

Gal-or

Overview

In 1900 Henri Bénard heated whale oil in a pan and noted a system of hexagonal convection cells. Lord Rayleigh in 1916 analyzed this in terms of the instability of a fluid heated from below. Since that time Rayleigh–Bénard convection has been taken as the classic example of thermal convection, and the hexagonal planform has been considered to be typical of convective patterns at the onset of thermal convection. Fifty years went by before it was realized that Bénard's patterns were actually driven from above, by surface tension, not from below by an unstable thermal boundary layer. Experiments showed the same style of convection when the fluid was heated from above, cooled from below or when performed in the absence of gravity. This confirmed the top-down surface-driven nature of the convection which is now called Marangoni or Bénard–Marangoni convection.

Although it is not generally recognized as such, mantle convection is a branch of the newly renamed *science of complexity*. Plate tectonics may be a *self-driven far-from-equilibrium system* that organizes itself by dissipation in and between the plates, the mantle being a passive provider of energy and material. Far-from-equilibrium systems, particularly those in a gravity field, can locally evolve toward a high degree of order. Plate tectonics was once regarded as passive motion of plates on top of mantle convection cells but it now appears that continents and plate tectonics organize the flow in the mantle. But mantle convection and plate tectonics involve more than geometry and space-filling considerations. The mantle is a heat engine, controlled by the laws of thermodynamics. One can go just so far without physics. Conservation of mass and energy are involved, as are balancing of forces. Although the mantle behaves as a fluid, mineral physics principles and classical solid-state physics are needed to understand this fluid. The effect of pressure suppresses the role of the lower thermal boundary layer at the core–mantle–boundary (CMB) interface.

The slow uplift of the surface of the Earth in response to the removal of an ice cap or drainage of a large lake changes the shape of the Earth and the geoid; this is not only proof of the fluid-like behavior of the mantle but also provides the data for estimating its viscosity. In contrast to everyday experience mantle convection has some unusual characteristics. The container has spherical geometry. The 'fluid' has stress-, pressure- and temperature-dependent properties. It is cooled from above and from within (slabs) and heated from within (radioactivity) and from below (cooling of the core and crystallization of the inner core). The boundary conditions and heat sources change with time. Melting and phase changes contribute to the buoyancy and provide additional heat sources and sinks. Mantle convection is driven partly by plate motions and partly by chemical buoyancy. The boundaries are deformable rather than rigid. None of these characteristics are fully treated in numerical calculations, and we are therefore woefully ignorant of the style of convection to be expected in the mantle. The cooling plates may well organize and drive mantle convection, as well as themselves. A mantle with continents on top will convect differently from one with no continents.

The theory of convection in the mantle cannot be decoupled from the theories of solids and petrology. The non-Newtonian rheology, the pressure and temperature sensitivity of viscosity, thermal expansion, and thermal conductivity, and the effects of phase changes and compressibility make it dangerous to rely too much on the intuition provided by oversimplified fluid-dynamic calculations or laboratory experiments. There are, however, some general characteristics of convection that transcend these details. Technical details of normal or classical thermal convection can be found in textbooks on mantle convection. Plate tectonics and mantle motions, however, are far from normal thermal convection.

Geochemists consider convection and stirring to be equivalent. They use convecting mantle as convenient shorthand for what they consider to be the homogenous upper mantle. The underlying assumption is that midocean ridge basalts, known for their chemical homogeneity, must come from a well-stirred mantle reservoir.

Generalities

SOFFE systems are extraordinarily sensitive to boundary and initial conditions. The corollary is that small differences between computer or laboratory simulations, or between them and the mantle, can completely change the outcome. The effect of pressure suppresses the role of the lower thermal boundary layer (TBL) at the core–mantle–boundary (CMB) interface. The state of stress in the lithosphere defines the plates, plate boundaries and locations of midplate volcanism. Fluctuations in stress, due to changing boundary conditions, are responsible for global plate reorganizations and evolution of volcanic chains. In Rayleigh–Bénard convection, by contrast, temperature fluctuations are the important parameters. In plume theory, plates break where heated or uplifted by hot buoyant upwellings. Ironically, the fluid flows in the experiments by Bénard, which motivated the Rayleigh theory of thermal convection, were driven by surface tension, i.e. stresses at the surface.

Computer simulations of mantle convection have not yet included a self-consistent thermodynamic treatment of the effect of temperature, pressure, melting and volume on the physical and thermal properties; understanding of the 'exterior' problem (the surface boundary condition) is in its infancy. Plate tectonics itself is implicated in the surface boundary condition. Sphericity, pressure and the distribution of radioactivity break the symmetry of the problem and the top and bottom boundary conditions play quite different roles than in the simple calculations and cartoons of mantle dynamics and geochemical reservoirs. Conventional (Rayleigh–Bénard) convection theory may have little to do with plate tectonics. The research opportunities are enormous.

The history of ideas

Convection can be driven by bottom heating, top or side cooling, and by motions of the boundaries. Although the role of the surface boundary layer and *slab-pull* are now well understood and the latter is generally accepted as the prime mover in plate tectonics, there is a widespread perception that active hot upwellings from deep in the interior of the planet, independent of plate tectonics, are responsible for 'extraordinary' events such as plate reorganization, continental break-up, extensive magmatism and events far away from current plate boundaries. Active upwellings from deep in the mantle are viewed as controlling some aspects of surface tectonics and volcanism, including reorganization, implying that the mantle is not passive. This is called the `plume mode of mantle convection`. This has been modeled by the injection of hot fluids into the base of a tank of motionless fluid.

Numerical experiments show that mantle convection is controlled from the top by continents, cooling lithosphere, slabs and plate motions and that plates not only drive and break themselves but can control and reverse convection in the mantle. Studies of the `time dependence in 3D spherical mantle convection with continental drift` show the extreme sensitivity to changes of conditions and give results quite different from simpler simulations. Supercontinents and other large plates generate spatial and temporal temperature variations. The migration of continents, ridges and trenches cause a constantly changing surface boundary condition, and the underlying mantle responds passively. Plates break up and move, and trenches roll back because of forces on the plates and interactions of the lithosphere with the mantle. Density variations in the mantle are, by and large, generated by plate tectonics itself by slab cooling, refertilization of the mantle, continental insulation; these also affect the forces on the plates. Surface plates are constantly evolving and reorganizing although major global reorganizations are infrequent. Plates are mainly under lateral compression although local regions having horizontal least-compressive axes may be the locus of dikes and volcanic chains. The Aegean plate is an example of a 'rigid' plate collapsing, or falling apart, because of changes in stress conditions.

The mantle is generally considered to convect as a single layer (whole mantle convection), or at most two. However, the mantle is more likely to convect in multiple layers as a result of gravitational sorting during accretion, and the density difference between the mantle products of differentiation.

Instabilities

Rayleigh–Taylor (RT) instabilities form when a dense, heavy fluid occurs above a low-density fluid, such as a layer of dense oil placed, carefully, on top of a layer of water. Two plane-parallel layers of immiscible fluid are stable, but the slightest perturbation leads to release of potential energy, as the heavier material moves down under the (effective) gravitational field, and the lighter material is displaced upwards. As the instability develops, downward-moving dimples are quickly magnified into sets of interpenetrating RT fingers or plumes. This process is evident not only in many examples, from boiling water to weather inversions. In mantle geophysics, plumes are often modeled by inserting a light fluid into a tank of a static higher density fluid. This is meant to mimic the instability of a hot basal layer. In the later situation, the instability develops naturally and the density contrast is limited. In the injection experiment, the density contrast is imposed by the experimenter, as is the scale of the upwelling. There is a difference between upwellings of intrinsically hot basal layers and intrinsically light chemical layers. The former case sets up the lateral temperature gradients that are the essence of thermal convection.

Rise of deep diapirs

Delamination and sinking of garnet pyroxenite cumulates, sinking of slabs, and upwelling of mantle at ridges are important geodynamic processes. Diapiric ascent, melt extraction and crystal settling are important processes in igneous petrology. Basic melts apparently separate from magma chambers, or rising diapirs, at depths as great as 90 km and possibly greater. Eclogite sinkers may equilibrate at upper mantle or transition zone depths; they then warm up and rise.

The basic law governing ascent and settling, `Stokes' Law`, expresses a balance between gravitational and viscous forces,

$$V = 2\Delta\rho g R^2/9\eta$$

where R is the radius of a spherical particle or diapir, $\Delta\rho$ is the density contrast, η is the dynamic viscosity and V is the terminal velocity. This equation can be applied to the rising or sinking of blobs through a mantle or a magma chamber, with modifications to take into account non-spherical objects and non-Newtonian viscosity. Additional complications are introduced by turbulence in the magma chamber and finite yield strengths.

Diapirs are usually treated as isolated spheres or cylinders rising adiabatically through a static mantle. Because of the relative slopes of the geotherm, and the melting curve, diapirs become more molten as they rise. At some point, because of the increased viscosity or decreased density contrast, ascent is slowed and cooling, crystallization and crystal settling can occur. The lithosphere serves as a viscosity or strength barrier, and the crust serves as a density barrier. Melt separation can therefore be expected to occur in magma chambers at shallow depths. In a convecting mantle the actual temperatures (adiabatic or subadiabatic) diverge from the melting point as depth increases. In a homogenous mantle, melting can therefore only occur in the upper parts of the rising limbs of convection cells or in thermal boundary layers. The additional buoyancy provided by melting contributes to the buoyancy of the ascending limbs. Although the melts will attempt to rise relative to the adjacent solid matrix, they are embedded in a system that is itself rising and melting further. If broad-scale vertical convection is fast enough, diapirs can melt extensively without fractionating. Fertile, low-melting-point patches, such as eclogite, can melt extensively if surrounded by subsolidus peridotite.

The stresses and temperatures in the vicinity of rising plumes or diapirs are high, and these serve to decrease the mantle viscosity; thus rapid ascent is possible. In order to achieve observed magma temperatures and large degrees of partial melting, allowing for specific and latent heats, melting probably initiates at depths of order 200 km under oceanic ridges and large volcanic provinces, assuming that the mantle is mainly peridotite. The solidus temperature of dry peridotite at this depth is at least 2100 °C. One question is, how fast can material rise between about 200 and 90 km, and is the material at 90 km representative of the deeper mantle source region or has it been fractionated upon ascent? Viscosities in silicates are very stress- and temperature-dependent, and diapirs occur in regions of the mantle that have higher than normal stresses and temperatures. Diapiric emplacement itself is a high-stress process and occurs in regions where mantle convection may have oriented crystals along flow lines. Diapirs may rise rapidly through such low-viscosity material. A 50 km partially molten diapir at a depth of 200 km can rise at a rate of about 40 cm/s. Kimberlites travel an order of magnitude faster still. Crystal settling velocities in magmas are of the order of cm/s. It appears therefore that deep diapirs can rise rapidly enough to entrain both melt and crystals. At depth the melt content and the permeability are low, and melt segregation may be very slow. The fertile low-melting point blobs may be encased in relatively impermeable subsolidus peridotite and can therefore melt extensively as they rise.

In a chemically stratified mantle, for example residual peridotite over eclogite or fertile peridotite, there is a conductive thermal boundary between the convecting layers. In such a region the thermal gradient is in excess of the melting gradient, and melting is likely to initiate at this depth. Eclogite has a melting temperature about 200 °C below that of dry peridotite and melting of fertile blobs is also likely between TBLs. Partial melting causes a reduction in density, and a Rayleigh–Taylor instability can develop. Material can be lifted out of the eclogite or piclogite – eclogite plus peridotite – layer by such a mechanism and extensive melting occurs during ascent to the shallower mantle. At shallow depths peridotite elevated adiabatically from greater depths can also melt and magma mixing is likely, particularly if the diapir is trapped beneath thick

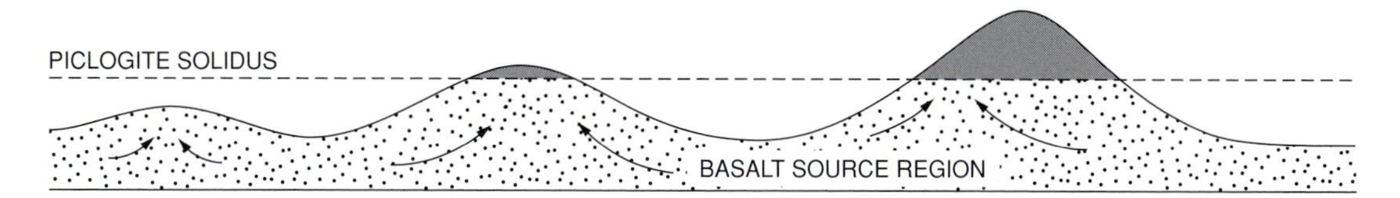

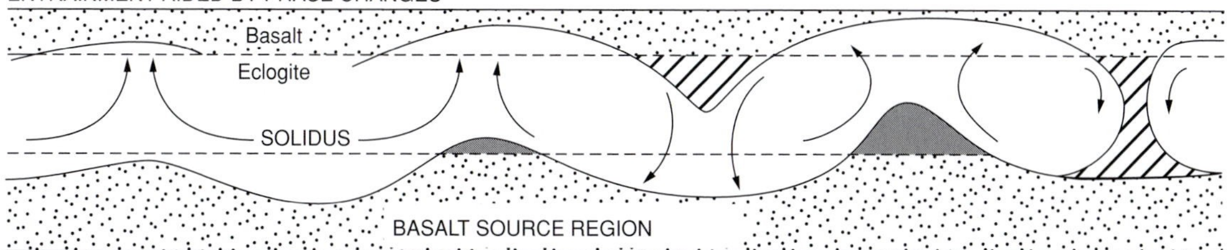

Fig. 7.1 Methods of removing material from a deep, dense layer. In a chemically layered mantle, the interface between layers is highly deformed. This may cause phase changes and melting, and a reduction in density. The deeper layer may also be entrained. If the deep layer is eclogite, from subduction or delamination, it will have a lower melting temperature than ambient mantle and may rise because of melt-induced buoyancy. Adiabatic decompression leads to further melting.

lithosphere. In mid-plate environments, such as Hawaii and other midplate hotspots, melts will cool, fractionate and mix with other melts prior to eruption. Such a mechanism seems capable of explaining the diversity of hotspot (ocean island, continental flood) basalts and midocean ridge magmas.

Material can leave a deep, dense source region by several mechanisms.

(1) Melting in the thermal boundary because of the high thermal gradient compared with melting point gradients.
(2) Melting, or phase changes, due to adiabatic ascent of hotter regions of a layer and crossing of phase boundary.
(3) Entrainment of material by adjacent convecting layers.

Some of these mechanisms are illustrated in Figure 7.1.

The *potential temperature* of the mantle is the temperature of the mantle adiabat if it were to ascend directly to the surface. The potential temperature of the mantle is usually between 1300 and 1400 °C, averaged over large areas. The adiabatic temperature gradient for the solid upper mantle is approximately 0.3–0.5 °C/km. The adiabatic gradient becomes smaller at very high pressures because the thermal expansivities of solids are smaller at high pressures. The adiabatic gradient for liquids (~1 °C/km) is higher than that for solids because the thermal expansivities of liquids are greater than solids.

Melting occurs where the geotherms intersect the mantle solidus. For a dry peridotite mantle, the geotherm (conduction gradient) in a region of high surface heatflow, 100 mW/m^2, intersects the solidus at ~30 km depth. Surface heat flows of 100 mW/m^2 or greater occur only at or near midocean ridges. A geotherm with 40 mW/m^2 surface heatflow, characteristic of the old interiors of continents, never intersects the dry peridotite solidus, implying that partial melting does not occur beneath continental interiors. In suture belts and along arcs there may be low-melting point constituents such as eclogite in the shallow mantle. High heatflow, however, in part, represents intrusion into the plate or thinning of the thermal boundary layer, rather than intrinsically high mantle temperatures.

We can induce melting at low temperatures if we flux the mantle with basalt, eclogite, CO_2 or water, and plate tectonics does all of these things. If portions of the mantle upwell rapidly, either passively in response to spreading at ridges or displacement by sinking slabs, decompressional partial melting occurs. Decompressional melting occurs at greater depths for higher mantle temperatures or lower melting temperatures, allowing higher degrees of partial melting. Any form of adiabatic decompression can give rise to partial melting if the solidus is crossed. Partial melting can be initiated when the plate is thinned by extension or by delamination of the lower crust or removal of the lower part of the TBL. Partial melting can thus occur by an increase in temperature, by adiabatic decompression, or by the depression of the solidus by the presence of eclogite, CO_2 or water. These processes can occur in a variety of tectonic environments, depending upon the background thermal state. Regions of high heat flow, such as midocean ridges and highly extended continental-crustal regions, are characterized by geotherms with steep temperature gradients such that the base of the thermal boundary layer lies well within the partial melting P-T field. In this case, partial melting occurs where the mantle adiabat intersects the solidus. Regions characterized by low heat flow, such as the stable interiors of continents, are not prone to melting as cold geotherms never intersect the solidus of dry peridotite. Partial melting of the mantle wedge overlying subducting slabs occurs because the peridotite solidus has been depressed by the addition of volatiles. Delaminated lower continental crust also lowers the melting point of the mantle that it sinks into.

Forces

Plate tectonic and convective motions represent a balance between driving forces and resisting forces. Buoyancy is the main driving force but there are a variety of resisting forces. Negative buoyancy is primarily created by cooling at the surface. Positive buoyancy is created by heating and melting.

The creation of new plates at ridges, the subsequent cooling of these plates, and their ultimate subduction at trenches introduce forces that drive and break up the plates. They also introduce chemical and thermal inhomogeneities into the mantle. Plate forces such as *ridge push* – a misnomer for the pulling force created by a cooling plate – *slab pull* and *trench suction* are basically gravitational forces generated by cooling plates. They are resisted by transform fault, bending and tearing resistance, mantle viscosity and *bottom drag*. If convection currents dragged plates around, the bottom drag force would be the most important. However, there is no evidence that this is a strong force, and even its sign is unknown (driving or resisting drag). The thermal and density variations introduced into the mantle by subduction also generate forces on the plates.

The pull of subducted slabs – the slab-pull force – is thought to be the main driver of the motions of Earth's tectonic plates and the motions beneath the plates. A slab mechanically attached to a subducting plate can exert a direct pull on the plate; a detached slab may drive a plate by causing a flow in the mantle that exerts a shear traction on the base of the plate. A cold slab can also set up thermal gradients that exert forces on the plates. Slab pull forces may account for about half of the total driving force on plates. Slabs in the deeper mantle are supported by viscous mantle forces and they may reach density equilibration.

Mantle convection may also be driven primarily by descent of dense slabs of subducted oceanic lithosphere. Slab forces cause both subducting and overriding plates to move toward subduction zones and they are also responsible for the migration of trenches and ridges. Ridges and trenches are stationary in the mantle reference frame only in very idealized symmetric cases. Cooling plates also exert forces that cause the plates to move away from ridges and toward subduction zones. One can alternatively think of mantle convection as the passive response to plate-tectonic stresses and thermal gradients created by plate tectonics and lithospheric architecture. Although both the plates and the underlying mantle are parts of the same convecting system it is useful to think of where most of the buoyancy and dissipation in the system resides. In 'normal' convection most of the energy comes from outside the system (bottom heating) and leaves at the top, and the buoyancy (via thermal expansion) and dissipation (viscosity) are distributed internally. In the mantle, much of the energy is provided from within (radioactivity) and much of the buoyancy and dissipation occurs in the surface layer; the upper and lower TBLs are not symmetric;

melting and the redistribution of radioactive elements are important.

Because of the high viscosity of the deep mantle the warm regions are semi-permanent compared to features in the upper mantle. The large viscosity contrast means that the various layers are more likely, on average, to be thermally coupled than shear coupled. From a tomographic point of view, this means that some mantle structures may appear to be continuous even if the mantle is stratified.

Energy

The energy for convection is provided by the decay of the radioactive isotopes of uranium, thorium and potassium and the cooling and crystallization of the Earth. This heat is removed from the interior by the upwelling of hot, buoyant material to the top of the system where it is lost by conduction and radiation. One can also view mantle convection as the result of cooling of the surface layer, which then sinks and displaces warmer material upward. In this view, the mantle is passive. The buoyancy is provided by thermal expansion and phase changes including partial melting. In a chemically layered Earth the heat is transferred by convection to internal thermal boundary layers across which the heat is transferred by the slow process of conduction. Thermal boundary layers at the base of convecting systems warm up and can also become unstable, generating hot upwellings or plumes. Adiabatically ascending hot upwellings, either passive (responding to slab sinking) or active (plumes), are likely to cross the solidus in the upper mantle, thereby buffering the temperature rise and magnifying the buoyancy. Because of the divergence, with pressure, of melting curves and the adiabat, deep buoyant plumes are subsolidus.

In the simplest model of convection in a homogenous fluid heated from below, hot material rises in relatively thin sheets and spreads out at the surface where it cools by conduction, forming a cold surface thermal boundary layer that thickens with time. Eventually the material achieves enough negative buoyancy to sink back to the base of the system where it travels along the bottom, heating up with time until it achieves enough buoyancy to rise. This gives the classical cellular convection with most of the motion and temperature change occurring in thin boundary layers, which surround the nearly isothermal or adiabatic cores. It has not escaped the attention of geophysicists that midocean ridges and subducting slabs resemble the edges of a convection cell and that the oceanic lithosphere thickens as a surface boundary layer cooling by conduction.

The planform of convection depends on the Rayleigh number and the boundary conditions (BC). At moderate Rayleigh number and constant uniform BC three-dimensional patterns result that resemble hexagons or spokes in plan view. Upwellings and downwellings can be different in shape. At very high Rayleigh number the fluid can become turbulent or chaotic. Whole mantle convection with constant properties and no pressure effects would be characterized by a high Rayleigh number and chaotic, well-stirred convection. Layered convection with pressure dependent properties can have very low Rayleigh numbers and an unmixed, heterogenous, mantle.

In an internally heated fluid the heat cannot be completely removed by narrow upwellings. The whole fluid is heating up and becoming buoyant, so very broad upwellings result. On the other hand, if the fluid has a stress-dependent rheology, or a component of buoyancy due to phase changes such as partial melting, then the boundary layers can become thinner. A temperature-dependent rheology can force the length scales of the surface boundary to be larger than the bottom boundary layer. The lower boundary layer, having a higher temperature and, possibly, undergoing phase changes to lighter phases, can go unstable and provide upwellings with a smaller spacing than the downwellings. The effect of pressure and phase changes on the rheology may reinforce or reverse this tendency. Finally, the presence of accumulations of light material at the surface continents can affect the underlying motion. Subduction, for example, depends on more than the age of the oceanic lithosphere. Convection in the mantle is therefore unlikely to be a steady-state phenomenon. Collection of dense material at the base of the mantle, or light material at the top of the core, may help explain the unusual characteristics of D″.

Although convection in the mantle can be described in general terms as thermal convection, it differs considerably from convection in a homogenous Newtonian fluid, heated from below, with constant viscosity and thermal expansion. The temperature dependence of viscosity gives a 'strong' cold surface layer. This layer must break or fold in order to return to the interior. When it does, it can drag the attached 'plate' with it, a sort of surface tension that is generally not important in normal convection. The deformation also introduces dissipation, a role played by internal viscosity in normal convection. In addition, light crust and depleted lithosphere serve to decrease the average density of the cold thermal boundary layer, helping to keep it at the surface. Buoyant continents and their attached, probably also buoyant, lithospheric roots move about and affect the underlying convection. The stress dependence of the strain rate in solids gives a stress-dependent viscosity. This concentrates the flow in highly stressed regions; regions of low stress flow slowly. Mantle minerals are anisotropic, tending to recrystallize with a preferred orientation dictated by the local stresses. This in turn gives an anisotropic viscosity, probably with the easy flow direction lined up with the actual flow direction. The viscosity controlling convection may therefore be different from the viscosity controlling postglacial rebound.

The core–mantle boundary region

The TBL at the base of the mantle generates a potentially unstable situation. The effects of pressure increase the thermal conductivity, decrease the thermal expansion and increase the viscosity. This means that conductive heat transfer from below is more efficient than at the surface, that temperature increases have little effect on density and that any convection will be sluggish. Although the amount of heat coming out of the core may be appreciable, it is certainly less ($\sim$10%) than that flowing through the surface. The net result is that lower-mantle upwellings take a long time to develop and they must be very large in order to accumulate enough buoyancy to overcome viscous resistance. The spatial and temporal scale of core–mantle–boundary instabilities are orders of magnitude larger than those at the surface. This physics is not captured in laboratory simulations or calculations that adopt the popular Boussinesq approximation. Pressure also makes it easier to irreversibly chemically stratify the mantle. A small intrinsic density difference due to subtle changes in chemistry can keep a deep layer trapped since it requires such large temperature increases to make it buoyant. Layered-mantle convection is the likely outcome.

Dimensionless scaling relations

The theory of convection is littered with dimensionless numbers named after prominent dead physicists. The importance of these numbers to Earth scientists is that they tell us what kinds of experiments and observations may be relevant to the mantle. Experiments and calculations that are in a different parameter space from the mantle are not realistic. Atmospheric thunderheads and smoke-stack plumes cannot be used as analogs to what might happen in the mantle.

The relative importance of conduction and convection is given by the Peclet number

$$Pe = vL/\kappa$$

where v is a characteristic velocity, L a characteristic length and κ the thermal diffusivity,

$$\kappa = K/\rho C_P$$

expressed in terms of conductivity, density and specific heat at constant pressure. The Peclet number gives the ratio of convected to conducted heat transport. For the Earth Pe is about 10^3 and convection dominates conduction. For a much smaller body (L small), conduction would dominate; this is an example of the scale as well as the physical properties being important in the physics. There are regions of the Earth, however, where conduction dominates, such as at the surface where the vertical velocity vanishes.

Dynamic similarity depends on two other non-dimensional parameters: the Grashof number, which involves the buoyancy forces and the resisting forces,

$$Gr = g\alpha\Delta T L^3/\nu^2$$

and the Prandtl number ($\mathrm{Pr} = \nu/\kappa$) where ν is the kinematic viscosity. Only when both of these are the same in two geometrically similar situations can the flow patterns be expected to be same. In general, the numbers appropriate for the mantle cannot be duplicated in the laboratory. The Prandtl number for the mantle is essentially infinite, $\sim 10^{23}$. Inertial forces can be ignored in mantle convection.

The Rayleigh number,

$$Ra = \mathrm{Gr}\ \mathrm{Pr}$$
$$Ra = g\alpha \Delta T L^3/\nu\kappa$$

is the ratio between thermal driving and viscous dissipative forces, and is proportional to the temperature difference, ΔT, and the cube of the scale of the system. It is a measure of the vigor of convection due to thermally induced density variations, $\alpha \Delta T$, in a fluid of viscosity ν operating in a gravity field g. This is for a uniform fluid layer of thickness d with a superadiabatic temperature difference of $\alpha \Delta T$ maintained between the top and the bottom. If the fluid is heated internally, the $\alpha \Delta T$ term is replacd by the volumetric heat production. Convection will occur if Ra exceeds a critical value of the order of 10^3. For large Ra the convection and heat transport are rapid. The Rayleigh number depends on the scale of convection as well as the physical properties.

If L is taken to be the depth of the mantle one obtains very large Ra. However, if convection is layered, the scale drops and Ra can become very small. At high pressure, the combination of properties in Ra also drives Ra down.

The Nusselt number gives the relative importance of convective heat transport compared with the total heat flux:

$$\mathrm{Nu} = \text{total heat flux across the layer/conducted heat flux in the absence of convection}$$
$$= QL/K\,\Delta T$$

where Q is the rate of heat transfer per unit unit area, and L and ΔT are the length and temperature difference scales.

The Nusselt number is the ratio of the actual heat flux to the flux that would occur in a purely conducting regime (so it expresses the efficiency of convection for enhancing heat transfer). For an internally heated layer Nu is the ratio of the temperature drops across the layer with and without convection or equivalently, the ratio of the half-depth of the layer to the thermal boundary layer thickness.

The Prandtl number, $\mathrm{Pr} = \nu/\kappa$ varies from about 10^{-2} in liquid metals to 1 for most gases, and slightly more than 1 for liquids such as water and oil. Pr for the mantle is about 10^{24}, which means that the viscous response to a perturbation is instantaneous relative to the thermal response. If one changes a boundary condition, or inserts a crack into a plate, the effect is felt immediately by the whole system. Thermal perturbations, however, take time to be felt. The square root of Pr gives the ratio of the thicknesses of the mechanical boundary layer (MBL) to the thermal boundary layer (TBL). If the container, or the mantle is smaller than this, then convection will organize itself so as to have a small number of upwellings and downwellings. The limiting case is rotation of the whole fluid.

One cannot take one's experience with smoke plumes in the atmosphere or hot plumes in boiling pots of water and apply it to the mantle. One must look at systems with comparable Rayleigh and Prandtl numbers. Narrow plumes are characteristic of high Rayleigh number, low Prandtl number flows. The mantle is the opposite. There are no large velocity gradients in high Pr fluids.

The Reynolds number is defined by $\mathrm{Re} = UL/v$, for a fluid of kinematic viscosity v, flowing with speed U past a body of size L. In aerodynamics, it characterizes similarities between flows with the same Reynolds number. Turbulence at very high Reynolds numbers is expected to be controlled by inertial effects (as viscosity passively smoothes out the smallest scales of motion), as seen for flows around an aircraft or car. The Reynolds number can also be written

$$\mathrm{Re} = \mathrm{Pe}/\mathrm{Pr}$$

For typical plate tectonic rates and dimensions, $\mathrm{Re} \sim 10^{-21}$. For $\mathrm{Re} \ll 1$ inertial effects are negligible, and this is certainly true for the mantle. Re is important in aerodynamics and hydrodynamics but not in mantle dynamics. Inertia can be ignored in plate tectonics. In free thermal convection Re and the velocity are functions of Pr and Ra; they are not independent parameters.

Convection can be driven by heating from below or within, or by cooling from above. The usual case treated is where the convection is initiated by a vertical temperature gradient. When the vertical increase of temperature is great enough to overcome the pressure effect on density, the deeper material becomes buoyant and rises. An adiabatic gradient simply expresses the condition that the parcel of fluid retains the same density contrast as it rises. Horizontal temperature gradients can also initiate convection. Convection can be driven in a tank of fluid where the side-walls differ in temperature. There is no critical Rayleigh number in this situation. Lateral temperature gradients can be caused by the presence of continents or variations in lithosphere thickness such as at fracture zones or between oceans and continents. For high Ra and Pr most of the temperature contrast occurs across narrow boundary layers.

In the case of natural convection, velocity is not imposed but is set by buoyancy effects. A central issue is to find a relationship between a temperature difference applied to the system and the corresponding heat flux. Fundamental studies are often concerned with Rayleigh–Bénard convection of a fluid layer heated from below and cooled from above, and where temperature is the only control on density. Natural systems are not so ideal. At very high Ra, the velocity and the transported heat flux are expected to become independent of viscosity and heat conductivity, which is reached in large-scale systems such as the atmosphere.

The Grüneisen parameter can be regarded as a nondimensional incompressibility

$$\gamma = \alpha K_s/\rho C_p = \alpha K_T/\rho C_s$$

It is important in compressible flow calculations. This effect is different from the effect of compression on physical and thermal properties.

The density scale height in a convecting layer is

$$h_d = \delta z/\delta \ln \rho = \gamma C_p/\alpha g = K_s/\rho g$$

where z is the radial (vertical) coordinate.

The dimensionless dissipation number, Di, is

$$\mathrm{Di} = hg\alpha/C_p$$

where h is the thickness of the convecting layer. If h is the thickness of the mantle Di is about 0.5. One hesitates to assign someone's name to this number, even a dead physicist. The dissipation number divided by the Grüneisen ratio is the ratio of the thickness of the convecting layer to the density-scale height or $h_d g\rho/K_s$. When Di is large, the assumption of incompressible flow is not valid. Nevertheless, the incompressible mass conservation equation is usually adopted in mantle convection studies. Compression also changes the physical properties of the mantle, the Rayleigh number and the possibility of chemical stratification.

The buoyancy ratio is

$$B = \Delta\rho_c/\rho\alpha\Delta T$$

where $\Delta\rho_c$ is the intrinsic chemical density contrast between layers. When this is small we have purely thermal convection but when it is large the dense components can no longer be entrained and chemical layering results. Pressure serves to decrease α and therefore to stabilize chemically stratified convection. In discussions of layered mantle convection B is the most important parameter.

Rayleigh-like numbers

The first order questions of mantle dynamics include:

(1) Why does Earth have plate tectonics?
(2) What controls the onset of plate tectonics, the number, shape and sizes of the plates, the locations of plate boundaries and the onset of plate reorganization?
(3) What is the organizing principle for plate tectonics; is it driven or organized from the top or by the mantle? What, if anything, is *minimized*?

Surprisingly, these are not the questions being addressed by mantle geodynamicists or computer simulations.

Marangoni or Bénard–Marangoni convection is controlled by a dimensionless number,

$$M = \sigma\Delta TD/\rho\nu\kappa$$

where σ is the temperature derivative of the surface tension, S, ΔT is the temperature difference, D is the layer depth, and ρ, ν and κ are the density, kinematic viscosity and thermal diffusivity of the layer. Fluid is drawn up at warm regions of the surface and flows toward cell boundaries where it returns to the interior. Surface tension forces replace thermal buoyancy which appears in the Rayleigh number. Systems with large Rayleigh or Marangoni numbers are far from conductive equilibrium; they are SOFFE systems.

In a fluid cooled from above, even without surface tension, the cold surface layer becomes unstable and drives convection in the underlying fluid when the local Rayleigh number of the thermal boundary layer (TBL) exceeds a critical value. Like Marangoni convection, this type of convection is driven from the top. Cold downwelling plumes are the only active elements; the upwellings are passive, reflecting mass balance rather than thermal instabilities. Plate tectonics, to a large extent, is driven by the unstable surface thermal boundary layer and therefore resembles convection in fluids which are cooled from above.

Pressure decreases α and increases ν and κ so it is hard to generate buoyancy or vigorous small-scale convection at the base of the mantle. In addition, heat flow across the CMB is about an order of magnitude less than at the surface so it takes a long time to build up buoyancy. In contrast to the upper TBL (frequent ejections of narrow dense plumes), the lower TBL is sluggish and does not play an active role in upper mantle convection.

There are additional surface effects. Lithospheric architecture and slabs set up lateral temperature gradients that drive small-scale convection. For example, a newly opening ocean basin juxtaposes cold cratonic temperatures of about 1000 °C at 100 km depth with asthenospheric temperatures of about 1400 °C. This lateral temperature difference, ΔT, sets up convection, the vigor of which is characterized by the Elder number,

$$\mathrm{El} = g\alpha \Delta T L^3/\kappa\nu$$

where L is a characteristic horizontal dimension, e.g. the width of a rift or an ocean basin or the distance between cratons, and ν is now the viscosity of the asthenosphere. This kind of small-scale convection has been called `EDGE`, for `edge-driven gyres and eddies`. Convective flows driven by this mechanism can reach 15 cm/year and may explain volcanism at the margins of continents and cratons, at oceanic and continental rifts, and along fracture zones and transform faults. Shallow upwellings by this mechanism are intrinsically 3D and may create such features as Iceland and Bermuda.

The role of pressure in mantle convection

Pressure decreases interatomic distances in solids and this has a strong nonlinear effect on such properties as thermal expansion, conductivity and viscosity, all in the direction of making it difficult for small-scale thermal instabilities to form in deep planetary interiors. Convection is sluggish and large-scale at high pressure. The Boussinesq approximation, widely used in geodynamics calculations, assumes that density, or volume (V), is a function of temperature (T) but that all other properties are independent of T, V and pressure (P), even those that are functions of V. This approximation, although thermodynamically (and algebraically) inconsistent, is widely used to analyze laboratory convection and is also used in mantle convection simulations. Sometimes this approximation is supplemented with a depth-dependent viscosity or with T-dependence of parameters other than density. It is preferable to use a thermodynamically self-consistent approach. To first order, the properties of solids depend on interatomic distances, or lattice volumetric strain, and to second order on what causes the strain (T, P composition, crystal structure). This is the basis of `Birch's Law`, the `seismic equation of state`, various laws of `corresponding states` and the `quasiharmonic approximation`.

Volume as a scaling parameter

As far as physical properties are concerned, the main effects of pressure, temperature and phase changes are via volume changes. The thermal, elastic and rheological properties of solids depend on interatomic distances, or lattice volumetric strain, and are relatively indifferent as to what causes the strain (T, P or crystal structure). Intrinsic temperature effects are those that occur at constant volume. The quasiharmonic approximation is widely used in mineral physics but not in seismology or geodynamics where less physically sound relationships are traditionally used.

A parameter that depends on P, T, phase (ϕ) and composition © (within limits, e.g. constant mean atomic weight) can be expanded as

$$M(P, T, \phi, ©) = M(V) + \varepsilon$$

where ε represents higher-order intrinsic effects at constant molar V. Lattice dynamic parameters and thermodynamic and anharmonic parameters are interrelated via V.

Beyond Boussinesq

The effect of volume changes on thermodynamic properties are determined by dimensionless parameters. Scaling parameters for volume-dependent properties can be written as power laws or as logarithmic volume derivatives about the reference state;

Lattice thermal conductivity	$-\mathrm{d}\ln\kappa_L/\mathrm{d}\ln V \sim 4$
Bulk modulus	$-\mathrm{d}\ln K_T/\mathrm{d}\ln V \sim 4$
Thermal expansivity	$-\mathrm{d}\ln\alpha/\mathrm{d}\ln V \sim -3$

Volume changes in laboratory convection experiments are small, so the changes of thermal parameters associated with volume changes are small, with the possible exception of viscosity. The Boussinesq approximation ignores these effects. The specific volume at the base of the mantle is 64% of that at the top. Compression, composition and phase changes, and to some extent, temperature, are all involved. When the above numbers are multiplied by $\Delta V/V$ the changes in physical properties are non-negligible for the mantle. Although volume scalings such as the Debye theory and the quasiharmonic approximation are strongly grounded in classical physics they have not been implemented in mantle convection codes.

Top-down tectonics

One can think of mantle convection as having various origins. The mantle is cooled from above; instability of the cold upper TBL is a *top-down* mechanism, which is basically plate tectonics. Heating the mantle from below is a *bottom-up* or thermal plume mechanism. Lithospheric architecture provides a *sideways* or EDGE mechanism that is lacking in Rayleigh–Bénard convection, or in simple fluids with simple boundary conditions. Internal heating generates time-dependent upwellings, an *inside-out* mechanism. *Delamination* is a *bottoms-off* thermo-chemical mechanism that does not involve the whole outer shell. Cooling of the surface and the motions of plates and plate boundaries, and their effect on the underlying mantle, constitute the main, or large-scale, mode of planetary convection. *Small-scale* convection takes the form of gyres and eddies, rolls and sprouts; these are secondary effects of plate tectonics. *Edge-driven* convection, and stress variations and cracks in the plates are all consequences of plate tectonics and offer explanations of volcanic chains and volcanism that are not at plate boundaries. Lateral variations in temperature, melting temperature, density and fertility of the upper mantle are also consequences of plate tectonics; recycling, continental insulation and slab cooling can explain variations in volcanic output from place to place. So-called *midplate volcanism, melting anomalies* and *hotspots* can be consequences of plate tectonics, and do not require high mantle temperatures or deep fluid dynamic instabilities.

Early views of plate tectonics treated plates as responding passively to mantle convection. Plates and continents drifted about passively on the surface. Ridges were the upwellings and slabs were the downwellings. Narrow hot upwellings were generally held responsible for 'hotspots,' 'hotlines' and 'hotspot tracks'; giant upwellings or plume heads were held responsible for large igneous provinces and continental break-up and for influencing plate motions. These ideas developed from the tacit assumptions that the lithosphere is rigid and uniform, the upper mantle is isothermal, generally subsolidus, and homogenous, and that locations of volcanoes are controlled by mantle temperature and convection, not the stress state of the lithosphere.

It was then recognized that plates could drive themselves and could also organize the underlying mantle convection. Plates and the architecture of the lithosphere provided the template and stress conditions for midplate magmatism and tectonics, phenomena not obviously related to plate tectonics or plate boundaries in the context of rigid plates. Variations in the thickness of the crust and the ages of continents, and the cooling of oceanic plates, set up lateral temperature gradients which can be just as important in driving mantle convection as the non-adiabatic temperature gradients in TBLs. Secular cooling of the Earth maintains a surface TBL; cooling from above can initiate and maintain mantle convection and plate tectonics. Although the mantle is, to some extent, heated from within, and from below, it is basically a system that is driven from the top.

The tectonic plate system can be viewed as an open, far-from-equilibrium, dissipative and self-organizing system that takes matter and energy from the mantle and converts it to mechanical forces (ridge push, slab pull), which drive the plates. 'Subducting slabs, delamination and cratonic roots cool the mantle and create pressure and temperature gradients that drive mantle convection. The plate system thus acts as a template to organize mantle convection. In contrast, in the conventional view, the lithosphere is simply the surface boundary layer of mantle convection and the mantle is the self-organizing dissipative system. If most of the buoyancy and dissipation – the alternative to mantle viscosity – is provided by the plates while the mantle simply provides heat, gravity, matter, and an entropy dump, then plate tectonics is a candidate for a self-organized system, in contrast to being organized by mantle convection or heat from the core. Stress fluctuations in such a system cause global reorganizations without a causative convective event in the mantle. Changes in stress affect plate permeability and can initiate or turn off fractures, dikes and volcanic chains. The mantle itself need play no active role in plate tectonic 'catastrophes.'

The traditional view of mantle geodynamics and geochemistry is that magmatism, and phenomena such as continental break-up and plate reorganization, are due to convection currents in the mantle, and the importation of core heat, via plumes, into the upper mantle. Mantle convection by-and-large controls itself, and can experience massive overturns called `mantle avalanches`. But mantle dynamics may be almost entirely a top-down system and it is likely that mantle convection of various scales is controlled by plates and plate tectonics, not vice versa. The surface boundary layer is the active element, the 'convecting mantle' is the passive element. When a plate tectonic and continental template is placed on top of the convecting system, it organizes the convective flow and the plates themselves become the dissipative self-organized system.

Plate driven flow

Marangoni convection is driven by surface tension. Since surface tension is isotropic, the fluid flows radially from regions of low surface tension to the cell boundaries, which are hexagonal in planform, where linear downwellings form. The equivalent surface force in mantle convection is the *ridge-push–slab-pull* gravitational force which has the same units as surface tension. Since plates are not fluids the forces are not isotropic. Plates move from ridge to trench, pulling up material at diverging regions, which are the equivalent of the centers of Bénard–Marangoni hexagons, and inserting cold material at subduction zones. The other difference between Marangoni and plate-driven convection is that plates are held together by lateral compression and fail in lateral extension. Cell boundaries are convergent and elevated and are regions of compressive stress in Marangoni convection.

The plate-tectonic equivalent of the `Marangoni number` can be derived by replacing surface tension by plate forces. I define the plate tectonic or Platonic number

$$\mathrm{Pl} = g\alpha \Delta T L^2 / U\mathcal{D}$$

where L is a characteristic length (e.g. ridge-trench distance) and U is plate velocity. $\mathcal{D}$ is a dissipation function, which accounts for plate deformation, intraplate resistance and mantle viscosity. It is a rheological parameter. The `roles` of plate bending and fault strength at subduction zones may be as important as mantle viscosity in mantle dynamics and in controlling the cooling of the mantle. When the only resisting forces are lithospheric bending we have the dimensionless lithosphere number

$$\mathrm{Pl} = g\alpha\Delta Tr^3/\kappa\nu_1$$

where r is the radius of curvature of the bend and ν_1 is the lithospheric viscosity. The thermal evolution of an Earth with strong subduction zones is quite different from one with a completely fluid mantle.

When plate interactions are involved we also need coupling parameters across plates such as transform fault resistance and normal stress. In the plate-tectonic system the plates (and slabs) account for much of both the driving force and dissipation (see Figure 4.8) and in this respect they play the role of the convecting fluid in Rayleigh–Bénard convection (internal sources of both buoyancy and viscous dissipation). Both the buoyancy and dissipative stresses affect the whole system. The plate sizes, shapes and velocities are self-controlled and should be part of the solution rather than input parameters. Even the rheology of the surface material may be self-controlled rather than something you can look up in a handbook. *Foams* and *grains* are examples of *fragile* or *soft* materials that to some extent control their own fates. The behavior of rocks, and probably the lithosphere, is controlled by damage rheology and complex feedback processes involving fracture and friction rather than fluid mechanics.

Layered mantle convection

If pressure is ignored it requires about a 6% increase in *intrinsic density* for a deep mantle layer to be stable against a temperature-induced overturn. Plausible differences in density between the silicate products of accretional differentiation, which are intermediate in density between the crust and the core, are about 1 or 2%, if the variations are due to changes in silicon, aluminum, calcium and magnesium. Changes in iron content can give larger variations. Such density differences have been thought to be too small to stabilize stratification. However, when pressure is taken into account chemical stratification is likely; dense layers become trapped, although the effect on seismic velocities can be slight. These are therefore *stealth* layers or reservoirs which are below the ability of seismic waves to detect, except by special techniques.

The lower parts of the mantle are now at high pressure and low temperature compared with accretional conditions. Chemically distinct dense material, accumulated at the base of the mantle, must become very hot in order to become buoyant, because of the very low thermal expansivity at lower-mantle pressures. Scaling relations also show that only very large features would accumulate enough buoyancy to rise.

Although the viscosity of the mantle increases with depth, because of pressure, and although the lower mantle may also have a higher viscosity, because the mineralogy is different and stresses may be lower, this is not sufficient to prevent whole-mantle convection. If the viscosity of the upper mantle is less than that of the lower mantle, circulation will be faster and there is more opportunity to recirculate crustal and upper-mantle material through the shallow melting zone. Differentiation processes would therefore change the composition of the upper mantle, even if the mantle were chemically homogenous initially. The separation of light and dense material, and low-melting-point and high-melting-point material, however, probably occurred during accretion and the early high-temperature history of the Earth. The differentiation will be irreversible if the recycled products of this differentiation (basalt, eclogite, depleted peridotite, continental crust) are unable to achieve the densities of the lower mantle or the parts of the upper mantle through which they must pass.

The magnitude of the Rayleigh number is a measure of the vigor of convection and the distance from static equilibrium. Most geodynamic discussions assume *Ra* to be 10^6 to 10^8. Using parameters appropriate for the base of the mantle yields a value of 4000. If the lower 1000 km of the mantle acts as an isolated layer, because of high intrinsic density, *Ra* drops to 500. The critical Rayleigh number in a spherical shell is about 10^3. The implication of these results are far-reaching. Instabilities at the base of the mantle must be sluggish and immense not narrow or plume-like. These inferences are consistent with lower mantle tomography and make it more plausible than previously thought for the mantle to be chemically stratified. Deep dense layers need only be a fraction of a percent denser that the overlying layers in order to be trapped. This is a consequence of low α (coefficient of thermal expansion) and the difficulty of creating buoyancy with available-temperature variations and heat sources. The gravitational differentiation of the deep mantle may be irreversible and ancient. The widespread belief that the mantle convects as a unit (whole mantle convection, deep slab penetration) or in accessible layers is based on non-Boussinesq calculations or experiments. A corner-stone of the standard model of mantle geochemistry is that the upper mantle is homogenous and is therefore vigorously convecting. The considerations in this section make this unlikely. If the mantle is heterogenous in chemistry and melting point the geochemical motivation for deep-mantle plumes disappears.

Part III

Radial and lateral structure

Descend into the crater of Yocul of Sneffels, Which the shade of Scartaris caresses, before the kalends of July, Audacious traveler, and you will reach the center of the Earth. I did it.

Arne Saknussemm

Overview

The Australian seismologist Keith Bullen introduced the nomenclature for the subdivisions of the Earth's interior. Table 8.1 gives these subdivisions. The lower mantle, starting at 1000-km depth, includes Regions D' and D". The latter is the only designation in common use today. Using his nomenclature, the lithosphere and the low-velocity zone are in Region B. The 650 km discontinuity is in Region C – the Transition Region – rather than being the boundary between the upper and lower mantles. The transition Region extends from 410 to 1000 km depth. The upper boundary is primarily a phase change and the lower boundary may be a chemical change and a geodynamic barrier.

Standard geochemical and geodynamic models of the mantle involve one or two large vigorously convecting regions. Petrological models of the mantle tend to be more complex. High-resolution seismic techniques involving reflected and converted phases show about 10 discontinuities in the mantle, not all of which are easily explained by solid-solid phase changes. They also show some deep low-velocity zones that may be eclogite layers.

It is increasingly clear that the upper mantle is heterogenous in all parameters at all scales. The parameters include seismic scattering potential, anisotropy, mineralogy, major and trace element chemistry, isotopes, melting point and temperature. An isothermal homogenous upper mantle, however, has been the underlying assumption in much of mantle geochemistry for the past 35 years. Derived parameters such as degree and depth of melting and the age and history of mantle 'reservoirs' are based on these assumptions. There is now evidence for major element, mineralogical, trace element and isotopic heterogeneity, on various scales (grain size to hemispheric) and for lateral variations in temperature and melting point.

The large-scale features of the upper mantle are well known from global tomographic studies. The mantle above 200–300 km depth correlates very well with known tectonic features. There are large differences between continents and oceans, and between cratons, tectonic regions, back-arc basins and different age ocean basins.

High-velocities appear beneath cratons – *archons*. Continental low-velocities appear in tectonically extending regions such as the Red Sea rift and in backarcs – *tectons*. Lithospheric thickening and asthenospheric thinning with age are evident beneath the oceans.

Low-velocity zones occur beneath ridges, tectonic regions, Yellowstone, and other places at depths less than 200 km. Yellowstone is not a particularly prominent anomaly when placed in the context of the western North American upper mantle, and does not extend below 200 km depth. At depths greater than 200 km there are low-velocity zones beneath India, Iceland and some ridges and back-arc basins. Significant features include a widespread and pronounced low-velocity zone beneath the western United States, and high-velocity anomalies associated with subducting slabs.

The upper mantle scatters seismic energy, indicating that it is heterogenous on the scale of seismic waves, ~10 km. At depths between 800 and 1000 km there is good correlation of seismic velocities with inferred regions of past subduction. Below the Repetti discontinuity, at about 1000 km depth – Bullen's lower mantle – the mantle is relatively homogenous and uncorrelated with surface processes. D" is heterogenous and may be chemically distinct from D' and C.

Although geodynamicists and geochemists are concentrating on one- and two-layer models of the mantle, high-resolution seismic techniques suggest that it is actually multilayered or laminated.

Chapter 8

Let's take it from the top: the crust and upper mantle

ZOE: Come and I'll peel off.
BLOOM: *(feeling his occiput dubiously with the unparalleled embarrassment of a harassed pedlar gauging the symmetry of her peeled pears)* Somebody would be dreadfully jealous if she knew.

James Joyce, Ulysses

The broad-scale structure of the Earth's interior is well known from seismology, and knowledge of the fine structure is improving continuously. Seismology not only provides the structure, it also provides information about the composition, mineralogy, dynamics and physical state. A 1D seismological model of the Earth is shown in Figure 8.1. Earth is conventionally divided into crust, mantle and core, but each of these has subdivisions that are almost as fundamental (Tables 8.1 and 8.2). Bullen subdivided the Earth's interior into shells, from A (the crust) through G (the inner core). The lower mantle, starting at 1000 km depth, is the largest subdivision, and therefore it dominates any attempt to perform major-element mass balance calculations. The crust is the smallest solid subdivision, but it has an importance far in excess of its relative size because we live on it and extract our resources from it, and, as we shall see, it contains a large fraction of the terrestrial inventory of many elements.

The crust

The major divisions of the Earth's interior – crust, mantle and core – have been known from seismology for about 80 years. These are based on the reflection and refraction of P- and S-waves. The boundary between the crust and mantle is called the Mohorovicic discontinuity (M-discontinuity or Moho for short) after the Croatian seismologist who discovered it in 1909. It separates rocks having P-wave velocities of 6–7 km/s from those having velocities of about 8 km/s. The term 'crust' has been used in several ways. It initially referred to the brittle outer shell of the Earth that extended down to the asthenosphere ('weak layer'); this is now called the lithosphere ('rocky layer'). Later it was used to refer to the rocks occurring at or near the surface and acquired a petrological connotation. Crustal rocks have distinctive physical properties that allow the crust to be mapped by a variety of geophysical techniques. Strictly speaking, the crust and the Moho are seismological concepts but petrologists speak of the 'petrological Moho,' which may actually occur in the mantle! The Moho may represent a transition from mafic to ultramafic rocks or to a high-pressure assemblage composed predominately of garnet and clinopyroxene. It may thus be a chemical change or a phase change or both, and differs from place to place. The lower continental crust can become denser than the mantle, and may founder and sink into the mantle.

The present surface crust represents 0.4% of the Earth's mass and 0.6% of the silicate Earth. It

Table 8.1 Bullen's regions of the Earth's interior

Region	Depth Range (km)		
A	0	33	*continental crust*
B	33	410	*upper mantle*
	220		*Lehmann discontinuity*
C	410	1000	*transition region*
	650		*discontinuity*
D'	1000	2700	*lower mantle*
	1000		*Repetti discontinuity*
D''	2700	2900	*transition region*
E	2900	4980	*outer core*
F	4980	5120	*transition region*
G	5120	6370	*inner core*

Table 8.2 Summary of Earth structure

Region	Depth (km)	Fraction of Total Earth Mass	Fraction of Mantle and Crust
Continental crust	0–50	0.00374	0.00554
Oceanic crust	0–10	0.00099	0.00147
Upper mantle	10–400	0.103	0.153
Transition (TZ)	400–650	0.075	0.111
Deep mantle	650–2890	0.492	0.729
Outer core	2890–5150	0.308	—
Inner core	5150–6370	0.017	—

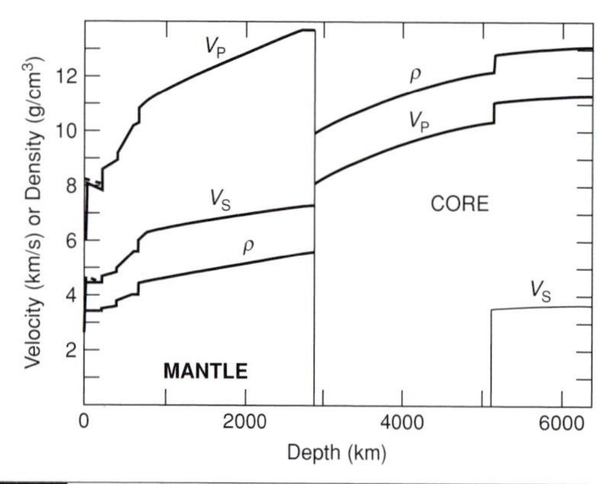

Fig. 8.1 The Preliminary Reference Earth Model (PREM). The model is anisotropic in the upper 220 km. Dashed lines are the horizontal components of the seismic velocity (after Dziewonski and Anderson, 1981).

contains a very large proportion of incompatible elements (20–70%), depending on element. These include the heat-producing elements and members of a number of radiogenic-isotope systems (Rb–Sr, U–Pb, Sm–Nd, Lu–Hf) that are commonly used in mantle geochemistry. Thus the continental crust factors prominently in any mass-balance calculation for the Earth as a whole and in estimates of the thermal structure of the Earth (Figure 8.2). (`Rudnick crust`).

The Moho is a sharp seismological boundary and in some regions appears to be laminated. There are three major crustal types – continental, transitional and oceanic. Oceanic crust generally ranges from 5–15 km in thickness and comprises 60% of the total crust by area and more than 20% by volume. In some areas, most notably near oceanic fracture zones, the oceanic crust is as thin as 3 km. Sometimes the crust is even absent, presumably because the underlying mantle is cold or infertile, or ascending melts freeze before they erupt. Oceanic plateaus and aseismic ridges may have crustal thicknesses greater than 20 km. Some of these appear to represent large volumes of material generated at oceanic spreading centers or triple junctions, and a few seem to be continental fragments. Although these anomalously thick crust regions constitute only about 10% of the area of the oceans, they may represent more than 25% of the total volume of the oceanic crust. They are generally attributed to hot regions of the mantle but they could also represent fertile regions of the mantle or transient responses to lithospheric extension. Islands, island arcs and continental margins are collectively referred to as *transitional crust* and range from 15–30 km in thickness. Continental crust generally ranges from 30–50 km thick, but thicknesses up to 80 km are reported in some active convergence regions. Older regions show a sharp cutoff in crustal thickness at 50 km; this may be the depth of the basalt–eclogite phase change and the depth at which over-thickened crust founders or delaminates. Based on geological and seismic data, the main rock type in the upper continental crust is granodiorite or tonalite in composition. The lower crust is probably diorite, garnet granulite and amphibolite. The average

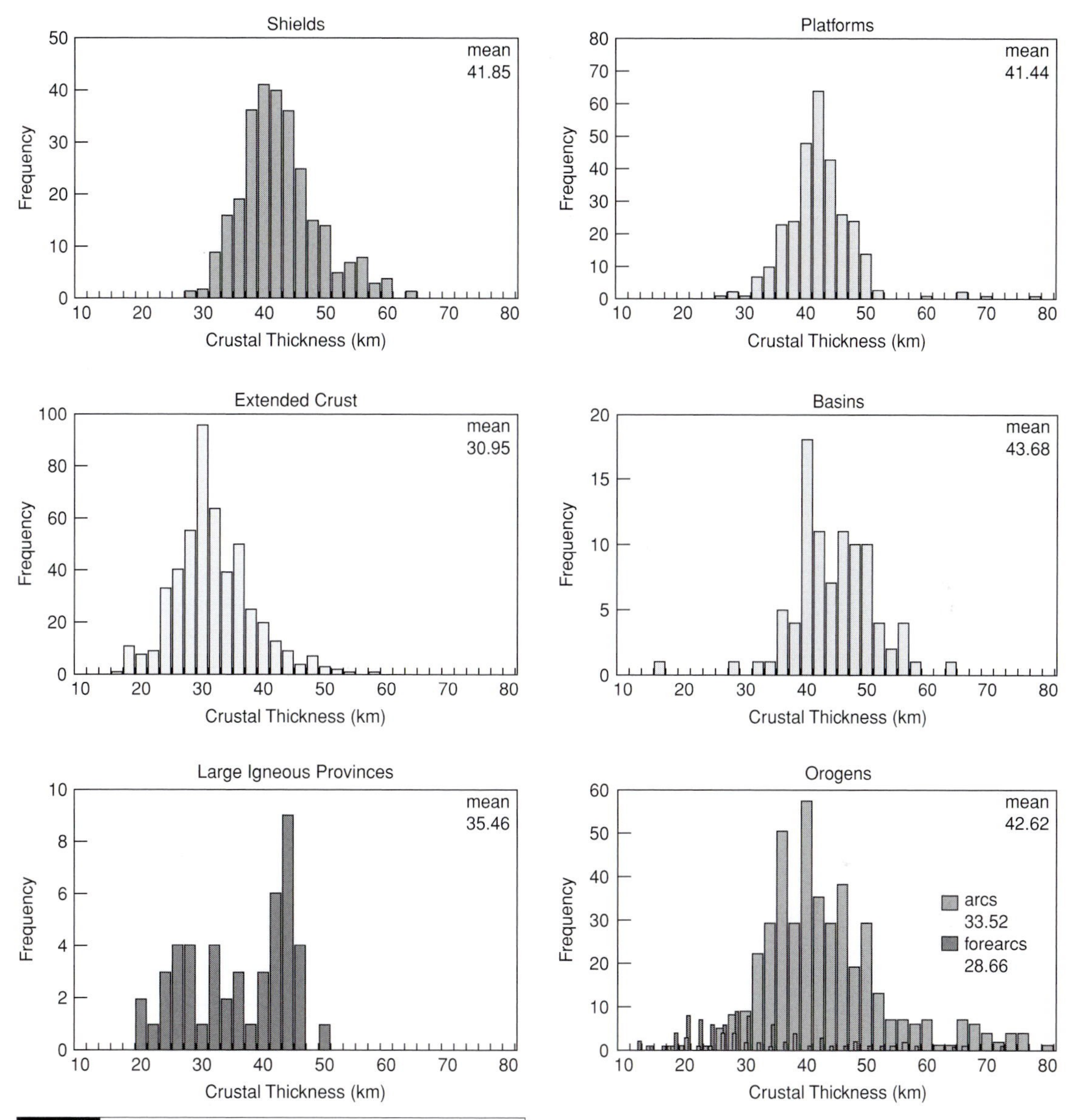

Fig. 8.2 Crustal thickness histograms for various tectonic provinces. Note the cutoff near 50 km thickness. Thicker crust exists in some mountain belts but apparently does not last long (after Mooney *et al.*, 1998).

composition of the continental crust is thought to be similar to andesite or diorite.

The terrestrial crust is unusually thin compared with the Moon and Mars and compared with the amount of potential crust in the mantle. This is related to the fact that crustal material – on a large body – converts to dense garnet-rich assemblages at relatively shallow depth. The maximum theoretical thickness of material with crust-like physical properties is about 50–60 km, although the crust may temporarily achieve somewhat greater thickness because of the sluggishness of phase changes at low temperature and in dry rocks. An Earth model based on cosmic abundances of the elements could, in principle, have a basaltic crust 200 km thick. It is usually considered that the Earth's extra crustal material is well mixed into the mantle, or that it was never extracted from parts of the mantle.

Table 8.3 | Crustal minerals

Mineral	Composition	Range of Crustal Abundances (vol. pct.)
Plagioclase		31–41
Anorthite	$Ca(Al_2Si_2)O_8$	
Albite	$Na(Al,Si_3)O_8$	
Orthoclase		9–21
K-feldspar	$K(Al,Si_3)O_8$	
Quartz	SiO_2	12–24
Amphibole	$NaCa_2(Mg,Fe,Al)_5\ [(Al,Si)_4O_{11}]_2(OH)_2$	0–6
Biotite	$K(Mg,Fe^{2+})_3(Al,Si_3)O_{10}(OH,F)_2$	4–11
Muscovite	$KAl_2(Al,Si_3)O_{10}(OH)$	0–8
Chlorite	$(Mg,Fe^{2+})_5Al(Al,Si_3)O_{10}(OH)_8$	0–3
Pyroxene		
Hypersthene	$(Mg,Fe^{2+})SiO_9$	0–11
Augite	$Ca(Mg,Fe^{2+})(SiO_3)_2$	
Olivine	$(Mg,Fe^{2+})_2SiO_4$	0–3
Oxides		~2
Sphene	$CaTiSiO_5$	
Allanite	$(Ce,Ca,Y)(Al,Fe)_3(SiO_4)_3(OH)$	
Apatite	$Ca_5(PO_4,CO_3)_3(F,OH,Cl)$	
Magnetite	$FeFe_2O_4$	
Ilmenite	$FeTiO_2$	

Alternatively, there may be layers or blobs of eclogite in the mantle, representing crust that was previously near the Earth's surface. Such fertile blobs are an alternative to the plume model for anomalous volcanism.

Feldspar (K-feldspar, plagioclase) is the most abundant mineral in the crust, followed by quartz and hydrous minerals (such as the micas and amphiboles) (Table 8.3). The minerals of the crust and some of their physical properties are given in Table 8.4. A crust composed of these minerals will have an average density of about 2.7 g/cm^3. There is enough difference in the velocities and V_p/V_s ratios of the more abundant minerals that seismic velocities provide a good mineralogical discriminant. One uncertainty is the amount of serpentinized ultramafic rocks in the lower crust since serpentinization decreases the velocity of olivine to crustal values. In regions of over-thickened crust, the lower portions can have abundant garnet and therefore high density and high seismic velocity, comparable to mantle values. The Moho in these cases can actually be due to the gabbro–eclogite phase change, rather than to a chemical change. In these cases, *crustal thickness* is not a good proxy for mantle temperature or the amount of melt that has been extracted from the mantle.

Table 8.4 | Average crustal abundance, density and seismic velocities of major crustal minerals

Mineral	Volume %	ρ (g/cm^3)	V_p (km/s)	V_s (km/s)
Quartz	12	2.65	6.05	4.09
K-feldspar	12	2.57	5.88	3.05
Plagioclase	39	2.64	6.30	3.44
Micas	5	2.8	5.6	2.9
Amphiboles	5	3.2	7.0	3.8
Pyroxene	11	3.3	7.8	4.6
Olivine	3	3.3	8.4	4.9

The floor of the ocean, under the sediments, is veneered by a layer of tholeiitic basalt that was generated at the midocean ridge systems. The pillow basalts that constitute the upper part of the oceanic crust extend to an average depth of 1–2 km and are underlain by sheeted dikes, gabbros and olivine-rich cumulate layers. The average composition of the oceanic crust is much more MgO-rich and lower in CaO and Al_2O_3 than the surface MORB. The oceanic crust rests on a depleted harzburgite layer of unknown thickness. In certain models of crustal genesis, the harzburgite layer is complementary to the crust and is therefore about 24 km thick if an average crustal thickness of 6 km and 20% melting are assumed. Oceanic plateaus and aseismic ridges have thick crust (>20 km in places), and the corresponding depleted layer would be more than 120 km thick if the simple model is taken at face value. Since depleted peridotites, including harzburgites and dunite, are less dense than fertile peridotite, or eclogite, several cycles of plate tectonics, crust generation and subduction would fill up the shallow mantle with harzburgite. Oceanic crust and oceanic plateaus may in part be deposited on ancient, not contemporaneous, buoyant underpinnings.

Table 8.5 Estimates of the chemical composition of the crust (wt.%)

Oxide	Oceanic Crust (1)	Continental Crust (2)	Continental Crust (3)
SiO_2	47.8	63.3	58.0
TiO_2	0.59	0.6	0.8
Al_2O_3	12.1	16.0	18.0
Fe_2O_3	—	1.5	—
FeO	9.0	3.5	7.5
MgO	17.8	2.2	3.5
CaO	11.2	4.1	7.5
Na_2O	1.31	3.7	3.5
K_2O	0.03	2.9	1.5
H_2O	1.0	0.9	—

(1) Elthon (1979).
(2) Condie (1982).
(3) Tayor and McLennan (1985).

Estimates of the composition of the oceanic and continental crust are given in Table 8.5; another estimate that includes the trace elements is given in Table 8.6. Note that the continental crust is richer in SiO_2, TiO_2, $A1_2O_3$, Na_2O and K_2O than the oceanic crust. This means that the continental crust is richer in quartz and feldspar and is intrinsically less dense than the oceanic crust. The mantle under stable continental-shield crust has seismic properties that suggest that it is less dense than mantle elsewhere. The elevation of continents is controlled primarily by the density and thickness of the crust and the intrinsic density and temperature of the underlying mantle.

It is commonly assumed that the seismic Moho is also the petrological Moho, the boundary between sialic or mafic crustal rocks and ultramafic mantle rocks. However, partial melting, high pore pressure and serpentinization can reduce the velocity of mantle rocks, and increased abundances of olivine, garnet and pyroxene can increase the velocity of crustal rocks. High pressure also increases the velocity of mafic rocks, by the gabbro–eclogite phase change, to mantle-like values. The increase in velocity from 'crustal' to 'mantle' values in regions of thick continental crust may be due, at least in part, to the appearance of garnet as a stable phase. The situation is complicated further by kinetic considerations. Garnet is a common metastable phase in near-surface intrusions such as pegmatites and metamorphic teranes. On the other hand, feldspar-rich rocks may exist at depths greater than the gabbro-eclogite equilibrium boundary if temperatures are so low, or the rocks so dry, that the reaction is sluggish.

The common assumption that the Moho is a chemical boundary is in contrast to the position taken with regard to other mantle discontinuities. It is almost universally assumed that the major mantle discontinuities represent equilibrium solid-solid phase changes in a homogenous mantle. A notable exception to this view is the geochemical model that attributes the 650 km discontinuity to a boundary separating the depleted 'convecting upper mantle' from the undegassed primitive lower mantle. This is strictly an assumption; there is no evidence in support of this view. It should be kept in mind that chemical boundaries may occur elsewhere in

Table 8.6 | Composition of the bulk continental crust, by weight*

SiO_2	57.3 pct.	Co	29 ppm	Ce	33 ppm
TiO_2	0.9 pct.	Ni	105 ppm	Pt	3.9 ppm
Al_2O_3	15.9 pct.	Cu	75 ppm	Nd	16 ppm
FeO	9.1 pct.	Zn	80 ppm	Sm	3.5 ppm
MgO	5.3 pct.	Ga	18 ppm	Eu	1.1 ppm
CaO	7.4 pct.	Ge	1.6 ppm	Gd	3.3 ppm
Na_2O	3.1 pct.	As	1.0 ppm	Tb	0.6 ppm
K_2O	1.1 pct.	Se	0.05 ppm	Dy	3.7 ppm
Li	13 ppm	Rb	32 ppm	Ho	0.78 ppm
Bc	1.5 ppm	Sr	260 ppm	Er	2.2 ppm
B	10 ppm	Y	20 ppm	Tm	0.32 ppm
Na	2.3 pct.	Zr	100 ppm	Yb	2.2 ppm
Mg	3.2 pct.	Nb	11 ppm	Lu	0.30 ppm
Al	8.41 pct.	Mo	1 ppm	Hf	3.0 ppm
Si	26.77 pct.	Pd	1 ppb	Ta	1 ppm
K	0.91 pct.	Ag	80 ppb	W	1 ppm
Ca	5.29 pct.	Cd	98 ppb	Re	0.5 ppm
Sc	30 ppm	In	50 ppb	Ir	0.1 ppb
Ti	5400 ppm	Sn	2.5 ppm	Au	3 ppb
V	230 ppm	Sb	0.2 ppm	Tl	360 ppb
Cs	185 ppm	Cs	1 ppm	Pb	8 ppb
Mn	1400 ppm	Ba	250 ppm	Bi	60 ppb
Fe	7.07 pct.	La	16 ppm	Th	3.5 ppm
				U	0.91 ppm

Taylor and McLennan (1985). *Major elements as oxides or elements.

the mantle. It is hard to imagine how the Earth could have gone through a high-temperature accretion and differentiation process and maintained a homogenous composition throughout.

It is probably not a coincidence that the maximum crustal thicknesses are close to the basalt–eclogite boundary. Eclogite is denser than peridotite, at least in the shallow mantle, and will tend to fall into normal mantle, thereby turning a phase boundary (basalt–eclogite) into a chemical boundary (basalt–peridotite). Some eclogites are less dense than the mantle below 410 km depth and may settle there, again turning a phase boundary into a chemical boundary.

Both the crust and the underlying lithosphere have intrinsic densities that increase with depth. That is, the outer shells of Earth are chemically stratified, probably as a result of early differentiation processes. This situation is in contrast to the bulk of the mantle, which is usually – and probably erroneously – treated as one or two homogenous layers.

Seismic velocities in the crust and upper mantle

Seismic velocities in the crust and upper mantle are typically determined by measuring the transit time between an earthquake or explosion and an array of seismometers. Crustal compressional wave velocities in continents, beneath the sedimentary layers, vary from about 5 km/s at shallow depth to about 7 km/s at a depth of 30–50 km. The lower velocities reflect the presence of pores and cracks more than the intrinsic velocities of the rocks. At greater depths the pressure closes cracks and the remaining pores are fluid-saturated. These effects cause a considerable increase in velocity. A typical crustal velocity range at depths greater than 1 km is 6–7 km/s. The corresponding range in shear velocity is about 3.5–4.0 km/s. Shear velocities can be determined from both body waves and the dispersion of short-period surface waves. The top of the mantle under continents usually has velocities in the range 8.0–8.2 km/s for compressional waves and 4.3–4.7 km/s for shear waves.

The compressional velocity near the base of the oceanic crust usually falls in the range 6.5–6.9 km/s. In some areas a thin layer at the base of the crust with velocities as high as 7.5 km/s has been identified. The oceanic upper mantle has a P-velocity (Pn) that varies from about 7.9 to 8.6 km/s. The velocity increases with oceanic age, because of cooling, and varies with azimuth, due to crystal orientation or to dikes and sills. The fast direction is generally close to the inferred spreading direction. The average velocity is close to 8.2 km/s, but young ocean has velocities as low as 7.6 km/s. Tectonic regions also have low velocities. The shear velocity increases from about 3.6–3.9 to 4.4–4.7 km/s from the base of the crust to the top of the mantle.

Ophiolite sections found at some continental margins are thought to represent upthrust or obducted slices of the oceanic crust and upper mantle. These sections grade downward from pillow lavas to sheeted dike swarms, intrusives, pyroxene and olivine gabbro, layered gabbro and peridotite and, finally, harzburgite and dunite (Figure 8.3). Laboratory velocities in these rocks are given in Table 8.7. There is good agreement between these velocities and those actually observed in the oceanic crust and upper mantle. The sequence of extrusives, intrusives and cumulates is consistent with what is expected at a midocean-ridge magma chamber.

The velocity contrast between the lower crust and upper mantle is commonly smaller beneath young orogenic areas (0.5–1.5 km/s) than beneath cratons and shields (1–2 km/s). Continental rift systems have thin crust (less than 30 km) and low Pn velocities (less than 7.8 km/s). Thinning of the crust in these regions appears to take place by thinning of the lower crust. In island arcs the crustal thickness ranges from about 5 km to 35 km. In areas of very thick crust such as in the Andes (70 km) and the Himalayas (80 km), the thickening occurs primarily in the lower crustal layers. Oceanic crust also seems to thicken by increasing the thickness of the lower layer. Paleozoic orogenic areas have about the same range of crustal thicknesses and velocities as platform areas.

Table 8.7 Density, compressional velocity and shear velocity in rock types found in ophiolite sections

Rock Type	ρ (g/cm^3)	V_P (km/s)	V_S (km/s)	Poisson's Ratio
Metabasalt	2.87	6.20	3.28	0.31
Metadolerite	2.93	6.73	3.78	0.27
Metagabbro	2.95	6.56	3.64	0.28
Gabbro	2.86	6.94	3.69	0.30
Pyroxenite	3.23	7.64	4.43	0.25
Olivine gabbro	3.30	7.30	3.85	0.32
Harzburgite	3.30	8.40	4.90	0.24
Durite	3.30	8.45	4.90	0.25

Salisbury and Christensen (1978), Christensen and Smewing (1981).

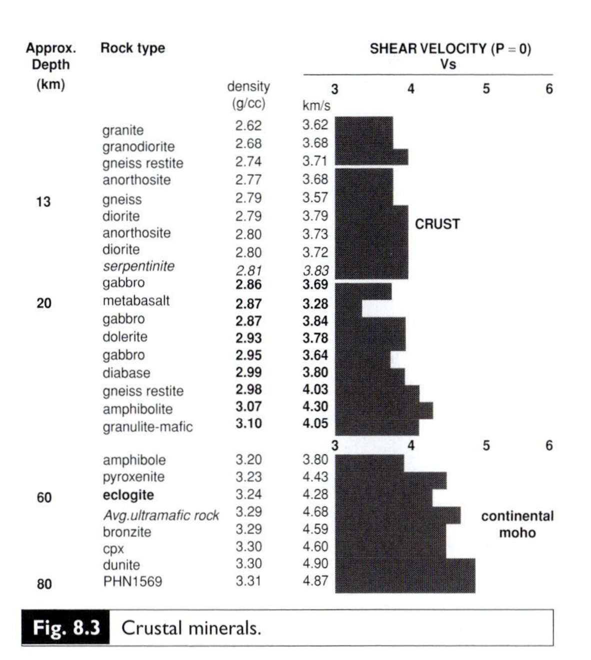

Fig. 8.3 Crustal minerals.

The seismic lithosphere or LID

The top of the mantle is characterized, in most places, by a thin high-velocity layer, a seismic lid.

This seismic lithosphere is not the same as the plate or the thermal boundary layer. It is defined solely on the basis of seismic velocities. In some places, e.g. Basin and Range, it is absent, perhaps due to delamination. The thickness of the LID corresponds roughly to the rheological or elastic lithosphere, e.g. the apparently strong or coherent upper layer overlying the *asthenosphere*. The *thermal boundary layer* is typically twice as thick, at least in oceanic regions. The high-velocity roots under cratons are chemically buoyant and are probably olivine-rich and FeO-poor. They extend to 200–300 km depth. They are long-lived because they are buoyant, cold, dry, high-viscosity and protected from plate-boundary interactions by the surrounding *mobile belts*.

Uppermost mantle compressional wave velocities, Pn, are typically 8.0–8.2 km/s, and the spread is about 7.9–8.6 km/s. Some long refraction profiles give evidence for a deeper layer in the lithosphere having a velocity of 8.6 km/s. The *seismic lithosphere*, or LID, appears to contain at least two layers. Long refraction profiles on continents have been interpreted in terms of a laminated model of the upper 100 km with high-velocity layers, 8.6–8.7 km/s or higher, embedded in 'normal' material. Corrected to normal conditions these velocities would be about 8.9–9.0 km/s. The P-wave gradients are often much steeper than can be explained by self-compression. These high velocities require oriented olivine or large amounts of garnet. The detection of 7–8% azimuthal anisotropy for both continents and oceans suggests that the shallow mantle at least contains oriented olivine or oriented cracks, dikes or lens. In some places, the lithosphere may have formed by the stacking of subducted slabs, another mechanism for creating anisotropy.

Typical values of V_p and V_s at 40 km depth, when corrected to standard conditions, are 8.72 km/s and 4.99 km/s, respectively. Short-period surface-wave data implies STP – standard temperature and pressure – velocities of 4.48–4.55 km/s and 4.51–4.64 km/s for 5-My-old and 25-My-old oceanic lithosphere. A value for V_p of 8.6 km/s is sometimes observed near 40 km depth in the oceans. This corresponds to about 8.87 km/s at standard conditions. These values can be compared with 8.48 and 4.93 km/s for olivine-rich aggregates. Eclogites are highly variable but can have V_p and V_s as high as 8.8 and 4.9 km/s in certain directions and as high as 8.61 and 4.86 km/s as average values. The above suggests that corrected velocities of at least 8.6 and 4.8 km/s, for V_p and V_s, respectively, occur in the lower lithosphere; this requires substantial amounts of garnet, about 26%. The density of such an assemblage is about 3.4 g/cm^3. The lower lithosphere may therefore be gravitationally unstable with respect to the underlying mantle, particularly when it is cold. The upper mantle under shield regions, on the other hand, is consistent with a very olivine-rich peridotite which is buoyant and therefore stable relative to 'normal' mantle.

Anisotropy of the upper mantle is a potentially useful petrological constraint, although it can also be caused by organized heterogeneity, such as laminations or parallel dikes and sills or aligned partial melt zones, and stress fields. Recycled material may also arrange itself so as to give a fabric to the mantle. The uppermost mantle under oceans exhibits an anisotropy of about 7%. The fast direction is in the direction of spreading, and the magnitude of the anisotropy and the high velocities of P-arrivals suggest that oriented olivine crystals control the elastic properties. Pyroxene exhibits a similar anisotropy, whereas garnet is more isotropic. The preferred orientation is presumably due to the emplacement or freezing mechanism, the temperature gradient or to nonhydrostatic stresses. A peridotite layer at the top of the oceanic mantle is consistent with the observations.

The anisotropy of the upper mantle, averaged over long distances, is much less than the values given above. Shear-wave anisotropies of the upper mantle average 2–4%. Shear velocities in the LID vary from 4.26–4.46 km/s, increasing with age; the higher values correspond to a lithosphere 10–50 My old. This can be compared with shear-wave velocities of 4.3–4.9 km/s and anisotropies of 1–5% found in relatively unaltered eclogites, at laboratory frequencies. The compressional velocity range in unaltered samples is 7.6–8.7 km/s, reflecting the large amounts of garnet.

Garnet and clinopyroxene may be important components of the lithosphere and upper mantle. A lithosphere composed primarily of peridotite does not satisfy the seismic data. The lithosphere, therefore, is not just cold asthenosphere or a pure thermal boundary layer. The roots of cratons, however, do seem to be composed mainly of cold olivine but they are buoyant in spite of the cold temperatures. They are therefore unlikely to fall off. They are probably depleted residual after basalt extraction, and are probably garnet- and FeO-poor.

It is likely that the upper mantle is laminated, with the volatiles and melt products concentrated toward the top. As the lithosphere cools, underplated basaltic material is incorporated onto the base of the plate, and as the plate thickens it eventually transforms to eclogite, yielding high velocities and increasing the thickness and mean density of the oceanic plate (Figure 8.4). Eventually the lower part of the plate becomes denser than the underlying asthenosphere, and conditions become appropriate for subduction or delamination.

The thickness of the seismic lithosphere, or high-velocity LID, is about 150–250 km under the older continental shields. A thin low-velocity zone (LVZ) at depth, as found from body-wave studies, however, cannot be well resolved with long-period surface waves. The velocity reversal between about 150 and 200 km in shield areas is about the depth inferred for kimberlite genesis, and the two phenomena may be related.

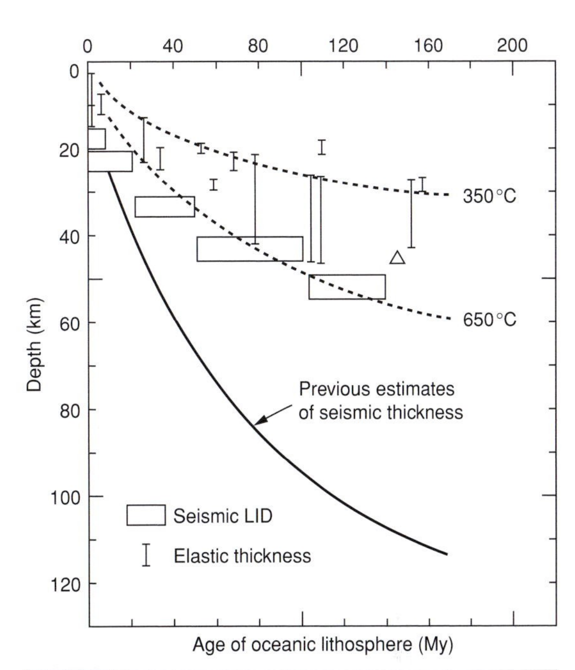

Fig. 8.4 The thickness of the lithosphere as determined from flexural loading studies and surface waves. The upper edges of the open boxes gives the thickness of the seismic LID (high-velocity layer, or seismic lithosphere). The lower edge gives the thickness of the mantle LID plus the oceanic crust (Regan and Anderson, 1984). The LID under continental shields is about 150–250 km thick.

There is very little information about the deep oceanic lithosphere from body-wave data. Surface waves have been used to infer a thickening with age of the oceanic lithosphere to depths greater than 100 km (Figure 8.4). However, when anisotropy is taken into account, the thickness may be only about 50 km for old oceanic lithosphere. This is about the thickness inferred for the 'elastic' lithosphere from flexural bending studies around oceanic islands and at trenches. This is not the same as the thickness of the thermal boundary layer (TBL) or the thickness of the plate.

The seismic velocities of some upper-mantle minerals and rocks are given in Tables 8.9 and 8.10. Garnet and jadeite have the highest velocities, clinopyroxene and orthopyroxene the lowest. Mixtures of olivine and orthopyroxene (the peridotite assemblage) can have velocities similar to mixtures of garnet–diopside–jadeite (the eclogite assemblage). Garnet-rich assemblages, however, have velocities higher than orthopyroxene-rich assemblages.

The V_p/V_s ratio is greater for the eclogite minerals than for the peridotite minerals. This ratio plus the anisotropy are useful diagnostics of mantle mineralogy. High velocities alone do not necessarily discriminate between garnet-rich and olivine-rich assemblages. Olivine is very anisotropic, having compressional velocities of 9.89, 8.43 and 7.72 km/s along the principal crystallographic axes. Orthopyroxene has velocities ranging from 6.92 to 8.25 km/s, depending on direction. In natural olivine-rich aggregates (Table 8.11), the maximum velocities are about 8.7 and 5.0 km/s for P-waves and S-waves, respectively. With 50% orthopyroxene the velocities are reduced to 8.2 and 4.85 km/s, and the composite is nearly isotropic. Eclogites are also nearly isotropic. For a

Table 8.8 | Density and shear velocity of mantle rocks

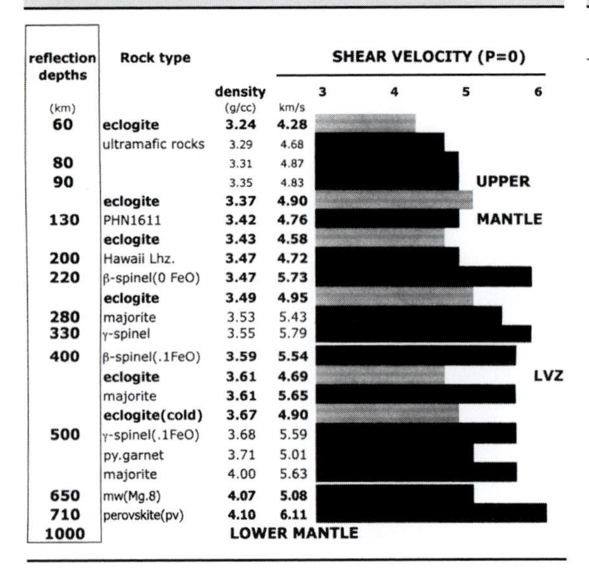

reflection depths (km)	Rock type	density (g/cc)	SHEAR VELOCITY (P=0) km/s
60	eclogite	3.24	4.28
	ultramafic rocks	3.29	4.68
80		3.31	4.87
90		3.35	4.83
	eclogite	3.37	4.90
130	PHN1611	3.42	4.76
	eclogite	3.43	4.58
200	Hawaii Lhz.	3.47	4.72
220	β-spinel(0 FeO)	3.47	5.73
	eclogite	3.49	4.95
280	majorite	3.53	5.43
330	γ-spinel	3.55	5.79
400	β-spinel(.1FeO)	3.59	5.54
	eclogite	3.61	4.69
	majorite	3.61	5.65
	eclogite(cold)	3.67	4.90
500	γ-spinel(.1FeO)	3.68	5.59
	py.garnet	3.71	5.01
	majorite	4.00	5.63
650	mw(Mg.8)	4.07	5.08
710	perovskite(pv)	4.10	6.11
1000	LOWER MANTLE		

Table 8.9 | Densities and elastic-wave velocities in upper-mantle minerals

Mineral	ρ (g/cm^3)	V_p (km/s)	V_s (km/s)	V_p/V_s
Olivine				
Fo	3.214	8.57	5.02	1.71
Fo_{93}	3.311	8.42	4.89	1.72
Fa	4.393	6.64	3.49	1.90
Pyroxene				
En	3.21	8.08	4.87	1.66
En_{80}	3.354	7.80	4.73	1.65
Fs	3.99	6.90	3.72	1.85
Di	3.29	7.84	4.51	1.74
Jd	3.32	8.76	5.03	1.74
Garnet				
Py	3.559	8.96	5.05	1.77
Al	4.32	8.42	4.68	1.80
Gr	3.595	9.31	5.43	1.71
Kn	3.85	8.50	4.79	1.77
An	3.836	8.51	4.85	1.75
Uv	3.85	8.60	4.89	1.76

Sumino and Anderson (1984).

Table 8.10 | Densities and elastic-wave velocities of upper-mantle rocks

Rock	ρ	V_p	V_s	V_p/V_s
Garnet lherzolite	3.53	8.29	4.83	1.72
	3.47	8.19	4.72	1.74
	3.46	8.34	4.81	1.73
	3.31	8.30	4.87	1.70
Dunite	3.26	8.00	4.54	1.76
	3.31	8.38	4.84	1.73
Bronzitite	3.29	7.89	4.59	1.72
	3.29	7.83	4.66	1.68
Eclogite	3.46	8.61	4.77	1.81
	3.61	8.43	4.69	1.80
	3.60	8.42	4.86	1.73
	3.55	8.22	4.75	1.73
	3.52	8.29	4.49	1.85
	3.47	8.22	4.63	1.78
Jadeite	3.20	8.28	4.82	1.72

Clark (1966), Babuska (1972), Manghnani and Ramananotoandro (1974), Jordan (1979).

given density, eclogites tend to have lower shear velocities than peridotite assemblages.

The 'standard model' for the oceanic lithosphere assumes 24 km of depleted peridotite – complementary to and forming contemporaneously with the basaltic crust – between the crust and the presumed fertile peridotite upper mantle. There is no direct evidence for this hypothetical model. It is a remnant of the pyrolite model. The lower oceanic lithosphere may be much more basaltic or eclogitic than in this simple model, and it might not have formed contemporaneously. Buoyant, refractory and olivine-rich lithologies may have accumulated in the shallow mantle for billions of years (Gyr). Basalts can pond beneath plates that are under lateral compression.

The perisphere

The cratonic lithosphere is chemically buoyant relative to the underlying mantle, a result of melt depletion. The shallow mantle elsewhere may

Table 8.11 Anisotropy of upper-mantle rocks

Mineralogy	Direction	V_p	V_{s_1}	V_{s_2}	V_p/V_s	
		Peridotites				
100 pct. ol	1	8.7	5.0	4.85	1.74	1.79
	2	8.4	4.95	4.70	1.70	1.79
	3	8.2	4.95	4.72	1.66	1.74
70 pct. ol,	1	8.4	4.9	4.77	1.71	1.76
30 pct. opx	2	8.2	4.9	4.70	1.67	1.74
	3	8.1	4.9	4.72	1.65	1.72
100 pct. opx	1	7.8	4.75	4.65	1.64	1.68
	2	7.75	4.75	4.65	1.63	1.67
	3	7.78	4.75	4.65	1.67	1.67
		Eclogites				
51 pct. ga,	1	8.476		4.70		1.80
23 pct. cpx,	2	8.429		4.65		1.81
24 pct. opx	3	8.375		4.71		1.78
47 pct. ga,	1	8.582		4.91		1.75
45 pct. cpx	2	8.379		4.87		1.72
	3	8.30		4.79		1.73
46 pct. ga,	1	8.31		4.77		1.74
37 pct. cpx	2	8.27		4.77		1.73
	3	8.11		4.72		1.72

Manghnani and Ramananotoandro (1974), Christensen and Lundquist (1982).

Table 8.12 Measured and estimated properties of mantle minerals

Mineral	ρ (g/cm^3)	V_p (km/s)	V_s (km/s)	V_p/V_s
Olivine ($Fa_{.12}$)	3.37	8.31	4.80	1.73
β-Mg_2SiO_4	3.63	9.41	5.48	1.72
γ-Mg_2SiO_4	3.72	9.53	5.54	1.72
Orthopyroxene ($Fs_{.12}$)	3.31	7.87	4.70	1.67
Clinopyroxene ($Hd_{.12}$)	3.32	7.71	4.37	1.76
Jadeite	3.32	8.76	5.03	1.74
Garnet	3.68	9.02	5.00	1.80
Majorite	3.59	9.05*	5.06*	1.79*
Perovskite	4.15	10.13*	5.69*	1.78*
($Mg_{.19}Fe_{.21}$)O	4.10	8.61	5.01	1.72
Stishovite	4.29	11.92	7.16	1.66
Corundum	3.99	10.86	6.40	1.70

* Estimated.

Duffy and Anderson (1988), Weidner (1986).

also be enriched in refractory and buoyant products of mantle differentiation such as olivine and orthopyroxene (the mantle perisphere). The base of this region may be related to the Lehmann discontinuity.

Perisphere is derived from the prefix *peri-* meaning *all around, surrounding, near by*. *Peri* in Persian folklore is a supernatural being descended from fallen angels or supernatural fairies and excluded from paradise until penance is done. The perisphere is mainly buoyant refractory *peridotite*. It may be enriched in the large-ion lithophile (LIL) elements, probably as a result of extraction of these elements from slabs. A shallow enriched layer is one alternative to a primordial deep layer.

There is evidence that two or three distinct lithologies contribute to the petrogenesis of observed magmas. These probably correspond to recycled components such as oceanic crust, delaminated lower continental crust and peridotite.

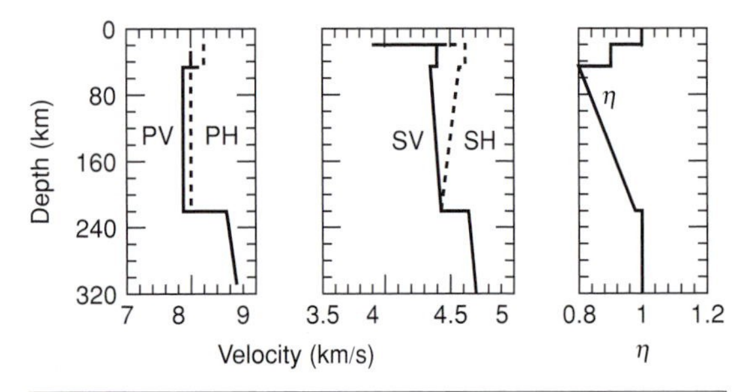

Fig. 8.5 Velocity-depth profiles for the average Earth, as determined from surface waves (Regan and Anderson, 1984). From left to right, the graphs show P-wave velocities (vertical and horizontal), Swave velocities (vertical and horizontal) and an anisotropy parameter (1 represents isotropy).

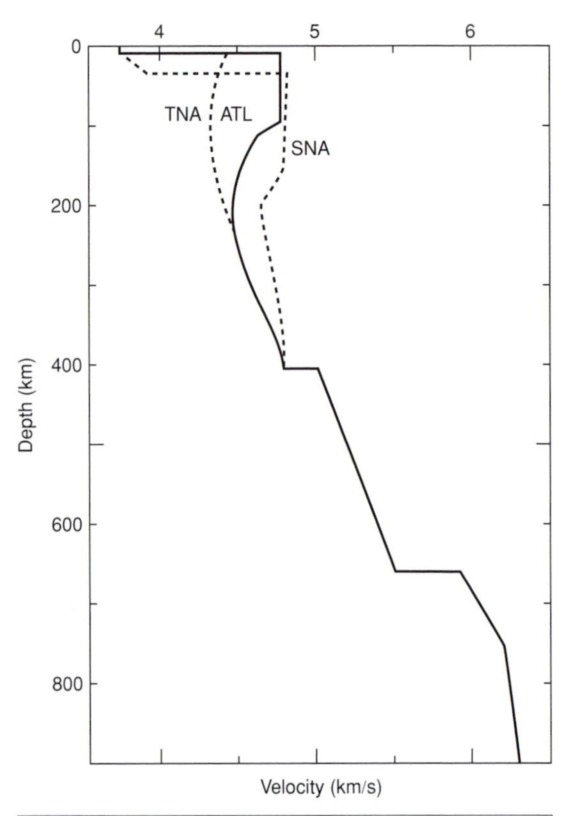

Fig. 8.6 Shear-wave velocity profiles for various tectonic provinces; TNA is tectonic North America, SNA is shield North America, ATL is north Atlantic. It is difficult to resolve the small variations below 400 km depth (after Grand and Helmberger, 1984a).

The low-velocity zone or LVZ

A region of diminished velocity or negative velocity gradient in the upper mantle was proposed by Beno Gutenberg in 1959. Earlier, just after isostasy had been established, it had been concluded that a weak region underlay the relatively strong lithosphere. This has been called the *asthenosphere*. The discovery of a low-velocity zone strengthened the concept of an asthenosphere.

Most models of the velocity distribution in the upper mantle include a region of high gradient between 250 and 350 km depth. Lehmann (1961) interpreted her results for several regions in terms of a discontinuity at 220 km (sometimes called the Lehmann discontinuity), and many subsequent studies give high-velocity gradients near this depth. Although the global presence of a discontinuity, or high-gradient, region near 220 km has been disputed, there is now appreciable evidence, from reflected phases, for its existence. The situation is complicated by the extreme lateral heterogeneity of the upper 200 km of the mantle. This region is also low Q (high attenuation) and anisotropic. Some upper mantle models are shown in Figures 8.5 and 8.6.

Various interpretations have been offered for the upper mantle low-velocity zone. This is undoubtedly a region of high thermal gradient, the boundary layer between the near surface where heat is transported by conduction and the deep interior where heat is transported by convection. If the temperature gradient is high enough, the effects of pressure can be overcome and velocity can decrease with depth. It can be shown, however, that a high temperature gradient alone is not an adequate explanation. Partial melting and dislocation relaxation both cause a large decrease in velocity. Water and CO_2 decrease the solidus and the seismic velocities. For partial melting to be effective the melt must occur, microscopically, as thin grain boundary films or, macroscopically, as dikes or sills, which also are very small compared with seismic wavelengths. Melting experiments suggest that melting occurs at grain corners and is more likely to occur in interconnected tubes. This also seems to be required by electrical conductivity data. However, slabs, dikes and sills act macroscopically as thin films for long-wavelength seismic waves.

High attenuation is associated with relaxation processes such as grain boundary relaxation, including partial melting, and dislocation relaxation. Small-scale heterogeneity such as slabs and dikes scatter seismic energy and this mimics intrinsic anelasticity. Allowance for anelastic dispersion increases the inferred high-frequency velocities in the low-velocity zone determined by free-oscillation and surface-wave techniques, but partial melting is still required to explain the regions of very low velocity. Allowance for anisotropy results in a further upward revision for the velocities in this region, as discussed below. This plus the recognition that subsolidus effects, such as dislocation relaxation, can cause a substantial decrease in velocity has complicated the interpretation of seismic velocities in the shallow mantle because one wants to compare results with laboratory data. Velocities in tectonic regions and under some oceanic regions, however, are so low that partial melting is implied. In most other regions a subsolidus mantle composed of oriented olivine-rich aggregates can explain the velocities and anisotropies to depths of about 200 km. Global tomographic inversions involve a laterally heterogenous velocity and anisotropy structure to depths as great as 400 km.

Mantle tomography averages the seismic velocity of the mantle over very long wavelengths and travel distances. Sampling theory tells us that extremes of velocity are averaged out in this procedure. As paths get shorter and shorter the variance goes up and sections of the path become much slower or much faster than inferred from global tomography. The minimum shear velocity found along ridges and backarc basins – and probably elsewhere – in high-resolution studies, is smaller than inferred from tomography. If global shear velocities approach the minimum shear velocities in solid rocks, then a reasonable variance of 5% placed on top of this probably means that the low-velocity regions require some melting.

The rapid increase in velocity below 220 km may be due to chemical or compositional changes (e.g. loss of water or CO_2) or to transition from relaxed to unrelaxed moduli. The latter explanation will involve an increase in Q, and some Q models exhibit this characteristic. However, the resolving power for Q is low, and most of the seismic Q data can be satisfied with a constant-Q upper mantle, at least down to 400 km.

Tomographic results show that the lateral variations of velocity in the upper mantle are as pronounced, and abrupt, as the velocity variations that occur with depth. Thus, it is misleading to think of the mantle as a simple layered or 1D system. Lateral changes are, of course, expected in a convecting mantle because of variations in temperature and anisotropy due to crystal orientation. They are also expected from the operation of plate tectonics and crustal processes. In particular, lithologic heterogeneity is introduced into the mantle by subduction and lower crustal delamination. But phase changes and partial melting are more important than temperature and this is why the major lateral changes are above about 400 km depth.

The geophysical data (seismic velocities, attenuation, heat flow) are consistent with partial melting in the low-velocity regions in the shallow mantle. This explanation, in turn, suggests the presence of volatiles in order to depress the solidus of mantle materials, or a higher-temperature mantle than is usually assumed. The top of the low-velocity zone may mark the crossing of the geotherm with the wet solidus of peridotite, or the solidus of a peridotite–CO_2 mix. Its termination would be due to (1) a crossing in the opposite sense of the geotherm and the solidus, (2) the absence of water, CO_2 or other volatiles or (3) the removal of water into high-pressure hydrous phases or escape of CO_2. In all of these cases the boundaries of the low-velocity zone would be expected to be sharp. Small amounts of melt or fluid (about 1%) can explain the velocity reduction if the melt occurs as thin grain-boundary films. Considering the wavelength of seismic waves, magma-filled dikes and sills, rather than intergranular melt films, would also serve to decrease the seismic velocity by the appropriate amount. Slabs that are thin and hot at the time of subduction may stack up in the shallow mantle. The basaltic parts will melt since the solidus of basalt is lower than the ambient temperature of the mantle. At seismic wavelengths this will have the same effect as intergranular melt films. Anelasticity and anisotropy of the upper mantle may also be due to these mega-scale effects rather than due to crystal physics, dislocations, oriented crystals and so on.

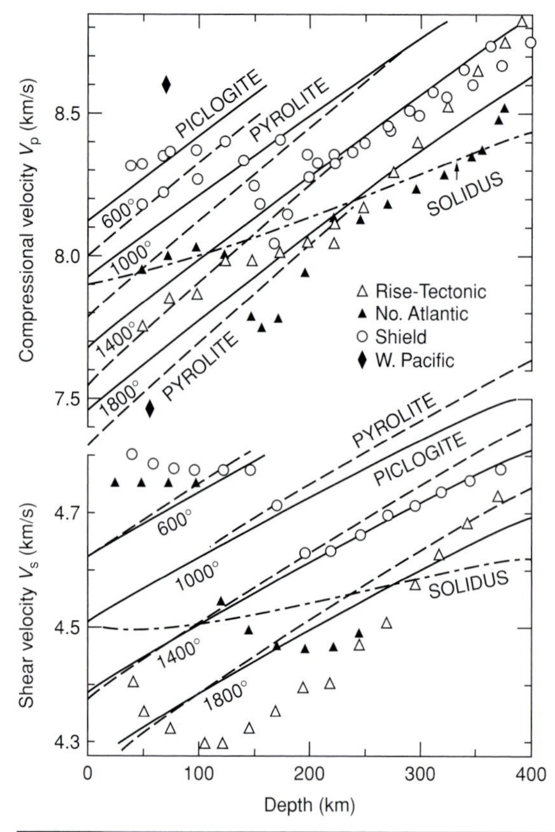

Fig. 8.7 Compressional and shear velocities for two petrological models, pyrolite and piclogite, along various adiabats. The temperatures (°C) are for zero pressure. The portions of the adiabats below the solidus curves are in the partial melt field. The seismic profiles are for two shields (Given and Helmberger, 1981; Walck, 1984), a tectonic-rise area (Grand and Helmberger, 1984a; Walck, 1984), and the North Atlantic region (Grand and Helmberger, 1984b).

The melting that is inferred for the lower velocity regions of the upper mantle may be initiated by adiabatic ascent from deeper levels. The high compressibility and high iron content of melts means that the density difference between melts and residual crystals decreases with depth. High temperatures and partial melting tend to decrease the garnet content and thus to lower the density of the mantle. Buoyant diapirs from depths greater than 200 km will extensively melt on their way to the shallow mantle. Therefore, partially molten material as well as melts can be delivered to the shallow mantle. The ultimate source of some basaltic melts may be below 300–400 km even if melt–solid separation does not occur until shallower depths. Low-melting point materials may be introduced into the shallow mantle from above, and heat up by conduction from ambient mantle. Upwellings do not have to initiate in thermal boundary layers.

Although a strong case can be made for localized or regional partial melting in the shallow mantle, the average seismic velocity over long paths, such as are used in most tomographic studies, may imply subsolidus conditions. If the partial melt zones are of the order of tens to hundreds of kilometers in lateral dimensions, and tens of kilometers thick, then they will serve to lower the average velocity, and perhaps to introduce anisotropy, in global tomographic models. There are also compositional effects to be considered. Eclogite, for example, has lower shear velocities than peridotite at depths between about 100 and 600 km. Eclogite, however, also has a melting point that is lower than ambient mantle temperature.

Absorption and the LVZ

Elastic-wave velocities are independent of frequency only for a nondissipative medium. In a real solid dispersion must accompany absorption. The effect is small when the seismic quality factor Q is large or unimportant if only a small range of frequencies is being considered. Even in these cases, however, the measured velocities or inferred elastic constants are not the true elastic properties but lie between the high-frequency and low-frequency limits or the so-called 'unrelaxed' and 'relaxed' moduli.

The magnitude of the effect depends on the nature of the absorption band and the value of Q. When comparing data taken over a wide frequency band, the effect of absorption can be considerable, especially considering the accuracy of present body-wave and free-oscillation data. The presence of physical dispersion complicates the problem of inferring temperature, chemistry and mineralogy by comparing seismic data with high-frequency ultrasonic data. Anelasticity as well as anharmonicity is involved in the temperature dependence of seismic velocity.

Figures 8.8 and 8.9 show calculations for seismic velocities for different mineral assemblages.

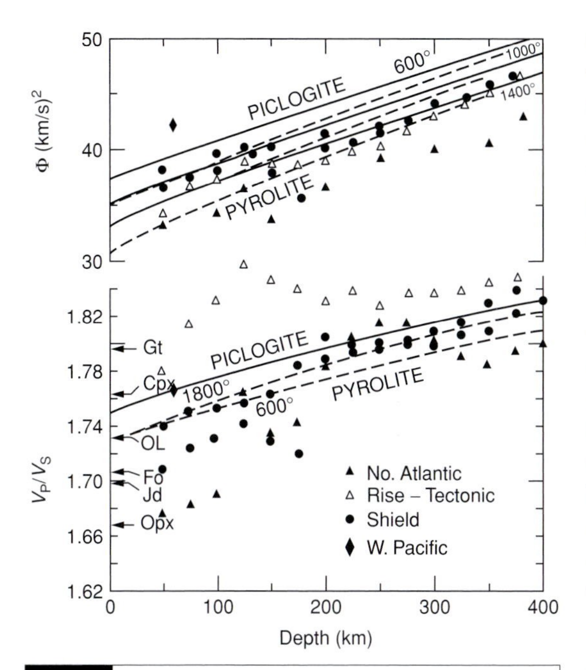

Fig. 8.8 Seismic parameters for two petrological models and various seismic models. Symbols and sources are the same as in Figure 8.7. V_p/V_s ratios for various minerals are shown in the lower panel. The high V_p/V_s ratio for the rise-tectonic mantle is consistent with partial melting in the upper mantle under these regions.

Pyrolite is a garnet peridotite composed mainly of olivine and orthopyroxene. Piclogite is a clinopyroxene- and garnet-rich aggregate with some olivine. Note the similarity in the calculated velocities. Below 200 km the seismic velocities under shields lie near the 1400 °C adiabat. Above 150 km depth the shield lithosphere is most consistent with cool olivine-rich material. The lower velocity regions have velocities so low that partial melting or some other high-temperature relaxation mechanism is implied. The adiabats falling below the solidus curves are predicted to fall in the partial-melt field. At depth, eclogite and piclogite have lower shear velocities and lower melting temperatures than peridotites. Low shear-velocity regions may be subducted or delaminated crust.

The 410 km discontinuity

The seismic discontinuity at 410 km depth is generally attributed to the phase transition of $(MgFe)_2SiO_4$ from the olivine to the β-spinel – wadsleyite – structure (Table 8.12). The best fitting mineralogy at this depth contains less than 50% olivine. The velocity jump at 410 km is too small to accommodate all the olivine (ol) and orthopyroxene (opx) in a pyrolite mantle converting to *spinel* and majorite; there must be substantial gt and cpx.

In some places there is a shear-wave velocity drop of about 5% on top of the 410 km discontinuity. This low-velocity zone has a variable thickness ranging up to perhaps 90 km. This may be due to a dense partial-melt layer, in which the solidus has been reduced by the presence of eclogite or CO_2. Eclogite itself, at depth, has a lower shear velocity than peridotite and even cold eclogite can be a low-shear velocity zone at 400 km.

Anisotropy

Rayleigh and Love wave data are often 'inconsistent' in the sense that they cannot be fit simultaneously using a simple isotropic model. This has been called the Love-wave–Rayleigh-wave discrepancy, and attributed to anisotropy. There are now many studies of this effect which has also been called polarization or radial anisotropy, or transverse isotropy. Independent evidence for anisotropy in the upper mantle is now strong (e.g. from receiver-functions amplitudes and shear-wave splitting). Radial anisotropy is very strong in the central Pacific. Anisotropy can be caused by crystal orientation or by a fabric of the mantle caused by slabs, dikes and sills. Seismic waves have such long wavelengths that it is immaterial whether the effect is due to centimeter- or tens-of-km-size features. Similarly, anelasticity may be due to km-size scatters, rather than to cm-sized dislocations.

The transition region (TR)

The mantle transition region, Bullen's region C is defined as that part of the mantle between 410 km and 1000 km (the Repetti discontinuity). The lower mantle was defined by Bullen as the mantle below 1000 km depth but more recently it is often taken as the mantle below the 650 km seismic discontinuity. This has caused immense confusion regarding whether slabs sink into the lower mantle. The 400 km discontinuity is mainly due to the olivine-spinel phase change, considered as an equilibrium phase boundary in a homogenous mantle. The seismic velocity jump, however, is smaller than predicted for this phase change. The orthopyroxene–garnet reaction leading to a garnet solid solution is also complete near this depth, possibly contributing to the rapid increase of velocity and density at the top of the transition region. Some eclogites are less dense than the beta-form of olivine and may be perched at 400 km depth. For these reasons the 400 km discontinuity should not be referred to as the olivine–spinel phase change. If the discontinuity is as small as in recent seismic models, then the olivine content of the mantle may be lower than in the shallow mantle. This would make sense since olivine is a buoyant product of mantle differentiation and would tend to accumulate at the top of the mantle.

The TR as described by Bullen is a diffuse region of high seismic wave-speed gradient extending from 410 to 1000 km. In Bullen's nomenclature the lower mantle (Region D) started at 1000 km. Birch suggested that the Repetti discontinuity near 1000 km marked the top of the lower mantle and that high seismic wave-speed gradients are caused by polymorphic phase changes. The early models of Jeffreys and Gutenberg were smooth and had high wave-speed gradients without abrupt discontinuities, but in the 1960s it was discovered that there are abrupt jumps in seismic velocity at depths of approximately 400 and 650 km. During that decade, various investigations, including detailed studies of the travel times of both first- and later-arriving body waves, seismic array measurements of apparent velocities, observations of reflected waves such as precursors to the core phase *P′P′*, and analysis of the dispersion of fundamental and higher-mode surface waves, all confirmed the existence of the discontinuities, which define the transition Z are, TZ.

Thermodynamic considerations have been used to argue that the discontinuities are abrupt phase changes of, mainly, olivine to the spinel crystal structure, and then to a 'post-spinel' phase, not chemical changes as in the standard geochemical models, and that the deeper one has a negative Clapyron slope. This means that cold subducting material of the same composition as the surrounding mantle would depress the 650 km discontinuity, inhibiting vertical motion, and would change to the denser phase only after warming up or being forced to greater depth. Material of different composition and intrinsic density can be permanently trapped at phase boundaries. Geochemical, and many convection, models assume that the 650 km phase change separates the 'depleted convecting upper mantle' from the 'primordial undegassed lower mantle.' Geodynamic modelers assume that if the 650 km discontinuity is not a chemical change then there can be no deeper chemical change and the mantle is chemically homogenous. The TZ thus holds the key to whether there is whole-mantle or layered-mantle convection.

In the transition zone the stable phases are garnet solid-solution, β- and γ-spinel and, possibly, jadeite. Garnet solid-solution is composed of ordinary garnet and SiO_2-rich garnet (majorite). The extrapolated elastic properties of the spinel forms of olivine are higher than those observed (Figures 8.9 and 8.10). The high velocity gradients throughout the transition zone imply a continuous change in chemistry or phase, or in lithology (eclogite vs. peridotite). Appreciable garnet is implied in order to match the velocities. A spread-out phase change involving clinopyroxene (diopside(di) plus jadeite(jd)) transforming to Ca-rich majorite(mj) can explain the high velocity gradients.

Unusually low temperatures, as expected in the vicinity of a downgoing slab, will warp the 410 km discontinuity up by about 8 km per 100 K, and the 650 km discontinuity down by about 5 km per 100 K. The Clapyron slopes are uncertain but most estimates for the total thickening/thinning of the TZ lie in the range 12–17 km per 100 degrees. This assumes a purely olivine mineralogy, which is unrealistic.

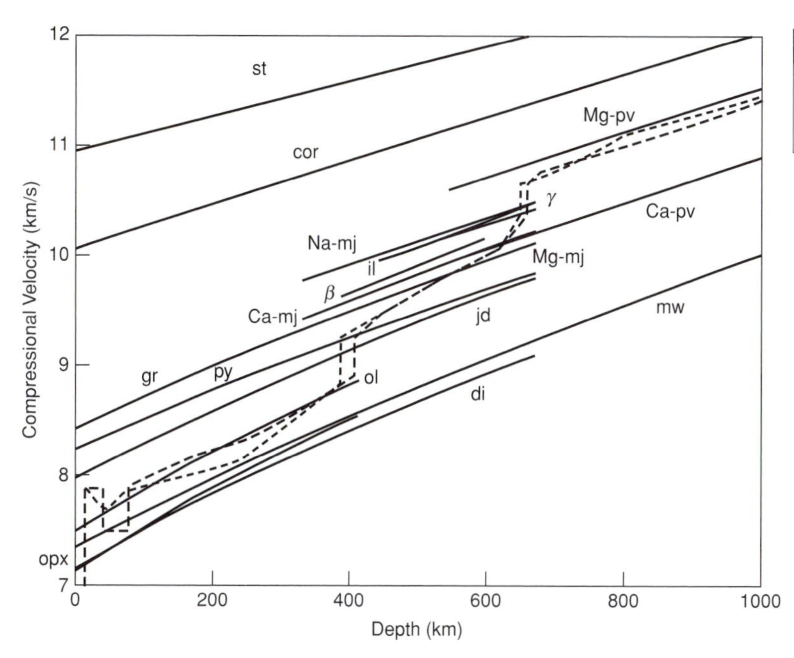

Fig. 8.9 Calculated compressional velocity versus depth for various mantle minerals along a 1400 °C adiabat.

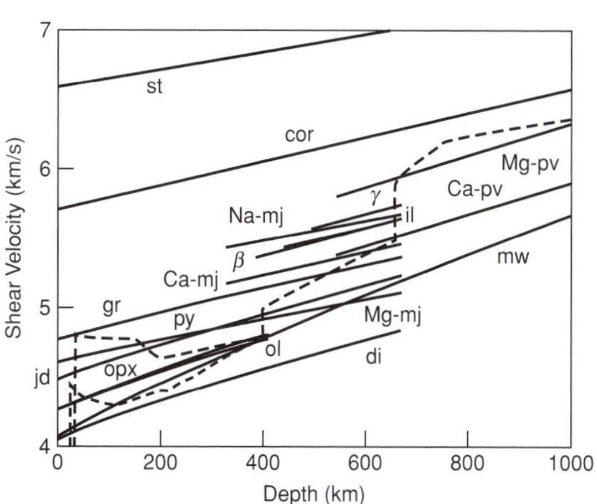

Fig. 8.10 Calculated shear velocity versus depth for various mantle minerals along a 1400 °C adiabat. 'Majorite' (mj), 'perovskite' (pv) and 'ilmenite' (il) are structural, not mineralogical terms. The dashed lines are two representative seismic profiles (after Duffy and Anderson, 1989).

The average thickness of the TZ is ~242 km; uncertainties are typically 3%. Typical thicknesses beneath high-heat-flow areas are 220–230 km. The topology of the relevant phase diagrams predicts antisymmetry in the directions of deflection of the discontinuities for the cases of both cold downwellings and hot upwellings. However, the depths of the 410 and 650 km discontinuities are uncorrelated on a global scale. They also sometimes show steps, which makes interpretation in terms of temperature not straightforward. Some places show rapid lateral changes in TZ thickness that may indicate non-thermal effects. It must therefore be borne in mind that temperature may not be the only control. The expected effects of temperature on the depths of the discontinuities are also based on uncertain laboratory calibrations, may be in error by a factor of two, and chemical effects may be stronger than generally supposed. In spite of these complications, TZ thickness may still prove to be a useful thermometer and an important part of any plan to map lateral variations of temperature in the mantle.

There is no global correlation between TZ thickness and the locations of surface hotspots and the large lower-mantle low-velocity regions. Transition zone thickness is normal beneath southern Africa (245 km) and the East African Rift and Afar (244 km), which are underlain by the postulated 'superplumes' and the postulated Afar plume. Transition-zone thickness beneath hotspots is generally within the range for normal oceans and often close to the global average.

The observed topography on the discontinuities does not seem to be explicable by thermal effects to the extent expected. The observations are consistent with the decorrelation of seismic anomalies between the upper and lower mantles observed both in tomographic images and revealed by matched filtering using plate/slab reconstructions. A few anomalies appear to extend from the surface through the TZ and into the deep mantle; it would be interesting to calculate if they are more numerous than would be expected by chance.

Detailed study of some specific regions have yielded surprises. The thinnest TZ region, 181 km thick, is found in Sumatra, where a thick accumulation of cold slabs is thought to exist, which would thicken the TZ. In western USA, the thickness of the TZ varies from 220 to 270 km, with 20 to 30 km relief on each discontinuity, and no correlation with surface geology, topography or between the discontinuities [mantleplumes].

There is evidence from scattering of seismic waves and plate-tectonic–tomographic correlations that there may be a chemical boundary near 1000-km depth. Much of modern mantle geochemistry is based on the conjecture that the 650-km phase change is also a major chemical change, and that this is the boundary between the upper and lower mantles. Geodynamic models assume that below 650 km depth, the mantle radioactivity is high. Mantle geodynamics is also based on the assumption that if slabs can penetrate the phase-change region they will sink to the core–mantle boundary. The transition zone of the upper mantle therefore continues to be a critical region for investigation.

The Repetti discontinuity

A layered convection model with a chemical interface near 900 km at the base of Bullen's Region C explains the geoid and dynamic topography. The evidence for stratification includes the mismatch between tomographic patterns and spectra between various depth regions, and evidence for slab flattening. There is a good correlation between subducted slabs and seismic tomography in the 900–1100 km depth range. The mantle does not become radially homogenous and adiabatic until about 800 km depth. A variety of evidence suggests that there might be an important geodynamic boundary, possibly a barrier to convection, and a thermal boundary at a depth of about 900—1000 km, the bottom of the transition region.

Chemical boundaries, in contrast to most phase-change boundaries, will not be flat, as assumed in some layered convection models, and will have little impedance contrast. The latter inference is based on plausible compositional differences between various lower-mantle assemblages. Complications between 650 and 1300 km depth in the mantle are perhaps related to slab trapping or thermal coupling and undulations in the Repetti discontinuity, the top of Region D.

Chapter 9

A laminated lumpy mantle

What can be more foolish than to think that all this rare fabric of heaven and earth could come by chance, when all the skill of art is not able to make an oyster!

Anatole France

Even so, it was 99,970,000 years getting ready . . . because God first had to make the oyster. You can't make an oyster out of nothing, nor you can't do it in a day. You've got to start with a vast variety of invertebrates, belemnites, trilobites, jebusites, amalekites, and that sort of fry, and put them into soak in a primary sea and observe and wait what will happen. Some of them will turn out a disappointment; the belemnites and the amalekites and such . . . but all is not lost, for the amalekites will develop gradually into encrinites and stalactites and blatherskites, and one thing and another . . . and at last the first grand stage in the preparation of the world for man stands completed; the oyster is done. Now an oyster has hardly any more reasoning power than a man has, so it is probable this one jumped to the conclusion that the nineteen million years was a preparation for him. That would be just like an oyster.

Mark Twain

Gravitation structures the earth in concentric shells, or geospheres, according to their specific gravity.

Gal-or

Standard geochemical and geodynamic models of the mantle involve one or two layers. Global tomographic models tend to be fairly simple; they show continental roots, slabs, shallow mid-ocean ridge structures and a few very large features in the deep mantle. Regional and high-resolution seismological models of the mantle are more complex. High-resolution seismic techniques involving reflected and converted phases show about 10 discontinuities in the mantle, not all of which are easily explained by solid–solid phase changes. They also show some deep low-velocity zones, which may be eclogite layers. The upper mantle is highly attenuating, anisotropic and scatters short period seismic energy.

The opposite extreme of a well-stirred homogenous mantle is a mantle that is stratified by intrinsic density. Convection can be expected to homogenize the mantle *if* the various components do not differ much in intrinsic density, of the order of 2 or 3%, depending on depth. The Earth itself is stratified by composition and density (atmosphere, hydrosphere, crust, mantle, core) and the crust and upper mantle are stratified as well. The region at the base of the mantle – D″ – appears to be iron-rich and intrinsically dense. There may be a buoyant refractory (melt depleted) discontinuous layer at the top of the mantle – the *perisphere*. The perisphere may never get cold enough to subduct and D″ may never get hot enough to rise. These are only the most obvious candidates for chemical layers; internal layers will likely be subtle and they need not be continuous or flat.

Figure 9.1 shows the shear velocity in a variety of rocks and minerals, at STP (standard temperature and pressure) arranged according to increasing zero-pressure density. This represents a stably stratified system. Many of the chemically distinct layers differ little in seismic properties and sometimes a denser layer has lower seismic velocity than a less dense overlying layer. The densities range from 2.6 to 4.2 g/cm^3; the density scale is monotonic but nonlinear. Several estimates of depth are given, calibrated according to estimates of mantle uncompressed density. Given enough time in a low enough viscosity mantle this is the stable density stratification. Note that shear velocity is not a monotonically increasing function of density. A stable density stratification has an irregular complex shear velocity structure. Even if the mantle achieves this stable stratification it will not be permanent. The different lithologies have different melting points and thermal properties, phase changes and can rise or sink as the temperature changes. Figure 9.2 is a similar plot, with some of the minerals and rocks identified.

Eclogites occur at various depths because they come in a variety of compositions; eclogite is not a uniform rock type. Arclogites are garnet clinopyroxenites that occur as xenoliths in arc magmas. The deeper eclogite layers in the figures are low-velocity zones relative to similar density rocks. Cold dense eclogite melts as it warms up to ambient mantle temperature, and becomes buoyant. The stable stratification of a chemically zoned mantle is only temporary. This kind of mantle convects but it is a different kind

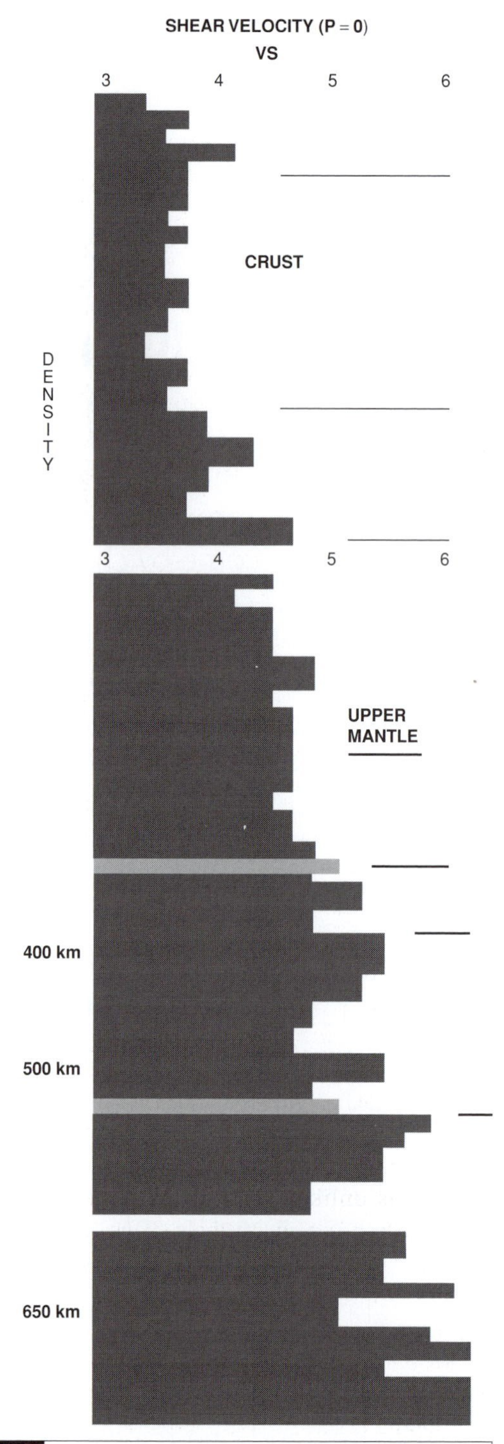

Fig. 9.1 Chemical stratification of the mantle if mantle rocks and minerals arrange themselves in the gravity field according to intrinsic density (density increases downward but is not tabulated). The velocities (horizontal axis) and densities (vertical axis) are appropriate for Standard Temperature and Pressure (STP) conditions.

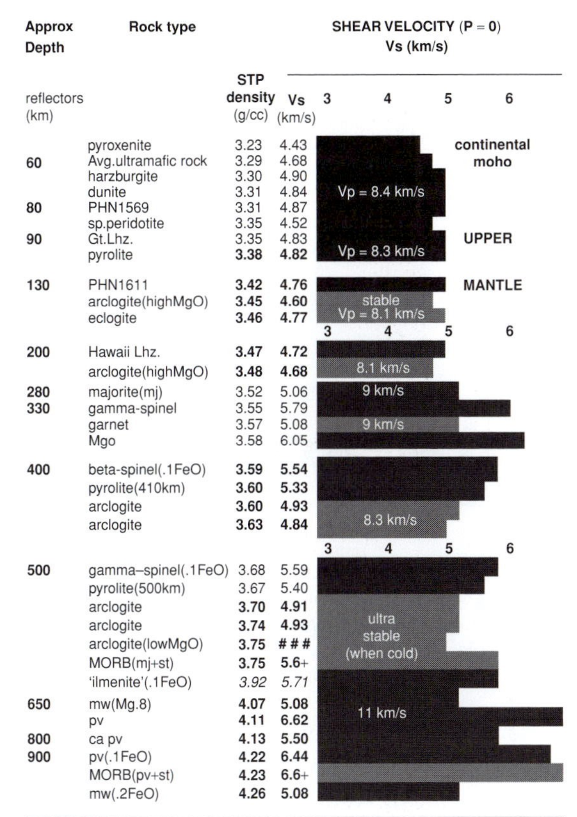

Fig. 9.2 Same as Figure 9.1 with additional information and fewer rock types.

of convection than the homogenous mantle usually treated by convection modelers or *geodynamicists*. Convection in the mantle is mainly driven by the differences in density between basalt, melt and eclogite. Note that sinking eclogite can be trapped above the various mantle phase changes, giving low-velocity zones. Although mantle stratification is unlikely to be as extreme or ideal as Figure 9.1 it is also unlikely to be as extremely homogenous or well-mixed as often assumed. Crustal type reflection seismology is required to see this kind of structure.

Recycled oceanic crust, one kind of eclogite, will have a particularly high density if it can sink below about 720 km because the high silica content of MORB gives a large stishovite content to MORB–eclogite. On the other hand, cumulate gabbros, the average composition of the oceanic crust and delaminated continental crust have much lower silica contents and this reduces their high-pressure densities. The controversy regarding the fate of eclogite in the mantle involves this point.

The arrangement in Figure 9.2 approximates the situation in an ideally chemically stratified mantle. The densities of peridotites vary from about 3.3 to 3.47 g/cm^3 while measured and theoretical eclogite densities range from 3.45 to 3.75 g/cm^3. The latter is comparable to the inferred STP density near 650 km and about 10% less dense than the lower mantle. The lower density eclogites (high-MgO, low-SiO_2) have densities less than the mantle below 410 km and will therefore be trapped at that boundary, even when cold.

There are several things to note. Eclogites come in a large variety of compositions, densities and seismic velocities. There is not a one-to-one correlation of seismic velocity and density in mantle rocks, and shear velocity is not a monotonic increasing function of density or depth. Some chemically distinct layers have similar seismic velocities. The velocities are quantized at about the 4% level, a typical variation observed in the shallow mantle globally, and under hotspots in particular; such variations on the slow side are usually attributed to partial melting or high-temperature. The shear-velocity quantum step is equivalent to a temperature variation of 1000 °C at constant pressure and about the size of the correction to be made to STP velocities to account for ambient mantle temperature and pressure. Pressure and temperature effects may change the ordering and the velocity and density jumps at depth. Eclogite can settle to various levels, depending on composition; the eclogite bodies that can sink to greater depths because of their density have low-velocity compared to similar-density rocks. Some eclogites have densities intermediate to the low- and high-pressure asemblages at the various peridotite phase boundaries (410 km, 500 km, 650 km); they will be trapped at these boundaries, affecting the seismic properties, and changing them from the ideal phase-change conditions. Cold eclogites with STP densities between 3.45 and 3.6 g/cm^3 may be trapped above the olivine-beta-spinel phase boundary near 410 km depth, giving a low-velocity zone (LVZ) at this depth. The observations of a LVZ atop the 410 km discontinuity are usually interpreted in terms of partial melting. A perched eclogite fragment will heat up by conduction from the surrounding hotter mantle and will eventually melt.

MORB–eclogite contains stishovite at high pressure and may sink below 500 km. Some low-MgO-arc eclogites have comparable densities. Note that the deeper eclogite layers form substantial low-velocity zones. The stratification shown in the figure is only temporary. Subducted or delaminated material warms up by conduction of heat inward from ambient mantle. Eclogites have much lower melting points than peridotites and will eventually heat up and rise, even if they are not in a TBL, creating a sort of *yo-yo* or *lava-lamp tectonics*. The *ilmenite* form of garnet and enstatite is stable at low temperature but will convert to more buoyant phases as it warms up. Whole mantle convection or vigorous mantle convection, and entrainment, are not necessary in order to bring fertile material into the shallow mantle.

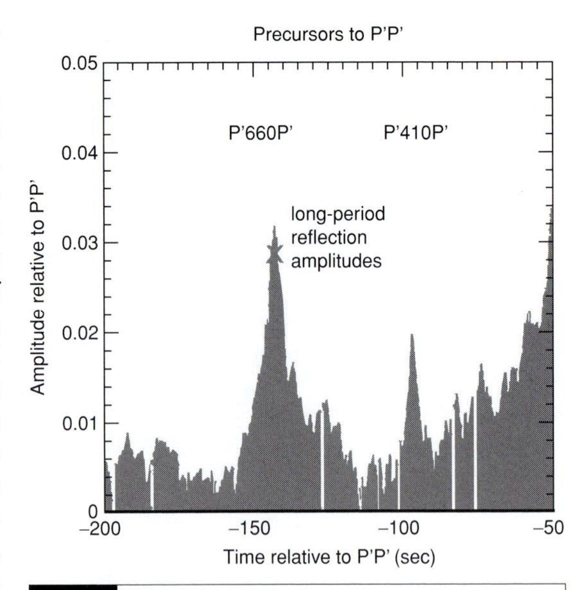

Fig. 9.3 Stacked seismograms of P′P′ precursors showing the major seismic discontinuities and many minor ones (Xu *et al.*, 2003).

Multiple mantle discontinuities

The classical 1D seismological models of the mantle include a TR between 400 and 1000 km – separating the upper and lower mantles – that was attributed to phase-changes. Early investigators, using reflected phases and breaks in teleseismic travel time curves identified many discontinuities and some of the early seismic models consisted of layers rather than smooth variations with depth. Whitcomb and Anderson (1970), using precursors to the seismic phase P′P′, identified about six seismic discontinuities in the mantle; a major one near 630 km depth and others at 280, 520, 940, 410 and 1250 km. Although the attention of geochemists and geodynamicists has been focused on the better known 410 and 650 km features, there are about 10 discontinuities in the mantle that have now been identified by seismologists by a variety of high-resolution or correlation techniques. Systematic searches for mantle discontinuities have yielded reflections or conversions from depths of 220, 320, 410, 500–520, 650–670, 800, 860, 1050, 1150–1160 and 1320 km. A survey of many such studies shows that most of the reflections and conversions occur in the depth intervals of 60–90, 130–170, 200–240, 280–320, 400–415, 500–560, 630–670, 800–940, 1250–1320 and 2500–2700 km. Some of these are probably chemical in nature and some apparently are either highly variable in depth or are independent scatterers (the reports of reflections or scatterers between 800 and 1320 km may all be due to a single highly irregular interface). The shallow features may represent boundaries between depleted and fertile peridotites and partial melt zones. Deeper ones may represent slabs that have been trapped at various depths because they are too buoyant to sink further.

A number of possible underside reflections are evident in Figure 9.3. Deuss and Woodhouse (2002) found a large number of reflections from a depth of 220 km beneath both oceans and continents (see Figure 9.4). Altogether, there are more reflections reported below 650 km and between 410 and 650 km than are reported at 410 and 650 km. These are not so evident in reflection histograms because they occur over a wide range of depths as expected for chemical interfaces.

Deep low-velocity zones

Figure 9.2 predicts the presence of low-velocity zones in a petrologically stratified mantle. Some eclogites create a LVZ having velocities about 2–5% lower than the surrounding mantle or mantle of the same density. The deeper eclogite LVZs

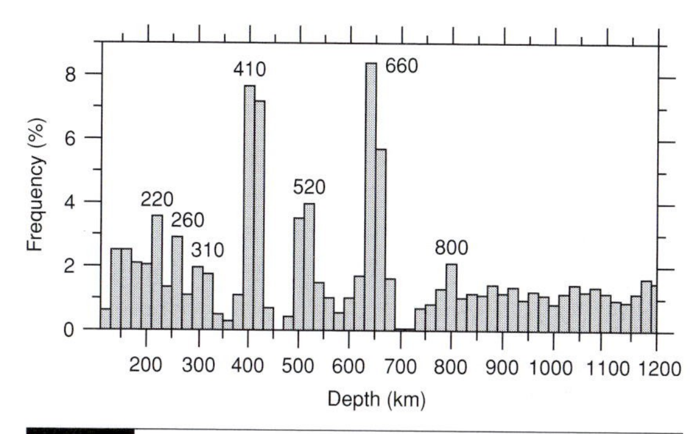

Fig. 9.4 Robust reflections from the mantle (Deuss and Woodhouse, 2002).

are about 9% slower. Mixing with normal mantle will reduce these differences; heating and melting will increase them. The main point is that lateral and radial reductions in seismic velocity of order 2–10% can have a simple petrological explanation. Shallow LVZ may be, in part, due to adiabatic upwelling of displaced asthenosphere but this also need not be particularly hot.

Velocity reversals, or low-velocity zones (LVZ) have been identified in regional studies at depths near 100, 185, 380, 410, 460–480, 570–600, 610 and 720 km (Nolet & Zielhuis, 1994; Vinnik *et al.*, 2003). The velocity reduction in these LVZ is generally between 2–5%. These LVZ are almost invariably attributed to the effects of water, partial melting or high temperature. These LVZ are in addition to those that occur in the upper 200 to 350 km in tectonic and volcanic regions such as Yellowstone, Iceland, western North America and near oceanic ridges. The LVZ that occur just above the major phase-change boundaries at 410 and 650 km are particularly instructive since these are the places where one expects to find barriers to certain kinds of subducted or delaminated materials.

Tomographic studies suggesting that some slabs cross the 650 km mantle discontinuity do not imply that all do. The transition zone may act as a petrological filter. Recycled material can also be trapped at other depths – deeper and shallower; thick, cold slabs can sink further and take longer to warm up; younger slabs or those with thick crust tend to underplate continents. The dry and depleted residual phases – peridotites and eclogites – equilibrate at various depths and the removed material metasomatizes the shallow mantle (the mantle wedge, the perisphere and the plate). Young oceanic plates, delaminated lower crust, subducted seamount chains and plateaus thermally equilibrate and melt at depths different from older thicker plates.

The 650 km discontinuity, with its negative Clapyron slope, is a temporary barrier to cold sinking material of the same composition, but such material may eventually break through. A different material, with higher pressure phase-changes, e.g. eclogite, can be stranded by phase-changes in peridotite. Eclogite can density-equilibrate at depths above 600 km (Figure 9.2). Chemical discontinuities, even those with very small density jumps, can be a barrier – or filter – to through-going convection.

Delaminated continental crust is a particularly potent source of mantle heterogeneity, low-velocity zones and melting anomalies; it starts out warmer and equilibrates faster than subducted oceanic crust. It is also low in SiO_2, which means it has more buoyancy below some 400 km where subducting MORB may have a high dense stishovite component. Large fertile low melting-point blobs trapped in the upper mantle may be responsible for 'melting anomalies' and LVZs. These recycling, filtering and sampling processes can explain many geochemical observations while avoiding the pitfalls associated with isolated mantle reservoirs and deep penetration of all slabs and all components.

High-resolution seismological techniques involving reflected and converted phase and scattering are starting to reveal the real complexity of the mantle. Abrupt seismic discontinuities are not necessarily isotope or reservoir boundaries and the deeper layers are not necessarily accessible to surface volcanoes. Plate tectonics and geochemical cycles may be entirely restricted to the upper ~1000 km, where thermal expansion is high and melting points, viscosity and thermal conductivity are low.

The seismic velocities of plausible materials in the mantle differ little from one another, even if the density contrasts are adequate to permanently stabilize the layering against convective overturn. Since chemical discontinuities can be almost invisible to 1D seismology, compared with phase-changes, and since even small chemical density contrasts can stratify the mantle, the possibility must be kept in mind that there may be multiple chemical layers in the mantle, some of which may be subtle. The major seismic discontinuities in the mantle are due to mineralogical and phase changes, not chemical changes, but this does not rule out a chemically heterogenous mantle.

Chemical discontinuities?

Mantle peridotites and eclogites have a variety of compositions, densities and seismic velocities. If the density differences are large enough, the mantle can become chemically stratified. Eclogites have higher seismic velocities and densities than peridotites at the same pressure and temperature, but they have much lower velocities than peridotites and peridotitic assemblages of the same density. After delamination sufficiently large eclogitic blobs will sink into olivine-rich mantle that has higher seismic velocities, and higher melting temperatures. Arc eclogites are predicted to have lower densities and velocities than transition zone minerals and may be trapped in and above the transition region. The cold ultramafic portions of the lithosphere have high seismic velocities and moderately high density. When sufficiently cold, they can become denser than the upper part of the subplate mantle. The delamination model predicts that regions possibly undergoing delamination, such as the Sierras, Yellowstone, and possibly most arc and rifted areas will be underlain by low-melting-point fertile material, and high seismic velocity curtains. At most depths, even subsolidus eclogite is predicted to have lower seismic velocities than the surrounding mantle, even if it is dense and sinking. If it is volatile-rich, or if it warms up to the solidus, it will have even lower seismic velocities. These seismic features may easily be mistaken for plumes. The delaminated mafic material eventually warms up and melts, creating fertility spots in the mantle. Such spots become buoyant as garnet is removed by melting. Melting anomalies may be due to these fertile spots rather than hotspots.

Many eclogites are seismically fast because they have a large proportion of garnet and much of the pyroxene is in the form of jadeite. Some garnet–pyroxenite xenoliths, however, have large amounts of clinopyroxene and very little jadeite, and, technically, are not eclogites. Such garnet pyroxenites, which can be called arclogites, are slower than common eclogites. They have higher V_P/V_S ratios than peridotites. Velocity decreases do not necessarily imply hot mantle.

In the upper mantle, residual dunites and eclogites have shear velocities that vary mainly from 4.60 to 4.95 km/s, while normal peridotites and lherzolites typically vary from 4.5 to 4.9 km/s, all at standard temperature and pressure (STP). Xenoliths from the Sierran root fall into two categories with inferred densities of 3.45–3.48 g/cm^3 and 3.6–3.75 g/cm^3 respectively. The associated STP shear velocities are lower than 4.68 km/s and higher than 4.83 km/s respectively. Low-density eclogites have densities comparable to peridotites and Hawaiian lherzolites and should sink to depths no greater than about 400 km, even when cold. The high-density low-MgO arc eclogites should density-equilibrate in the lower part of the transition region, 500–650 km, where they will show up as low-velocity anomalies. These could be confused with hot plumes, even if they are cold dense sinkers!

Eclogites and dunites in the surface TBL, or delaminated, will have slightly higher seismic velocities than the warmer peridotites and lherzolites that they displace. Eclogites are much denser than dunites and are more likely to delaminate, on their own, and sink into the underlying mantle. Delaminated material will displace asthenospheric material, causing adiabatic decompression melting and a lowering of the shear velocity. Between 60 and 200 km eclogites can have seismic velocities that are lower or higher (particularly when cold) than mantle peridotites of similar density; the denser (or colder) ones may show up as high velocity curtains. Below about 400 km eclogites are typically much lower velocity than transition zone minerals and, in fact, they are also less dense than much of the mantle below 500 km. If the mantle is close to its normal (peridotitic) solidus, then eclogitic blobs will eventually heat up to above their solidi.

Pyroxenites and eclogites can have radioactivities up to 10% of those in the continental crust. This is a non-negligible heat source, particularly since the amount of basaltic/eclogitic material in the mantle is about ten times the amount of continental crust. But the major heating of eclogitic blobs is due to conduction of heat from ambient mantle and this is more rapid for the more deeply equilibrated fragments or layers. The lower crustal cumulates are embedded in a cold thermal boundary layer and the cooling is more efficient than the radioactive heating. The residence time is also short, ~30 Myr. At depth, heat conduction and radioactivity work in the same direction and it is probable that eclogite sinkers can be brought back up to their melting temperature in hundreds of millions of years, about the residence times of oceanic plates at the surface. Sinking oceanic slabs can also displace warm eclogite layers and this is in addition to intrinsic density considerations. Figure 9.2 assumes that all materials are at the same temperature and pressure. The eclogites will actually be at higher homologous temperatures - temperatures relative to the melting point - and many will be above the melting point at normal mantle temperatures.

Thermal implications

In a chemically stratified mantle the thermal structure is not simply a thermal boundary layer over an adiabatic interior. Higher temperature gradients and higher interior temperatures can be expected. On the other hand, the sinking of cold material to various depths serves to cool off the mantle and give rise to negative temperature gradients. The mantle is cooled both by subducting slabs and by delamination events. In the plate-tectonic cycle the cold subducting slabs occur mainly at subduction zones, but these migrate about, allowing much of the mantle to be cooled by this process. Crustal delamination is also expected to occur at island arcs but also at sutures and midplate locations far away from current subduction. In summary, in a chemically heterogenous mantle one does not expect a monotonic increase of temperature, along an adiabat, below a thermal boundary layer. High conduction gradients will be interspersed with low or negative temperature gradients in the vicinity of delaminated crust and subducted slabs.

The high temperatures at shallow depths probably mean that the shallow mantle is stripped of melts, leaving behind buoyant peridotitic residues. An infertile olivine-rich mantle is predicted (the perisphere concept). In such a mantle, the fertile components will either be in the surface thermal boundary layer or in deeper eclogite layers. Current estimates of mantle temperatures in the transition zone are based on adiabatic extrapolations from depths of about 100 km and are well below the melting point of eclogite. On the other hand, geophysical estimates of mantle temperature tend to be much higher than petrological estimates, and eclogites tend to have higher U and Th contents than peridotites. They can also be displaced upwards by subducting and delaminating material. The dense majorite portions of eclogite exsolve pyroxene as they warm up or decompress, contributing to the decrease of negative buoyancy.

Chapter 10

The bowels of the Earth

The lower mantle

I must be getting somewhere near the centre of the earth. Let me see: that would be four thousand miles down. I think-

Alice

The traditional lower mantle starts near 800–1000 km where the radial gradient of the seismic velocities becomes small and smooth. This is Bullen's Region D. The 1000-km depth region appears to be a fundamental geodynamic interface, perhaps a major-element chemical and a viscosity interface. Some authors take the lower mantle to start just below the major mantle discontinuity near 650 km. The depth of this discontinuity varies, perhaps by as much as 40 km and is variously referred to as the '660 km discontinuity' or '670 km discontinuity'; the average depth is 650 km. In detailed Earth models there is a region of high velocity gradient for another 50–100 km below the discontinuity. This is probably due to phase changes, but it could represent a chemical gradient. Plate reconstructions show that past subduction zones correlate with high-velocity regions of the mantle near 800–1000 km depth. The 'lower mantle proper' therefore does not start until a depth well below the 650 km boundary, more in agreement with the classical definition. Below this depth the lower mantle is relatively homogenous until about 300 km above the core-mantle boundary. If there are chemical discontinuities in the mantle the boundaries will not be at fixed depths. These clarifications are needed because of the controversy about whether slabs penetrate into the lower mantle or whether they just push down a discontinuity, and where the boundary of the lower mantle really is.

The *midmantle*, or *mesosphere*, extends from 1000 to 2000 km. This is the blandest part of the mantle. Estimates of its composition range from pure $MgSiO_3$ *perovskite* to chondritic – in major refractory elements – to pyrolite, within the uncertainties. It can be Si- and Fe-rich compared with the upper mantle, or identical to the upper mantle. Geochemical models assume that it is *undegassed* and *unfractionated* cosmic material, but there is no support for this conjecture. The lower mantle must be quite different from that which appears in the `standard model of mantle geochemistry`, and the geodynamic models that are based on it.

Composition of the lower mantle

The probable *mineralogy* of the deep mantle is known from high-pressure mineral physics – *squeezing* – experiments. The mineralogy is simple in comparison to that of the crust and upper mantle, consisting of $(Mg,Fe)SiO_3$–Al_2O_3 orthorhombic *perovskite*, $CaSiO_3$ cubic *perovskite*, and $(Mg,Fe)O$ magnesiowüstite. No Al-rich phases, except for *perovskites*, are considered to exist in the main part of lower mantle. The total basaltic content of the mantle is only about 6% and most of this is probably in the upper mantle and transition region. If most of the crustal and basaltic material has been sweated out of the lower mantle, then it will be low in Ca, Al, U, Th and K, among many other things. The lower mantle would then be mainly oxides of Si, Mg and Fe. The uncertain spin-state and oxidation-state of Fe introduces a bit of spice into lower mantle mineralogy.

The *composition* of the lower mantle is another story. Most plausible compositions have similar properties. Candidates for the dominant rock types in various deep-mantle layers are so similar in seismic properties that standard methods of seismic petrology fail. Small differences in density, however, can irreversibly stratify the mantle so it is methods based on density, impedance, anisotropy, dynamic topography, pattern recognition, scattering and convective style that must be used, in addition to seismic velocity. Visual inspection of color tomographic cross-sections cannot reveal subtle chemical contrasts.

Several methods have been used to estimate the composition of the lower mantle from seismic data but they are all non-unique and require assumptions about temperature gradients, temperature and pressure derivatives, equations of state and homogeneity. Perhaps the most direct method is to compare shock-wave densities at high pressure of various silicates and oxides with seismically determined densities. There is a trade-off between temperature and composition, so this exercise is non-unique. Materials of quite different compositions, say (Mg,Fe)SiO_3 (*perovskite*) and (Mg,Fe)O, can have identical densities, and mixtures involving different proportions of MgO, FeO and SiO_2 can satisfy the density constraints. In addition, the density in the Earth is not as well determined as such parameters as the compressional and shear velocities. The mineralogy and composition of the lower mantle are hard to determine since plausible combinations of perovskite and magnesiowüstite ranging from chondritic to pyrolite have similar elastic properties when FeO and temperature are taken as free parameters. But they can differ enough in density to allow chemical stratification that is stable against overturn. Oxide mixtures, such as MgO + SiO_2 (stishovite), can have densities, at high pressure, similar to compounds such as *perovskite* having the same stoichiometry.

It can be shown that a chondritic composition for the lower mantle gives satisfactory agreement between shockwave, equation of state and seismic data, for the most plausible lower mantle temperature. The SiO_2 content of the lower mantle may be closer to chondritic than pyrolitic. If the lower mantle falls on or above the 1400 °C adiabat, then chondritic or pyroxenitic compositions are preferred. If temperatures are below the 1200 °C adiabat, then more olivine (*perovskite* plus (MgFe)O) can be accommodated. A variety of evidence suggests that the higher temperatures are more appropriate. The temperature gradient in the lower mantle can be subadiabatic or superadiabatic. Attempts to estimate composition assume chemical and mineralogical homogeneity and adiabaticity but the problem is still indeterminate. A variety of chemical models can be made consistent with the geophysical data but the actual chemical composition of the lower mantle is unknown, except within very broad limits. Equation-of-state modeling is much too blunt a tool to 'prove' that the lower mantle has the same, or different, chemistry as the upper mantle.

Internal chemical boundaries in the mantle, in contrast to phase boundaries, and the surface, Moho and core–mantle boundaries, must exhibit enormous variations in depth, because of the low density contrast. This plus the low predicted seismic impedance means that compositional boundaries are difficult to detect, even if they are unbreachable by mantle convection. They are *stealth boundaries*.

Low-spin $Fe^{2}+$

Fe undergoes a spin-transition at high pressure with a large reduction in ionic radius and a probable increase in the bulk modulus and seismic velocities. The transition may be spread out over a large depth interval The major minerals in the deep mantle are predicted to be almost Fe-free perovskite [$MgSiO_3$] and Fe-rich magnesiowüstite, (Mg,Fe)O. This has several important geodynamic implications. Over time, the dense FeO-rich material may accumulate, irreversibly, at the base of the mantle, and, in addition, may interact with the core. The lattice conductivity of this iron-rich layer will be high and the radiative term should be low. A thin layer convects sluggishly (because of the h^3 term in the Rayleigh number) but its presence slows down the cooling of the mantle and the core. The overlying FeO-poor layer may have high radiative conductivity, because of high T and transparency, and have high viscosity and low thermal expansivity, because of P effects on volume. This part of the mantle will also convect sluggishly. If it represents about one-third of the mantle (by depth) it will have a Rayleigh number about 30 times less than Rayleigh numbers based on whole mantle convection and orders of magnitude less than Ra based on $P = 0$ properties.

It is likely than some of the Fe in the lower mantle is low-spin and some is high-spin with the proportions changing with depth. The oxidation state of Fe in the lower mantle is also likely to be different than in the shallow mantle. These considerations complicate the interpretation of lower mantle properties and the geodynamics and melting point of the deep mantle. It is certainly dangerous to fit a single equation of state to the whole lower mantle or to argue that seismic data requires the lower mantle to be homogenous, or the same as the upper mantle.

Region D″

The lowermost mantle, Bullen's Region D″ is a region of generally low seismic gradient and increased scatter in seismic travel times and amplitudes. Lay and Heimberger (1983) found a shear-velocity jump of 2.8% in this region that may vary in depth by up to 40 km. They concluded that a large shear-velocity discontinuity exists about 280 km above the core, in a region of otherwise low velocity gradient. There appears to be a lateral variation in the velocity increase and sharpness of the structure, but the basic character of the discontinuity seems to be well established.

Because the core is a good conductor and has low viscosity, it is nearly isothermal. Lateral temperature variations can be maintained in the mantle, but they must converge at the base of D″. This means that temperature gradients are variable in D″. In some places, in hotter mantle, the gradient may even be negative. Regions of negative shear velocity gradient in D″ are probably regions of high temperature gradient and high heat loss from the core. It is plausible that the layers at the base of the mantle interact with the core and therefore differ in composition from the rest of the mantle.

Lateral heterogeneity

D″ may represent a chemically distinct region of the mantle. If so it will vary laterally, and the discontinuity in D″ will vary considerably in radius, the hot regions being elevated with respect to the cold regions. A chemically distinct layer at the base of the mantle that is only marginally denser than the overlying mantle would be able to rise into the lower mantle when it is hot and sink back when it cools off. The mantle–core boundary, being a chemical interface, is a region of high thermal gradient, at least in the colder parts of the lower mantle.

Seismic observations suggest the presence of broad seismic velocity anomalies in the deep mantle. The nature of these anomalies is inconsistent with purely thermal convection and suggests the existence of large-scale chemical heterogeneities in the lower mantle. The anticorrelation between bulk sound speed and shear wave velocity anomalies in the lowermost mantle suggests the presence of chemical density heterogeneities.

The core

In the first place please bear in mind that I do not expect you to believe this story. Nor could you wonder had you witnessed a recent experience of mine when, in the armor of blissful and stupendous ignorance, I gaily narrated the gist of it to a Fellow of the Royal Geological Society. . . . The erudite gentleman in whom I confided congealed before I was half through! – it is all that saved him from exploding – and my dreams of an Honorary Fellowship, gold medals, and a niche in the Hall of Fame faded into the thin, cold air of his arctic atmosphere.

But I believe the story, and so would you, and so would the learned Fellow of the Royal Geological Society, had you and he heard it from the lips of the man who told it to me

Edgar Rice Burroughs

A molten iron-rich core appeared early in Earth history, the evidence being in the remnant magnetic field and isotopic record of ancient rocks. This in turn implies a short high temperature accretion for the bulk of the Earth, with perhaps a drawn out accretionary tail to bring in noble gases and other volatile elements and to salt the upper mantle with siderophile elements that would otherwise be in the core. The long-standing controversy regarding a drawn-out (100 milllion years) versus a rapid (~1 Myr) terrestrial accretion appears to be resolving itself in favor of the shorter time scales and a high-temperature origin. The core is approximately half the radius of the Earth and is about twice as dense as the mantle. It represents 32% of the mass of the Earth. A large dense core can be inferred from the mean density and moment of inertia of the Earth, and this calculation was performed by Emil Wiechert in 1891. The existence of stony meteorites and iron meteorites had earlier led to the suggestion that the Earth may have an iron core surrounded by a silicate mantle. The first seismic evidence for the existence of a core was presented in 1906 by Oldham, although it was some time before it was realized that the core does not transmit shear waves and is therefore probably a fluid. It was recognized that the velocity of compressional waves dropped considerably at the core-mantle boundary. Beno Gutenberg made the first accurate determination of the depth of the core, 2900 km, in 1912, and this is remarkably close to current values. The core-mantle-boundary is referred to as the Gutenberg discontinuity and as the CMB.

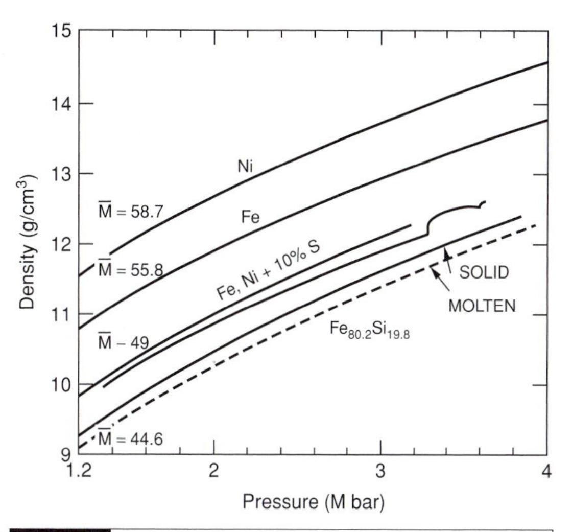

Fig. 10.1 Estimated densities of Fe, Ni and some Fe-rich alloys compared with core densities. The estimated reduction in density due to melting is shown (dashed line) for one of the alloys.

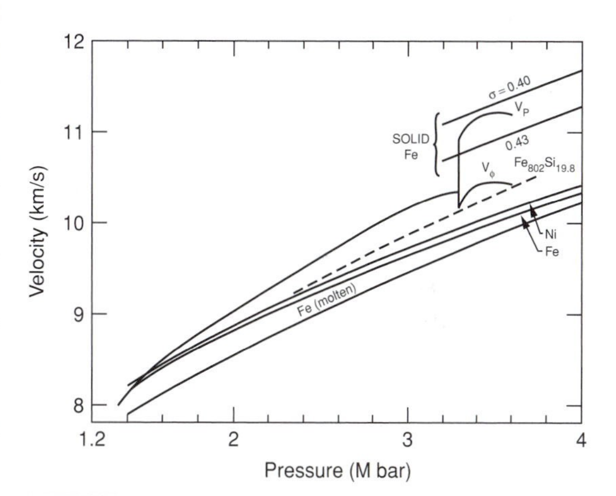

Fig. 10.2 Compressional wave velocities in the outer core and compressional and bulk sound speeds in the inner core compared to estimates for iron and nickel. Values are shown for Poisson ratios in the inner core.

Although the idea that the westward drift of the magnetic field might be due to a liquid core goes back 300 years, the fluidity of the core was not established until 1926 when Jeffreys pointed out that tidal yielding required a smaller rigidity for the Earth as a whole than indicated by seismic waves for the mantle. It was soon agreed by most that the transition from mantle to core involves both a change in composition and a change in state. Subsequent work has shown that the boundary is extremely sharp. There is some evidence for variability in depth, in addition to hydrostatic ellipticity. Variations in lower-mantle density and convection in the lower mantle can cause at least several kilometers of relief on the core-mantle boundary. The outer core has extremely high Q and transmits P-waves with very low attenuation. The elastic properties and density of the core are consistent with an iron-rich alloy (Figures 10.1 and 10.2). Evidence that the outer core is mainly an iron-rich fluid also comes from the magnetohydrodynamic requirement that the core be a good electrical conductor.

Although the outer core behaves as a fluid, it does not necessarily follow that temperatures are above the liquidus throughout. It would behave as a fluid even if it contained 30% or more of suspended particles. All we know for sure is that at least part of the outer core is above the solidus or eutectic temperature and that the outer core, on average, has a very low rigidity and low viscosity. Because of the effect of pressure on the liquidus temperature, a homogenous core can only be adiabatic if it is above the liquidus throughout. An initially homogenous core with an adiabatic temperature profile that lies between the solidus and liquidus will contain suspended particles that will tend to rise or sink, depending on their density. The resulting core will be on the liquidus throughout and will have a radial gradient in iron content. The core will be stably stratified if the iron content increases with depth.

The inner core

In 1936 Inge Lehmann used seismic data from the *core shadow* to infer the presence of a higher velocity inner core. Although no waves have yet been identified that have traversed the inner core unambiguously as shear waves, indirect evidence indicates that the inner core is solid (Birch, 1952). Julian and others (1972) reported evidence for PKJKP, a compressional wave in the mantle and outer core that traverses the inner core as a shear wave. Observation of PKJKP is difficult and claimed observations are controversial.

Early free-oscillation models gave very low shear velocities for the inner core, 2 to 3 km/s; more recent models give shear velocities in the inner core ranging from 3.46 to 3.7 km/s, in the range of crustal values. The boundary of the inner core is also extremely sharp. The Q of the inner core is relatively low, and appears to increase with depth. The high Poissons ratio of the inner core, 0.44, has been used to argue that it is not a crystalline solid, or that it is near the melting point or partially molten or that it involves an electronic phase change. However, Poissons ratio increases with both temperature and pressure and is expected to be high at inner core pressures, particularly if it is metallic. Some metals have Poisson's ratios of 0.43 to 0.46 even under laboratory conditions.

The solid inner core is the most remote and enigmatic part of our planet, and except for the crust, is the smallest 'official' subdivision of Earth's interior. Only a few seismic waves ever reach it and return to the surface. The inner core is a small target for seismologists and seismic waves are distorted by passing though the entire Earth before reaching it. The inner core is isolated from the rest of Earth by the low viscosity fluid outer core and it can rotate, nod, wobble, precess, oscillate, and even flip over, only loosely constrained by the surrounding shells. Its existence, size and properties constrain the temperature and mineralogy near the center of the Earth. Among its anomalous characteristics are low rigidity and viscosity (compared to other solids), bulk attenuation, extreme anisotropy and super-rotation (or deformation).

The inner core has a radius of 1222 km and a density about 13 g/cm^3. Because of its small size, it is difficult to determine a more accurate value for density. It represents about 1.7% of the mass of the Earth. The density and velocity jumps at the inner-core–outer-core boundary are large enough, and the boundary is sharp enough, so that the boundary is a good reflector of short-period seismic energy. The inner core is seismically anisotropic; compressional wave speeds are 3–4% faster along the Earth's spin axis.

The main constraint on composition and structure is the compressional velocity and anisotropy. From seismic velocities and cosmic abundances we know that it is mainly composed of iron–nickel crystals, and the crystals must exhibit a large degree of common orientation. The inner core is predicted to have very high thermal and electrical conductivity, a non-spherical shape, frequency-dependent properties and it may be partially molten. It may be essential for the existence of the magnetic field and for polarity reversals of this field. Freezing of the inner core and expulsion of impurities is likely responsible for powering the geodynamo.

Within the uncertainties the inner core may be simply a frozen version of the outer core, Fe_2O or FeNiO, pure iron or an iron–nickel alloy.

If the inner core froze out of the outer core, then the light alloying element may have been excluded from the inner core during the freezing or sedimentation process. An inner core growing over time could therefore cause convection in the outer core and may be an important energy source for maintaining the dynamo. The possibility that the outer core is below the liquidus, with iron in suspension, presents an interesting dynamic problem. The iron particles will tend to settle out unless held in suspension by turbulent convection. If the composition of the core is such that it is always on the iron-rich side of the eutectic composition, the iron will settle to the inner-core–outer-core boundary and increase the size of the solid inner core. Otherwise it will melt at a certain depth in the core. The end result may be an outer core that is chemically inhomogenous and on the liquidus throughout. The effect of pressure on the liquidus and the eutectic composition may, however, be such that solid iron particles can form in the upper part of the core and melt as they sink. In such a situation the core may oscillate from a nearly chemically homogenous adiabatic state to a nearly chemically stratified unstable state. Such complex behavior is well known in other nonlinear systems. The apparently erratic behavior of the Earth's magnetic field may be an example of chaos in the core, oscillations being controlled by nonlinear chemistry and dynamics.

Since the outer core is a good thermal conductor and is convecting, the lateral temperature gradients are expected to be quite small. The mantle, however, with which the outer core is in contact, is a poor conductor and is convecting much less rapidly. Seismic data for the lowermost mantle indicate large lateral changes in velocity and, possibly, a chemically distinct layer of variable thickness. Heat can only flow across the core-mantle boundary by conduction. A thermal boundary layer, a layer of high temperature gradient, is therefore established at the base of the colder parts of the mantle. That in turn can cause small-scale convection in this layer if the thermal gradient and viscosity combine to give an adequately high Rayleigh number. It is even possible for material to break out of the thermal boundary layer, even if it is also a chemical boundary, and ascend into the lower mantle above D″. The lateral temperature gradient near the base of the mantle also affects convection in the core. This may result in an asymmetric growth of the inner core. Hot upwellings in the outer core will deform and possibly erode or dissolve the inner core. Iron precipitation in cold downwellings could serve to increase inner-core growth rates in these areas. These considerations suggest that the inner-core boundary might not be a simple surface in rotational equilibrium.

The orientation of the Earth's spin axis is controlled by the mass distribution in the mantle. The most favorable orientation of the mantle places the warmest regions around the equator and the coldest regions at the poles. Insofar as temperatures in the mantle control the temperatures in the core, the polar regions of the core will also be the coldest regions. Precipitation of solid iron is therefore most likely in the axial cylinder containing the inner core.

Formation of the inner core

There are two processes that could create a solid inner core. (1) Core material was never completely molten and the solid material coalesced into the solid inner core, and (2) the inner core solidified due to gradual cooling, increase of pressure as the Earth grew, and the increase of melting temperature with pressure. It is possible that both of these processes occurred; that is, there was an initial inner core due to inhomogenous accretion, incomplete melting or pressure freezing and, over geologic time, there has been some addition of solid precipitate. The details are obviously dependent on the early thermal history, the abundance of aluminum-26 and the redistribution of potential energy. The second process is controlled by the thermal gradient and the melting gradient. The inner core is presently 5% of the mass of the core, and it could either have grown or eroded with time, depending on the balance between heating and cooling. Whether or not the core is thermally stable depends on the distribution of heat sources and the state of the mantle. If all the uranium and thorium is removed with the refractories to the lowermost mantle, then the only energy sources in the core are cooling, a growing inner core and further gravitational separation in the outer core.

In the *inhomogenous accretion model* the early condensates, calcium–aluminum-rich silicates, heavy refractory metals, and iron accreted to form the protocore. The early thermal history is likely to be dominated by aluminum-26, which could have produced enough heat to raise the core temperatures by 1000 K and melt it even if the Earth accreted 35 My after the Allende meteorite, the prototype refractory body. Melting of the protocore results in unmixing and the emplacement of refractory material (including uranium, thorium and possibly 26 Al) into the lowermost mantle. Calculations of the physical properties of the refractory material and normal mantle suggest that the refractories would be gravitationally stable in the lowermost mantle but would have a seismic velocity difference of a few percent.

Mantle–core equilibration

Upper mantle rocks are extremely depleted in the siderophile elements such as cobalt, nickel, osmium, iridium and platinum, and it can be assumed that these elements have mostly entered the core. This implies that material in the core had at one time been in contact with material currently in the mantle, or at least the upper mantle. Alternatively, the siderophiles could have experienced preaccretional separation, with the iron, from the silicate material that formed the mantle. In spite of their low concentrations, these elements are orders of magnitude more abundant than expected if they had been partitioned into core material under low-pressure equilibrium conditions. The presence of iron in the mantle would serve to strip the siderophile elements out of the silicates. The magnitude of the partitioning depends on the oxidation state of the mantle. The 'overabundance' of siderophiles in the the upper mantle is based primarily on observed partitioning between iron and silicates in meteorites. The conclusion that has been drawn is that the entire upper mantle could never have equilibrated with metallic iron, which subsequently settled into the core. Various scenarios have been invented to explain the siderophile abundances in the mantle; these include rapid settling of large iron blobs so that equilibration is not possible or a late veneer of chondritic material that brings in siderophiles after the core is formed. The siderophiles are not fractionated as strongly as one would expect if they had been exposed to molten iron. Some groups of siderophiles occur in chondritic ratios in upper mantle rocks.

The highly siderophile elements (Os, Re, Ir, Ru, Pt, Rh, Au, Pd) strongly partition into any metal that is in contact with a silicate. These elements are depleted in the crust–mantle system by almost three orders of magnitude compared with cosmic abundances but occur in roughly chondritic proportions. If the mantle had been in equilibrium with an iron-rich melt, which was then completely removed to form the core, they would be even more depleted and would not occur in chondritic ratios. Either part of the melt remained in the mantle, or part of the mantle, the part we sample, was not involved in core formation and has never been in contact with the core. Many of the moderately siderophile elements (including Co, Ni, W, Mo and Cu) also occur in nearly chondritic ratios, but they are depleted by about an order of magnitude less than the highly siderophile elements. They are depleted in the crust–mantle system to about the extent that iron is depleted. These elements have a large range of metal–silicate partition coefficients, and their relatively constant depletion factors suggest, again, that the upper mantle has not been exposed to the core or that some core-forming material has been trapped in the upper mantle. It is not clear why the siderophiles should divide so clearly into two groups with chondritic ratios occurring among the elements within, but not between, groups. The least depleted siderophiles are of intermediate volatility, and very refractory elements occur in both groups.

Light element in the core

The core's density is about 10% less than that of Fe (or Fe–Ni alloy) at core conditions (Figure 10.1) and thus there is a significant amount of an element or element mixture having a lower atomic number than Fe in the core. There is evidence to suggest that some, but not all, of this density deficient is due to the presence of sulfur. The S content of the core may be of the order of ~2 weight%, based on meteorites and mantle chemistry. H, C, O and Si, and very high temperature, are additional candidates for explaining core density.

Radioactive elements in the core?

Most of the large ion lithophile elements such as K, U and Th are undoubtedly in the crust and mantle, and were probably placed there during accretion. Nevertheless, the presence of radioactive elements in the Earth's core is often suggested in order to power the geodynamo or to explain where the volatile elements are, in the Earth. The cosmochemical argument for K in the core is based on the presence of potassium sulfide in enstatite chondrites. Enstatite chondrites also contain other more abundant sulfides, including CaS, (Mn,Fe)S, and (Mg,Fe)S and substantial concentrations of REE. Although one may wish to place some K in the core, there are associated consequences that exclude this possibility. Likewise it is not possible that U or Th or both are in the core, with our present understanding of crystal chemistry and solubilities at high temperature and pressure. The possibility of a nuclear reactor in the core, however, has been proposed.

Core formation

Density stratification explains the locations and relative mass of the crust, mantle and core. The inner core is likely also the result of chemical stratification although the effect of pressure on the melting point would generate a solid inner core even if it were chemically identical to the outer core. Low-density materials are excluded when solidification is slow so the inner core may be purer and denser than the outer core. As the inner core crystallizes and the outer core cools, the material held in solution and suspension will plate out, or settle, at the core–mantle boundary and may be incorporated into the lowermost mantle. Analogous processes contribute to sedimentation in deep ocean basins. The mantle is usually treated as a chemically homogenous layer but this is unlikely. Denser silicates, possibly silicon- and iron-rich, also gravitate toward the lower parts of the mantle. Crustal and shallow mantle materials were sweated out of the Earth as it accreted and some were apparently never in equilibrium with core material. The effect of pressure on physical properties implies that the mantle and core probably irreversibly stratified upon accretion and that only the outer shells of the mantle participate in surface processes such as volcanism and plate tectonics.

Geophysical data require rapid accretion of Earth and early formation of the core. Until recently this has been at odds with accretional theory and isotopic data but now these disciplines are also favoring a contracted time scale. A variety of isotopes have recently confirmed short time intervals between the formation of the solar system and planetary differentiation processes. This has bearing on the age of the inner core and its cooling history.

There are three quite different mechanisms for making a planetary core. In the homogenous accretion hypothesis the silicates and the metals accrete together but as the Earth heats up the heavy metals percolate downwards, eventually forming large dense accumulations that sink rapidly toward the center, taking the siderophile elements with them. In the heterogenous accretion hypothesis the refractory condensates from a cooling nebula, including iron and nickel, start to form the nucleus of a planet before the bulk of the silicates and volatiles are available. The late veneer contributes low-temperature condensates and gases, including water, from far reaches of the solar system. Finally, large late impacts can efficiently and rapidly inject their metallic cores toward the center of the impacted planet, and trigger additional separation of iron from the mantle. The Moon is a byproduct of one of these late impacts. The material in the core may therefore have multiple origins and a complex history. In addition to its age and growth rate, other issues regarding the inner core involve density, temperature, texture and internal energy sources.

Chapter 11

Geotomography: heterogeneity of the mantle

You all do know this mantle.

Shakespeare (Julius Caesar)

We must now admit that the Earth is not like an onion. It is time for some lateral thinking. One-dimensional radial variations in the mantle were responsible for what are now the standard one- and two-layer models of geodynamics and mantle geochemical reservoirs. The lateral variations of seismic velocity and density are as important as the radial variations. The shape of the Earth tells us this directly but provides little depth resolution. The long-wavelength geoid tells us that lateral density variations – and probable chemical variations – occur at great depth. Heat flow tells us that there are pronounced shallow variations in heat productivity, structure and physical properties. Lateral variations in the mantle affect the orientation of Earth in space and convection in the mantle and core. This property of the Earth is known as *asphericity*. It is best studied with seismic tomography. In the following chapters we further recognize that the Earth is neither elastic nor isotropic; it is *anelastic* and *anisotropic*. Long-wavelength lateral variations are revealed by global tomography. High-resolution seismic studies and scattering of high-frequency seismic waves complement the long-wavelength studies but are not consistent with the simple dynamic and chemical models based on older 1D or long-wavelength studies. Scattering may contribute to the anisotropy and attenuation of seismic waves in the upper mantle, and may help resolve the fates of recycled materials and the question of homogeneity of the upper mantle.

Tomography

Seismic tomography can be used to infer the three-dimensional structure of the Earth's interior, and, under certain conditions, the mineralogy. It is more difficult to infer temperature and composition. Although seismology and tomography are quantitative sciences, the resulting maps and cross-sections are often interpreted in a visual or intuitive way. Conclusions based on visual inspection of color tomographic cross-sections [see `mantleplumes`] can be called *Qualitative chromo-tomography* or QCT. Quantitative tomographic interpretations involve `probabilistic` tomography `maps of chemical heterogeneities, understanding how the power is distributed in the wavenumber domain`, anisotropy and derived quantities such as V_p/V_s ratios, correlations, spectral densities vs. depth, matched filters, statistics and so on. Anelasticity, anharmonicity and mineral physics and geodynamic constraints, are also used in quantitative interpretations of tomographic models. Quite often a tomographic cross-section is misleading because of ray coverage, color saturation, cropping, data selection, bleeding and smearing. There are also artifacts associated with source, receiver and finite frequency effects, lateral refraction, anisotropy and projection. Global tomography gives only the long-wavelength components of heterogeneity. Seismic scattering and high-frequency reflection experiments provide details of small-scale structure.

Preliminaries

In mantle tomography the number of parameters that one would like to estimate far exceeds the information content (the number of independent data points) of the data. It is therefore necessary to decide which parameters are best resolved by the data, what is the resolution, or averaging length, which parameters to hold constant, how to treat unsampled areas, and how the model should be parameterized (for example as layers or smooth functions, isotropic or anisotropic). In addition, there are a variety of corrections that might be made (such as crustal thickness, water depth, elevation, ellipticity, attenuation). The resulting models are as dependent on these assumptions and corrections as they are on the quality and quantity of the data. This is not unusual in science; data must always be interpreted in a framework of assumptions, and the data are always, to some extent, incomplete and inaccurate. In the seismological problem the relationship between the solution, or the model, including uncertainties, and the data can be expressed formally. The effects of the assumptions and parameterizations, however, are more obscure, but these also influence the solution. The hidden assumptions are the most dangerous. For example, most seismic modeling assumes perfect elasticity, isotropy, geometric or finite-frequency optics and linearity. To some extent all of these assumptions are wrong, and their likely effects must be kept in mind. These artifacts and assumptions do not show up in color tomographic cross-sections, which also usually do not indicate which parts of the mantle are unsampled and have been filled in by smoothing. Published cross-sections are usually selected, oriented and cropped to make a certain point. Thus, it is easy for nonspecialists to accept these cross-sections as data and to overinterpret them. Tomography is best used as a hypothesis tester rather than as a definitive and unique mapper of heterogeneity and mantle temperature.

Qualitative vs. quantitative tomography

Body-wave tomography is a powerful but imperfect tool. Travel-time tomography, used alone, is particularly limited. The results depend crucially on the ray geometry, which is constrained by the geometry of earthquakes and seismic stations (Vasco *et al.*, 1995), and – to a lesser extent – the details of the mathematical techniques employed (Spakman and Nolet, 1988; Spakman *et al.*, 1989; Shapiro and Ritzwoller, 2004). Seismic ray coverage is sparse and spotty. Moreover, the visual appearance of displayed results depends on the color scheme, reference model, cropping and cross-sections chosen. The resulting images contain artifacts that appear convincing (Spakman *et al.*, 1989). Even further difficulties result from the limitations of present algorithms, which cannot correct completely for finite-frequency, source and anisotropic effects and can certainly not constrain regions where there are little data. Global tomographic models have little detailed resolution. High-frequency reflection, scattering and coda studies paint a much more complex picture of the mantle than is available from long-period waves.

Visual, or intuitive, interpretations of tomographic images, qualitative chromo-tomography (QCT) and the association of color with temperature have had more crossdisciplinary impact than quantitative and statistical analysis of the tomographic results. To a geochemist a color cross-section provides overwhelming tomographic evidence that at least some of the subducting lithospheric plates are currently reaching the core–mantle boundary. The visual impressions of tomographic images, their implications regarding isotopic as well as major element recycling, and the plausibility or implausibility of layering and survival of heterogeneities in a convecting mantle are now at the heart of perceived conflicts of geophysics and geochemistry and mantle convection. It has recently become possible to quantify tomographic models, to resolve density and to separate the effects of temperature and composition (Ishii and Tromp, 2004; Trampert *et al.*, 2004). The basic assumption in many tomographic interpretations, that low seismic velocity is always a proxy for high temperature and low density, is not valid. There is no correlation in properties between the upper mantle, midmantle and lower mantle and no evidence for either deep slab penetration or continuous plume-like low-velocity upwellings (Becker and Boschi, 2002; Ishii and Tromp, 2004; Trampert *et al.*, 2004); this is consistent with chemical stratification (Wen and Anderson, 1995, 1997).

Color 2D cross-sections are particularly ambiguous; although certainly vivid – and impressive to nonspecialists – they are not 'overwhelming, compelling or convincing' evidence for mantle dynamics or composition; they can be overinterpreted. They cannot do justice to the information content of a typical tomographic study. There are also issues of physical, geodynamic and petrological interpretations; are 'blue' regions of the lower mantle, even if real, unambiguous indicators of cold, dense material that started at the Earth's surface?

Geodynamic interpretations of tomographic models based on quantitative analyses of tomography (Scrivner and Anderson, 1992; Wen and Anderson, 1995; Masters *et al.*, 2000; Gu *et al.*, 2001; Anderson, 2002a; Ishii and Tromp, 2004; Trampert *et al.*, 2004) are quite different from the interpretations of QCT and color images [Probabilistic Tomography Maps of Chemical Heterogeneities]. Very few slab-like features appear to be dense; hardly any hotspots appear to be underlain by buoyant material in the lower mantle. Long wavelength chemical heterogeneities, however, exist throughout the mantle (Ishii and Tromp, 1999, 2004).

Tomographic images are often interpreted in terms of an assumed velocity–density–temperature correlation, e.g. high shear-velocity (blue) is attributed to cold dense slabs, and low shear velocity (red) is interpreted as hot rising low-density blobs. There are many factors controlling seismic velocity and some do not involve temperature or density. Cold, dense regions of the mantle can have low shear velocities (e.g. Presnall and Gudfinnsson, 2004; Trampert *et al.*, 2004). Changes in composition or crystal structure can lower the shear-velocity and increase the bulk modulus and/or density, as can be verified by checking any extensive tabulation of elastic properties and densities of minerals. Ishii and Tromp (2004), for example, found negative correlations between velocity and density in the upper mantle.

For reviews of the situation regarding seismic modeling – including uncertainties and limitations – see Dziewonski (2005); Boschi and Dziewonski (1999); Vasco *et al.* (1994) and Ritsema *et al.* (1999). The bottom line is that the mantle is not similar to the one- and two-layer 1D structures that underlie the standard models of mantle geodynamics and geochemistry and temperature is not the sole parameter controlling seismic velocities.

Spectrum of heterogeneity

Understanding how the power is distributed in the wavenumber domain and the distribution of the spectral power as a function of depth in the mantle is an extremely valuable diagnostic in the interpretation of tomographic models and assessing the applicability of various mantle convection models [www.mantleplumes.org/Convection.html]. Tanimoto (1991) was the first to point out the significance of the predominance of the large-scale heterogeneity in the Earth. Su and Dziewonski (1991, 1992) showed that tomographic power is approximately constant up to degree 6, but then decreases rapidly. Mantle convection simulations must satisfy this constraint.

Approaches

There are several approaches for interpreting global seismic data. The regionalized approach divides the Earth into tectonic provinces and solves for the velocity of each. Application of this technique shows that shields are the highest velocity regions at shallow depths but convergence regions are faster at greater depth, suggesting that cold subducted material is being sampled (Nakanishi and Anderson, 1983, 1984a,b). Convergence regions, on average, are slow at short periods, due to high temperatures and melting at shallow mantle depths. The regionalization approach is necessary when the data are limited or when only complete great-circle or long-path data are available. In the latter case the velocity anomalies cannot be well isolated.

In the regionalized models it is assumed that all regions of a given tectonic classification have the same velocity. This is clearly oversimplified. It is useful, however, to have such reference maps in order to define anomalous regions. In fact, all shields are not the same, and velocity does not increase monotonically with age within oceanic regions. The region around Hawaii, for example, is faster than equivalent-age ocean elsewhere at shallow depth and slower at somewhat greater depth. From ScS data we know that the average seismic velocity and attenuation in the mantle under Hawaii are normal. On average, hotspots are not associated with anomalous upper mantle, although visual correlations are often claimed with the deep mantle.

There are now many models of the velocity, attenuation and anisotropy – radial and azimuthal – of the upper mantle.

[tomography upper mantle anisotropy] [mantle tomographic maps] [global seismic struture maps] [bullard.esc.cam.ac.uk/~maggi/Physics_Earth_Planet/Lecture_7/colour_figures.html]

Global surface wave tomography

The most complete global maps of seismic heterogeneity of the upper mantle are obtained from surface waves [global tomographic images]. By analyzing the velocities of Love and Rayleigh waves, of different periods, over many great circles, small arcs and long arcs, it is possible to reconstruct the radial and lateral velocity and anisotropy variations. Although global coverage is possible, the limitations imposed by the locations of seismic stations and of earthquakes limit the global spatial resolution to about 1000 km. The raw data consist of amplitude variations, phase delays, travel times, or average group and/or phase velocity over many arcs. These averages can be converted to images using techniques similar to medical tomography. Body waves have better resolution, but coverage, particularly for the upper mantle, is poor. The best global maps of mantle structure combine surface waves, higher modes, body waves and normal modes. Even the early surface-wave studies indicated that the upper mantle was extremely inhomogenous and anisotropic. Shield paths are fast, oceanic paths are slow, and tectonic regions are also slow. The most pronounced differences are in the upper 200 km, but substantial differences between regions extend to about 400 km.

The mantle above 200–300 km depth correlates very well with known tectonic features. There are large differences between continents and oceans, and between cratons, tectonic regions, back-arc basins and different age ocean basins. High velocities appear beneath all Archean cratons. Platforms are variable. At depths between 800 and 1000 km there is good correlation of seismic velocities with inferred regions of past subduction. Hypothetical hot, narrow mantle upwellings – the elusive mantle plumes – do not, in general, show up consistently in mantle tomography.

Over most of the Earth, long-period Rayleigh- and Love-wave dispersion curves are 'inconsistent' in the sense that they cannot be fit simultaneously using a simple isotropic model. This has been called the Love-wave—Rayleigh-wave discrepancy and attributed to anisotropy. There are now many studies of this effect – also called polarization or radial anisotropy, or transverse isotropy. Independent evidence for anisotropy in the upper mantle is strong (e.g. from receiver functions amplitudes and shear wave splitting).

Love waves are sensitive to the SH (horizontally polarized shear waves) velocity of the shallow mantle, above about 300 km for most studies. The slowest regions are at plate boundaries, particularly triple junctions. Slow velocities extend around the Pacific plate and include the East Pacific Rise, western North America, Alaska-Aleutian arcs, Southeast Asia and the Pacific-Antarctic Rise. Parts of the Mid-Atlantic Rise and the Indian Ocean Rise are also slow. The Red Sea-Gulf of Aden–East Africa Rift (Afar triple junction) is one of the slowest regions. The upper-mantle velocity anomaly in this slowly spreading region is as pronounced as under the rapidly spreading East Pacific Rise. Since it also shows up for S-delays and long-period Rayleigh waves, this is a substantial and deep-seated anomaly.

Rayleigh waves are sensitive to shallow Primary or compressional (P) velocities and SV (vertically polarized shear wave, or polarized in the plane of the ray) velocities from about 100

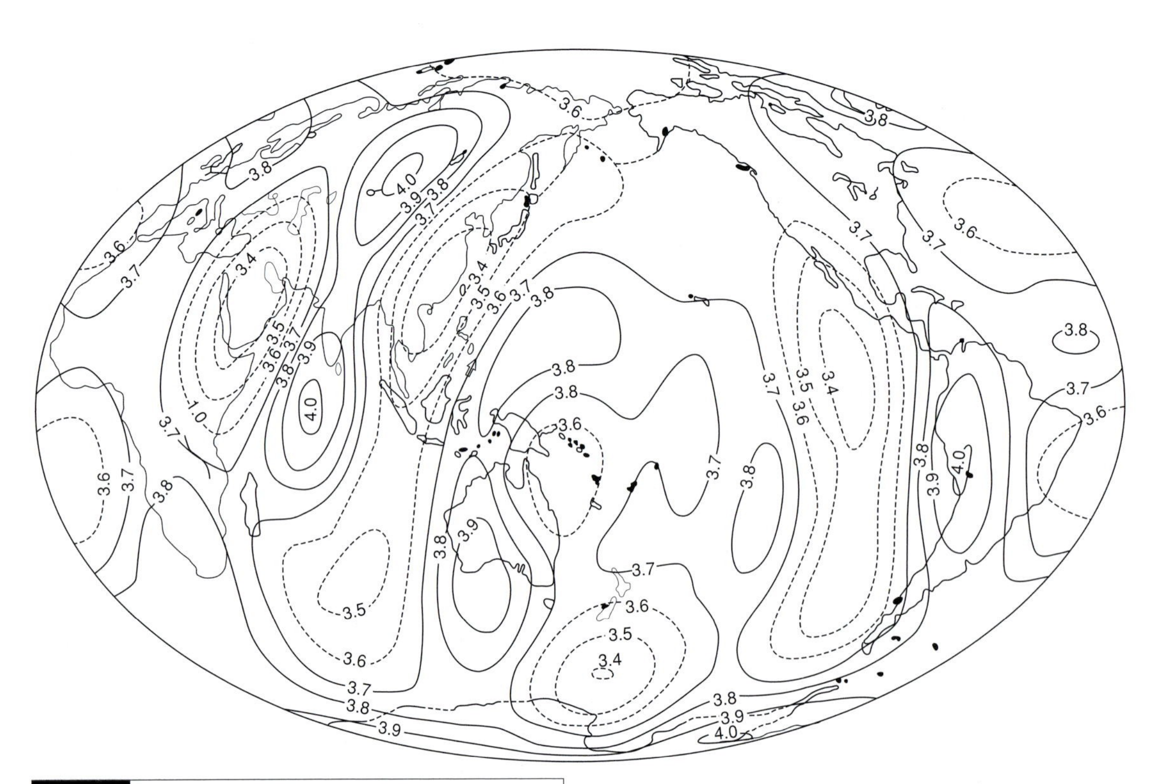

Fig. 11.1 Group velocity of 152-s Rayleigh waves (km/s). Tectonic and young oceanic areas are slow (dashed), and continental shields and older oceanic areas are fast. High temperatures and partial melting are responsible for low velocities. These waves are sensitive to the upper several hundred kilometers of the mantle (after Nakanishi and Anderson, 1984a). This was the first seismic indication of a profound mantle feature under the Afar.

to 600 km. The fastest regions are the western Pacific, western Africa and the South Atlantic. Western North America, the Red Sea area, Southeast Asia and the North Atlantic are the slowest regions. The velocities in the South Atlantic and the Philippine Sea plate are faster than shields.

The overall pattern of velocity variations shows a general correlation with surface tectonics. The lowest velocity regions are located in regions of extension or active volcanism: the southeastern Pacific, western North America, northeast Africa centered on the Afar region, the central Atlantic, the central Indian Ocean, Kerguelen–Indian Ocean triple junction, western North America centered on the Gulf of California, the northeast Atlantic, the Tasman Sea–New Zealand–Campbell Plateau and the marginal seas in the western Pacific. The fastest regions are the western Pacific, New Guinea–western Australia–eastern Indian Ocean, west Africa, northern Europe and the South Atlantic. A high-velocity region is located in the north central Pacific, centered near the Hawaiian swell, suggesting that the swell is not a thermal anomaly. `Mid-Atlantic ridge tomography` shows low-velocities especially near triple junctions in the North and South Atlantic; these show up particularly well when the reference model is ORM, the Oceanic Reference Model. [`www.mantleplumes.org/Seismology.html`]

High-velocity regions in continents generally coincide with Precambrian shields and Phanerozoic platforms (northwestern Eurasia, western and southern parts of Africa, eastern parts of North and South America, and Antarctica). Low-velocity regions in or adjacent to continents coincide with tectonically active regions, such as the Middle East centered on the Red Sea, eastern and southern Eurasia, eastern Australia, and western North America and island arcs or back-arc basins such as the southern Alaskan margin, the Aleutian, Kurile, Japanese, Izu-Bonin, Mariana,

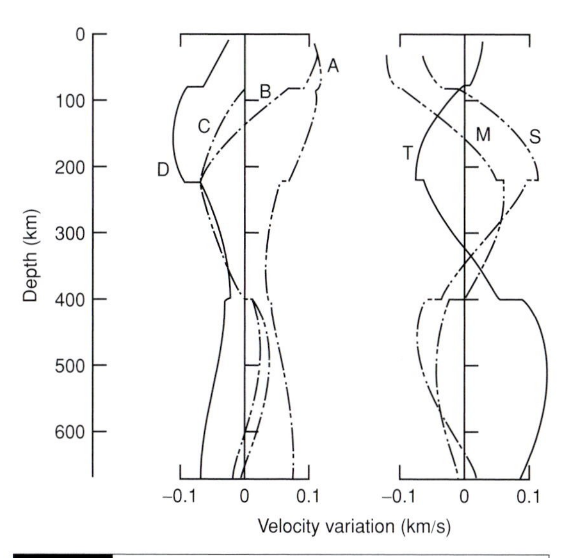

Fig. 11.2 Variation of the SV velocity with depth for various tectonic provinces. A–D, oceanic age provinces ranging from old, A, to young, D; S, continental shields; M, mountainous areas; T, trench and island-arc regions. These are regionalized results.

Ryukyu, Philippine, Fiji, Tonga, Kermadec and New Zealand arcs.

Maps of surface-wave velocity (Nakanishi and Anderson, 1983, 1984a,b) provide the most direct display possible of the lateral heterogeneity of the mantle. The phase and group velocities can be obtained with high precision and with relatively few assumptions. In general, the shorter period waves, which sample only the crust and shallow mantle, correlate well and as expected with surface tectonics. The longer-period waves, which penetrate into the transition region (400–1000 km), correlate with past subduction zones but less well with current surface tectonics.

Regionalized inversion results

Shields (S) are faster than all other tectonic provinces except old ocean from 100 to 250 km depth. Below 220 km the velocities under shields decrease, relative to average Earth, and below 400 km shields, on average, are among the slowest regions. At all depths beneath shields the velocities averaged over hundreds of km can be accounted for by reasonable mineralogies and temperatures without any need to invoke partial melting. Trench and marginal sea regions (T), on the other hand, are relatively slow above 200 km, probably indicating the presence of a partial melt, and fast below 400 km, probably indicating the presence of cold subducted lithosphere. The large size of the tectonic regions and the long wavelengths of surface waves require that the anomalous regions at depth are much broader than the sizes of slabs or the active volcanic regions at the surface. This is consistent with broad passive upwellings under young oceans and abundant piling up of slabs under trench and old ocean regions. The latter is evidence for layered-mantle convection and the cycling of oceanic plates into the transition region.

Shields and young oceans are still evident at 250 km. At 350 km the velocity variations are much suppressed. Below 400 km, most of the correlation with surface tectonics has disappeared, in spite of the regionalization, because shields and young oceans are both slow, and trench and old-ocean regions are both fast. Most oceanic regions have similar velocities at depth. Shields do not have higher velocities than some other tectonic regions below 250 km and definitely do not have 'roots' extending throughout the upper mantle or even below 400 km, as in the original `tectosphere` hypothesis. In high-resolution body-wave studies, subshield velocities drop rapidly at 150 km depth, although velocities remain relatively high to about 390 km.

Spherical harmonic inversion

An alternative way to analyze tomographic data is through a spherical harmonic expansion that ignores the surface tectonics. This provides a less biased way to assess the depth extent of tectonic features but the results are similar. In both cases, the unsampled regions are essentially filled in by interpolation. The major tectonic features correlate well with the shear velocity above 50 km. Shields and old oceans are fast. Young oceanic regions and tectonic regions are slow. The slowest regions are centered near the midocean ridges, back-arc basins and the Red Sea. The hotspot province in the south Pacific is slow at shallow depths, but the shallow mantle in the north-central Pacific, including Hawaii, is fast.

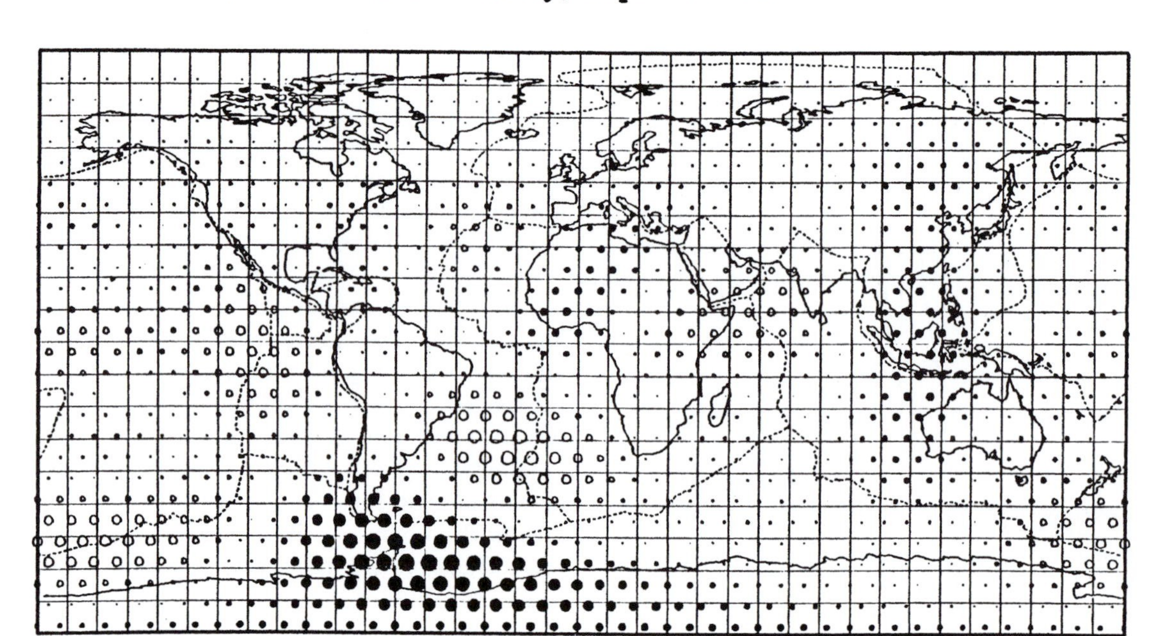

Fig. 11.3 Map of SV velocity at ~250 km depth from sixth-order spherical harmonic representation of Nataf *et al.* (1984). Note the slow regions associated with the midocean ridges. The fastest regions are in the far south Atlantic, some subduction areas and northwest Africa. Other Surface wave tomography maps can be accessed on the web.

The central Pacific and the northeastern Indian Ocean are fast. The Arctic Ocean between Canada and Siberia is slow. At 250 km (see Figure 11.3), the shields are less evident than at shallower depths and than in the regionalized model. On the other hand, the areas containing ridges are pronounced low-velocity regions. The central Pacific and the Red Sea are slow. The prominent low-velocity region in the central Pacific roughly bounded by Hawaii, Tahiti, Samoa and the Caroline Islands is the Polynesian Anomaly. This feature may be related to extension and a possible breaking up of the Pacific plate. The highest velocity anomalies are in the far south Atlantic, north-west Africa to southern Europe and the eastern Indian Ocean to southeast Asia and are not confined to the older continental areas.

The Red Sea–Afar anomaly extends to at least 400 km, but there is no evidence for a thermal thinning of the transition region. The very slow velocities associated with western North America die out by 300 km. Other data show that the Yellowstone anomaly extends to only 200 km, and below that depth Yellowstone is no more anomalous than the rest of western North America. Most hotspots, in fact, have low-velocity anomalies that are cut-off at depth (Anderson, 2005) [scoring hotspots].

By 400 or 450 km depth the pattern seen at shallow depths changes dramatically. The mantle beneath most ridges is fast. The Polynesian Anomaly, although shifted, is still present. Global seismic structure maps show that the South Pacific Superswell is generally underlain by low velocities but do not correlate well with it nor are the velocities lower than elsewhere. Eastern North America and/or the western North Atlantic are slow. Most of South America, the South Atlantic and Africa are fast. The north-central Pacific is slow. Most hotspots are above faster-than-average parts of the mantle at this depth. The fast regions under the Atlantic,

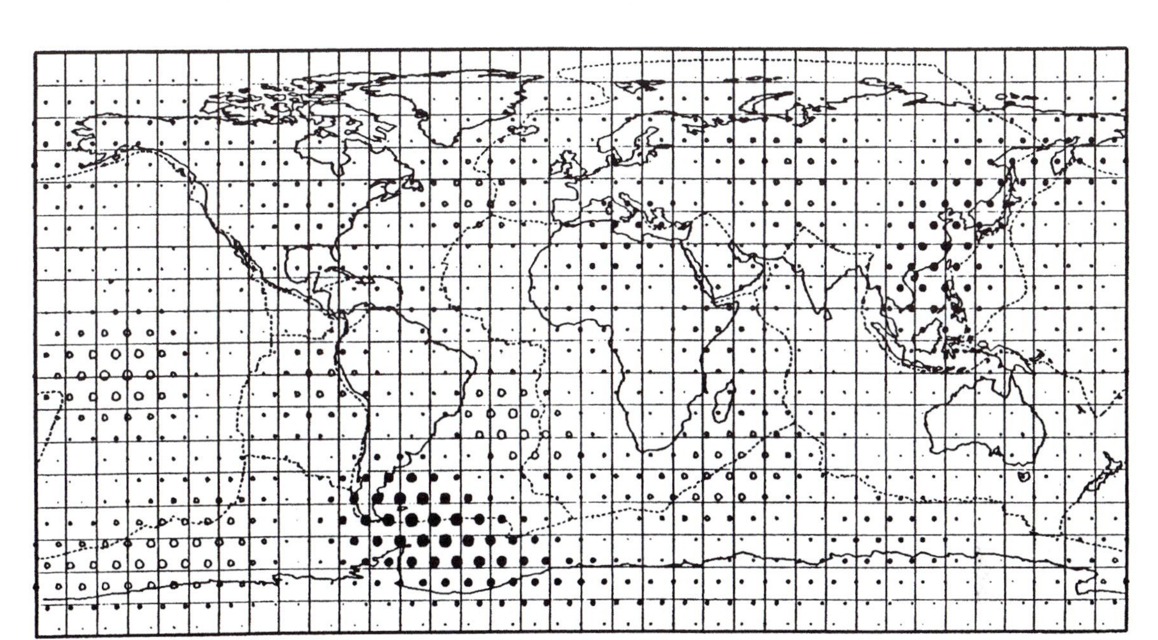

Fig. 11.4 Map of SV velocity at ~340 km depth. A prominent low-velocity anomaly shows up in the central Pacific (the Polynesian Anomaly). The fast anomalies under eastern Asia, northern Africa and the South Atlantic may represent mantle that has been cooled by subduction.

western North America, the western Pacific and south of Australia may be sites of subducted or over-ridden oceanic lithosphere. Prominent slow anomalies are under the northern East Pacific Rise and in the northwest Indian Ocean. The central Atlantic and the older parts of the Pacific are fast.

The upper-mantle shear velocities along three great-circle paths are shown in Figures 11.5 to 11.7. The open circles are slower than average regions of the mantle. These show the effects of ridges and cratons.

Quantitative tectonic correlations

Cratonic 'roots' (*archons*), thickening oceanic plates and subducting slabs are the first-order contributors to tomography above about 400 km depth. In order to investigate the fate of slabs, the structure of ridges and hotspots and the style of mantle convection, one can calculate residual maps by excluding from the tomography the first-order effects of conductive cooling of oceanic plates, deep cratonic roots, and partial melting or cooling caused by subducted lithosphere (Wen and Anderson, 1995). The good correlations between residual tomography in the transition zone (400–650 km) with 0–30 Ma subduction can be explained by slab accumulation in this region. The correlations between seismic velocities in the transition region and the subduction during earlier periods are poor. This may indicate that slabs reside near the 650 km discontinuity for only a certain period of time. This discontinuity may not be the place where long-lived-mantle convective stratification takes place, as often assumed, and it may not be a boundary between chemically distinct regions of the mantle. Mantle convection may be decorrelated closer to 900 km, near a recently rediscovered mantle discontinuity (the Repetti discontinuity). This may be a chemical boundary. If so, we expect it to be highly variable in depth.

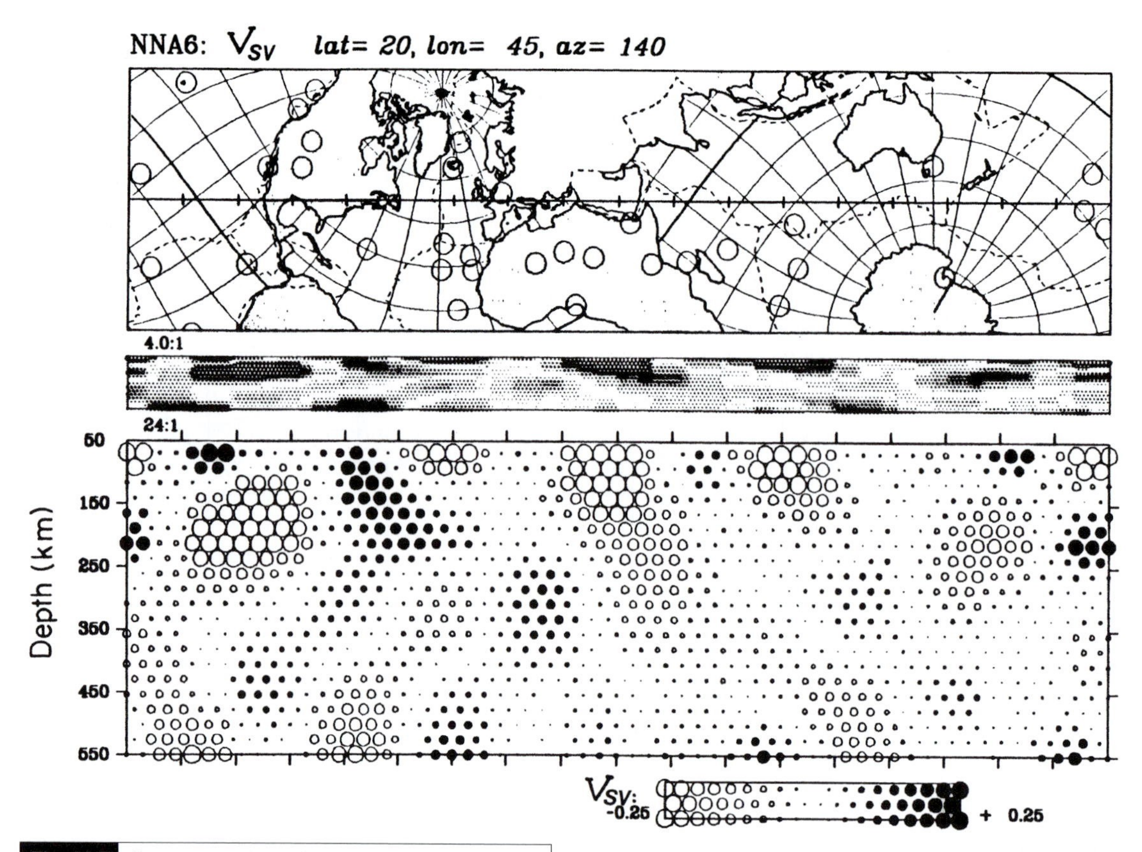

Fig. 11.5 Shear wave velocity from 50 to 550 km depth along the great-circle path shown. Cross-sections are shown with two vertical exaggerations. Note the low-velocity regions in the shallow mantle below Mexico, the Afar and south of Australia and the asymmetry of the North Atlantic. Velocity variations are much more extreme at depths less than 250 km than at greater depths. The circles on the map represent hotspots.

There is no clear global, statistically significant, correlation between surface hotspots or swells and mantle structure below some 400 km.

Slabs and tomography

The relationship between subduction and seismic tomography has been studied widely. Scrivner and Anderson (1992) correlated subduction positions since the breakup of Pangea, with seismic tomography depth by depth throughout the whole mantle. They found the best correlations in the transition zone region. Ray and Anderson (1994) found good correlations between integrated slab locations since Pangea breakup and fast velocities in the depth range 220–1022 km. Wen and Anderson (1995) quantified the slab flux by estimating the subducted volume in the hotspot reference frame and correlated it with seismic tomography throughout the mantle. They found significant correlations in the depth interval 900–1100 km and attributed these to the accumulation of subducted lithosphere in this region. Correlations were found in the upper mantle and transition zone for more recent subduction.

Some slabs may be stopped at the 650 km discontinuity for a period of time, while some slabs may penetrate to 1000 km. The residual tomography shows high velocity beneath the Kurile, Japan, Izu-Bonin, Mariana, New Hebrides and Philippine trenches. This implies that the subducted slabs may accumulate beneath these trenches at the 650 km discontinuity. The good correlations between the 0–30 Ma subduction and residual tomography can be explained by the accumulation of slabs beneath the Kurile, Japan,

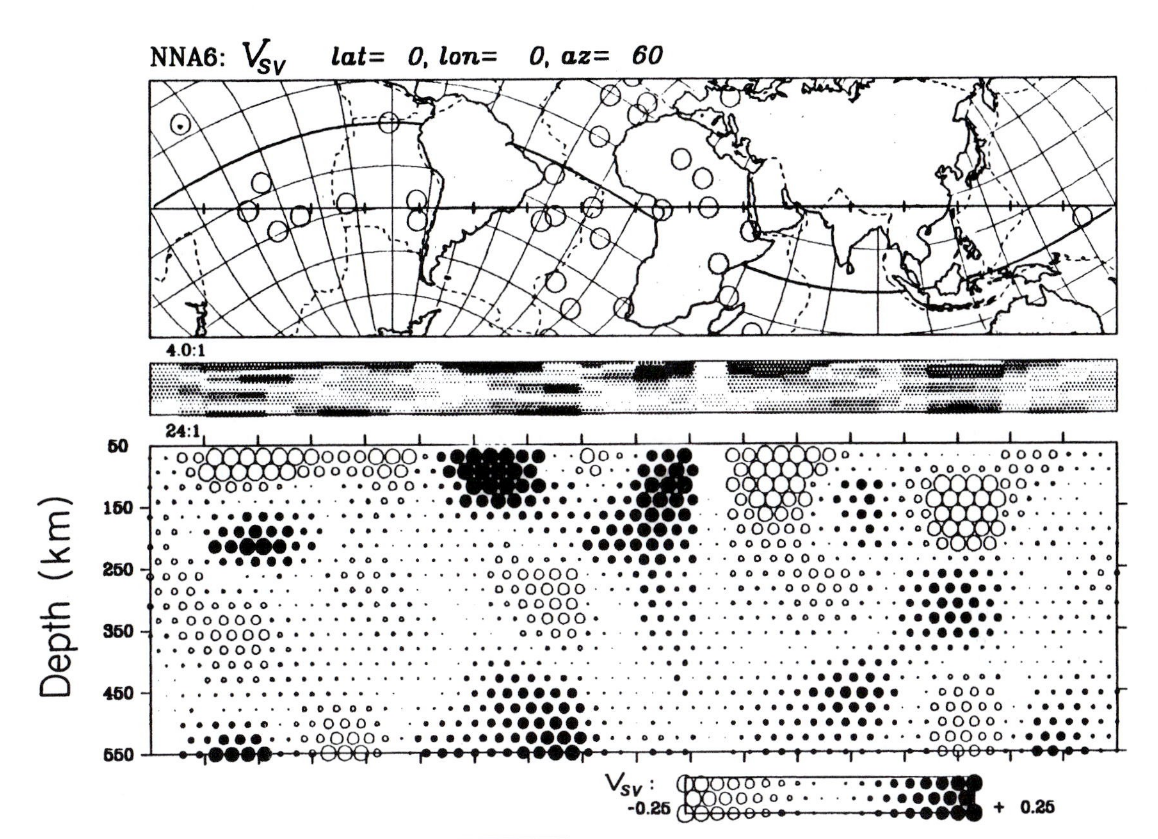

Fig. 11.6 Shear wave velocity in the upper mantle along the cross-section shown. Note low velocities at shallow depth under the western Pacific backarc basins, replaced by high velocities, presumably slabs, at greater depth. The eastern Pacific is slow at all depths. The Atlantic is fast below 400 km, possibly due to accumulated slabs. The fast regions above 200 km are cratonic roots.

Izu-Bonin, Mariana, New Hebrides and Philippine trenches in the transition zone.

Cratons

Quantitative correlations between cratons and tomography confirm the impressions one gets from visual interpretations (Polet and Anderson, 1995; Wen and Anderson, 1997; Becker and Boschi, 2002).

Hotspots

The relationship between hotspots and seismic tomography has been investigated extensively, with mixed results. Visually, some LVZ in the mantle appear to correlate with hotspots but these are not statistically significant. Excellent correlations at spherical harmonic degree 2 are found in the lower mantle and at degree 6 in the upper mantle; the latter is controlled, in part, by the distribution of cratons and trenches. Some have interpreted these results in terms of degree 2 convection in the lower mantle and degree 6 dominated convection in the upper mantle.

The overall distribution of hotspots correlates with very long wavelength low seismic velocities in the upper mantle and in the deep lower mantle (1700 km CMB) but hotspot locations are no better correlated with lower mantle tomography than are ridge locations. Global correlations are very poor in the mid-mantle, below 1000 km depth, and at shorter wavelengths and decrease very rapidly into the transition zone. Some hotspots correlate with fast velocities in the 400–500 km interval.

There are strong negative correlations between hotspot positions and the 0–180 Ma reconstructed slab locations; hotspots apparently do not originate in mantle that has been cooled or blocked by slab. Normal mantle, cooled by subduction will, of course, correlate with the subduction history. But this mantle will also correlate with hotspots, even if only 'normal' mantle is present. Some hotspots correlate with some tomographic maps at some depths [scoring hotspots].

Chemical stratification

The possible chemical stratification of mantle convection is an issue of current interest, which I will keep coming back to. It is hard to prove or disprove simply by looking at tomographic cross-sections. The endothermic nature of the phase change at 650 km depth may cause slab flattening and delay slab penetration but it cannot prevent large masses of cold material – of the same composition as the surrounding mantle – from episodically cascading across the boundary. The combination of a negative Clapyron slope at 650 km, a chemical barrier near 1000 km, and an increase of viscosity with depth may serve to confine the plate tectonic cycle and recycling to the outer 1000 km of the mantle.

Below some – not all – subduction zones, some – not all – seismic tomographic cross-sections contain bands of fast anomalies that extend into the mantle below 650 km. These visual anomalies are generally interpreted as slabs penetrating through the 650 km seismic discontinuity, and as evidence in support of 'whole-mantle convection flow' (Grand *et al.*, 1997; Bijwaard *et al.*, 1998). However, thermal coupling between two flow systems separated by an impermeable interface provides an alternative explanation of the tomographic results.

The dynamical interpretations of the geoid show that the model with an impermeable interface at 650 km or deeper can satisfy the data equally well as the whole-mantle model (Wen and Anderson, 1997) and some resistance to up- and downgoing flows can reduce the amplitudes of surface dynamic topography (Thoraval *et al.*, 1995). If there is an impermeable interface at 650 km, cold subducted lithosphere that has accumulated above the boundary at 660 km can cool the underlying material and initiate a downwelling instability into the lower mantle. In such a case, it would be difficult on the basis of seismic tomography alone to discriminate between 'thermal slabs' (with no material exchange between the upper and lower mantles) and slabs penetrating into the lower mantle. Thermal coupling could be an alternative explanation of the tomographic results.

Correlations with heat flow and geoid

Love-wave velocities are well correlated with heat flow and, therefore, with surface tectonics. Shields are areas of low heat flow and exhibit high Love-wave velocities, in spite of the thick low-velocity crust. From the Love-wave data one would predict that southeast Asia and the Afar region are characterized by high heat flow.

The geoid has a weak correlation with surface-wave velocities, consistent with a deep origin for the causative mass anomalies. The correlation between surface-wave velocities and the geoid is weak.

Regions of generally faster than average velocity occur in the western Pacific, the western part of the African plate, Australia–southern Indian Ocean, part of the south Atlantic, north-eastern North America–western North Atlantic and northern Europe. These are all geoid lows. Dense regions of the mantle that are in isostatic equilibrium generate geoid lows. High density and high velocity are both consistent with cold mantle. The above regions may be underlain by cold subducted material. Geoid highs occur near Tonga-Fiji, the Andes, Borneo, the Red Sea, Alaska, the northern Atlantic and the southern Indian Ocean. These are generally slow regions of the mantle and are therefore presumably hot. The upward deformation of boundaries counteracts the low density associated with the buoyant material, and for uniform viscosity the net result is a geoid high (Hager, 1983). An isostatically compensated column of low-density material also generates a geoid high because of the elevation of the surface.

Features of the geoid having wavelengths of about 4000 to 10 000 km are generated in the upper mantle. Geoid anomalies of this wavelength generally have an amplitude of about 20 to 30 m. An isostatically compensated density anomaly of 0.5% spread over the upper mantle would give geoid anomalies of this size. It

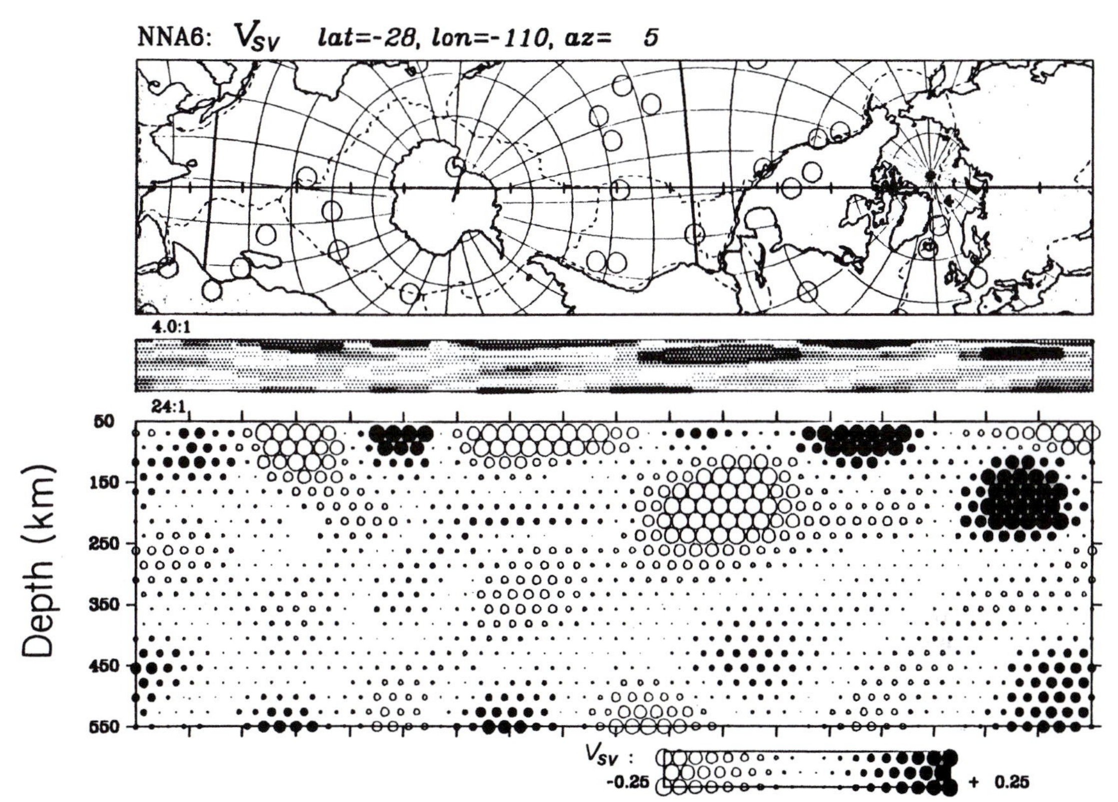

Fig. 11.7 Cross-section illustrating the low upper-mantle shear velocities under midocean ridges and western North America. Stable regions are relatively fast in the shallow mantle.

therefore appears that a combination of slabs and broad thermal anomalies in the upper mantle can explain the major features of the degree 4–9 geoid. The longer wavelength part of the geoid, degrees 2 and 3, correlates with seismic velocities in the deeper part of the mantle. Figure 11.8 shows the actual distribution of Love-wave phase velocities and that computed from the geoid assuming a linear relationship between velocity and geoid height. Most subduction regions are slow at short periods, presumably because of the presence of back-arc basins and hot, upwelling material above the slab.

Slabs are colder than normal mantle and therefore they can be denser. Intrinsically denser minerals also occur in the slab because of temperature-dependent phase changes. The phase-change effect leverages the role of temperature with the result that slabs confined to the upper mantle can explain the magnitude of the slab-related geoid. Dense delaminated lower continental crust may also contribute to density and velocity anomalies in the upper mantle.

Azimuthal anisotropy

The velocities of surface waves depend on position and on the direction of travel. If an adequately dense global coverage of surface-wave paths is available, then azimuthal anisotropy of the upper mantle, as well as lateral heterogeneity can be studied (Tanimoto and Anderson, 1984, 1985, Forsyth and Vyeda, 1975). Global azimuthal anisotropy may correlate with convective motions in the mantle; it is a further constraint on mantle geodynamics.

Azimuthal anisotropy can be caused by oriented crystals or a consistent fabric caused by, for example, dikes or convective rolls in the shallow mantle. In either case, the azimuthal variation of seismic-wave velocity is telling us something about convection in the mantle.

Figure 11.9 is a map of the azimuthal results for 200-s Rayleigh waves. The lines are oriented in the maximum velocity direction, and the length of the lines is proportional to the anisotropy. The

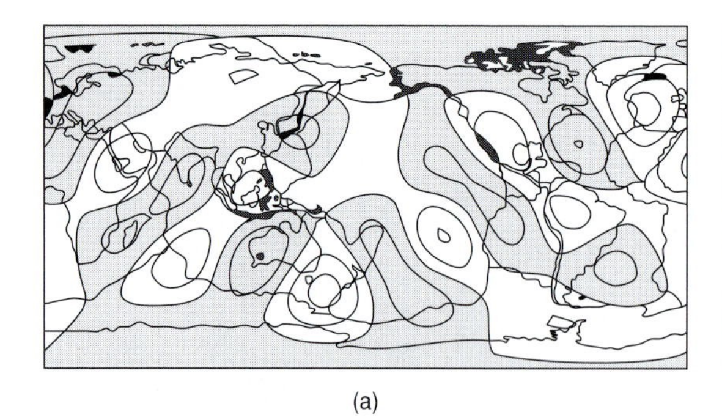

(a)

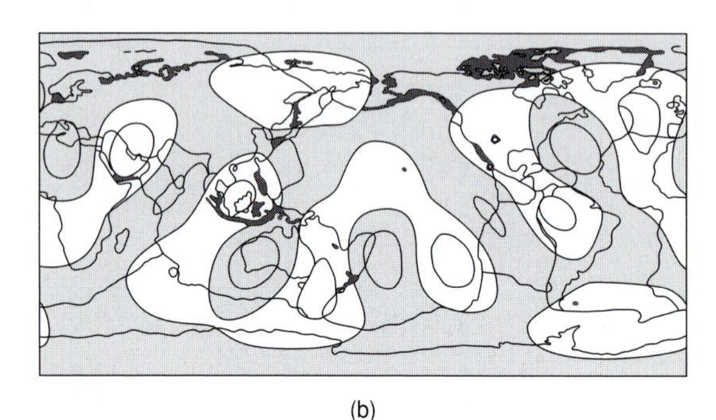

(b)

Fig. 11.8 The intermediate wavelength geoid is controlled by processes in the upper mantle (slabs, slow asthenosphere). (a) The $l = 6$ component of a global spherical harmonic expansion of Love-wave phase velocities. These are sensitive to shear velocity in the upper several hundred kilometers of the mantle. Note that most shields are fast (gray areas), and oceanic and tectonic regions are slow (white areas). Hot regions of the upper mantle, in general, cause geoid highs because of thermal expansion and uplift of the surface and internal boundaries. Tectonic and young oceanic areas are generally elevated over the surrounding terrain. (b) Phase velocity computed from the $l = 6$ geoid, assuming a linear relationship between geoid height and phase velocity. Note the agreement between these two measures of upper-mantle properties (Tanimoto and Anderson, 1985).

azimuthal variation is low under North America and the central Atlantic, between Borneo and Japan, and in East Antarctica. Maximum velocities are oriented approximately northeast-southwest under Australia, the eastern Indian Ocean, and northern South America and east-west under the central Indian Ocean; they vary under the Pacific Ocean from north-south in the southern central region to more northwest-southeast in the northwest portion. The fast direction is generally perpendicular to plate boundaries. There is little correlation with plate

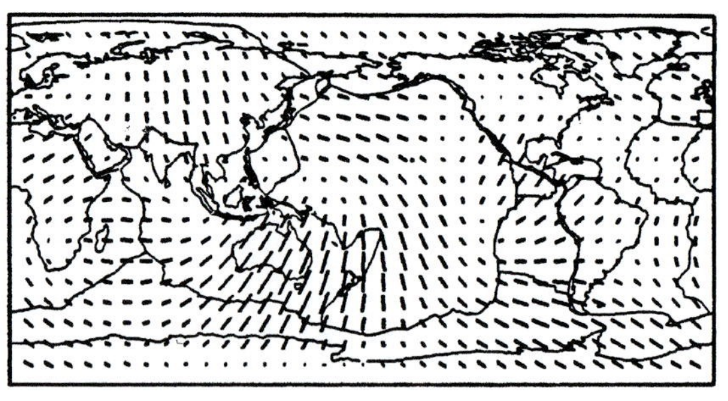

Fig. 11.9 Azimuthal variation of phase velocities of 200-s Rayleigh waves (expanded up to $l = m = 3$) (after Tanimoto and Anderson, 1984, 1985).

motion directions, and little is expected since 200-s Rayleigh waves are sampling the mantle beneath the lithosphere. Pn velocity correlates well with spreading direction.

Lateral heterogeneity from body waves

Because of the irregular distribution of seismic stations and seismic events, one cannot determine tomographic anomalies on a globally uniform basis from body waves. This limitation is fundamental and cannot be cured by improvements in ray theory, reference models or inversion techniques. Many body-wave tomographic maps and cross-sections simply reflect the ray coverage and regions where the reference model has been perturbed. If there is inadequate coverage, the reference model is unchanged but this does not mean it is right, or that contrasts with adjacent regions are correct. In contrast to surface-wave studies, however, the anomalies can be fairly well localized geographically although the depth extent is ambiguous. Average properties along rays or ray bundles can be determined accurately. Cross-sections oriented along the great-circle paths between many sources and receivers are the most useful. Tomographic cuts in random directions across body-wave models that do not include many sources and receivers can be misleading. In principle, the most reliable tomographic studies use both body waves and surface waves, and correct for frequency effects.

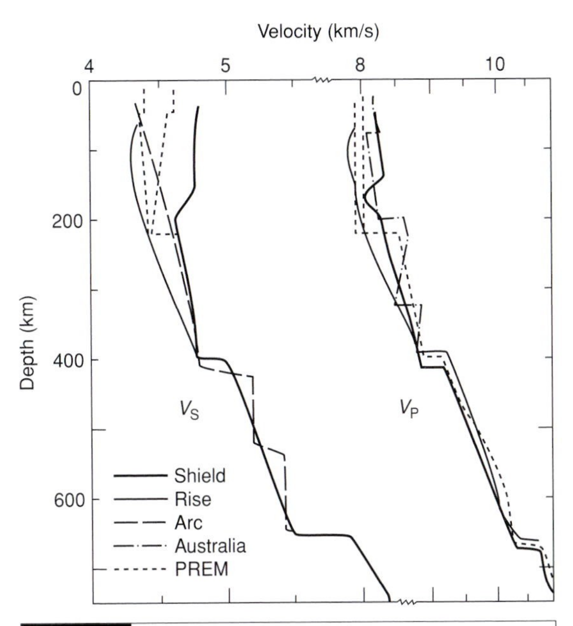

Fig. 11.10 V_s and V_p in various tectonic provinces. Note the large lateral variations above 200 km and the moderate variations between 200 and 400 km. The reversal in velocity between 150 and 200 lcrn under the shield area may indicate that this is the thickness of the stable shield plate. Models from Grand and Helmberger (1984a,b) and Walck (1984).

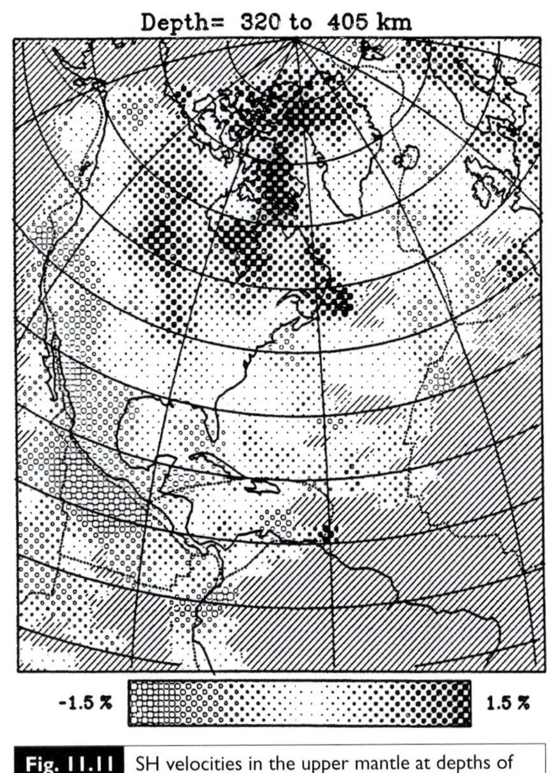

Fig. 11.11 SH velocities in the upper mantle at depths of 320 to 405 km (after Grand, 1986).

The highest resolution body-wave studies, involving the use of travel times, apparent velocities, amplitudes and waveform fitting, have provided details about upper-mantle velocity structures in several tectonic regions. Figure 11.10 shows some results. Note that low velocities extend to depths of about 390 km for the tectonic and oceanic structures. These regional studies confirm the general features of the global surface-wave studies. Recent results for the tectonic province of SW USA are available [`Ristra/ristra.html`].

Although the largest variations (of the order of 10%) in seismic velocity occur in the upper 200 km of the mantle, the velocities from 200 to about 400 km under oceanic and tectonic regions are slightly less (on the order of 4% on average) than under shields. The question then arises, what is the cause of these deeper velocity variations? Is the continental plate 400 km thick or are the velocities between 150–200 and 400 km beneath shields appropriate for 'normal' subsolidus mantle?

Body-wave tomography of the lower mantle

The large lateral variations of seismic velocity in the upper mantle make it difficult to detect the smaller variations in the lower mantle and deep small-scale structure. Long-wavelength velocity and density variations in the lower mantle are easier to detect from tomography and have more influence on the geoid and orientation of the Earth than comparable variations in the upper mantle. `Body-wave tomography of the lower mantle` has revealed features that are similar to the low-order components of the geoid (Hager and Clayton, 1989). The polar regions are fast, and the equatorial regions, in general, are slow. The slowest regions are

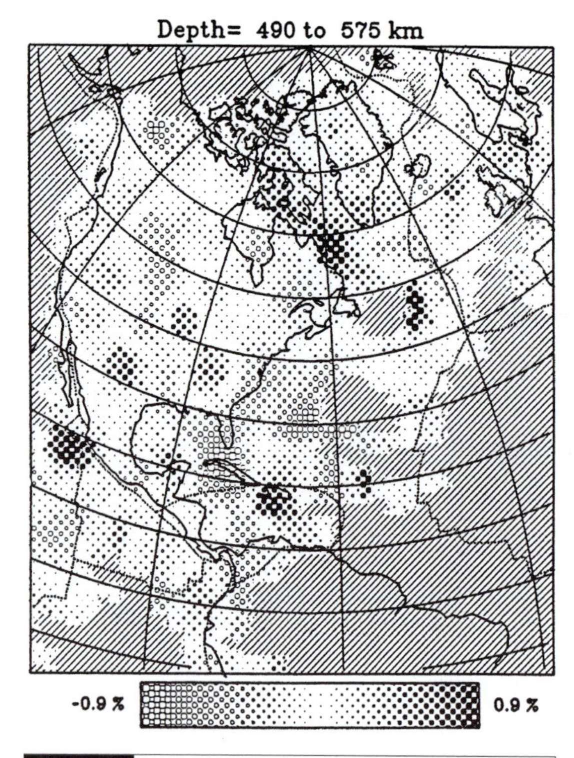

Fig. 11.12 SH velocities at depths of 490 to 575 km (after Grand, 1986).

centered near the long-wavelength geoid highs, which occupy the central Pacific and the North Atlantic through Africa to the southwestern Indian Ocean. The slow-shear-velocity regions were originally thought to represent hotter than average mantle but the bulk modulus and density are higher than average, ruling out a thermal explanation. The range of the long wavelength velocity variations is much less than in the shallow mantle but there are small ultra-low velocity regions near the core that have 10–20% velocity reductions.

Slow regions of the lower mantle occur under the mid Indian Ocean Ridge, the East Pacific Rise, western North America and South Africa. Fast regions occur under Australia, China, South America and northern Pacific. Generally, there is lack of radial continuity in the lower mantle. Large-scale convection-like features are not evident. The fastest regions are Siberia, south Africa, south of South America and the northeast Pacific. At mid-to-lower mantle depths (Figures 11.17 and 11.18), the slowest regions are southeast Africa and adjacent oceans, the Cape Verde–Canaries–Azores region of the Atlantic, and the equatorial Pacific. The fast regions are in the north polar regions, North America, eastern Indian Ocean and the South Atlantic. Most continental regions are fast. Near the base of the lower mantle, the prominent low-velocity regions are under Africa, Brazil and the south-central Pacific. The fast regions are Asia, the North Atlantic, the northern Pacific and Antarctica. The locations of hotspots do not correlate with the slower regions at the base of the mantle. [global seismic structure maps, scoring hotspots].

Superplumes?

The so-called Pacific and African lower mantle superplumes, identified visually in global tomographic cross-sections, apparently are dense (Ishii and Tromp, 2004) and must have a chemical origin; they are not thermally buoyant, as has often been proposed. In the lower part of the mantle, thermal buoyancy is weak compared to chemical buoyancy because thermal expansivity decreases with pressure. Purely thermal upwellings are expected to have low bulk modulus, low compressional velocity and low density. This is not the case (Ishii and Tromp, 2004, Trampert *et al.*, 2004) for the large lower-mantle features. The large-scale features have the appropriate dimensions to be thermal in nature but resemble more a chemically dense layer at high pressure, i.e., large-scale marginally-stable domes with large relief. These domes may have neutral density, or be slowly rising or sinking. In any case, they will affect the geoid, the dynamic surface topography, and the relief on other chemical boundaries. D″ would then be a very dense – probably iron-rich – layer, and the overlying 'layer', which is called D′, would be a less dense region trapped between D″ and the rest of the lower mantle. Stratification may have been established during accretional melting of the Earth by downward drainage of dense melts and residual refractory phases, and iron partitioning into phases that may include post-perovskite, low-spin iron-rich oxides and sulfides and intermetallic compounds. The large low-shear-velocity features are more appropriately called 'domes,' a geologically descriptive term, than 'megaplumes' or

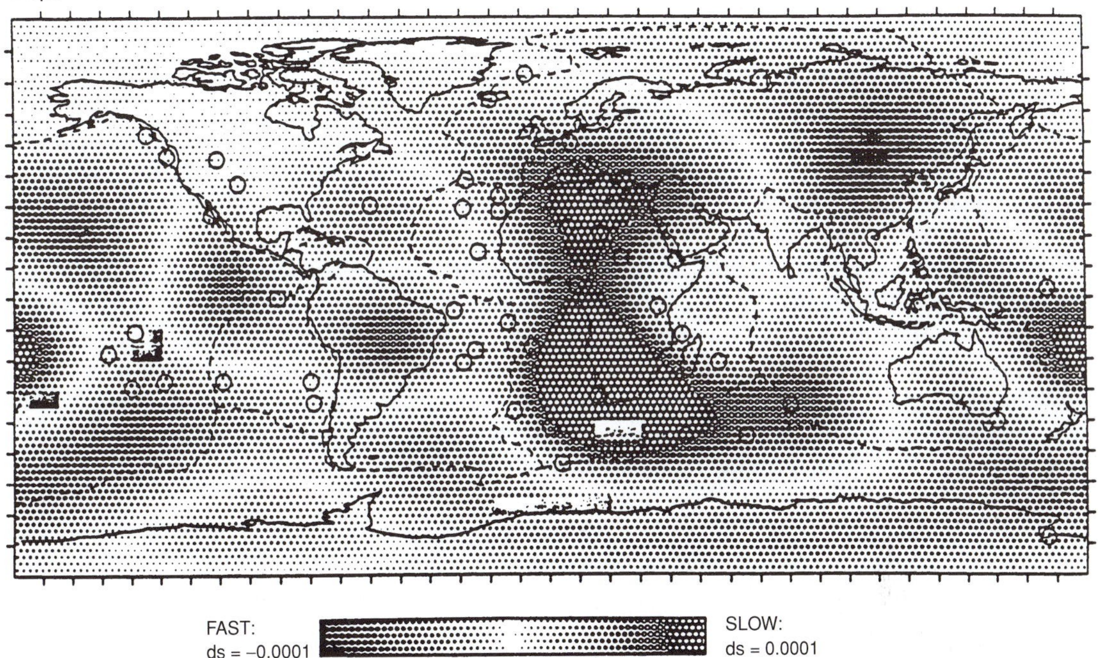

Fig. 11.13 Compressional velocity at 2500 km depth. Note the large low-velocity zone under Africa. This was called a superplume by later workers but it has high density and high bulk modulus; it is a compositional feature.

'superplumes,' which implies a thermal, active upwelling with low-density and low-bulk modulus. The existence of large lateral changes in chemistry and lithology makes suspect all attempts to infer composition and temperature from 1D mantle models such as PREM.

Seismic scattering

Because of resolution limitations global seismic models are smooth and heterogeneities of order 1 to 10 km are not imaged. High-frequency techniques, however, are available to map small-scale heterogeneities throughout the mantle. Early measurements of precursors to multi-reflected core phases showed that the upper 300 km of the mantle was complex, giving both coherent and incoherent reflections (Whitcomb and Anderson, 1970). Recent results confirm the multiple reflections in the mantle (Figure 9.3). Scattering is minor in the lower mantle compared with the upper mantle and the upper 200 km contain more or stronger scatterers than the transition region data (Baig and Dahlen, 2004; Shearer and Earle, 2004). Small scale structures (~8 km) in the upper 200 km appear to be an order of magnitude greater than elsewhere. The strongly scattering upper mantle is also strongly anisotropic, a possible manifestation of some organized heterogeneity, and high attenuation, also consistent with scattering, although other explanations for these are available.

Controversial evidence for strong seismic scattering in the lower mantle has been used to support the whole mantle convection model but newer and more powerful techniques allow a different conclusion. It appears that the upper mantle is the stronger scatterer of seismic energy and the lower mantle – below 1000 km depth – is rather bland except in D″. The same features that scatter energy may also be responsible for the pronounced anisotropy and anelasticity of the upper mantle.

Color images

Experienced geodynamicists are fond of saying that one mantle convection simulation is worse than none. This is true in all branches of non-linear mechanics and thermal convection. Complex systems are sensitive to changes in initial and boundary conditions and one must feel out a large part of parameter space before one understands the system. The same is true for one or two tomographic images. Although a picture may be worth a thousand words a single color tomographic cross-section can be worse than none. Although quantitative and statistical interpretations of tomographic models are required to fully capitalize on their information content, much can be learned by looking at hundreds of tomographic maps and cross-sections from different groups. Numerous color images of various tomographic studies can be found on www.mantleplumes.org/Penrose/BookChapterPDFs/RitsemaWebSuppl_Accepted.pdf and by searching for global seismic structure maps, mantleplumes, caltech tomography Ritsema ORM, and tomography maps mantle, and searching Google Images with the key words tomography or cross sections combined with global, lower mantle, upper mantle, transition zone, mantle, Iceland, Yellowstone, Hawaii, Harvard, Berkeley, Ritsema, Harvard, Grand and so on. At the time of writing there are also images on [geophysics.nmsu.edu/jni/research], [bullard.esc.cam.ac.uk/~maggi/Physics_Earth_Planet/Lecture_7/colour_figures.html] and www.gps. caltech.edu. See also the book, *Plates, Plumes and Paradigms*, Foulger *et al.* (2005).

Part IV

Sampling the Earth

Chapter 12

Statistics and other damned lies

The remark attributed to Disraeli would often apply with justice and force: 'There are three kinds of lies: lies, damned lies, and statistics'.

Mark Twain

['lies, damn lies – and statistics' (sic) – usually attributed to Twain (a lie) – or to Disraeli (a damn lie), as Twain took trouble to do; Lord Blake, Disraeli's biographer, thinks that this is most unlikely (statistics)].

Overview

Semantics, rhetoric, logic and assumptions play a large role in science but are usually relegated to specialized books and courses on philosophy and paradigms. Conventional wisdom is often the controlling factor in picking and solving problems. This chapter is a detour into issues that may be holding up efforts to develop a *theory of the Earth*, one that is as paradox free as possible.

A large part of petrology, geochemistry and geophysics is about sampling the Earth. `Sampling theory` is a branch of statistics. If the mantle is blobby on a kilometer scale, then individual rocks will exhibit large scatter – or variance – but the mean of a large number of samples will eventually settle down to the appropriate mean for the mantle; the same mean will be achieved in one fell swoop by a large volcano sampling a large volume of the same mantle. If the outputs of many large volcanoes are plotted on a histogram, the spread will be much smaller than from the rock samples, even though the same mantle is being sampled. Midocean ridges sample vast volumes of the mantle and mix together a variety of components. They are the world's largest blenders. Seamount and ocean island volcanoes sample much smaller volumes. This is a situation where the `central limit theorem` (CLT) applies.

The central limit theorem

The `central limit theorem and the law of large numbers` state that variably sized samples from a heterogenous population will yield the same mean but will have variances that decrease as n (the sample size) or V (the volume of the sampled region) increases. Small sample sizes are more likely to have extreme values than samples that blend components from a large volume [`CLT isotopes`]. One expects that melt inclusions in a lava sample will have greater isotopic diversity than that sampled in whole rocks.

An amazing thing about the central limit theorem is that no matter what the shape of the original distribution – bimodal, uniform, exponential, multiple peaks – the sampling distribution of the mean approaches a normal distribution. Furthermore, a normal distribution is usually approached very quickly as n increases;

n is the *sample size* for each mean and is not the number of samples. The *sample size* is the number of measurements that goes into the computation of each mean. In the case of sampling by a volcano of a heterogenous mantle, the volume of the sample space is equivalent to the number of discrete samples that are averaged; the volcano does the averaging in this case.

Suppose the mantle has a variety of discrete components, or blobs, with distinctive isotope ratios. The probability density function has, say, five peaks; call them DM, EM1, EM2, HIMU and Q. The CLT states that a sufficiently large sample, or average, from this population, will have a narrow Gaussian distribution. MORB has this property; it has nothing to do with the homogeneity of the mantle or convective stirring. MORB is best viewed as an *average*, not a reservoir or component.

If one has several large datasets there are statistical ways to see if they are drawn from the same population. If two datasets have different means and different standard deviations they may still be drawn from the same population. But if the datasets have been *filtered*, *trimmed* or *corrected*, then statistics cannot be applied. Often these corrections involve some hypothesis about what the data should look like. This is a very common error in seismology and geochemistry; data can be selected, filtered and discarded for various reasons but there are formal and statistically valid ways of doing this.

The problems of scale

Volcanoes sample and average large volumes of the mantle. It is not necessary that the source region correspond to a rock type familiar in hand-specimen size. The 'grains' of the source may be kilometres in extent, and the melt from one source region may rise and interact with melts from a different lithology, or smaller degree melts from the same lithology. The properties of individual crystals are greatly washed out in the average. The hypothetical rock types – pyrolite and piclogite – and hypothetical reservoirs – MORB, FOZO – may only exist as averages, rather than as distinct 10 cm by 10 cm 'rocks' that might be found in a rock collection. Magmas average over a large volume of the mantle, just as river sediments can average over a large area of a continent. Similarly, seismic waves average over tens to hundreds of km and see many different kinds of lithologies; they do not see the extremes in properties of the material that they average.

Anisotropy of the mantle may be due to anisotropy of individual mantle minerals or due to large-scale organized heterogeneity. A 1-second P-wave has a wavelength of about 10 km and therefore does not usually care whether it is microphysics or macrophysics that gives the fabric. Similar considerations apply to seismic anelasticity. Geophysics is as much a matter of composites, and averaging, as it is a science of crystallography and mineral physics. The problem of averaging occurs throughout the sciences of petrology, geochemistry and geophysics. Much of deep Earth science is about unravelling averages. In seismology this is called *tomography*.

The scale becomes important if it becomes big enough so that gravity takes over from diffusion. A large blob behaves differently, over time, than a small impurity, because of buoyancy. If the heterogeneities are large, a *fertile blob* can be confused with a *hot plume*.

Scrabble statistics

The central limit theorem can be illustrated with the game of Scrabble, which has a well-defined distribution function of letters or scores (Figure 12.1). Put all the letter tiles into a can and draw them out repeatedly. Plot the number of times each letter is pulled out on a cumulative histogram. The histogram is very ragged; this is what the population – the world of letters – looks like. Now play a game with only two-letter combinations – words – allowed, and plot the average scores. Continue this process with three- and four-letter words (but do not filter out the offensive words or words that do not make sense). The histograms get smoother and smoother and narrower and narrower. By the time one gets to three-letter words one already has nearly a Gaussian distribution, with very few average scores of 4 or 5, or 20 and higher. If this

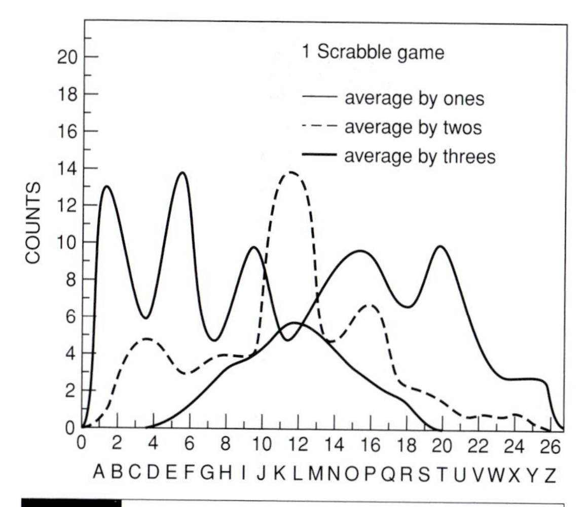

Fig. 12.1 The distribution of letters, and their assigned numerical values, in a Scrabble game. Also shown are average values for two and three-letter combinations, illustrating the smoothing effect of the CLT. One might conclude that the various combinations represent different bags, or reservoirs, of letters.

were a distribution of normalized helium isotopic ratios one might conclude that the population that was being sampled by four-letter words did not contain the extreme values that one- and two-letter words were sampling. The lower curve in Figure 12.1 approximates the helium isotope histogram for MORB; the other curves look more like OIB statistics.

SUMA – sampling upon melting and averaging

In dealing with the mantle, one is dealing with a heterogenous system – heterogenous in properties and processes. In geochemical box models the mantle is viewed as a collection of large, discrete and isolated *reservoirs*, with definite locations and compositions. In petrological models there are distinct *components* that can be intimately mixed. The *sampling process* usually envisaged is similar to using a dipper in a bucket of water. In models based on *sampling theory* the various products of mantle differentiation are viewed as averages, different kind of samples from a heterogenous population; a homogenous product does not imply a homogenous or well-mixed source. In the real world, sampling at volcanoes involves melting, recrystallization and incomplete extraction, as well as statistics.

Fast spreading oceanic ridges process large volumes of the mantle and involve large degrees of melting. A consequence of the central limit theorem is that the variance of samples drawn from a heterogenous population (reservoir) depends inversely on the sampled volume (Anderson, 2000b; Meibom & Anderson, 2003). The homogeneity of a sample population (e.g. all MORB samples) can thus simply reflect the integration effect of large volume sampling. The presumed homogeneity of the MORB source may thus be an illusion.

The 'MORB reservoir' is thought to be homogenous because some isotopic ratios show less scatter in MORB than in ocean-island basalts (OIB). The common explanation is that MORB are derived from a well-stirred, convecting part of the mantle while OIB are derived from a different, deeper reservoir. Alternatively, the homogeneity of MORB can be explained as a consequence of the sampling process. The standard, two-reservoir model of geochemistry is reinforced by questionable – from a statistical point of view – data filtering practices. Samples that are judged to be contaminated by plumes (i.e. OIB-like samples) are often removed from the dataset prior to statistical analysis. Sometimes the definition of plume influence is arbitrary. For example, isotopic ratios that exceed an arbitrary cutoff may be eliminated from the dataset. In this way, the MORB dataset is forced to appear more homogenous than it really is. This method is commonly applied to $^3He/^4He$. Despite this, various ridges still have different means and variances, and these depend on spreading rate and ridge maturity.

The *upper mantle, sub-continental lithosphere*, and *lower mantle* are usually treated as distinct and accessible *geochemical reservoirs*. There is evidence, however, for ubiquitous small- to moderate-scale heterogeneity in the upper mantle, referred to as the statistical upper mantle assemblage (SUMA). This heterogeneity is the result of processes such as inefficient melt extraction and long-term plate tectonic recycling of sedimentary and crustal *components*. The SUMA concept derives from the CLT and contrasts with the idea of a convectively homogenized MORB mantle reservoir, and different reservoirs for OIB where homogenization of the source is achieved by convective stirring and mixing. In contrast, in the SUMA model, the isotopic compositions of MORB and OIB are the outcome of homogenization during sampling, by partial melting and magma mixing. The primary homogenization process *sampling upon melting and averaging*, SUMA, does not require the participation of distinct (e.g. lower mantle) reservoirs to explain OIB compositions. Statistical distributions of lithologic components and sampling theory replace the concept of distinct, isolated geochemical reservoirs, and extensive solid-state convective stirring prior to sampling. In sampling theory terms, SUMA is the heterogenous population to be sampled – the probability density function. MORB represents a large scale sample, or average, from this population; near-ridge seamounts are a smaller-scale sample; grain boundaries, fluid-inclusions and melt-inclusions are very small-scale samples. MORB is uniform and does not have the extremes of composition because it is large-scale average of the sampled mantle. Mathematical and statistical treatments of isotopic heterogeneity of basalts and upper mantle assemblages are starting to replace the static reservoir and convective stirring concepts.

Bayesian statistics

The use of prior probabilities and subjective constraints external to the dataset is known as bayesian statistics [Harold Jeffreys bayesian statistics]. Bayesian reasoning is common in geophysical inversion problems and in the statistical treatments of isotope data. For example, it is often assumed that there are two populations in isotopic data – the MORB reservoir, corresponding to the 'convecting degassed upper mantle', and the OIB reservoir, assumed to be an isolated, more primitive, less-degassed, more variable reservoir in the lower mantle. Data are corrected or filtered to remove the influence of contamination or pollution by materials from the 'wrong' reservoir, and then statistics is applied. In seismology, a *prior* or *reference* model is adopted and perturbations are made to this to satisfy new datasets.

Which kind of statistical approach is preferable in these situations? Bayesian methods have a long and controversial history. Bayesian reasoning has emerged from an intuitive to a formal level in many fields of science. *Subjective probability* was developed to quantify the plausibility of events under circumstances of uncertainty. Bayes' theorem is a natural way of reasoning, e.g. www.ipac.caltech.edu/level5/March01/Dagostini/Dagostini2.html.

There is an apparent contradiction between rigorous normal statistics and the intuition of geologists. Geologists' intuition resembles the less familiar bayesian approach; conclusions should not deviate too much from prior beliefs. On the other hand, a common objection to bayesian statistics is that science must be objective – there is no room for belief or prejudice. Science is not a matter of religion. But scientific beliefs and assumptions are always there, but often well hidden.

There are good reasons for applying bayesian methods to geological problems and geochemical datasets. First of all, geochemical data are often ratios, they cannot be negative. Conclusions should not depend on whether a ratio or its inverse is analysed. In the absence of information to the contrary it can be assumed that all values of 0 to infinity are equally probable in the underlying distribution (i.e. the magma source – or mantle – prior to sampling and averaging). Ruling out negative values already is a prior constraint. Sampling of a heterogenous source, according to the central limit theorem (CLT) will yield a peaked distribution that, in the limit of a large sample volume, is *normal* or *log-normal*. Many geochemical samples can be considered to have sampled fairly large volumes. These considerations are more critical for $^3He/^4He$ than for heavier isotopes since the spread of values about the mean is larger, and median values are not far from 0. Similarly, histograms of seismic wave travel-time residuals, which can be negative, or heat flow values, depend on the distance the rays travel or the number of samples averaged used to characterize a heat flow province. Large volcanoes and global tomography average large volumes of the mantle.

Distributions are commonly asymmetric and skewed. Medians of isotope data are more robust measures than the arithmetic means, with which they commonly disagree. Log-normal distributions are more appropriate than linear Gaussian distributions for many geophysical and geochemical datasets. These are relatively mild applications of bayesian reasoning. Stronger bayesian priors would involve placing a low prior probability on certain ranges of values of the parameter being estimated, or on external parameters.

Basaltic volcanism is by nature an integrator of the underlying source. All volcanoes average, to a greater or lesser extent, the underlying heterogeneities. To determine the true heterogeneity of the mantle, samples from a large variety of environments are required, including fast and slow spreading ridges, small off-axis seamounts, fracture zones, new and dying ridges, various ridge depths, overlapping spreading centers, melt-starved regions, unstable ridge systems such as back-arc ridges, and so on. These regions are often avoided, as being *anomalous*. Various materials enter subduction zones, including sediments, altered oceanic crust and peridotite, and some of these are incorporated into the upper mantle. To the extent that anomalous materials are excluded, or anomalous regions left unsampled, the degree of true intrinsic heterogeneity of the mantle will be unknown. In essence, one must sample widely, collecting specimens that represent different degrees of melting and different source volumes. The data can be weighted or given low probability in the bayesian approach.

The main distinguishing feature of the bayesian approach is that it makes use of more information than the standard statistical approach. Whereas the latter is based on analysis of 'hard data', i.e. data derived from a well-defined observation process, bayesian statistics accommodates 'prior information' which is usually less well specified and may even be subjective. This makes bayesian methods potentially more powerful, but also imposes the requirement for extra care in their use. In particular, we are no longer approaching an analysis in an 'open-minded' manner, allowing the data to determine the result. Instead, we input 'prior information' about what we think the answer is before we analyse the data! The danger of subjective bayesian priors, if improperly applied, is that prior beliefs become immune to data.

Fallacies

Logic, argument, rhetoric and fallacies are branches of philosophy; science started out as a branch of philosophy. *Scientific truth* is now treated differently from *logical truth*, or *mathematical truth*, and statistics. But in the search for scientific truth it is important not to make logical errors, and not to make arguments based on logical fallacies. The rules of logical inference form the foundation of good science.

A fallacy is an error in reasoning and differs from a factual or statistical error. A fallacy is an 'argument' in which the premises given for the conclusion do not provide the necessary support. Many of the paradoxes in Earth science, as discussed in the following chapters, are the result of poor assumptions or erroneous reasoning (`logical paradoxes`) [`mantleplumes fallacies`].

The following chapters discuss some of the controversies and paradoxes in Earth science. Some of these can be traced to assumptions and fallacies. The following are a few examples of logical fallacies;

Midocean-ridge basalts are homogenous; therefore their mantle source is homogenous; therefore it is vigorously convecting, therefore it is well stirred. (There are four logical fallacies in that sentence, and at least one factual error.)

Midocean-ridge basalts are derived from the upper mantle; ocean-island basalts are not midocean-ridge basalts; they therefore are derived from the lower mantle.

Seismic velocities and density decrease with temperature; therefore regions of the mantle that have low seismic velocities are hotter and less dense than other regions.

Some other well known fallacies are categorized below, with examples. The examples are picked from an ongoing debate about the source of oceanic magmas and `large igneous provinces`.

Circulus in demonstrando and affirming the consequent

Absolute plate motions agree well with the fixed hotspot hypothesis; **therefore** . . .

Argumentum ad populum

For many geoscientists, the mantle-plume model is as well established as plate tectonics.

False dilemma and bifurcation

A limited number of options (usually two) is given, while in reality there are more options (or perhaps only one). A false dilemma is an illegitimate use of the 'or' operator. Putting issues or opinions into 'black or white' terms is a common instance of this fallacy.

Melting anomalies (oceanic islands, hotspots) are due to either high absolute temperatures or high degrees of melting

Plumes either come from the core–mantle boundary or the transition region.

Red herring and fallacy of irrelevant conclusion

The upwelling mantle under Hawaii must also be 200 to 300 K hotter than the surrounding mantle to achieve the required large melt fractions . . . such hot rock must come from a thermal boundary layer . . . the core–mantle boundary is the most likely source, unless there is another interface within the mantle between compositionally distinct layers.

These are requirements of the plume hypothesis, not general requirements. Large melt volumes have a variety of explanations and thermal boundary layers are not essential for creating upwellings.

The ratio fallacy and the slippery slope fallacy

The chemistry and isotopic composition of hotspot lavas indicate that the hotspots sample a part of the mantle distinct from that sampled by midocean-ridge basalts. High ratios imply high ^{3}He contents, therefore an ancient undegassed reservoir, therefore the deep mantle.

Fallacy of irrelevant conclusion

Numerical simulations of plumes reproduce many of the geophysical observations, such as the rate of magma production and the topography and gravity anomalies produced by plume material as it spreads beneath the lithosphere. Therefore, . . .

Fallacy of irrelevant conclusion, affirming the consequent and permissivity

Theoretical and laboratory studies of fluids predict that plumes should form in the deep Earth because the core is much hotter than the mantle. Therefore hotspots are caused by plumes from the core–mantle boundary.

Confusion of 'should' with 'do' or 'must'.

Ignoratio elenchi and circulus in demonstrando

Continental flood basalts (CFB) erupt a million cubic kilometers of basalt or more in 1 million years or less. Therefore plumes erupt a million cubic kilometers of basalt or more in 1 million years or less.

This circular argument is often used as a basis for supporting a plume model for CFB.

Some invalid deductive arguments are so common that logicians have tabulated them, and given some of them names; the following are just a few:

If P then Q
Q
Therefore, P

If P then Q
Not-P
Therefore, Not-Q

If P then Q
Therefore, if Not-P then Not-Q

The following statements can all be written in these forms:

We have a midocean-ridge basalt (MORB). It is from the upper mantle. We have a sample from the upper mantle. It is therefore MORB.

We have a sample that is not MORB. It therefore cannot come from the upper mantle.

These are all variants of the *modus moron fallacy*.

Hypothesis testing

In discussing the early history of quantum mechanics, Werner Heisenberg noted how difficult it is when we try to push new ideas into an old system of concepts belonging to an earlier philosophy and attempt to put new wine into old bottles. *Paradigms* differ from *theories* and *hypotheses* in that they represent a culture, a way of viewing the world. They come with sets of assumptions and dogmas that are shared by the community, but which are generally not stated. The `standard models of geodynamics and mantle geochemistry` combine the independent paradigms of *rigid plate tectonics, the convective mantle* and accessible but isolated *reservoirs*. These, in turn, spawned the *plume paradigm*. The assumptions hidden in '*the convective mantle*' are that the upper mantle is homogenous, it is vigorously stirred by convection, and that MORB is homogenous and requires a homogenous source.

Seismic tomography uses inverse methods to convert seismic data to models of the interior. Because the sampling is so sparse and the data represent large-scale averages, and because the local properties of the interior are unknown, tomography is best used as a *hypothesis tester* rather than a method for determining definitive and unique structures. `Probabilistic tomography maps` are one way to display the results. The parameters are all given *prior probabilities*, as in bayesian statistics. Tomography is a form of large-scale averaging. Magmatism is another.

The technique of testing a hypothesis by assuming the opposite, is called *reductio ad absurdum*. A simpler more powerful theory often emerges when we drop the adjectives – such as *rigid, accessible, homogenous* – and reverse the assumptions. This technique is useful if an alternative to a popular reigning paradigm is not yet fully developed. One can at least test whether a standard model is more successful than its opposite, and whether paradoxes are the results of assumptions.

Language: a guide to the literature

We think only through the medium of words – The art of reasoning is nothing more than a language well arranged. Thus, while I thought myself employed only in forming a Nomenclature, and while I proposed to myself nothing more than to improve the chemical language, my work transformed itself by degrees, without my being able to prevent it, into a treatise upon the Elements of Chemistry.

. . . we cannot improve the language of any science without at the same time improving the science itself; neither can we, on the other hand, improve a science, without improving the language or nomenclature which belongs to it. However certain the facts of any science may be, and, however just the ideas we may have formed of these facts, we can only communicate false impressions to others, while we want words by which these may be properly expressed.

Antoine Lavoisier

The study of the Earth's interior involves a large number of parameters, some much more important than others, and a specialized language. Once we leave the familiar parameter range of the surface of the Earth and of the laboratory, our intuition often fails. Temperature is one of the key parameters of the Earth's interior but it is often over-used and abused; it is not an explanation of everything. But what could be more natural than to assume that volcanic islands are the result of hotter than average mantle? Or to assume that low seismic velocities in the mantle are caused by high mantle temperatures? The words *hotspots* and *upwelling thermal plumes* have been applied to these features. But the mantle is heterogenous; it is variable in composition and melting point and is not simply a homogenous ideal fluid with variable temperature. Pressure increases with depth and this changes the role and importance of temperature.

The study of the Earth's interior also involves a number of assumptions, often unstated. The unwary reader can be misled by semantics and assumptions and the following is a guide to both the parameters and assumptions that are used in the following chapters.

Temperature

A rise in temperature generally decreases the density and the seismic velocity and it is usually assumed that temperature increases with depth. It does not follow that seismic velocity is a proxy for temperature and density. Rates of magmatism and crustal thicknesses are also not proxies for temperature.

The largest temperature variations in the mantle are across conduction boundary layers, plates and slabs; the high-temperature gradient allows the internal heat to be conducted away or the surrounding heat to be conducted in. The temperature rise across the upper thermal boundary layer (TBL) is about 1400 °C. Geophysical and petrological data are consistent with a variation of temperature of about 200 °C at any given depth in the mantle, averaged over several hundred km.

Petrologically inferred temperature variations in the upper mantle are much lower than geophysically plausible estimates for the temperature change across the core–mantle boundary – CMB. This implies that deep mantle material does not get into the upper mantle or contribute to surface volcanism.

There are trade-offs between composition and temperature. Equally satisfactory fits to seismic data can be obtained using a range of plausible compositions and temperatures. If the mantle is homogenous in composition and if the lower mantle composition is the same as estimates of upper mantle peridotites, then the lower mantle is cold and iron-poor. On the other hand, if the lower mantle has chondritic Mg/Si ratios then the same data require a hot and iron-rich lower mantle. In most cases, the inferred geotherms are subadiabatic.

In the calculation of seismic velocity, there are several temperature effects to be considered; intrinsic, extrinsic, anharmonic, anelastic, quantum and Arrhenius or exponential. In some of these, the associated density effect is small.

Potential temperature

If a parcel of material at depth is brought to the surface on an adiabatic path (rapid upwelling) it defines the potential temperature of the mantle at that point. Potential temperatures of the mantle vary by about 200 °C. Such variations in temperature are also required to cause the mantle to convect and are the result of plate tectonic processes such as subduction cooling and continental insulation.

Homologous temperature

The homologous temperature is the absolute temperature divided by 'the melting temperature.' Many physical properties depend on this scaled temperature rather than on the absolute temperature. The homologous temperature varies from place to place in the mantle and this may be confused with variations of absolute temperature. This confusion has led to the hotspot hypothesis. There are other parameters that also control the locations and rates of excess magmatism at places called melting anomalies or hotspots.

Melting temperature

Rocks begin to melt at the solidus and are completely molten at the liquidus. The addition of water and CO_2 and other 'impurities' such as potassium decrease the solidus temperature. The materials in the mantle have quite different melting temperatures; eclogite, for example, melts at temperatures about 200 K below the melting point of peridotite. Partially molten eclogite can be denser and colder than unmelted peridotite. Variations of seismic velocity in the upper mantle can be due to variations in lithology and melting temperature, in addition to variations in absolute temperature. A cold dense sinking eclogite blob – perhaps a bit of delaminated lower continental crust – can have low seismic velocities and high homologous temperature.

Adiabatic gradient

The temperature gradient in the mantle is fundamental in prescribing properties, such as the composition, density, seismic velocity, viscosity, thermal conductivity and electrical conductivity. It is also essential in discussions of whole mantle vs. layered or stratified mantle convection. An adiabatic temperature gradient is commonly assumed. The adiabatic gradient is achieved by a homogenous self-compressed solid. A fluid heated from below will not convect until the radial temperature gradient exceeds the adiabatic gradient by a given amount. The adiabat does not define a unique temperature; it defines a gradient. The average temperature variation with depth is called the geotherm. Its basic form is almost always assumed to consist of adiabatic regions where temperatures rises slightly with depth (approximately 0.3 to 0.5 °C/km), and of narrow thermal boundary layers (TBL) where temperatures increase rapidly (approximately 5 to 10 °C/km) over a depth of a few hundred kilometers. If the mantle were entirely heated from below, was not experiencing secular cooling and if the mantle were a fluid with constant properties, the horizontally averaged temperature gradient would be adiabatic. Material brought up rapidly from depth without cooling by conduction or heating by radioactivity will follow an adiabatic path. The mantle geotherm is not, in general, along the adiabat; departures from an adiabat are due to secular cooling, internal radioactive heat production and the cooling from below by material that sank from the upper cold TBL. As a volume element rises through the mantle its temperature decreases in response to adiabatic decompression but at the same time its temperature increases due to internal heat released from radioactive decay. The actual temperature gradient in the mantle can be slightly subadiabatic. The mantle geotherm away from thermal boundary layers may depart by as much as 500 K from the adiabat.

Pressure

Pressure is not an intuitive parameter, at least in the pressure range of the deep mantle. For geophysical purposes it is the compression, or the volume, that controls physical properties. The *specific volume*, or *inverse density*, of the mantle decreases by about 30% from the top to the bottom and most physical properties are strong non-linear functions of volume. This is ignored in most simulations of mantle convection and the insight gained from these simulations is therefore restricted. Laboratory simulations of mantle convection cannot model this pressure effect. Since mantle dynamics is an example of a *far-from-equilibrium system* that is sensitive to small perturbations in initial or boundary conditions, or changes in parameters, it is difficult, or impossible, even under the best of conditions, to generalize from a small number of simulations or experiments.

Volume

Equations of state can be cast into forms that involve P and T, or V and T. Many physical properties depend on P and T only in-so-far as the specific volume is changed. The intrinsic effect of T is often a small perturbation. This is not true for radiative conductivity, rigidity and viscosity. Various laws of corresponding states, Debye theory, and the quasi-harmonic approximation are cast in terms of inter-atomic distances or volume. Usually, volume refers to specific volume. In discussions of sampling theory, volume will refer to the size of the sampled domain.

Composition

Density, seismic velocity, melting temperature, fertility and so on are affected by changes in composition or lithology. The mantle is a multi-component system. Yet it is often assumed that the mantle is homogenous and that variations in seismic velocity and magmatism are controlled only by variations in temperature.

Phase changes

Phase changes are responsible for most of the seismic velocity and density increases at 410- and 650-km depth. There are other seismic reflectors and scatterers in the mantle and some of these may be due to chemical or lithologic changes. Chemical boundaries are generally harder to detect than phase boundaries because they are more variable in depth and reflectivity or impedance, and their existence is not controlled by P and T. The absence of features similar to 410- and 650-km discontinuities does not imply that there are no chemical variations in the mantle or that the mantle is not chemically stratified.

Ratios

Geophysics, fluid dynamics and geochemistry abound in dimensionless ratios such as the V_P/V_S ratio, Poisson's ratio, Rayleigh number and isotopic ratios such as $^3He/^4He$. Ratios are handy but they do not constrain absolute values of the numerator and denominator. For example, high $^3He/^4He$ ratios do not imply high ^3He-contents, or an undegassed reservoir. They also cannot be treated the same as other quantities in mixing calculations and in statistics. Means of ratios can be misleading. The robust measures of central tendency for ratios are the *geometric mean* and the *median*, these are invariant to inversion of the ratio. *Arithmetic means* of ratios are widely used, even in this book, but they can be misleading.

Convection vs. stirring

Stirring is one way to homogenize a fluid. Gravity and centrifuging are good ways to de-homogenize a fluid. Thermal convection is not necessarily a good homogenizer. The mantle is stirred by plate motions and subduction but these also are not efficient homogenizers. Homogeneity of basalts is more likely to be a result of the central limit theorem; the averaging out of heterogeneity by the sampling process. Large-scale averages of the properties of the upper mantle tend toward Gaussian distributions with small variance. Smaller-scale averages tend to have more outliers and these have given rise to ideas about different reservoirs. The *convecting mantle* is geochemical jargon for the *upper mantle*, the *MORB-reservoir* or for the *source* of homogenous basalts. A homogenous product does not imply a homogenous source.

Plume

This is a catch-all word in mantle studies, although it has a precise fluid dynamic meaning. In tomographic studies it means *red*, a region that has relatively *slow seismic velocities*. In isotope geochemistry it means any composition other than MORB, as in *the plume component*. In marine geophysics it is any *anomalously shallow region*. In geology it often means any region except ridges and arcs with extensive magmatism. In geophysics it means a hot buoyant upwelling from a deep unstable thermal boundary layer, heated from below. [`do plumes exist`]

Chapter 13

Making an Earth

Thus God knows the world, because He conceived it in His mind, as if from the outside, before it was created, and we do not know its rule, because we live inside it, having found it already made.

Brother William of Baskerville

Overview

Attempts to estimate mantle composition fall into two broad categories.

Cosmochemical approaches take meteorites as the basic building blocks; these materials are processed or mixed in order to satisfy such constraints as core size, heat flow and crustal ratios of certain elements. These models constrain the bulk chemistry of the Earth rather than that of the mantle alone. The only primitive meteorites that match satisfactorily both the stable isotope and redox characteristics of the Earth are the enstatite chondrites. The actual material forming the Earth is unlikely to be represented by a single meteorite class. It may be a mixture of various kinds of meteorites and the composition may have changed with time. The oxidation state of accreting material may also have changed with time.

Petrological models take the present continental crust, basalts and peridotites as the basic building blocks. Since basalts represent melts, and peridotites are thought to be residues, some mixture of these should approximate the composition of the upper mantle. Peridotites are the main mantle reservoirs for elements such as magnesium, chromium, cobalt, nickel, osmium and iridium. The continental crust is an important reservoir of the crustal elements; potassium, rubidium, barium, lanthanum, uranium and thorium. Enriched magmas, such as kimberlites, are rare but they are so enriched in the crustal elements that they cannot be ignored. Thus, each of these components plays an essential role in determining the overall chemistry of the primitive mantle. An alternate approach is to search for the most 'primitive' ultramafic rock or component and attribute its composition to the whole mantle or the original mantle, or to take the most depleted MORB and attribute its source region to the whole upper mantle.

The earliest, and simplest, petrological models tended to view the mantle or the upper mantle as homogenous, and capable of providing primitive basalts by partial melting. When the petrological data are combined with isotopic and geophysical data, and with considerations from accretional calculations, a more complex multi-stage evolution is required. Similarly, simple evolutionary models have been constructed from isotope data alone that conflict with the broader database. There is no conflict between cosmochemical, geochemical, petrological and geophysical data, but a large conflict between models that have been constructed by specialists in these separate fields. The mantle is heterogenous on all scales and there are vast regions of the Earth that are unsampled today because they were isolated by gravitational stratification during accretion of the planet.

The composition of the crust and the upper mantle are the results of a series of melting and fractionation events, including the high-temperature accretion of the planet. Attempts to estimate upper-mantle chemistry usually start from the assumption that it initially was the same as `bulk silicate Earth` (BSE) and differs from it only by the extraction of the crust, or that the most depleted midocean-ridge basalts (MORB) plus their refractory residues constitute the entire upper mantle. Traditionally, geochemists have assumed that the lower mantle is still undifferentiated BSE. On the other hand, large-scale melting and differentiation upon accretion probably pre-enriched the upper mantle with incompatible elements, including the radioactive elements; the crust and the various enriched and depleted components sampled by current melting events were probably already in the upper mantle shortly after accretion and solidification.

Midocean-ridge basalts represent large degrees of melting of a large source volume, and blending of magmas having different melting histories. The central limit theorem explains many of the differences between MORB and other kinds of melts that sample smaller volumes of the heterogenous mantle. Observed isotopic arrays and mixing curves of basalts, including ocean-island basalts (OIB), can be generated by various stages of melting, mixing, melt extraction, depletion and enrichment and do not require the involvement of unfractionated, primitive or lower, mantle components. However, the first stage in building an Earth – the accretional stage – does involve large degrees of melting that essentially imparted an unfractionated – but enriched – chondritic REE pattern to the upper mantle. Small-degree melts from this then serve to fractionate LIL. Mass-balance and box-model calculations can go just so far in constraining the chemistry of the mantle. Geophysics provides information about physical properties and boundaries in the mantle.

Recycling of crust into the upper mantle is an important process. It is possible to estimate the composition of the `fertile upper mantle` by combining known components of the upper mantle – basalts, peridotites, recycled crust and so on – in such a way as to satisfy cosmic ratios of the lithophile refractory elements. Other elements are not necessarily concentrated into the crust and upper mantle. The MORB source is just part of the upper mantle and it is not the only LIL depleted part of the mantle. It is not necessarily convectively homogenized.

Attempts to establish an average composition for the upper mantle have focused on MORB because of the assumption the whole upper mantle is the MORB-reservoir. This procedure involves major assumptions about melt generation, melt transport and differentiation processes that have affected these melts, and the sources of non-MORB melts. The `depleted upper mantle`, that part of the mantle that is assumed to provide MORB by partial melting is variously called DUM, DM, DMM and the `convecting upper mantle`. Simplified mass balance calculations suggested to early workers that this depleted mantle constituted ~30% of the mantle; the 650–670-km discontinuity was adopted as the boundary between DUM and '`the primitive undepleted undegassed lower mantle`'. The starting condition for the upper mantle (UM) was taken as identical to `primitive mantle` (PM) and the present lower mantle (LM). There are many estimates of PM and BSE based on various cosmological and petrological considerations. The primitive upper mantle – crust plus DUM – is labeled PUM. It was further assumed that the upper mantle was vigorously convecting, well-stirred and chemically homogenous, and extended from the base of the plate to 650-km depth. Thus, this part of the mantle was also called the *convecting mantle*. The non-MORB basalts that occur at the initiation of spreading and at various locations along the global spreading system were attributed to plumes from the lower mantle.

Most or all of the mantle needs to be depleted and degassed to form the crust and upper mantle and the ^{40}Ar in the atmosphere (e.g. the first edition of Theory of the Earth, or TOE). Depletion of the upper mantle alone cannot explain the continental crust (CC); the MORB reservoir and the CC are not exactly complementary. There must be other components and processes beyond single-stage small-degree melt removal from part of the primordial mantle to form CC. There are other enriched components in the mantle, probably in the shallow mantle. Other depleted components or reservoirs, in addition to the MORB-source, are required by mass-balance calculations. The upper mantle cannot be treated as if its composition can be uniquely determined from the properties of depleted MORB – NMORB or DMORB – and depleted peridotites, continental crust, and an undifferential starting condition.

Both the cosmochemical and petrological approaches utilize terrestrial and meteoritic data. The common theme is that the Earth should have an unfractionated chondritic pattern of the refractory elements. This can be used as a formal a priori constraint in geochemical modeling of the composition of the Earth. This mass balance approach is consistent with the idea that most of the radioactive elements, and other crustal elements, are in the crust and upper mantle. Some investigators, however, decouple their models of the Earth from meteorite compositions. Two extreme positions have been taken: (1) the Earth is a unique body and is not related to material currently in the solar system; upper mantle rocks are representative of the whole mantle and Earth-forming material is not to be found in our meteorite collections; (2) only part of the Earth has been sampled and the upper mantle is not representative of the whole mantle. There are several variants of the second option: (1) the mantle is extensively differentiated and the deepest layers are the dense residues of this differentiation, which is irreversible; (2) only the upper mantle has been processed and differentiated; the deeper mantle is 'primordial'. There are numerous petrological reasons why the Earth's upper mantle should be distinct from any known type of meteorite and why no part of the mantle should have survived in a homogenous primitive state. However, some argue that it is more reasonable that the Earth accreted from material with major-element compositions that were distinct from primitive meteorite types than to accept chemical stratification and petrological differentiation of the mantle.

'Primitive mantle' or PM is the silicate fraction of the Earth, prior to differentiation and removal of the crust and any other parts of the present mantle that are the result of differentiation, or separation, processes. This is called bulk silicate Earth (BSE). In geochemical models, which were popular until very recently, it was assumed that large parts of the Earth escaped partial melting, or melt removal, and are therefore still 'primitive'. Some petrological models assumed that melts being delivered to the Earth's surface are samples from previously unprocessed material. It is difficult to believe that any part of the Earth could have escaped processing during the high-temperature accretional process. 'Primitive mantle', as used here, is a hypothetical material that is the sum of the present crust and mantle. Some petrological models assume that it is a mixture of the MORB-source and continental crust. 'Primitive magma' is a hypothetical magma, the parent of other magmas, which formed by a single-stage melting process of a parent rock and has not been affected by loss of material (crystal fractionation) prior to sampling. It is much more likely that magmas are the result of a multi-stage process and that they represent blends of a variety of melts from various depths and lithologies.

Petrological building blocks

Most mantle magmas can be matched by mixtures of depleted MORB (DMORB) and enriched components (Q). Midocean-ridge basalt (MORB) represents the most uniform and voluminous magma type and is often taken as an end-member for large-ion lithophile (LIL) concentrations and isotopic ratios in other basalts. The uniformity of MORB, however, may be the result of sampling, and the central limit theorem. In this case, MORB is a good average, but not a good end-member. The MORB source has been depleted by removal of a component – Q – that must be rich in LIL but relatively poor in Na, Al and Ca. Kimberlitic and some other enriched magmas have a complementary relationship to MORB.

The extreme enrichments of kimberlitic magmas in incompatible elements are usually attributed to low degrees of melting and/or metasomatized source compositions. The observed enrichment of kimberlitic magmas with rare earth elements (REE) can be explained in terms of melt migration through source rocks having the composition of normal mantle. The resulting saturated REE spectrum is practically independent of source mineral composition, which may explain the similarity of kimberlites from different geographic localities. Kimberlite is thus an important mantle component and can be used as such – component Q – in mass-balance calculations.

Chemical composition of the mantle

Considerations from cosmochemistry and the study of meteorites permit us to place only very broad bounds on the chemistry of the Earth's interior. These tell us little about the distribution of elements in the planet. Seismic data tell us a little more about the distribution of the major elements. General considerations suggest that the denser major elements will be toward the center of the planet and the lighter major elements, or those that readily enter melts or form light minerals, will be concentrated toward the surface. To proceed further we need detailed chemical information about crustal and mantle rocks. The bulk of the material emerging from the mantle is in the form of melts, or magmas. It is therefore important to understand the chemistry and tectonic setting of the various kinds of magmatic rocks and the kinds of sources they may have come from.

Midocean-ridge basalt represents the most uniform and voluminous magma type and is an end member for LIL concentrations and many isotopic ratios. This is usually taken as one of the *components* of the mantle, even though it itself is an average or a blend. Most mantle magma compositions can be approximated with a mixture of a depleted MORB-component and one or more enriched components, variously called EM1, EM2, HIMU, DUPAL and Q. The resulting magmas themselves are called NMORB, EMORB, OIB, AOB, CFB and so on. The refractory residue left after melt extraction – the *restite* – is usually considered to be a peridotite, dunite or harzburgite, all *ultramafic rocks* (UMR). All of the above, plus *continental crust* (CC), are candidate components for primitive mantle.

The MORB source appears to have been depleted by removal of a component that is rich in LIL but relatively poor in Na and the clinopyroxene-compatible elements (such as Al, Ca, Yb, Lu and Sc). Kimberlitic magmas have the required complementary relationship to MORB, and I adopt them in the following as a possible Q component. Some elements, such Nb, Ta, Ti and Zr are extraordinarily concentrated into specific minerals – rutile and zircon, for example – and estimates of these elements in rocks can be highly variable and dependent on the amount of these minerals. Peridotites and sulfides are the main carriers of elements such as magnesium,

Table 13.1 Estimates of average composition of the mantle

Oxide	(1)	(2)	(3)	(4)	(5)
SiO_2	45.23	47.9	44.58	47.3	45.1
Al_2O_3	4.19	3.9	2.43	4.1	3.9
MgO	38.39	34.1	41.18	37.9	38.1
CaO	3.36	3.2	2.08	2.8	3.1
FeO	7.82	8.9	8.27	6.8	7.9
TiO_2	—	0.20	0.15	0.2	0.2
Cr_2O_3	—	0.9	0.41	0.2	0.3
Na_2O	—	0.25	0.34	0.5	0.4
K_2O	—	—	0.11	0.2	(0.13)

(1) Jacobsen and others (1984): extrapolation of ultramafic and chondritic trends.
(2) Morgan and Anders (1980): cosmochemical model.
(3) Maaløe and Steel (1980): extrapolation of lherzolite trend.
(4) 20 percent eclogite, 80 percent garnet lherzolite (Anderson, 1980).
(5) Ringwood and Kesson (1976, Table 7): pyrolite adjusted to have chondritic CaO/Al_2O_3 ratio and Ringwood (1966) for K_2O.

chromium, cobalt, nickel, osmium and iridium, some of the so-called compatible elements. The continental crust is an important reservoir of potassium, rubidium, barium, lanthanum, uranium and thorium, some of the classical LIL elements. Thus, each of these components plays an essential role in determining the overall chemistry of the primitive mantle.

It is conventional to adopt a single hypothetical mix – lherzolite or harzburgite plus basalt – as the dominant silicate portion of the mantle; this has been called *pyrolite*, for *pyroxene–olivine rock*. An orthopyroxene-rich component (OPX) is also present in the mantle and is required if such major-element ratios as Mg/Si and Al/Ca ratios of the Earth are to be chondritic. Clinopyroxenites, rather than fertile peridotites, may be important source rocks for basalts. Some peridotites appear to have been enriched (metasomatized) by a kimberlite-like (Q) component. Seawater is an important repository of Cl, I and Br. The atmosphere may contain most of the heavier rare gases. Mixtures of the above components, plus continental crust, can be expected to give a first approximation to the composition of primitive mantle.

There may also be inaccessible reservoirs that do not provide samples for us to measure. The so-called `missing element and isotope paradoxes in geochemical box-models` suggest that some material is hidden away, probably in deep dense layers that formed during the accretion of the Earth. Ratios such as Ca/Al, Mg/Si, U/Pb and U/Nb and some isotope ratios imply that there is hidden or inaccessible material. The most obvious missing elements are iron and other siderophiles, such as Os and Ir. These are in the core. The missing silicon is probably in the perovskite-rich lower mantle. Other missing elements are S and C and other volatiles that left the Earth entirely or were never incorporated into it. There are numerous paradoxes associated with U and Th and their products – heat, Pb-isotopes, He-isotopes and Ne-isotopes. The obvious implication is that we are missing something; the mantle may be chemically stratified and we are sampling only the outer reaches, or we are ignoring certain components such as fluid-filled or melt inclusions, carbonatites and exotic minerals such as rutile, osmiridium etc. It is possible that certain rock types such as lower-crustal cumulates, carbonatites, recycled material and island-arc basalts are not added into the mix in appropriate quantities.

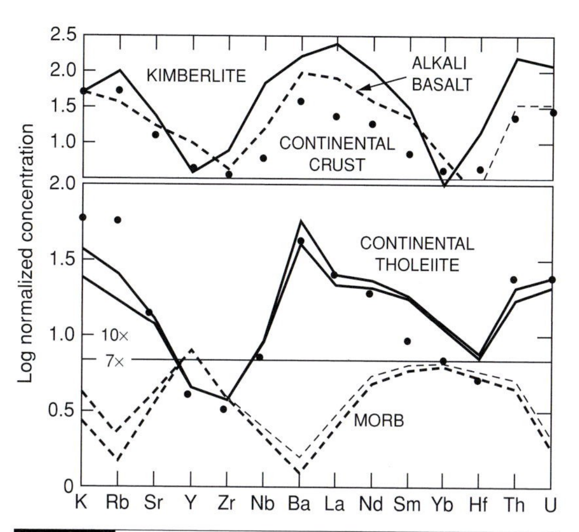

Fig. 13.1 Trace-element concentrations in the continental crust (dots), continental basalts and midocean-ridge basalts (MORB), normalized to average mantle compositions derived from a chondritic model. Note the complementary relationship between depleted basalts (MORB) and the other materials. MORB and continental tholeiites are approximately symmetric about a composition of 7 × C1. This suggests that about 14% of the Earth may be basalt. For other estimates, see text.

Figure 13.1 shows representative compositions of kimberlite, crust, MORB and ultramafic rock. For many refractory elements kimberlite and crust have a similar enrichment pattern. However, the volatile/refractory ratios are quite different, as are ratios involving strontium, hafnium, titanium, lithium, yttrium, ytterbium and lutetium. Kimberlite and MORB patterns are nearly mirror images for the refractory elements, but this is only approximately true for MORB and crust, especially for the HREE, and the small-ion–high-charge elements. MORB and kimberlite also represent extremes in their strontium and neodymium isotope compositions.

When LIL-rich materials (KIMB, lamproites) are mixed with a depleted magma (MORB), the resulting blend can have apparently paradoxical geochemical properties. For example, the hybrid magma can have such high La/Yb, Rb/Sr and Pb-isotope ratios that derivation from an enriched source is indicated, but the Sr- and Nd-isotope ratios imply derivation from an ancient depleted reservoir. Many island, island-arc and continental basalts have these characteristics. These apparently paradoxical results simply mean that ratios do not average as do concentrations. An additional complication is that the melt components, and the mixed hybrid, can experience crystal fractionation while mixing. Magma mixtures and evolved magmas do not necessarily define simple mixing lines.

The isotope zoo

A large number of components in MORB and OIB have been proposed and labeled. These are primarily based on isotopic characteristics. A short list is: EM1, EM2, HIMU, DUPAL, C, PHEM, PREMA, FOZO, LOMU, LONU and Q. The origin of the isotope species in the mantle zoo is a matter of considerable controversy. In the isotope geochemistry literature these are attributed to the lower mantle and to plumes but there is no evidence that this is the case. There is considerable evidence that all these species originated at the surface of the Earth, in the crust or in the lithosphere, and may have evolved in the shallow mantle. Various kinds of altered and recycled materials have been proposed as contributors or ancestors to these species. Among the suggestions are oceanic sediments, recycled oceanic crust with some pelagic sediments (EM1), continental sediments or dehydrated oceanic sediments (EM2), altered oceanic crust and sediments or ancient altered MORB (HIMU), delaminated lower continental crust (EM1, DUPAL), peridotites, and high ^{3}He/U and low U and Th bubbles in depleted lithosphere, restites or cumulates (FOZO). Some EM components of the mantle have oxygen isotopic compositions similar to MORB and peridotite xenoliths, ruling out a large contribution of subducted sediment and other sources that were proposed on the basis of Sr and Nd isotopes alone.

All of these species may co-exist in the shallow mantle, along with the sources of MORB; they do not have to be recycled into the deep lower mantle before they return to the surface. Observed magmas are blends or mixtures of components, or partial melts therefrom, with various isotopic signatures followed by variable degrees of crystal fractionation and reactions with the crust and mantle on the way to the surface. Trace-element and isotope ratios of mixtures can have anti-intuitive properties.

The recipes and mathematics of mixtures

Rocks, and the mantle, are composed of *components*. Let us call them A, B, C . . . The fraction of each component in the mix is x, y, z . . . The mixture itself, M, has properties that are the appropriate weighted average of the properties of the components. It is straightforward to construct a rock, or a mantle, or a fruitcake, if we know all the ingredients and the proportions in which they occur. It is not straightforward to infer the ingredients and the mixing ratios if we just have samples from M, or if we know the ingredients but not the proportions. Our inferences depend on assumptions and how we sample M. There are techniques, known as geophysical inverse theory, sampling theory and the central limit theorem (CLT) for approaching these kinds of problems.

Consider first, the fruitcake. The *components* are bits of fruit and dough. Every large bite is more-or-less the same. Every tiny bite is different, a different bit of fruit or a different combination. The big bites have very little variance in composition or taste, while the compositions (tastes) of the small bites have enormous scatter. This is a consequence of the CLT. One cannot infer the composition of the fruitcake from the homogenous big bites, one can only infer the average taste.

Consider next, the rock. A rock is composed of minerals, inclusions and intergranular material. By grinding up a large sample, one infers the average composition of the rock. Many such samples from a homogenous outcrop will give similar results. That is, there will be little scatter or variance. However, if very small samples are taken, there will be enormous variance, a reflection of the basic heterogeneity of rocks. However, the mean of all the small samples will be close to the means of the large samples. This is another consequence of the CLT. The homogeneity of the large samples does not mean that the rock is homogenous; it simply means that large-scale, large-volume averages have been taken. Multiple microprobe analyses of the same rock would tell a different story. Some small-scale samples would give extreme compositions, compositions not evident in any of the large-volume averages. This does not mean that they come from a different rock.

Finally, consider the mantle. We have homogenous samples of MORB from midocean ridges, which are enormous blending and averaging machines. Seamounts on the flanks of oceanic ridges provide smaller volume samples from the same mantle, and these exhibit enormous variance. Ocean-island basalts also exhibit a wide range of compositions, partly a reflection of the large-scale heterogeneity of the mantle and partly a reflection of the nature and scale of the sampling process. Sometimes the magmas bring up small fragments, or xenoliths, from the mantle. These are highly variable in composition and sometimes exhibit a greater range of compositions than one observes globally in basalts. Midocean-ridge basalts are blends of magmas and solid materials from huge volumes, and, as expected, they show little variance, at least in some properties. MORB is not a *component* or a *reservoir*; it is an average, and will be treated as such below.

Considering all of this, how does one infer the composition of the mantle? One approach is to assume that one knows the composition and volume of the major reservoirs such as the crust and the upper mantle and then to attribute discrepancies to the lower mantle or other hidden reservoirs. The pyrolite model assumes that the mantle is composed of known amounts and compositions of a mafic and an ultramafic component. A more objective approach is to make a list of all the materials that are known to enter and leave the mantle and construct a composite from these that satisfies certain constraints. These constraints might include the total volume of the crust (one of the components), the sum of the components is 1, and the ratios of some of the components should be chondritic. This is the approach taken in the following. It is clearly important to know something about the various components; magmas, ultramafic rocks, crust, kimberlites and so on. Most of the components are also mixtures of other components. In some cases, and for some elements, it may be important to consider mineral components, such as zircon and rutile, or rare magmas such as carbonatites. The reason for this is the same reason that one must include the crust in any mass balance in spite of the fact that it has a very small total volume. Some rocks have concentrations of some elements that exceed the crustal concentrations by an order of magnitude; one cannot ignore such rocks. For a complete mass balance one cannot even ignore the oceans and the atmosphere; iodine and chlorine, for example, are concentrated on the atmosphere and the biosphere.

In addition to all the above, one must always remember that isotopic and trace-element ratios do not average and mix as do concentrations or absolute abundances.

The inverse approach

The inverse approach is more widely used in geophysics than in geochemistry. The problem is one of inferring mantle chemistry from various products of differentiation, which themselves are averages. One alternative approach is to use petrological insight to pick the most representative constituents for the mantle, and to adjust their proportions to satisfy certain key constraints or assumptions.

The refractory elements can be assumed to be in the Earth in chondritic ratios but we do not know the absolute concentrations or their distribution. In order to estimate absolute concentrations, and the volatile and siderophile content of the mantle, we seek a linear combination of components that gives chondritic ratios for the refractory elements. We can then estimate such key ratios as Rb/Sr, K/U and U/Pb that involve non-refractory elements. In essence, we replace the five basic building blocks of mantle chemistry, ol,

Table 13.2 Chemical composition of mantle components (ppm, % or ppb).

	MORB	Ultramafic Rocks	KIMB	Crust	Picrite	OPX	Morgan and Anders (1980)
Li	9	1.5	25	10	7	1.5	1.85
F	289	7	1900	625	219	—	14
Na	15 900	2420	2030	26 000	12 530	—	1250
Mg*	5.89	23.10	16.00	2.11	10.19	21.00	13.90
Al*	8.48	2.00	1.89	9.50	6.86	1.21	1.41
Si*	23.20	21.80	14.70	27.10	22.85	22.60	15.12
P	390	61	3880	1050	308	—	1920
S	600	8	2000	260	452	—	14 600
Cl	23	1	300	1074	17	—	20
K	660	35	17 600	12 500	504	10	135
Ca*	8.48	2.40	7.04	5.36	6.96	1.32	1.54
Sc	37.3	17	15	30	32	12	9.6
Ti	5500	1000	11 800	4800	4375	800	820
V	210	77	120	175	177	60	82
Cr	441	2500	1100	55	956	2500	4120
Mn	1080	1010	1160	1100	1063	1010	750
Fe*	6.52	6.08	7.16	5.83	6.41	6.08	32.07
Co	53	105	77	25	66	105	840
Ni	152	2110	1050	20	642	2110	18 200
Cu	77	15	80	60	62	30	31
Zn	74	60	80	52	71	20	74
Ga	18	4	10	18	14	3	3
Ge	1.5	1.1	0.5	1.5	1.4	1.1	7.6
Se	0.181	0.02	0.15	0.05	0.14	0.006	9.6
Rb	0.36	0.12	65	42	0.30	0.04	0.458
Sr	110	8.9	707	400	85	4.5	11.6
Y	23	4.6	22	22	18	0.023	2.62
Zr	70	11	250	100	55	5.5	7.2
Nb	3.3	0.9	110	11	2.7	0.45	0.5
Ag	0.019	0.0025	—	0.07	0.015	0.001	0.044
Cd	0.129	0.255	0.07	0.2	0.103	0.01	0.016
In	0.072	0.002	0.1	0.1	0.055	—	0.002
Sn	1.36	0.52	15	2	1.15	0.52	0.39
Cs	0.007	0.006	2.3	1.7	0.01	0.002	0.0153
Ba	5	3	1000	350	4.50	2	4
La	1.38	0.57	150	19	1.18	0.044	0.379
Ce	5.2	1.71	200	38	4.33	0.096	1.01
Nd	5.61	1.31	85	16	4.54	0.036	0.69
Sm	2.08	0.43	13	3.7	1.67	0.009	0.2275
Eu	0.81	0.19	3	1.1	0.66	0.0017	0.079
Tb	0.52	0.14	1	0.64	0.43	0.0017	0.054
Yb	2.11	0.46	1.2	2.2	1.70	0.0198	0.229
Lu	0.34	0.079	0.16	0.03	0.27	0.0072	0.039
Hf	1.4	0.34	7	3	1.14	0.17	0.23

(cont.)

Table 13.2 | *(cont.)*

	MORB	Ultramafic Rocks	KIMB	Crust	Picrite	OPX	Morgan and Anders (1980)
Ta	0.1	0.03	9	2	0.08	—	0.023
Re†	1.1	0.23	1	1	0.88	0.08	60
Os†	0.04	3.1	3.5	5	0.81	3.1	880
Ir†	0.0011	3.2	7	1	0.80	3.2	840
Au†	0.34	0.49	4	4	0.38	0.49	257
Tl†	0.01	0.02	0.22	0.25	0.01	—	0.003 86
Pb	0.08	0.2	10	7	0.11	—	0.068
Bi	0.007	0.005	0.03	0.2	0.01	—	0.002 94
Th	0.035	0.094	16	4.8	0.05	—	0.0541
U	0.014	0.026	3.1	1.25	0.02	—	0.0135

*Results in percent.
†Results in ppb.
Anderson (1983a).

opx, cpx, gt and Q, or their low-pressure equivalents) with four composites or averages; peridotite, orthopyroxenite, basalt and Q. In practice, we can use two different ultramafic rocks (UMR and OPX) with different ol/opx ratios to decouple the ol+opx contributions. The chemistry of the components (MORB, UMR, OPX and crust) are given in Table 13.2.

Having measurements of m elements in n components where m far exceeds n, we can find the weight fraction x_j of each component, given the concentration C_j of the jth element in the ith component, that yields chondritic ratios of the refractory oxyphile elements. In matrix form,

$$C_{ij}x_j = kC_j \quad (1)$$

where C_j is the chondritic abundance of element j and k is a dilution or enrichment factor, which is also to be determined. The least-squares solution is

$$x_j/k = (C^{\mathrm{T}}C)^{-1}C^{\mathrm{T}}C_j \quad (2)$$

where C^{T} is the matrix transpose of C, with the constraints

$$\sum x_{\bar{r}} = 1 \text{ and } k = \text{constant} \quad (3)$$

Equation 1 then gives the mantle concentrations of the volatile and siderophile elements, elements not used in the inversion.

Schematically, primitive mantle can be written

$$\mathrm{PM} = \mathrm{MORB} + \mathrm{UMR} + \mathrm{CC} + \mathrm{OPX} + \mathrm{Q}$$

and the proportions of each component are found by the inversion.

The mixing ratios found from Equations 2 and 3 are UMR, 32.6%, OPX, 59.8%, MORB, 6.7%, crust, 0.555% and Q, 0.11%. This is a model composition for primitive mantle (PM) i.e. mantle plus crust. The Q component is equivalent to a global layer 3.6 km thick. The MORB component represents about 25% of the upper mantle. This solution is based on 18 refractory elements and, relative to chondrite (C1) abundances, yields k = 1.46.

The result is given in Table 13.3 under 'mantle plus crust.' This will be referred to as the bulk silicate Earth (BSE) composition. Concentrations normalized to Cl are also given. Note that the high-charge, small-ionic radius elements (Sc, Ti, Nb and Hf) have the highest normalized ratios and may be overestimated. Taking all the refractory oxyphile elements into account (23 elements), the normalized enrichment of the refractory elements in the mantle plus crust is 1.59. If five elements (Fe, S, Ni, Co and O) are removed from C1 to a core of appropriate size and density, the remaining silicate fraction will be enriched

Table 13.3 | Chemical composition of mantle (ppm, % or ppb)

	CI	Morgan and Anders (1980)	UDS	Upper Mantle	Lower Mantle	Mantle + Crust	Normalized to Morgan and Anders (1980)	CI
Li	2.4	1.85	7.48	3.49	1.50	2.09	0.76	0.87
F	90	14	259	91	1	28	1.39	0.31
Na	7900	1250	13 172	6004	359	2040	1.10	0.26
Mg*	14.10	13.90	9.80	18.67	21.31	20.52	1.00	1.46
Al*	1.29	1.41	6.95	3.65	1.33	2.02	0.97	1.57
Si*	15.60	15.12	23.00	22.20	22.48	22.40	1.00	1.44
P	1100	1920	387	170	9.06	57	0.02	0.05
S	19 300	14 600	458	158	1.19	48	0.0022	0.0025
Cl	1000	20	79	27	0.07	8	0.27	0.008
K	890	135	1356	475	14	151	0.76	0.17
Ca*	1.39	1.54	6.87	3.89	1.48	2.20	0.96	1.58
Sc	7.8	9.6	31.92	22	13	15	1.09	1.99
Ti	660	820	4477	2159	830	1225	1.01	1.86
V	62	82	176	110	63	77	0.63	1.24
Cr	3500	4120	907	1969	2500	2342	0.38	0.67
Mn	2700	750	1066	1029	1010	1016	0.91	0.38
Fe*	27.20	32.07	6.39	6.18	6.08	6.11	0.13	0.22
Co	765	840	64	91	105	101	0.08	0.13
Ni	15 100	18 200	611	1610	2110	1961	0.07	0.13
Cu	160	31	62	31	28	29	0.62	0.18
Zn	455	74	70	63	26	37	0.34	0.08
Ga	14	3	15	7	3	4	0.94	0.31
Ge	47	7.6	1.40	1.20	1.10	1.13	0.10	0.02
Se	29	9.6	0.14	0.06	0.01	0.02	0.0016	0.0008
Rb	3.45	0.458	3.32	1.19	0.05	0.39	0.57	0.11
Sr	11.4	11.6	109	42.3	5.2	16.2	0.94	1.42
Y	2.1	2.62	18.64	9.28	0.70	3.26	0.84	1.55
Zr	5.7	7.2	60	27	6	13	1.18	2.20
Nb	0.45	0.5	4.30	2.03	0.52	0.97	1.31	2.15
Ag	0.27	0.044	0.02	0.01	0.00	0.003	0.05	0.01
Cd	0.96	0.016	0.11	0.05	0.01	0.02	1.03	0.03
In	0.12	0.002	0.06	0.02	0.0003	0.006	2.13	0.05
Sn	2.46	0.39	1.34	0.79	0.52	0.60	1.04	0.24
Cs	0.29	0.0153	0.13	0.05	0.003	0.02	0.68	0.05
Ba	3.60	4	34	13	1.72	5.22	0.88	1.45
La	0.367	0.379	3.75	1.63	0.12	0.57	1.02	1.56
Ce	0.957	1.01	8.28	3.90	0.34	1.40	0.93	1.46
Nd	0.711	0.69	6.03	2.88	0.23	1.02	1.00	1.43
Sm	0.231	0.2275	1.90	0.92	0.07	0.32	0.96	1.40
Eu	0.087	0.079	0.70	0.362	0.03	0.13	1.10	1.48
Tb	0.058	0.054	0.44	0.241	0.02	0.09	1.09	1.51
Yb	0.248	0.229	1.72	0.88	0.09	0.32	0.95	1.30
Lu	0.038	0.039	0.27	0.144	0.02	0.06	0.96	1.46
Hf	0.17	0.23	1.30	0.66	0.20	0.33	0.98	1.96

(cont.)

Table 13.3 (*cont.*)

	CI	Morgan and Anders (1980)	UDS	Upper Mantle	Lower Mantle	Mantle + Crust	Normalized to Morgan and Anders (1980)	CI
Ta	0.03	0.023	0.28	0.11	0.004	0.04	1.10	1.24
Re†	60	60	0.89	0.45	0.10	0.21	0.0023	0.0034
Os†	945	880	1.08	2.43	3.10	2.90	0.0022	0.0031
Ir†	975	840	0.88	2.43	3.20	2.97	0.0024	0.0031
Au†	255	257	0.62	0.53	0.49	0.50	0.0013	0.002
Tl†	0.22	0.0039	0.03	0.02	0.00	0.01	1.28	0.033
Pb	3.6	0.068	0.60	0.33	0.03	0.12	1.19	0.033
Bi	0.17	0.0029	0.02	0.009	0.0007	0.0033	0.75	0.019
Th	0.051	0.0541	0.48	0.224	0.014	0.0765	0.96	1.50
U	0.014	0.0135	0.12	0.057	0.004	0.0196	0.98	1.40

*Results in percent.
†Results in ppb.
Anderson (1983a).

in the oxyphile elements by a factor of 1.48. This factor matches the value for k determined from the inversions.

The Earth's mantle can be chondritic in major and refractory-element chemistry if an appreciable amount of oxygen has entered the core. H_2O and CO_2 are not included in most mass balance calculations and unknown amounts of these may be in the upper mantle or escaped from the Earth.

Table 13.3 compares these results with cosmochemically based models. Both the volatile elements and the siderophile elements are strongly depleted in the crust–mantle system relative to cosmic abundances.

In pyrolite-type models, it is assumed that primitive mantle (PM) is a mix of basalt and peridotite and that one knows the average compositions of the basaltic (MORB or OIB) and ultramafic (UMR) components. These are mixed in somewhat arbitrary, predetermined, proportions; the results are comparable with volatile-free chondritic abundances for some of the major elements.

Tholeiitic basalts are thought to represent the largest degree of partial melting among common basalt types and to nearly reflect the trace-element chemistry of their mantle source. Tholeiites, however, range in composition from depleted midocean-ridge basalts to enriched ocean-island basalts (OIB) and continental flood basalts (CFB). Enriched basalts (alkali-olivine, OIB, CFB) can be modeled as mixtures of MORB and an enriched (Q) component that experience varying degrees of crystal fractionation prior to eruption. Ultramafic rocks from the mantle, likewise, have a large compositional range. Some appear to be crystalline residues after basalt extraction, some appear to be cumulates, and others appear to have been secondarily enriched in incompatible elements (metasomatized). This enriched component is similar to the Q component of basalts. Some ultramafic rocks are relatively 'fertile' (they can yield basalts upon partial melting), but they do not have chondritic ratios of all the refractory elements.

The above calculation, in essence, decomposes the basaltic component of the mantle into a depleted (MORB) and LIL-enriched (Q) component. These can be combined and compared with the basaltic component of other two-component models, such as pyrolite, by comparing 'undepleted basalt' (MORB +1.5% KIMB) with the basalts (Hawaiian tholeiite or 'primitive' oceanic tholeiite) used in the construction of pyrolite. The present model has an undepleted basaltic component (MORB + Q) of 914 ppm potassium, 0.06 ppm uranium and 0.27 ppm thorium. The basaltic components of pyrolite have much higher concentrations of these and other LIL. Refractory ratios such as U/Ca and Th/Ca for pyrolite are about 50% higher than chondritic because of the high LIL content of the arbitrarily chosen basalt. One disadvantage of two-component pyrolite-type models is that the final results are controlled by the arbitrary choice of components, including basalt, and compositions. Indeed, the various pyrolite models differ by an order of magnitude in, for example, the abundance of potassium.

By and large, there is excellent agreement with cosmochemical models (Table 13.3) and BSE found by the above inversion. In the present model, the alkalis lithium, potassium, rubidium and cesium are somewhat more depleted than in M&A, as are volatiles such as chlorine, vanadium and cadmium. The Rb/Sr and K/U ratios are correspondingly reduced. The elements that are excessively depleted (P, S, Fe, Co, Ni, Ge, Se, Ag, Re, Os, Ir and Au) are plausibly interpreted as residing in the core. Note that the chalcophiles are not all depleted. In particular, lead is not depleted relative to other volatiles such as manganese, fluorine and chlorine, which are unlikely to be concentrated in the core. The composition of primitive mantle, derived by the above approach, is given in Figure 13.2.

Some elements are extraordinarily concentrated into the crust. The above results give the following proportions of the total mantle-plus-crust inventory in the continental crust; rubidium, 58%; cesium, 53%; potassium, 46%; barium, 37%; thorium and uranium, 35%; bismuth, 34%; lead, 32%; tantalum, 30%; chlorine, 26%; lanthanum, 19% and strontium, 13%.

In addition, the atmospheric ^{40}Ar content represents 77% of the total produced by 151 ppm potassium over the age of the Earth. This has probably degassed from the crust and upper mantle and probably reflects 23% retention, rather than 23% primordial undegassed mantle. These results all point toward an extensively differentiated Earth and efficient upward concentration of the incompatible trace elements. It is difficult to imagine how these concentrations could be achieved if the bulk of the mantle is still primitive or unfractionated. If only the upper mantle provides these elements to the crust, one would require more than 100% removal of most of the LIL (U, Th, Bi, Pb, Ba, Ta, K, Rb and Cs). More likely, the whole mantle has contributed to crustal, and upper-mantle, abundances, and most of the mantle is strongly depleted and, probably, infertile. The crust, MORB reservoir and the Q component account for a large fraction of the incompatible trace elements. It is likely, therefore, that the lower mantle is depleted in these elements, including the heat producers potassium, uranium and thorium.

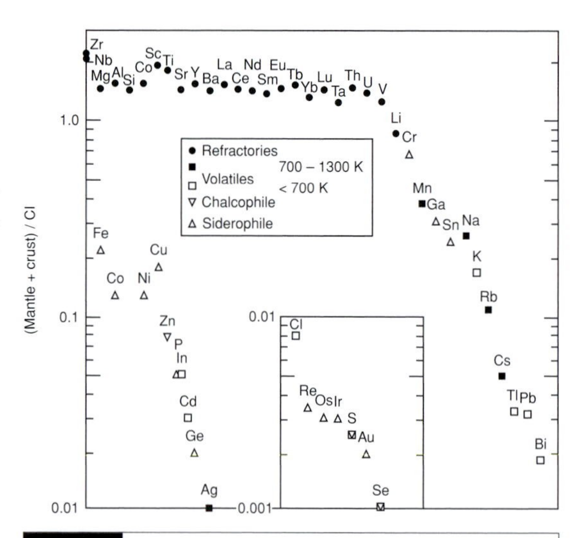

Fig. 13.2 Abundances of elements in 'primitive mantle' (mantle + crust) relative to C1, derived by mixing mantle components to obtain chondritic ratios of the refractory lithophile elements.

In an alternative approach we can replace UMR and OPX by their primary constituent minerals, olivine, orthopyroxene and clinopyroxene. The present mantle can then be viewed as a five-component system involving olivine, orthopyroxene, clinopyroxene, MORB (cpx and gt) and Q. In this case the LIL inventory of the primitive mantle is largely contained in four components: MORB, Q, clinopyroxene and CRUST. The results of this approach are: olivine, 33%, orthopyroxene, 48.7%, clinopyroxene, 3.7%, MORB, 14.0%, Q, 0.085% and CRUST, 0.555%.

Concentrations of certain key elements are sodium, 2994 ppm, potassium, 205 ppm,

Table 13.4 Elemental ratios.

	CI	Morgan and Anders*	UDS	Upper Mantle	Lower Mantle	Mantle + Crust	Normalized M & A*	Normalized CI
Rb/Sr	0.3026	0.0395	0.0304	0.0281	0.0101	0.024	0.61	0.08
K/U	63 571	10 000	11 429	8356	3552	7693	0.77	0.12
Sm/Nd	0.3249	0.3297	0.315	0.319	0.318	0.319	0.968	0.982
Th/U	3.64	4.01	4.08	3.94	3.62	3.98	0.97	1.07
U/Pb	0.0039	0.20	0.20	0.17	0.13	0.16	0.82	42
La/Yb	1.48	1.66	2.18	1.85	1.43	1.77	1.07	1.20
K/Na	0.11	0.11	0.10	0.08	0.04	0.07	0.69	0.66
Mg/Si	0.90	0.92	0.43	0.84	0.95	0.92	1.00	1.01
Ca/Al	1.08	1.09	0.99	1.07	1.12	1.09	1.00	1.01
Yb/Sc	0.032	0.024	0.054	0.04	0.007	0.021	0.87	0.65
Ce/Nd	1.35	1.46	1.37	1.35	1.49	1.37	0.94	1.02
Eu/Nd	0.12	0.11	0.12	0.13	0.13	0.13	1.10	1.03
Yb/Lu	6.53	5.87	6.26	6.10	4.77	5.80	0.99	0.89
Sr/Ba	3.17	2.90	3.17	3.14	2.99	3.11	1.07	0.98
U/La	0.038	0.036	0.03	0.035	0.032	0.034	0.97	0.90

*Model of Morgan and Anders (1980).

rubidium, 0.53 ppm, strontium, 25 ppm and cesium, 0.02 ppm. The alkalis are generally within 50% of the concentrations determined previously. The K/U ratio is 44% lower.

A four-component (crust, basalt, peridotite and Q) model for the upper mantle gives close to chondritic ratios for the refractory trace elements. The model gives predictions for volatile/refractory ratios such as K/U and Rb/Sr. It predicts that EMORB in the upper mantle can be up to 10% of NMORB.

For BSE a pyroxene-rich component is required in order to match chondritic ratios of the major elements. Such a component is found in the upper mantle and is implied by the seismic data for the lower mantle. The abundances in the mantle-plus-crust system (BSE) are 151 ppm potassium, 0.0197 ppm uranium and 0.0766 ppm thorium, giving a steady-state heat flow that implies that slightly more than half of the terrestrial heat flow is due to cooling of the Earth, consistent with convection calculations in a stratified Earth.

In summary, primitive mantle (PM) can be viewed as a five-component system; crust, MORB, peridotite, pyroxenite and Q (quintessence, the fifth essence) or, alternatively, as olivine, orthopyroxene, garnet plus clinopyroxene (or basalt) and incompatible element- and alkali-rich material (crust and kimberlite or LIL-rich magmas). Advantages of the inverse method are that the mixing proportions do not have to be fixed in advance and all potential components can be included; the inversion will decide if they are needed.

The upper mantle

The mass-balance method gives the average composition of the mantle but makes no statement about how the components are distributed between regions of the mantle. We can, however, estimate the composition of the MORB source region prior to extraction of crust and Q. The bulk of the mantle is depleted and refractory, UMR plus orthopyroxene (OPX). Pyroxenite, the most uncertain and to some extent arbitrary of the components, plays a minor role in the mass-balance calculations for the trace refractories and is required mainly to obtain chondritic ratios of major elements. Detailed discussion of the composition of the upper mantle is postponed until Chapter 26.

Table 13.5 | Elemental ratios in upper-mantle components

	MORB	Ultramafic Rocks	KIMB	Crust	Picrite	OPX	Morgan and Anders (1980)
Rb/Sr	0.0033	0.0135	0.0919	0.105	0.0035	0.0089	0.039
K/U	47 143	1346	5677	10 000	29 632	—	10 000
Sm/Nd	0.371	0.328	0.153	0.231	0.368	0.25	0.3297
Th/U	2.50	3.62	5.16	3.84	2.94	—	4.01
U/Pb	0.18	0.13	0.31	0.18	0.15	—	0.20
La/Yb	0.65	1.24	125	8.64	0.69	2.22	1.66
K/Na	0.84	0.01	8.67	0.48	0.04	—	0.11
Mg/Si	0.25	1.06	1.09	0.08	0.45	0.93	0.92
Ca/Al	1.00	1.20	3.72	0.56	1.01	1.09	1.09
Yb/Sc	0.057	0.027	0.08	0.073	0.053	0.0017	0.0239
Ce/Nd	0.93	1.31	2.35	2.38	0.95	2.67	1.46
Eu/Nd	0.14	0.15	0.04	0.07	0.14	0.05	0.11
Yb/Lu	6.21	5.82	7.50	7.33	6.18	2.75	5.87
Sr/Ba	22.00	2.97	1.18	1.14	18.83	3.00	2.90
U/La	0.01	0.46	0.021	0.066	0.014	—	0.036

We first find the composition of a possible picritic parent to MORB. The relation PICRITE = 0.75 MORB + 0.25 UMR gives the results tabulated under 'picrite' in Table 13.2 and 13.4.

The mixing ratios that are required to give a chondritic pattern for the refractory elements yield

$$\text{UDS} = 0.935\ \text{PICRITE} + 0.016\ \text{Q} + 0.00559\ \text{CRUST}$$

The composition of this reconstructed *Undepleted Source Region* is tabulated under UDS. This has also been called primitive upper mantle (PUM). UDS accounts for about 10% of the mantle. The remainder of the mantle above 670-km depth is assumed to be ultramafic rocks (UMR). This gives the composition tabulated under 'upper mantle' in Tables 13.3 and 13.4. This region contains 23.4% basalt (MORB). The resulting composition has element ratios (Table 13.5), which, in general, are in agreement with chondritic ratios. The La/Yb, Al/Ca and Si/Mg ratios, however, are too high. These are balanced by the deeper mantle in the full calculation. The solution for the mantle below 670-km depth, is 0.145 UMR and 0.855 OPX. This gives chondritic ratios for Mg/Si and Ca/Al for the Earth as a whole. The mixing

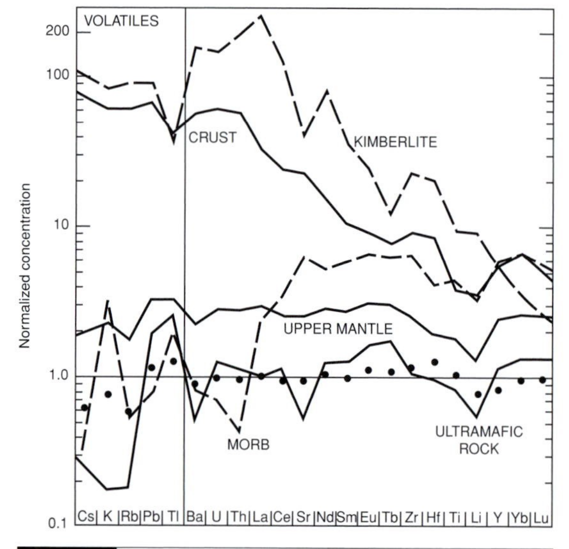

Fig. 13.3 Shows the normalized concentrations of the lithophile elements in the various components, upper mantle and mantle-plus-crust. The refractory elements in the upper mantle have normalized concentrations of about 3; this includes the crustal contribution. Since the upper mantle is about one-third of the whole mantle, a strongly depleted lower mantle is implied. Note that the upper mantle is depleted in lithium and titanium. These elements may be in the inaccessible lower mantle (below 1000 km). The composition of the upper mantle is discussed in more detail in Chapter 23. Dots are derived PM.

ratios will be slightly different if it is assumed that the upper mantle extends to 1000-km depth, which is the classical depth to the top of the lower mantle. An orthopyroxene-rich composition for the bulk of the mantle is expected with a chondritic model for the major elements, particularly if the upper mantle is olivine- and basalt-rich. At low pressure olivine and orthopyroxene are refractory phases and are left behind as basalt is removed. However, at high pressure the orthopyroxene-rich phases, majorite and perovskite, are both refractory and dense. If melting during accretion extended to depths greater than about 350-km, then the melts would be olivine-rich and separation of olivine from orthopyroxene can be expected.

Figure 13.3 shows the concentrations of the lithophile elements in the various components, all normalized to mantle equivalent concentrations. The refractory elements in the upper mantle have normalized concentrations of about 3. Since the upper mantle is about one-third of the whole mantle, a strongly depleted lower mantle is implied. The upper mantle is not necessarily homogenous. The basaltic fraction, as eclogite, may be in a separate layer. Seismic data are consistent with an eclogite-rich transition region.

Chapter 14

Magmas: windows into the mantle

If our eye could penetrate the Earth and see its interior from pole to pole, from where we stand to the antipodes, we would glimpse with horror a mass terrifying riddled with fissures and caverns

Tellus Theoria Sacra (1694), Thomas Burnet

Various kinds of magmas, ranging from midocean-ridge basalts (MORB) to kimberlite (KIMB) emerge from the mantle. Material is also recycled into the mantle; sediments, oceanic crust, delaminated continental crust, water and peridotite. Parts of the mantle are inaccessible to direct sampling and we can only infer their compositions by subtracting off the sampled components from what is thought to be the original composition. There is no assurance that we are currently receiving samples from all parts of the Earth, although this is the hope of many geochemists. There is reason to believe that the chemical stratification of the mantle is irreversible, and that there exist hidden 'reservoirs' (a bad term for a permanent repository). Nevertheless, the accretional stratification of the Earth may have placed most of the low-melting point lithophile elements within reach. In fact, one can construct a plausible compositional model for the mantle by isolating the dense refractory depleted products of differentiation in the deep mantle. This would make the Mg, Si and Fe contents of the mantle uncertain but these can be constrained by geophysics, mineral physics and cosmology. In fact, if one mixes together all the materials known to enter or leave the mantle, one can obtain cosmic ratios of the refractory elements except for Mg/Si and other elements likely to enter dense refractory residues of mantle differentiation. Likewise, this mix is deficient in the siderophile elements, which are plausibly in the core, and the volatile elements, which were probably excluded from the beginning.

Magmas are an important source of information about conditions and composition of the Earth's interior. The bulk composition, trace-element chemistry, isotope geochemistry and volatile content of magmas all contain information about the source region and the processes that have affected the magmas before their eruption. Mantle fragments, or xenoliths, found in these magmas tell us about the material through which the magmas have passed on the way to the surface. Representative compositions of various magmas are given in Table 14.1.

Three principal magma series are recognized: tholeiite, calc-alkaline and alkali. The various rock types in each series may be related by varying degrees of partial melting or crystal separation. The dominant rock type is tholeiite, a fine-grained dark basalt containing little or no olivine. Tholeiites are found in both oceanic and intraplate settings. Those formed at midocean ridges are low-potassium and LIL-depleted and relatively high in Al, while those found on oceanic islands and continents are generally LIL-enriched. Ridge tholeiites differ from continental and island tholeiites by their higher contents of Al and Cr, low contents of large-ion lithophile

Table 14.1 | Representative compositions of basalts and andesites.

	Tholeiite					Oceanic	Cont. Rift			
		Continental				Alkali	Alkali		Andesite	
	Ridge	Arc	Rift	Island	High-Al	Basalt	Basalt	Arc	Low-K	High-K
SiO_2	49.8	51.1	50.3	49.4	51.7	47.4	47.8	57.3	59.5	60.8
TiO_2	1.5	0.83	2.2	2.5	1.0	2.9	2.2	0.58	0.70	0.77
Al_2O_3	16.0	16.1	14.3	13.9	16.9	18.0	15.3	17.4	17.2	16.8
Fe_2O_3*	10.0	11.8	13.5	12.4	11.6	10.6	12.4	8.1	6.8	5.7
MgO	7.5	5.1	5.9	8.4	6.5	4.8	7.0	3.5	3.4	2.2
CaO	11.2	10.8	9.7	10.3	11.0	8.7	9.0	8.7	7.0	5.6
Na_2O	2.75	1.96	2.50	2.13	3.10	3.99	2.85	2.63	3.68	4.10
K_2O	0.14	0.40	0.66	0.38	0.40	1.66	1.31	0.70	1.6	3.25
Cr	300	50	160	250	40	67	400	44	56	3
Ni	100	25	85	150	25	50	100	15	18	3
Co	32	20	38	30	50	25	60	20	24	13
Rb	1	5	31	5	10	33	200	10	30	90
Cs	0.02	0.05	0.2	0.1	0.3	2	>3	~ 0.1	0.7	1.5
Sr	135	225	350	350	330	800	1500	215	385	620
Ba	11	50	170	100	115	500	700	100	270	400
Zr	85	60	200	125	100	330	800	90	110	170
La	3.9	3.3	33	7.2	10	17	54	3.0	12	13
Ce	12	6.7	98	26	19	50	95	7.0	24	23
Sm	3.9	2.2	8.2	4.6	4.0	5.5	9.7	2.6	2.9	4.5
Eu	1.4	0.76	2.3	1.6	1.3	1.9	3.0	1.0	1.0	1.4
Gd	5.8	4.0	8.1	5.0	4.0	6.0	8.2	4.0	3.3	4.9
Tb	1.2	0.40	1.1	0.82	0.80	0.81	2.3	1.0	0.68	1.1
Yb	4.0	1.9	4.4	1.7	2.7	1.5	1.7	2.7	1.9	3.2
U	0.10	0.15	0.4	0.18	0.2	0.75	0.5	0.4	0.7	2.2
Th	0.18	0.5	1.5	0.67	1.1	4.5	4.0	1.3	2.2	5.5
Th/U	1.8	3.3	3.8	3.7	5.9	6.0	8.0	3.2	3.1	2.5
K/Ba	105	66	32	32	12	28	16	58	49	68
K/Rb	1160	660	176	630	344	420	55	580	440	300
Rb/Sr	0.007	0.022	0.089	0.014	0.029	0.045	0.13	0.046	0.078	0.145
La/Yb	1.0	1.7	10	4.2	3.7	11	32	1.1	6.3	4.0

*Total Fe as Fe_2O_3. (After Candie, 1982.)

(LIL) elements (such as K, Rb, Cs, Sr, Ba, Zr, U, Th and REE) and depleted isotopic ratios (i.e. low integrated Rb/Sr, U/Pb etc. ratios). Tholeiitic basalts have low viscosity and flow for long distances, constructing volcanic forms of large area and small slope. The major- and trace-element differences between the tholeiitic and calc-alkaline suites can be explained by varying proportions of olivine, plagioclase and pyroxene crystallizing from a basaltic parent melt.

The world-encircling midocean-ridge system accounts for more than 90% of the material flowing out of the Earth's interior. The whole ocean floor has been renewed by this activity in less than 200 Ma. Hotspots represent less than 10% of the material and the heat flow. The presence of a crust and core indicates that the Earth is a differentiated body. Major differentiation was contemporaneous with the accretion of the Earth. The high concentrations of incompatible elements in the crust and the high argon-40 content of the atmosphere indicate that the differentiation and degassing has been relatively efficient. The presence of helium-3 in mantle magmas shows, however, that outgassing has not been 100% efficient. The evidence from helium and argon is not contradictory. Helium dissolves more readily in magma than argon and the heavy noble gases so we expect helium degassing to be less effective than argon degassing. We also have no constraint on the initial helium content of the mantle or the total amount that has degassed. So helium may be mostly outgassed and what we see is just a small fraction of what there was. The evidence for differentiation and a hot early Earth suggest that much of the current magmatism is a result of recycling or the processing of already processed material. The presence of helium-3 in the mantle suggests that a fraction of the magma generated remains in the mantle; that is, magma removal is inefficient.

Generalities

In most models of basalt genesis, it is assumed not only that olivine and orthopyroxene are contained in the source rock but that these are the dominant phases. Petrologic and isotope data alone, however, cannot rule out a source that is mainly garnet and clinopyroxene. The eclogite (garnet plus clinopyroxene) source hypothesis differs only in scale and melt extraction mechanism from the fertile peridotite hypothesis. In the fertile peridotite, or pyrolite, model the early melting components, garnet and clinopyroxene, are distributed as grains in a predominantly olivine-orthopyroxene rock. On a larger scale, eclogite might exist as pods or blobs in a peridotite mantle. Since eclogite is denser than peridotite, at least in the shallowest mantle, these blobs would tend to sink and coalesce. In the extreme case an isolated eclogite-rich layer might form below the lighter peridotite layer. Such a layer could form by crustal delamination, subduction or by crystal settling in an ancient magma ocean. Melts from such blobs or layers still interact with olivine and orthopyroxene. If eclogite blobs are surrounded by refractory peridotite they can extensively melt upon adiabatic decompression, without the melt draining out. The isotopic evidence for isolated reservoirs and the geophysical evidence for gross layering suggest that differentiation and chemical stratification may be more important in the long run than mixing and homogenization.

Fluids and small-degree melts are LIL-enriched, and they tend to migrate upward. Sediments and altered ocean crust, also LIL-enriched, re-enter the upper mantle at subduction zones. Thus there are several reasons to believe that the shallow mantle serves as a scavenger of incompatible elements, including the radioactive elements (U, Th and K) and the key tracers (Rb, Sr, Nd, Sm and, possibly, Pb and CO_2). The continental crust and lithosphere are commonly assumed to be the main repositories of the incompatible elements, but oceanic islands, island arcs and deep-seated kimberlites also bring LIL-enriched material to the surface. Even a moderate amount of LIL in the upper mantle destroys the arguments for a primitive lower mantle or the need for a deep radioactive-rich layer.

It is becoming increasingly clear that all magma are blends of melts from an inhomogenous mantle. In fact, the source of magma has been described as a `statistical upper mantle assemblage`, or SUMA. Various apparent inconsistencies between the trace element ratios and isotopic ratios in basalt can be understood if (1) partial melting processes are not 100% efficient in removing volatiles and incompatible elements from the mantle; (2) basalts are hybrids or blends of magmas from depleted and enriched components; and (3) different basalts represent different averages, or different volume sampling. In a chemically heterogenous mantle mixing or contamination is inevitable. Even with this mixing it is clear from the isotopes that there are four or five ancient components in the mantle, components that have had an independent history for much of the age of the Earth.

Some components in a heterogenous mantle melt before others. Material rising from one depth level advects high temperatures to shallow levels, and can cause melting from the material in the shallow mantle. It is not necessary that material melt itself *in situ*. Cold but fertile subducted or delaminated material will melt as it warms up to ambient mantle temperature; excess temperatures are not required. Thus, there are a variety of ways of generating *melting anomalies*.

In a stratified mantle, without transfer of material between layers, convection in a lower layer can control the location of melting in the overlying layer. In a homogenous mantle, the high temperature gradient in the thermal boundary layer between layers is the preferred source of magma genesis since the melting gradient is larger than the adiabatic gradient in a homogenous convecting fluid (Figure 14.5). In a homogenous mantle, if the melting point is not exceeded in the shallow mantle, then it is unlikely to be exceeded at greater depth since the melting point and the geotherm diverge with depth. Material can leave a deep source region by several mechanisms.

(1) Melting in the thermal boundary at the bottom or the top of a region because of the high thermal gradient.
(2) Melting, or phase changes, due to adiabatic ascent of hotter or lower melting point regions.
(3) Entrainment of material by adjacent convecting layers.
(4) Heating of fertile blobs by internal radioactivity or by conduction of heat from the surrounding mantle.

Some of these mechanisms are illustrated in Figure 7.1.

Midocean ridge basalts

The most voluminous magma type, MORB, has low LIL-content and depleted isotopic ratios, with very low variance, that is, they are homogeneous. The Rb/Sr, Nd/Sm and U/Pb ratios have been low since >1 Ga. Since these ratios are high in melts and low in residual crystals, the implication is that the MORB source is a cumulate or a crystalline residue remaining after the removal of a melt fraction or residual fluid. Extraction of very enriched, very small-degree melts (<1%) is implied. In some geochemical models the LIL-enriched melt fraction is assumed to efficiently leave the upper mantle (the depleted MORB-source in these models) and enter, and become, the continental crust. The continental crust is therefore regarded as the complement – often the only complement – to the depleted MORB source. Early mass-balance calculations based on this premise, and on just a few elements and isotopes, suggested that most of the mantle remains undepleted or *primitive*. The depleted reservoir was assumed to occupy most or all of the upper mantle. Since the continental crust is the only enriched reservoir in this three-reservoir model, magmas that have high LIL-contents and $^{87}Sr/^{86}Sr$ ratios are assumed to be contaminated by the continental crust, or to contain a recycled ancient crustal component.

A depleted reservoir (low in LIL, low in Rb/Sr, $^{87}Sr/^{86}Sr$ and so on) can still be *fertile*, i.e. it can provide basalts by partial melting. A clinopyroxenite or gabbro cumulate, for example, can be depleted but fertile. Similarly, an enriched reservoir can be infertile, being low in Ca, Al, Na and so on. The fact that most of the mantle must be depleted as implied by mass-balance calculations, does not mean that it is fertile or similar to the MORB reservoir. Most of the volume of the mantle is depleted infertile- or barren-refractory residue of terrestrial accretion, and most of it is magnesium *perovskite* in the lower mantle.

Ocean-island basalts

The trace element and isotope ratios of basalts from ocean islands (OIB) differ from otherwise similar MORB. There is still no general agreement on how such variations are produced; chemical and isotope variability of the mantle is likely to be present everywhere, not just in the source regions of OIB. Both geochemical and geophysical observations, and plate-tectonic processes, require the upper mantle to be inhomogenous, and a variety of mechanisms have been suggested to produce such inhomogeneities. Perhaps the most obvious model for generating a variably fertile and inhomogeneous mantle is subduction, delamination and incomplete melt extraction. Variability is generated by recycling of oceanic and continental crust, and of seamounts, aseismic ridges, oceanic islands and sediments, all of which are being transported into subduction zones. Recycling of the oceanic crust involves the greatest mass flux of these various components but this does not require that such material is the dominant source of OIB. A popular model involves transport of cold slabs, including the crust, to the core–mantle boundary (CMB) where they heat up and generate ~3000 km deep narrow plumes that come up under oceanic islands. Enrichment of the lower part of the lithosphere by upward migrating metasomatic melts and fluids, and inefficient extraction of residual melts could produce extensive fractionation, and also allow the isotopic anomalies to form within a region that was not being stirred by mantle convection. Many oceanic islands and volcanic chains – often called hotspots and hotspot tracks – are on pre-existing lithospheric features such as fracture zones, transform faults, continental sutures, ridges and former plate boundaries. Volcanism is also associated with regions of lithospheric extension and thinning, and swells. The cause and effect relations are often not obvious. Oceanic islands are composed of both alkali and tholeiitic basalts. Alkali basalts are subordinate, but they appear to dominate the early and waning stages of volcanism. The newest submarine mountain in the Hawaii chain, Loihi, is alkalic. Intermediate-age islands are tholeiitic, and the latest stage of volcanism is again alkalic. The volcanism in the Canary Islands in the Atlantic changes from alkalic to tholeiitic with time.

Ocean-island basalts, or OIB, are LIL-rich and have enriched isotopic ratios relative to MORB. Their source region is therefore enriched, or depleted parent magmas have suffered contamination en route to the surface. The larger oceanic islands such as Iceland and Hawaii generally have less enriched magmas than smaller islands and seamounts. A notable exception is Kerguelen, which may be built on a micro-continent. Other volcanic islands may involve continental crust in a less obvious way. The islands in the Indian ocean and the south Atlantic are from mantle that was recently covered by Gondwana and may contain fragments of delaminated continental crust.

Enriched magmas such as alkali-olivine basalts, OIB and nephelinites occur on oceanic islands and have similar LIL and isotopic ratios to continental flood basalts (CFB) and continental alkali basalts. Continental contamination is unlikely in these cases unless the upper mantle is polluted by delaminated lower continental crust. Enriched MORB (EMORB) is also common along the global spreading ridge system and on near-ridge seamounts. There is a continuum between depleted and enriched tholeiites (DMORB, NMORB, TMORB, EMORB, PMORB, OIB). Veins in mantle peridotites and xenoliths contained in alkali basalts and kimberlites are also commonly enriched and, again, crustal contamination is unlikely unless continental crust somehow gets back into the mantle. In many respects enriched magmas and xenoliths are also complementary to MORB (in LIL contents and isotopic ratios), suggesting that there is ancient enriched material in the mantle. Island-arc basalts are also high in LIL, $^{87}Sr/^{86}Sr$, $^{143}Nd/^{144}Nd$ and $^{206}Pb/^{204}Pb$, suggesting that there are shallow – and global – enriched components. Back-arc-basin basalts (BABB) are similar to MORB in composition and, if the depth of the low-velocity zone and the depths of earthquakes can be used as a guide, tap a source deeper than 150 km. Many BABBs are intermediate in chemistry to MORB and OIB. This and other evidence indicates that enriched components – or enriched reservoirs – may be as shallow or shallower than the depleted MORB reservoir. On average, because of recycling and crustal delamination, enriched and fertile components may be preferentially collected in the shallow mantle. However, they may be dispersed throughout much of the upper mantle. It is quite probable that the upper mantle is also lithologically stratified. This would show up as scattered energy in short-period seismograms.

The trace-element signatures of some ocean-island basalts and some ridge basalts are consistent with derivation from recycled lower-continental crust. Others are apparently derived from a reservoir that has experienced eclogite fractionation or metasomatism by melts or other fluids from an eclogite-rich source. Kimberlites are among the most enriched magmas. Although they are rare, the identification of a kimberlite-like component in enriched magmas means that they may be volumetrically more important, or more wide-spread, than generally appreciated. Simple mass balance calculations suggest that kimberlite-like components may account for up to 0.05% of the mantle, or about 10% of the crustal volume. A small amount of such enriched magma can turn a depleted MORB into an enriched melt.

Trace-element and isotopic patterns of OIB overlap continental flood basalts (CFB), and common sources and fractionation patterns can be inferred. Island-arc basalts (IAB) also share many of the same geochemical characteristics, suggesting that the enriched character of these basalts may be derived from the shallow mantle. This is not the usual interpretation. The geochemistry of CFB is usually attributed to continental contamination or to a plume head from the lower mantle, that of IAB to sediment and hydrous melting of the mantle wedge, and that of OIB to 'primitive' lower mantle. In all three cases the basalts have evolved beneath thick crust or lithosphere, giving more opportunity for crystal fractionation and contamination by crust, lithosphere and shallow mantle prior to eruption. CFB and continental island arcs are on continental or craton boundaries, or on old sutures, and may therefore tap over-thickened continental or arc crust.

Rhyolites and silicic LIPs

Silicic volcanism is a component of large igneous provinces (LIPs) such as continental flood basalt provinces related to continental breakup and assembly. A crustal source is the main difference in silicic and mafic LIP formation. Large degrees of crustal partial melting, implied by the large volumes of rhyolitic magma, are controlled by the water content and composition of the crust, and the thermal input from the mantle via basaltic magmas. Rhyolites are more common and voluminous on the continents and continental margins than in the ocean basins. Silicic volcanism is associated with most CFB provinces, and some oceanic plateaus (e.g. Kerguelen, Wallaby and Exmouth).

Silicic LIPs - SLIPs - have erupted volumes similar to CFB, but are produced over time periods extending up to 40 Myr. The Kerguelen and Ontong–Java oceanic plateaus were erupted, episodically, over similar time-spans (30–40 Myr). The generation of SLIPs and oceanic plateaus requires the sustained input of basaltic mantle melts rather than the transient impact of a hot plume head. Suitable stress conditions of the plate and fertility of the mantle may be more important than the absolute temperature of the mantle. The final stages of ocean closure at a convergent margin or suture and the presence of batholiths with cumulate roots serve to fertilize the mantle and lower the melting point of the mantle and the hydrated crust.

Apparently hot magmas

Magmas that are associated with so-called hotspots differ in various ways from what are thought to be *normal* mid-ocean ridge basalts (i.e. along *normal* portions of the ridge system). There is little evidence, however, that they are hotter than normal basalts. High MgO contents are usually considered to be a diagnostic of high mantle temperatures but such magmas are rare at hotspots. These magmas include the following.

Boninite: A high-MgO, low-alkali, andesitic rock with textures characteristic of rapid crystal growth. Boninites are largely restricted to Pacific island arcs, and are more abundant in the early stages of magmatism.

Picrite: A high-MgO basalt extremely rich in olivine and pyroxene. In some petrological schemes, picrite is the parent magma of other basalts, prior to olivine separation. Picrites are common at the base of continental flood basalt sequences. They occur in Hawaii, and, possibly, in Iceland. Sometimes they have had olivine added by crystal settling and are not true high-MgO melts.

Komatiite: An ultramafic lava with 18–32% MgO. The more highly magnesian varieties are termed *peridotitic komatiite*.

Meimechite: A high-MgO, TiO_2-rich ultramafic volcanic rock composed of olivine, clinopyroxene, magnetite and glass occurs in association with flood basalts, Ni–Cu–PGE-bearing plutons, and diamond-bearing kimberlite and carbonatite.

All of these, plus kimberlites and carbonatites, have been taken by many authors as diagnostics of deep mantle plumes. Ocean-island basalts, almost by definition, and certainly by convention, are considered to be hotspot magmas. Although island-arc magmas and oceanic-island magmas differ in their petrography and inferred pre-eruption crystal fractionation, they overlap almost completely in the trace elements and isotopic signatures that are characteristic of the source region. It is almost certain that island-arc magmas originate above the descending slab, although it is uncertain as to the relative contribution of the slab and the intervening mantle wedge. Water from the descending slab may be important, but most of the material is presumed to be from the overlying 'normal' mantle. Whether the subducted oceanic crust also melts below island arcs is controversial. It is curious that the essentially identical range of geochemical signatures found in oceanic-island and continental flood basalts have generated a completely different set of hypotheses, generally involving a 'primitive' lower-mantle plume source. If island-arc magmas and boninites originate in the shallowest mantle, it is likely that island and continental basalts do as well. Subduction, and bottom side erosion and delamination, rather than plumes, are probably the primary causes of shallow mantle enrichment.

The main diagnostic difference between some islands and most other magmas is the higher maximum values and the greater spread of the $^3He/^4He$ ratios of the former. High ratios are referred to as 'primitive,' meaning that the helium-3 is assumed to be left over from the accretion of the Earth and that the mantle is not fully outgassed. The word 'primitive' has introduced semantic problems since high $^3He/^4He$ ratios are assumed to mean that the part of the mantle sampled has not experienced partial melting. This assumes that (1) helium is strongly concentrated into a melt (relative to U and Th), (2) the melt is able to efficiently outgas, (3) mantle that has experienced partial melting contains no helium and (4) helium is not retained in or recycled back into the mantle once it has been in a melt. As a matter of fact it is difficult to outgas a melt.

Magmas must rise to relatively shallow depths before they vesiculate, and even under these circumstances gases are trapped in rapidly quenched glass or olivine cumulates. Although most gases and other volatiles are probably concentrated into the upper mantle by magmatic processes, only a fraction manages to get close enough to the surface to outgas and enter the atmosphere. High helium-3 contents relative to helium-4, however, require that the gases evolved in a relatively low uranium–thorium reservoir for most of the age of the Earth. Shallow depleted reservoirs such as olivine cumulates may be the traps for helium.

Komatiites

Komatiites are ultrabasic melts that occur mainly in Archean rocks. The peridotitic variety apparently require temperatures of the order of 1450–1500 °C and degrees of partial melting greater than 60–70% in order to form by melting of dry peridotite. At one time such high temperatures and degrees of melting were thought to be impossible. Even today, similar degrees of partial melting of eclogite, at much lower temperatures, are considered unlikely. At high pressure it is possible to generate MgO-rich melts from peridotite with smaller degrees of partial melting. It is even possible that olivine is replaced as the liquidus phase by the high-pressure majorite phase of orthopyroxene, again giving high-MgO melts. Komatiites may represent large degrees of melting of a shallow olivine-rich parent, small degrees of melting of a deep peridotite source, melting of a rock under conditions such that olivine is not the major residual phase, or melting of wet peridotite at much lower temperatures. At depths greater than about 200 km, the initial melts of a peridotite may be denser than the residual crystals. This may imply that large degrees of partial melting are possible and, in fact, are required before the melts, or the source region, become buoyant enough to rise. The ratios of some trace-elements in some komatiites suggest that garnet has been left behind in the source region or that high-pressure garnet fractionation occurred prior to eruption. The existence of komatiites appears to refute the common assumption that melt-crystal separation must occur at relatively small degrees of partial melting. The rarity of komatiites since Precambrian times could mean that the mantle has cooled, but it could also mean that a suitable peridotite parent no longer exists at about 300 km depth, or that the currently relatively thick lithosphere prevents their ascent to the surface. Diapirs ascending rapidly from about 300 km from an eclogite-rich source region would be almost totally molten by the time they reach shallow depths, and basaltic magmas would predominate over komatiitic magma.

Komatiites may be the result of particularly deep melting. By cooling and crystal fractionation komatiitic melts can evolve to less dense picritic and tholeiitic melts. Low-density melts, of course, are more eruptable. It may be that komatiitic melts exist at present, just as they did in the Precambrian, but, because of the colder shallow mantle they can and must cool and fractionate more, prior to eruption. In any case komatiites provide important information about the physics and chemistry of the upper mantle.

The major-element chemistry of komatiites and related magmas (komatiitic basalts) are more similar to modern mafic subduction magmas than to magmas thought to be produced by high temperatures. Komatiites with high SiO_2-contents show similarities to modern subduction magmas (boninites). SiO_2 contents of magmas generally decrease as the pressure of melting increases therefore high SiO_2 is difficult to explain in a plume scenario. Boninites have similarly high SiO_2 at high MgO contents because they are high-degree melts produced at shallow depths. In an anhydrous, high-temperature decompression melting regime, high degrees of melting begin at great depth; in the case of boninites, large extents of melting can occur at shallow depth because of high H_2O contents (Crawford *et al.*, 1989; Parman *et al.*, 2001).

The chemical similarities between boninites, basaltic komatiites and komatiites suggests that komatiites were produced by similar melting processes that produce modern boninites, with the primary difference being that the mantle was about 100 °C hotter in the Archean. Rather than being products of deep mantle thermal plumes, komatiites may be products of normal plate tectonic processes.

Komatiitic rocks are depleted in both light and heavy rare earth elements (LREE and HREE) relative to middle REE (MREE) and possess relatively high TiO2 even in the most LREE-depleted varieties. Uncontaminated komatiites and picrites have similar isotopic ratios overlapping MORB, indicating generation from a mantle source with a long-term depletion in LREE. Geochemical characteristics of the komatiite–picrite association, including REE and Nb/Y–Zr/Y systematics, indicate chemical heterogeneities in the source region. The high MgO contents of the rocks has been used to support a mantle plume model for their genesis but they do not have the enriched or primitive isotopic signatures expected for plumes and significant regional uplift before volcanism is lacking.

Hot mantle

Hot mantle intersects the solidus at different depths and it melts over a larger pressure interval and to greater extents than cooler mantle. This is why melting anomalies have been called *hotspots*, and why they are attributed to hot upwelling plumes. The melts produced at high-temperature have higher MgO and FeO and lower Na_2O, Al_2O_3 and SiO_2 and lower LIL-contents than melts of cooler mantle.

Archean volcanism produced high-MgO magmas called *komatiites* that seemed to fit the criteria for being the products of super-hot plumes. These magmas, however, may be produced by hydrous melting processes in the upper mantle, possibly associated with subduction (Parman and Grove, 2001). [mantleplumes]

Some modern subduction related magmas, boninites, show a large compositional overlap with Archean basaltic komatiites. Komatiites have a wide range of major and minor element composition. High-SiO_2 komatiites resemble modern boninite magmas that are produced by hydrous melting, while low-SiO_2 komatiites resemble more closely modern basalts produced by anhydrous decompression melting. If high MgO magmas represent deep plume sources then the isotopic and trace element patterns should resemble ocean island basalts, but they do not. The low-TiO_2 contents of komatiite magmas requires their source to have been depleted (i.e. to have had a melt extracted). The isotopic compositions of komatiites overlap those of depleted

MORB rather than being similar to those exhibited by OIB or hotspots.

Melting temperatures can be lowered by the presence of water, carbon dioxide, and eclogite. Boninites are high-degree melts produced at shallow depths. High degrees of melting in hot upwellings necessarily must begin at great depth; in the case of boninites, large extents of melting occur at shallow depth because the melting is facilitated by high water contents. Fertile mantle, such as eclogite, also melts at higher pressure and at lower temperature than peridotite, and can therefore have a long melting column, and produce large amounts of basalt.

Kimberlite

Some rocks are rare but are so enriched in certain key elements that they cannot be ignored in any attempt to reconstruct the composition of the mantle. Kimberlite is a rare igneous rock that is volumetrically insignificant compared with other igneous rocks. Kimberlite provinces themselves, however, cover very broad areas and occur in most of the world's stable craton, or shield, areas. Kimberlites are best known as the source rock for diamonds, which crystallized at pressures greater than about 50 kilobars. They carry other samples from the upper mantle that are the only direct samples of mantle material below about 100 km. Some kimberlites appear to have exploded from depths as great as 200 km or more, ripping off samples of the upper mantle and lower crust in transit. The fragments, or xenoliths, provide samples unavailable in any other way. Kimberlite itself is an important rock type that provides important clues as to the evolution of the mantle. It contains high concentrations of lithophile elements (Table 14.2) and higher concentrations of the most incompatible trace elements (Pb, Rb, Ba, Th, Ce, La) than any other ultrabasic rock. Kimberlite is also enriched in elements of ultramafic affinity (Cr and Ni).

Diamond-bearing kimberlites are usually close to the craton's core, where the lithosphere may be thickest. Barren kimberlites (no diamonds) are usually on the edges of the tectonically stable areas. Kimberlites range in age from Precambrian to Cretaceous. Some areas have been subjected to kimberlite intrusion over long periods of geological time, suggesting lithospheric control.

Table 14.2 | Kimberlite composition compared with ultrabasic and ultramafic rocks

Oxide	Average Kimberlite	Average Ultrabasic	Average Ultramafic
SiO_2	35.2	40.6	43.4
TiO_2	2.32	0.05	0.13
Al_2O_3	4.4	0.85	2.70
	6.8	—	—
FeO	2.7	12.6	8.34
MnO	0.11	0.19	0.13
MgO	27.9	42.9	41.1
CaO	7.6	1.0	3.8
Na_2O	0.32	0.77	0.3
K_2O	0.98	0.04	0.06
H_2O	7.4	—	—
CO_2	3.3	0.04	—
P_2O_5	0.7	0.04	0.05

Wederpohl and Muramatsu (1979).
Dawson (1980).

Compared to other ultrabasic rocks – lherzolites, dunites, harzburgites – kimberlites contain high amounts of K, Al, Ti, Fe, Ca, C, P and water. For most incompatible trace elements kimberlites are the most enriched rock type; important exceptions are elements that are retained by garnet and clinopyroxene. Carbonatites may be enriched in U, Sr, P, Zr, Hf and the heavy REEs relative to kimberlites. Since kimberlites are extremely enriched in the incompatible elements, they are important in discussions of the trace-element inventory of the mantle. Such extreme enrichment implies that kimberlites represent a small degree of partial melting of a mantle silicate or a late-stage residual fluid of a crystallizing cumulate layer. The LIL elements in kimberlite show that it has been in equilibrium with a garnet–clinopyroxene-rich source region, possibly an eclogite cumulate. The LIL contents of kimberlite and MORB are complementary. Removal of a kimberlite-like fluid from an eclogite cumulate gives a crystalline residue that has the required geochemical characteristics of the depleted source region that provides MORB. Adding a kimberlitic fluid to a depleted basalt can make it similar to enriched magmas.

Carbonatites and other ultramafic alkaline rocks, closely related to kimberlites, are widespread. They are probably common in the upper mantle, although they have been argued to come from the lower mantle because they are not MORB. Kimberlites provide us with a sample of magma that probably originated below about 200 km and, as such, contain information about the chemistry and mineralogy of the mantle in and below the continental lithosphere. Kimberlites are anomalous with respect to other trace-element enriched magmas, such as nephelinites and alkali basalts, in their trace-element chemistry. They are enriched in the very incompatible elements such as rubidium, thorium and LREE, consistent with their representing a small degree of partial melting or the final concentrate of a crystallizing liquid. The extreme enrichments of kimberlitic magmas in incompatible elements (e.g. compared with a typical undepleted mantle) are usually attributed to low degrees of melting and/or metasomatized source compositions. The observed enrichment of kimberlitic magmas with rare earth elements (REE) can be explained in terms of melt migration through source rocks having the composition of normal mantle.

Kimberlites are relatively depleted in the elements (Sc, Ti, V, Mn, Zn, Y, Sr and the HREE) that are retained by garnet and clinopyroxene. They are also low in silicon and aluminum, as well as other elements (Na, Ga, Ge) that are geochemically coherent with silicon and aluminum. This suggests that kimberlite fluid has been in equilibrium with an eclogite residue. Kimberlites are also rich in cobalt and nickel.

Despite their comparative rarity, disproportionately high numbers of eclogite xenoliths have been found to contain diamonds. Diamond is extremely rare in peridotitic xenoliths. Eclogitic garnets inside diamonds imply a depth of origin of about 200–300 km if mantle temperatures in this depth range are of the order of 1400–1600 °C. Seismic velocities in the transition region are consistent with piclogite (eclogite plus peridotite). The bulk modulus in the transition zone for olivine in its spinel forms is higher than observed.

Alkali basalts have LIL concentrations that are intermediate to MORB and kimberlite. Although kimberlite pipes are rare, there may be a kimberlite-like component (Q) dispersed throughout the shallow mantle. Indeed, alkali basalts can be modeled as mixtures of a depleted magma (MORB) and an enriched magma (kimberlite), as shown in Table 14.3. Peridotites with evidence of secondary enrichment may also contain a kimberlite-like component.

Carbonatite

Carbonatites are igneous rocks that typically occur in regions of lithospheric extension of thick continental plates, i.e. continental rift zones (e.g. East Africa) and on the margins of continental flood-basalt provinces (e.g. Parana-Etendeka, Deccan). They have also been found on ocean-islands on continental margins (e.g. Cape Verde and Canary Island carbonatites). The close spatial and temporal association of kimberlites and carbonatites suggests that these rocks are genetically related, although diamond-bearing kimberlites may come from deeper in the upper mantle. Kimberlites are rare on oceanic islands. This may be because the adiabatic ascent of the parent rock through the asthenosphere generates large-degree melting, while small-degree melts can be better preserved if the eruption is through cold continental mantle. Both carbonatites and kimberlites have enriched trace-element patterns, consistent with small-degree melts. Carbonatites have depleted isotopic signatures while some kimberlites are enriched. Kimberlites and carbonatites provide an opportunity to study low-degree partial melting in the mantle. This is important since the removal of low-degree partial melts seems to be the mechanism for depleting the mantle. Since much of this depletion apparently occurred in early Earth history, when the mantle was much hotter, the implication is that this melting and melt removal occurred at shallow depths, probably in the thermal boundary layer and the lithosphere. The solidi of CO_2-bearing rocks is much lower than volatile-poor rocks; this allows

Table 14.3 Trace-element chemistry of MORB, kimberlite and intermediate (alkali) basalts (ppm)

		Alkali Basalts					Kimberlite	
Element	MORB	EPR*	Australia	Hawaii	BCR†	Theoretical**	Average	Max
Sc	37	27	25	33	34–37	35	30	—
V	210	297	260	170	399	197–214	120	250
Co	53	41	60	56	38	57–61	77	130
Ni	152	113	220	364	15	224–297	1050	1600
Rb	0.36	13	7	24	47	10–45	65	444
Sr	110	354	590	543	330	200–289	707	1900
Y	23	37	20	27	37	23–28	22	75
Zr	70	316	111	152	90	97–133	250	700
Nb	3.3	—	24	24	14	19–48	110	450
Ba	5	303	390	350	700	150–580	1000	5740
La	1.38	21	24	23	26	16–31	150	302
Ce	5.2	46	47	49	54	31–57	200	522
Nd	5.6	28	24	23	29	16–26	85	208
Sm	2.1	6.7	6.8	5.5	6.7	3.1–4.8	13	29
Eu	0.8	2.1	2.1	2.0	2.0	1.1–1.4	3	6.5
Tb	0.5	1.2	0.92	0.87	1.0	0.6–0.7	1	2.1
Yb	2.1	3.4	1.6	1.8	3.4	2.0–2.1	1.2	2.0
Lu	0.34	0.49	0.28	0.23	0.55	0.32–0.34	0.16	0.26
Hf	1.4	5.3	2.8	3.9	4.7	2.2–4.3	7	30
Ta	0.1	9	1.8	2.1	0.9	1.3–2.5	9	24
Pb	0.08	—	4.0	5	18	1.6–5.1	10	50
Th	0.35	2.7	2.0	2.9	6.8	2.4–5.4	16	54
K	660	8900	4700	8800	1200	3200–4770	17 600	41 800
U	0.014	0.51	0.4	0.6	1.8	0.48–1.84	3.1	18.3

*EPR = East Pacific Rise.
†BCR = Basalt, Columbia River (USGS Standard Rock).
**Mixture ranging from 85 percent MORB plus 15 percent Average Kimberlite to 90 percent MORB plus 20 percent Max Kimberlite.

relatively cold rocks to have very low seismic velocities. The low-velocity zone in the upper mantle and the low-velocity regions under some so-called hotspots may be due to this rather than to high absolute temperatures.

There is no particular reason for supposing that carbonatites come from deep in the mantle, although their isotopic similarity with some OIB has led to suggestions that they come from the lower mantle. This is circular reasoning and is based on the assumption that only MORB originates in the upper mantle.

The isotopic components HIMU, EM1, FOZO and DMM (depleted MORB mantle) have been identified in carbonatites. HIMU may represent recycled crust, or lower continental crust, dehydrated and converted to eclogite. HIMU sources are enriched in U, Nb and Ta relative to MORB. HIMU islands include Mangaia, and St. Helena. EM1 has extremely low $^{143}Nd/^{144}Nd$ and very low lead isotopic ratios compared to other OIB components. Pitcairn and the Walvis Ridge are the type locations for EM1 basalts. EM1 sources may be recycled crust plus pelagic sediment or metasomatized continental lithosphere. The EM1 component has $\delta^{18}O$ similar to MORB sources and peridotite xenoliths, ruling out a large contribution of subducted sediment.

The most likely place to generate low-degree melting in a low-melting point mantle with CO_2 is in the outer reaches, probably in the conduction layer or above 200 km. The EM1, HIMU and FOZO components may all originate in the upper and lower crust, the sub-continental lithosphere and the shallow mantle. They can enter the source region of OIB and carbonatites by erosion, subduction or delamination. In the absence of subduction, the lower crustal delamination explanation is preferred. A wide-spread but sporadic, and shallow depleted FOZO source provides an explanation for carbonatites. The 'depleted mantle' signatures may be attributed to differentiation events in the mantle that produced depleted sub-continental lithosphere and continental crust. Carbonatite isotope data suggest that a shallow source in the lithosphere or perisphere could be FOZO mantle. It could have existed for billions of years as an approximately closed system if it is buoyant and of high viscosity.

Carbonatites may represent the initial melting of carbonated mantle peridotite and kimberlites may represent larger degrees of partial melting after the CO_2 is exhausted. There is also the possibility that carbonated eclogite produces carbonatitic melt under various depth conditions in the upper mantle and that this melt metasomatizes mantle peridotite. The source of the eclogite could be delaminated lower continental crust, or subducted oceanic crust.

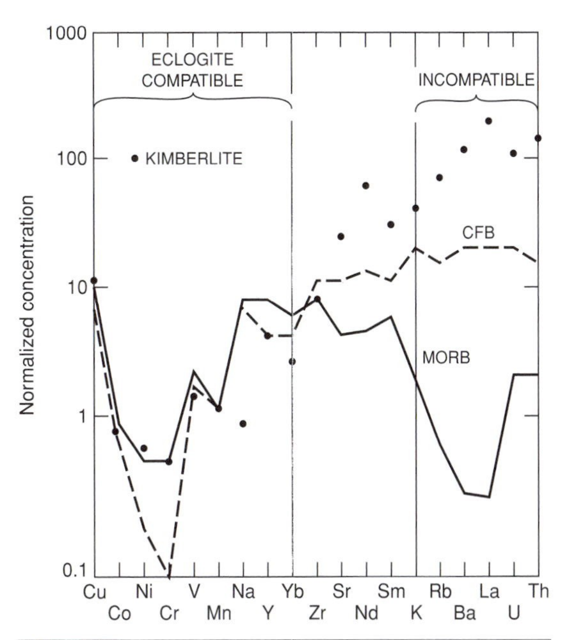

Fig. 14.1 Trace-element concentrations in MORB, continental flood basalts (CFB) and kimberlites. The elements to the right are incompatible in all major mantle phases (olivine, pyroxene and garnet) while those to the left are retained in the eclogite minerals (clinopyroxene and garnet). Note the complementary pattern between MORB and kimberlite and the intermediate position of CFB. Concentrations are normalized to estimates of mantle composition.

The Kimberlite–MORB connection

Kimberlites are enriched in the LIL elements, especially those that are most depleted in MORB. Figure 14.1 gives the composition of kimberlites, MORB and continental tholeiites. The complementary pattern of kimberlite and MORB is well illustrated as is the intermediate position of continental tholeiites (CFB). Note that kimberlite is not enriched in the elements that are retained by the eclogite minerals, garnet and clinopyroxene. This is consistent with kimberlite having been a fluid in equilibrium with subducted oceanic crust or an eclogite cumulate. If a residual cumulate is the MORB reservoir, the ratio of an incompatible element in kimberlite relative to the same element in MORB should be the same as the solid/liquid partition coefficient for that element. This is illustrated in Figure 14.2. The solid line is a profile of the MORB/kimberlite ratio for a series of incompatible elements. The vertical lines bracket the solid/liquid partition coefficients for garnet and clinopyroxene. Although MORB is generally regarded as an LIL-depleted magma and kimberlite is ultra-enriched in most of the incompatible elements, MORB is enriched relative to KIMB in yttrium, ytterbium and scandium, elements that have a high solid/melt partition coefficient for an eclogite residue. The trend of the MORB/KIMB ratio is the same as the partition coefficients, giving credence to the idea that enriched magmas, such as kimberlite, and MORB are genetically related.

The LIL content of continental tholeiites and alkali-olivine basalts suggest that they are mixtures of MORB and a melt from a more enriched part of the mantle, or blends of high-degree melts

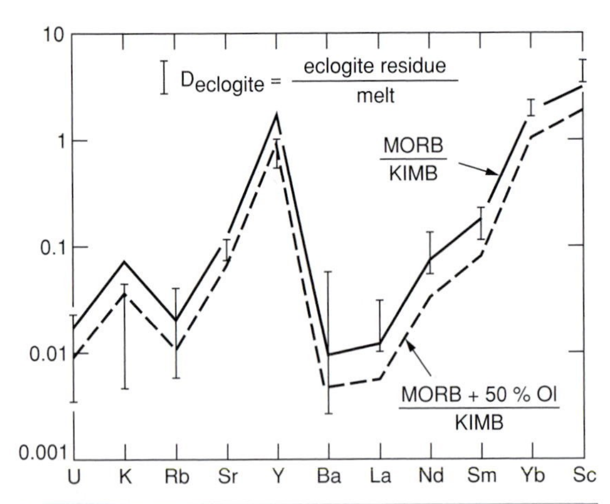

Fig. 14.2 Solid line is ratio of concentrations in MORB and kimberlite (KIMB). Vertical bars are solid/liquid partition coefficients for garnet plus clinopyroxene. The dashed line gives the ratio for MORB plus 50 percent olivine, a possible picrite parent magma for MORB. If the MORB or picrite source region is the crystalline residue remaining after removal of a kimberlitic fluid, the ratio of concentrations, MORB/KIMB or picrite/KIMB, should equal the solid–liquid partition coefficient, which depends on the crystalline (xl) phases in the residue.

and low-degree melts. If diapirs initiate in a deep depleted layer, they must traverse the shallow mantle during ascent, and cross-contamination seems unavoidable. The more usual model is that the whole upper is depleted and enriched magmas originate as plumes in the deep mantle.

The Kimberlite--KREEP Relation

KREEP is a lunar material having very high concentrations of incompatible elements (K, REE, P, U, Th). It is thought to represent the residual liquid of a crystallizing magma ocean. Given proposals of a similar origin for kimberlite, it is of interest to compare the composition of these two materials. An element by element comparison gives the remarkable result that for many elements (K, Cs, P, S, Fe, Ca, Ti, Nb, Ta, Th, U, Ba and the LREE) kimberlite is almost identical (within 40%) to the composition of KREEP (Figures 14.3 and 14.4). This list includes compatible and incompatible elements, major, minor and trace elements, and volatiles as well as refractories. KREEP is relatively depleted in strontium and europium, elements that have been removed from the KREEP source region by plagioclase fractionation. Kimberlite is depleted in sodium, HREE, hafnium, zirconium and yttrium, elements that are removed by garnet-plus-clinopyroxene fractionation. It appears that the differences between KIMB and KREEP are due to difference in pressure between the Earth and the Moon; garnet is not stable at pressures occurring in the lunar mantle. The plagioclase and garnet signatures show through such effects as the different volatile-refractory ratios of the two bodies and expected differences in degrees of partial melting, extent of fractional crystallization and other features. The similarity in composition extends to metals of varying geochemical properties and volatilities such as iron, chromium, manganese as well as phosphorus. KREEP is depleted in other metals (V, Co, Ni, Cu and Zn), the greater depletions occurring in the more volatile metals and the metals that are partitioned strongly into olivine. This suggests that olivine has been more important in the evolution of KREEP than in the evolution of kimberlite, or that cobalt, copper and nickel are more effectively partitioned into MgO-rich fluid such as kimberlite.

It appears from Figure 14.3 that KREEP and kimberlite differ from chondritic abundances in similar ways. Both are depleted in volatiles relative to refractories, presumably due to preaccretional processes. Strontium is less enriched than the other refractories, although this is much more pronounced for KREEP. The pronounced europium and strontium anomalies for KREEP are consistent with extensive plagioclase removal. The HREE, yttrium, zirconium and hafnium are relatively depleted in kimberlite, suggesting eclogite fractionation or a garnet-rich source region for kimberlite. The depletion of scandium simply indicates that olivine, pyroxene or garnet have been in contact with both KREEP and KIMB. The depletion of sodium in KIMB is also consistent with the involvement of eclogite in its history. The depletion of sodium in KREEP, relative to the other alkalis, is presumably due to the removal of feldspar, and the greater relative depletion of sodium in kimberlite therefore

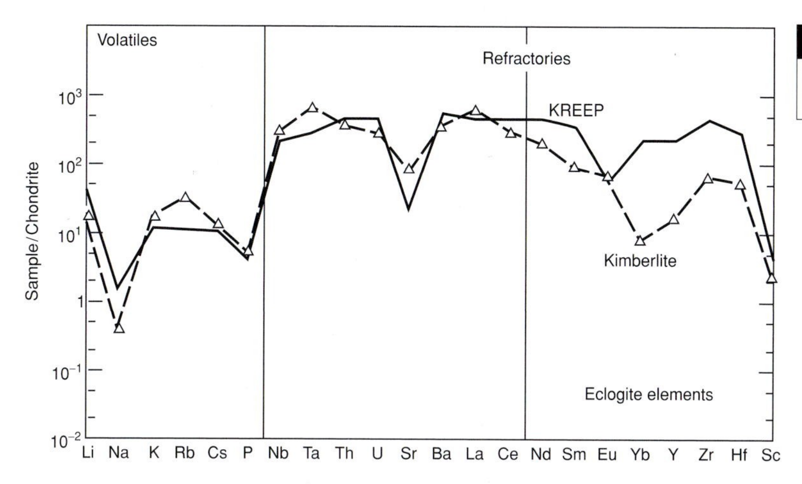

Fig. 14.3 Chondrite-normalized trace-element compositions for kimberlite and lunar KREEP.

requires another process, such as removal of a jadeite component by eclogite fractionation.

Chromium, manganese and iron are approximately one-third the chondritic level in both KREEP and kimberlite. This is about the level in the mantles of these bodies. Nickel and cobalt in kimberlite are about the same as estimated for the Earth's mantle. These elements are extremely depleted in KREEP, indicating that they have been removed by olivine or iron extraction from the source region.

The similarity of kimberlite and KREEP is shown in Figure 14.4. For many elements (such as Ca, Ba, Nd, Eu, Nb, Th, U, Ti, Li and P) the concentrations are identical within 50%. Kimberlite is enriched in the volatiles rubidium, potassium and sulfur, reflecting the higher volatile content of the Earth. Kimberlite is also relatively enriched in strontium and europium, consistent with a prior extraction of plagioclase from the KREEP source region.

Kimberlite and lunar KREEP are remarkably similar in their minor- and trace-element chemistry. The main differences can be attributed to plagioclase fractionation in the case of KREEP and eclogite fractionation in the case of kimberlite. KREEP has been interpreted as the residual fluid of a crystallizing magma. In a small body the Al-content of a crystallizing melt is reduced by plagioclase crystallization and flotation. In a magma ocean on a large body, such as the Earth, the Al_2O_3 is removed by sinking or residual garnet. Kimberlite is depleted in eclogite elements including HREE and sodium. Kimberlite may represent the late-stage residual fluid of a crystallizing terrestrial magma ocean. A buried eclogite-rich cumulate layer is the terrestrial equivalent of the lunar anorthositic crust.

Removal of a kimberlite-like fluid from a garnet-clinopyroxene-rich source region gives a crystalline residue that has the appropriate trace-element chemistry to be the reservoir for LIL-depleted magmas such as MORB. Enriched fluids permeate the shallow mantle and are responsible for the LIL-enrichment of island-arc and oceanic island basalts.

Alkali basalt

Continental alkaline magmatism may persist over very long periods of time in the same region and may recur along lines of structural weakness after a long hiatus. The age and thickness of the lithosphere play an important role, presumably by controlling the depth and extent of crystal fractionation and the ease by which the magma can rise to the surface.

The non-random distribution of alkaline provinces has been interpreted alternatively as hotspot tracks and structural weaknesses in the lithosphere. What have been called *hotspots* may be passive centers of volcanism whose location is determined by pre-existing zones of weakness, rather than manifestations of deep mantle plumes. Alkaline basalts both precede and follow abundant tholeiitic volcanism. In general alkaline basalts erupt through thick lithosphere

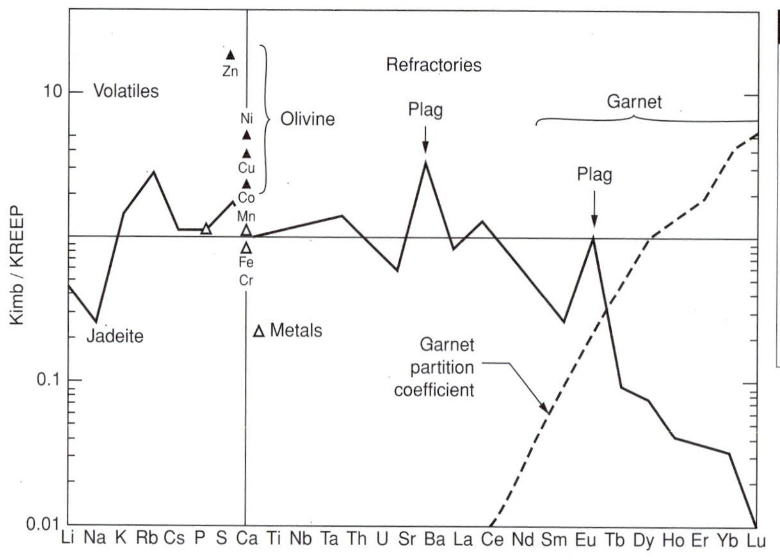

Fig. 14.4 Trace-element concentrations in kimberlite relative to KREEP. Elements that are fractionated into olivine, plagioclase, garnet and jadeite are noted. The solid/liquid partition coefficient for garnet is shown. If kimberlite has been in equilibrium with garnet, or an eclogite cumulate, the concentrations in kimberlite will be proportional to the reciprocal of the garnet partition coefficient.

and are early when the lithosphere is thinning (rifting) and late when the lithosphere is thickening (flanks of ridges, downstream from midplate volcanic centers).

On Kerguelen, the third largest oceanic island, and on Ascension Island, the youngest basalts are alkalic and overlie a tholeiitic base. In Iceland, tholeiites dominate in the rift zone and grade to alkali basalts along the western and southern shores. Thus, there appears to be a relation between the nature of the magmatism and the stage of evolution. The early basalts are tholeiitic, change to transitional and mixed basalts and terminate with alkaline compositions. It was once thought that Hawaii also followed this sequence, but the youngest volcano, Loihi, is alkalic and probably represents the earliest stage of Hawaiian volcanism. It could easily be covered up and lost from view by a major tholeiitic shield-building stage. Thus, oceanic islands may start and end with an alkalic stage.

In continental rifts, such as the Afar trough, the Red Sea, the Baikal rift and the Oslo graben, the magmas are alkalic until the rift is mature and they are then replaced by tholeiities, as in a widening oceanic rift. A similar sequence occurs in the Canary Islands and the early stages of Hawaii. The return of alkalics in the Hawaiian chain may occur when the islands drift away from a fertile blob. Seamounts on the flanks of midocean ridges, may be tapping the flanks of a mantle heterogeneity, where crystal fractionation and contamination by shallow mantle are most prevalent. In all cases lithospheric thickness appears to play a major role. This plus the long duration of alkalic volcanism, its simultaneity over large areas and its recurrence in the same parts of the crust argue against the simple plume concept.

Alkali-rich basalts generally have trace-element concentrations intermediate between MORB and highly alkalic basalts such as nephelinite. In fact, they can be treated as mixtures of MORB, or a picritic MORB parent, and nephelinite. Alkaline rocks are the main transporters of mantle inclusions.

Continental flood basalts

Continental flood basalts (CFB), the most copious effusives on land, are mainly tholeiitic flows that can cover very large areas and are typically 3–9 km in thickness. These are also called traps, plateau basalts and fissure basalts, and large igneous provinces (LIPs). Examples are the Deccan traps in India, the Columbia River province in the USA, the Parana basin in Brazil, the Karoo province in South Africa, the Siberian traps, and extensive flows in western Australia, Tasmania, Greenland and Antarctica. They have great similarity around the world. Related basalts are found in recent rifts such as East Africa and the Basin and Range province. Rift and flood basalts may both form during an early stage of continental rifting.

In some cases alkali-rich basalts occurring on the edge of rifts were erupted first, while tholeiites occur on the floor of the rift. Continental alkali basalts and tholeiites are highly enriched in LIL elements and have high $^{87}Sr/^{86}Sr$ ratios. There are many similarities between continental and ocean-island basalts. Fractional crystallization or varying degrees of partial melting can produce various members of the alkali series. The large range of isotopic ratios in such basalts indicates that their source region is inhomogeneous or that mixing between magma types is involved. The similarity between many continental basalts and ocean-island basalts indicates that involvement of the *upper* continental crust is not necessary to explain all continental flood basalts.

Continental flood basalts have been the subject of much debate. Their trace-element and isotopic differences from depleted midocean-ridge basalts have been attributed to their derivation from enriched mantle, primitive mantle, magma mixtures or from a parent basalt that experienced contamination by the continental crust. In some areas these flood basalts appear to be related to the early stages of continental rifting or to the proximity to an oceanic spreading center. In other regions there appears to be a connection with subduction, and in these cases they may be analogous to back-arc basins. In most isotopic and trace-element characteristics they overlap basalts from oceanic islands and island arcs.

Chemical characteristics considered typical of continental tholeiites are high SiO_2 and incompatible-element contents, and low MgO/FeO and compatible-element concentrations. The very incompatible elements such as cesium and barium can be much higher than in typical ocean island basalts. Many CFB, generally speaking, are in back-arc environments. The Columbia River flood basalts are not necessarily typical but they appear to have been influenced by the subduction of the Juan de Fuca plate beneath the North American plate and may have involved interaction with old enriched subcontinental mantle. The old idea that continental flood basalts are derived from a primordial 'chondritic' source does not hold up when tested against all available trace-element and isotopic data, even though some of these basalts have neodymium isotopic ratios expected from a primitive mantle.

The relationships between some continental basalts and oceanic volcanism are shown in Figure 14.5. The basalts in the hatched areas are generally attributed to hotspot volcanism. The dashed lines show boundaries that apparently represent sutures between fragments of continents.

Andesite

Andesites are associated primarily with island arcs and other convergent plate boundaries and have bulk compositions similar to the continental crust. Tholeiites and high-MgO magmas are also found in island arcs. The accretion of island arcs onto the edges of continents is a primary mechanism by which continents grow or, at least, maintain their area. Delamination of the lower mafic arc crust may be the main mechanism of crustal recycling and an important mechanism for creating fertile heterogeneities, or melting anomalies, in the mantle.

Andesites are lavas with >54 wt.% SiO_2. Andesites can, in principle, originate as primary melts from the mantle, as melts of continental crust, by contamination of melts by continental crust, by partial fusion of hydrous peridotite, by partial fusion of subducted oceanic crust, by differentiation of basalt by crystal fractionation, or by magma mixing of melts from a variety of sources. Andesites may involve interactions between mantle peridotite and partial melts of subducted basalt, eclogite and sediments. Arc magmas are enriched in thorium and lanthanum, and depleted in niobium and tantalum – relative to thorium and lanthanum – this may involve dehydration of subducted crust.

Melting of hydrous peridotite and eclogite and low-pressure differentiation of basalt

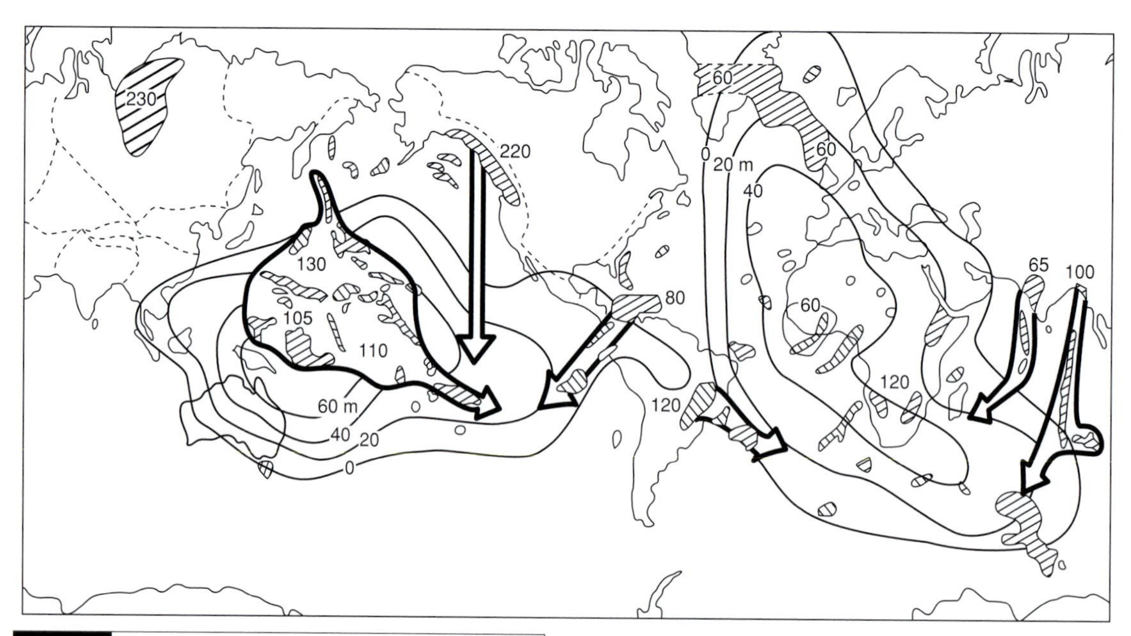

Fig. 14.5 Map of continental flood basalts, selected volcanic islands, and oceanic plateaus, with ages in millions of years. The period between 130 and 65 Ma was one of extensive volcanism and most of this occurred inside geoid highs (contours) and areas of high elevation. Some of the basalt provinces have drifted away from their point of origin (arrows). (See also Fig. 5.5.)

by crystal fractionation may all be involved. Andesites probably originate in or above the subducted slab and, therefore, at relatively shallow depths in the mantle. The thick crust in most andesitic environments means that shallow-level crystal fractionation and crustal contamination are likely. In their incompatible-element and isotopic chemistry, andesites are similar to ocean-island basalts and continental flood basalts. They differ in their bulk chemistry.

Calc-alkalic andesite production involves mixing of two end-member magmas, a mantle-derived basaltic magma and a crust-derived felsic magma. The bulk continental crust possesses compositions similar to calc-alkalic andesites. If continental crust is made at arcs, the mafic residue after extraction of felsic melts must be removed – delaminated from the initial basaltic arc crust – in order to form andesitic crust. Chemically modified oceanic materials and delaminated mafic lower crust materials, create chemical and melting anomalies in the shallow mantle and may be responsible for the magmatism associated with continental breakup and the formation of oceanic plateaus some 40–80 Myr after breakup.

Volcanic or island arcs typically occur 100–250 km from convergent plate boundaries and are of the order of 50–200 km wide. The volcanoes are generally about 100–200 km above the earthquake foci of the dipping seismic zone. Crustal thicknesses at island arcs are generally 20–50 km thick. The upper bound may be the basalt-eclogite phase change boundary and the level of crustal delamination. Andesitic volcanism appears to initiate when the slab has reached a depth of 70 km and changes to basaltic volcanism when the slab is no longer present or when compressional tectonics changes to extensional tectonics. Figure 14.6 shows that island-arc basalts (hatched) overlap ocean-island and continental basalts in isotopic characteristics. A similar mantle source is indicated, and it is probably shallow.

Back-arc-basin basalts (BABB) are similar to midocean-ridge basalts but in many respects they are transitional toward island-arc tholeiites and ocean-island basalts (Tables 14.4, 14.5 and 14.6). They are LIL-enriched compared with normal MORB and generally have LREE enrichment. High $^3He/^4He$ ratios have been found in some BABB, in spite of the fact that they are cut off from

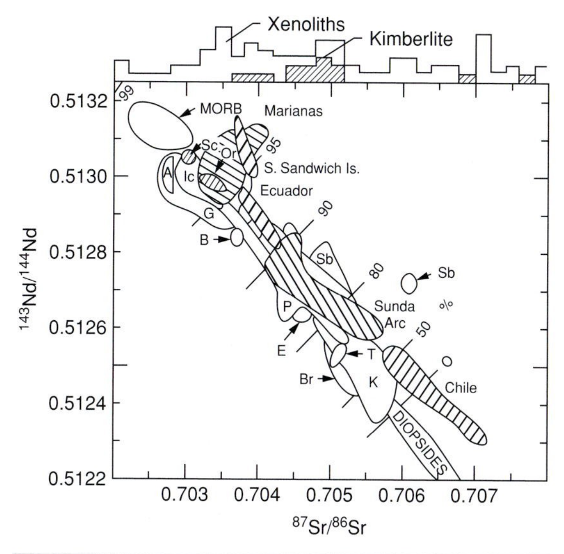

Fig. 14.6 Neodymium and strontium isotopic ratios for oceanic, ocean island, continental and island-arc (hatched) basalts and diopside inclusions from kimberlites. The numbers are the percentage of the depleted end-member if this array is taken as a mixing line between depleted and enriched end members. Mixing lines are flat hyperbolas that are approximately straight lines for reasonable choices of parameters. The fields of MORB and CFB correspond to >97% and 70–95%, respectively, of the depleted end-member. The enriched end-member has been arbitrarily taken as near the enriched end of Kerguelen (K) basalts. The most enriched magmas are from Kerguelen, Tristan da Cunha (T) and Brazil (Br). Other abbreviations are Sc (Scotia Sea), A (Ascension), Ic (Iceland), G (Gouch), Or (Oregon), Sb (Siberia), P (Patagonia) and E (Eifel). Along top are strontium isotopic data for xenoliths and kimberlites.

the lower mantle by a slab. Yellowstone is not in a classic back-arc basin but it has high $^3He/^4He$ in spite of the underlying thick U-rich crust and no deep low-velocity zone. BABB are similar to, or gradational toward, ocean-island basalts and other hotspot magmas.

In most respects active back-arc basins appear to be miniature versions of the major oceans, and the spreading process and crustal structure and composition are similar to those occurring at midocean ridges. The age–depth relationships of marginal seas, taken as a group, are indistinguishable from those of the major oceans. To first order, then, the water depth in marginal sea basins is controlled by the cooling, contraction and subsidence of the oceanic lithosphere, and the density of the mantle under the spreading centers is the same. The marginal basins of Southeast Asia, however, tend to be deeper than similar age oceanic crust elsewhere, by one-half to one kilometer, and to deepen faster with age. The presence of cold subducting material underneath the basins may explain why they are deeper than average and why they sink faster. The low-density material under the ridge axis may be more confined under the major midocean ridges. On average the upper 200–300 km of the mantle in the vicinity of island arcs and marginal basins has slower than average seismic velocities and the deeper mantle is faster than average, probably reflecting the presence of cold subducted material. The depth of back-arc basins is an integrated effect of the thickness of the crust and lithosphere, the low-density shallow mantle and the presumed denser underlying subducted material. It is perhaps surprising then that, on average, the depths of marginal basins are so similar to equivalent age oceans elsewhere. The main difference is the presence of the underlying deep subducting material, which would be expected to depress the seafloor in back-arc basins. Basalts in marginal basins with a long history of spreading are essentially similar to MORB, while basalts generated in the early stages of back-arc spreading have more LIL-enriched characteristics. A similar sequence is found in continental rifts (Red Sea, Afar) and some oceanic islands, which suggests a vertical zonation in the mantle, with the LIL-rich zone being shallower than the depleted zone. This is relevant to the plume hypothesis, which assumes that enriched magmas rise from deep in the mantle. The early stages of back-arc magmatism are the most LIL-enriched, and this is the stage at which the effect of hypothetical deep mantle plumes would be cut off by the presence of the subducting slab. Continental basalts, such as the Columbia River basalt province, are also most enriched when the presence of a slab in the shallow mantle under western North America is indicated. The similarity of the isotopic and trace-element geochemistry of island-arc basalts, continental flood basalts and ocean-island basalts and the slightly enriched nature of the back-arc-basin basalts all suggest that the enrichment occurs at shallow depths,

Table 14.4 Composition of marginal basin basalts compared with other basalts

Oxide	MORB	MORB	Mariana Trough	Scotia Sea	Lau Basin	Bonin Rift	North Fiji Plateau	Arc Tholeiites		Island Tholeiite
SiO_2	49.21	50.47	50.56	51.69	51.33	51.11	49.5	48.7	51.1	49.4
TiO_2	1.39	1.58	1.21	1.41	1.67	1.10	1.2	0.63	0.83	2.5
Al_2O_3	15.81	15.31	16.53	16.23	17.22	18.20	15.5	16.5	16.1	13.9
FeO	7.19	10.42	8.26	8.28	5.22	8.18	6.2	8.4	—	—
Fe_2O_3	2.21	—	—	—	2.70	—	3.9	3.4	11.8	12.4
MnO	0.16	—	0.10	0.17	0.15	0.17	0.1	0.29	0.2	—
MgO	8.53	7.46	7.25	6.98	5.23	6.25	6.7	8.2	5.1	8.4
CaO	11.14	11.48	11.59	11.23	9.95	11.18	11.3	12.2	10.8	10.3
Na_2O	2.71	2.64	2.86	3.09	3.16	3.12	2.7	1.2	1.96	2.13
K_2O	0.26	0.16	0.23	0.27	1.07	0.38	0.3	0.23	0.40	0.38
P_2O_5	0.15	0.13	0.15	0.18	0.39	0.18	0.1	0.10	0.14	—
H_2O		—	1.6	—	1.18	—	1.4	—	—	—
$^{87}Sr/^{86}Sr$	0.7023	0.7028	0.7028	0.7030	0.7036	—	—	—	—	—

Gill (1976), Hawkins (1977).

Table 14.5 Representative properties of basalts and kimberlites

Ratio	Midocean Ridge Basalt	Back-Arc Basin Basalt	Island-Arc Basalt	Ocean-Island Basalt	Alkali-Olivine Basalt	Kimberlite
Rb/Sr	0.007	0.019–0.043	0.01–0.04	0.04	0.05	0.08
K/Rb	800–2000	550–860	300–690	250–750	320–430	130
Sm/Nd	0.34–0.70	0.31	0.28–0.34	0.15–0.35	0.37	0.18
U/Th	0.11–0.71	0.21	0.19–0.77	0.15–0.35	0.27	0.24
K/Ba	110–120	33–97	7.4–36	28–30	13.4	7.5
Ba/Rb	12–24	9–25	10–20	14–20	15	15
La/Yb	0.6	2.5	2–9	7	14	125
Zr/Hf	30–44	43	38–56	44	41	36
Zr/Nb	35–40	4.8	20–30	14.7	2.7	2.3
$^{87}Sr/^{86}Sr$	0.7023	0.7027–0.705	0.703–0.706	0.7032–0.706	—	0.71
$^{206}Pb/^{204}Pb$	17.5–18.5	18.2–19.1	18.3–18.8	18.0–20.0	—	17.6–20.0

perhaps by contamination of MORB rising from a deeper layer. The high 3He/4He of Lau Basin basalts also suggests a shallow origin for 'primitive' gases.

Both back-arc basins and midocean ridges have shallow seafloor, high heat flow and thin sedimentary cover. The upper mantle in both environments has low seismic velocity and high attenuation. The crusts are typically oceanic and basement rocks are tholeiitic. The spreading center in the back-arc basins, however, is much less well defined. Magnetic anomalies are less coherent, seismicity is diffuse and the ridge crests appear to jump around. Although extension is undoubtedly occurring, it may be oblique to the arc and diffuse and have a large shear

Table 14.6 Representative values of large-ion lithophile and high-field-strength elements (HFSE) in basalts (ppm)

Element	Midocean-Ridge Basalt	Back-Arc-Basin Basalt	Island-Arc Basalt	Ocean-Island Basalt
Rb	0.2	4.5–7.1	5–32	5
Sr	50	146–195	200	350
Ba	4	40–174	75	100
Pb	0.08	1.6–3	9.3	4
Zr	35	121	74	125
Hf	1.2	2.8	1.7	3
Nb	1	25	2.7	8
Ce	3	32	6.7–32.1	35
Th	0.03	1–1.9	0.79	0.67
U	0.02	0.4	0.19	1.18

component. The existence of similar basalts at midocean ridges and back-arc basins and the subtle differences in trace-element and isotopic ratios provide clues as to the composition and depth of the MORB reservoir.

Trace-element modeling

An alternative to mixing known components to infer the composition of the depleted MORB mantle is to calculate the partitioning of elements with a particular model in mind. For example, instead of using MORB and Q, one can suppose that the MORB source was the result of a certain kind of melting process, and calculate the properties of the melt and the residue. This usually gives a more depleted result than the mixing method.

The trace-element contents of basalts contain information about the composition, mineralogy and depth of their source regions. When a solid partially melts, the various elements composing the solid are distributed between the melt and the remaining crystalline phases. The equation for equilibrium partial melting is simply a statement of mass balance:

$$C_m = \frac{C_o}{D(1-F)+F} = \frac{C_r}{D}$$

where C_m, C_o and C_r are the concentrations of the element in the melt, the original solid and the residual solid, respectively; D is the bulk distribution coefficient for the minerals left in the residue, and F is the fraction of melting. Each element has its own D that depends not on the initial mineralogy but on the residual minerals, and, in some cases, the bulk composition of the melt. For the very incompatible elements ($D \ll 1$); essentially all of the element goes into the first melt that forms. The so-called compatible elements ($D \gg 1$) stay in the crystalline residue, and the solid residual is similar in composition to the original unmelted material. The above equation is for equilibrium partial melting, also called batch melting. The reverse is equilibrium crystal fractionation, in which a melt crystallizes and the crystals remain in equilibrium with the melt. The same equations apply to both these situations. The effective partition coefficient is a weighted average of the mineral partition coefficients.

The Rayleigh fractionation law

$$C_m/C_o = F^{(D-1)}$$

applies to the case of instantaneous equilibrium precipitation of an infinitesimally small amount of crystal, which immediately settles out of the melt and is removed from further equilibration with the evolving melt. The reverse situation is called fractional fusion.

Nickel and cobalt are affected by olivine and orthopyroxene fractionation since D is much greater than 1 for these minerals. These are called compatible elements. Vanadium and scandium are moderately compatible. Rare-earth patterns are particularly diagnostic of the extent of garnet involvement in the residual solid or in the fractionating crystals. Melts in equilibrium with garnet would be expected to be low in HREE, Y and Sc. MORB is high in these elements; kimberlite is relatively low.

Melt-crystal equilibrium is likely to be much more complicated than the above equations imply. As semi-molten diapirs or melt packets rise in the mantle, the equilibrium phase assemblages change. The composition of a melt depends on its entire previous history, including shallow crystal fractionation. Melts on the wings of a magma chamber may represent smaller degrees of partial melting. Erupted melts are hybrids of low- and high-degree melts, shallow and deep melting, and melts from peridotites and eclogites.

Chapter 15

The hard rock cafe

There are three kinds of rocks: Ingenious, Metaphoric, and Sedentary, named after the three kinds of geologists.

Anon.

Ultramafic rocks

Ultramafic rocks (UMR) are composed chiefly of ferromagnesian minerals and have a low silicon content compared with the crust, mafic rocks and basalts. The term is often used interchangeably with *ultrabasic*; pyroxene-rich rocks are ultramafic but not ultrabasic because of their high SiO_2 content. Peridotites, lherzolite, dunite and harzburgite are specific names applied to ultramafic rocks that are chiefly composed of olivine, orthopyroxene, clinopyroxene and an aluminous phase such as plagioclase, spinel or garnet. Ultramafic rocks are dense and mainly composed of refractory minerals with high seismic velocities. Basic rocks, such as basalts, become dense at high pressure (for example, eclogite) and can have properties comparable to the more refractory peridotites. Some eclogites overlap basalts in their bulk chemistry. The relationships between these rocks are shown in Figure 15.1.

Peridotites can represent

(1) The refractory residue left after basalt extraction.
(2) Cumulates formed by the crystallization of a magma.
(3) Primitive mantle that can yield basalts by partial melting.
(4) Cumulates or residues that have been intruded by basalt.
(5) High-pressure or high-temperature melts.

Peridotites contain more than 40% olivine. They are divided into fertile or infertile (or barren). Fertile peridotites can be viewed as having an appreciable basaltic component. The terms 'enriched' and 'depleted' are often used interchangeably with 'fertile' and 'infertile' but have trace-element and isotopic connotations that are often inconsistent with the major-element chemistry. Table 15.1 gives compositions for representative ultramafic rocks.

Garnet lherzolites are composed mainly of olivine and orthopyroxene (Table 15.2). Olivine is generally in the range of 60–70 vol. % and orthopyroxene 30–50%. The average clinopyroxene and garnet proportions are about 5% and 2%, respectively.

The major oxides in peridotites and lherzolites generally correlate well (Figure 15.2 and Table 15.3). The lherzolite trend can be explained by variable amounts of clinopyroxene and garnet. Olivine- and orthopyroxene-rich rocks, presumably from the mantle, are found in foldbelts, ophiolite sections, oceanic fracture zones and, as xenoliths, in kimberlites and alkali-rich magmas. They are rare in less viscous magmas such as tholeiites. Olivine and orthopyroxene in varying proportion are the most abundant minerals in peridotites. These are dense refractory minerals, and peridotites are therefore generally thought

Table 15.1 Compositions of spinel and garnet lherzolites

Oxide	Spinel Lherzolite: Continental (avg. of 301)	Spinel Lherzolite: Oceanic (avg. of 83)	Garnet Lherzolite
SiO_2	44.15	44.40	44.90
Al_2O_3	1.96	2.38	1.40
FeO	8.28	8.31	7.89
MgO	42.25	42.06	42.60
CaO	2.08	1.34	0.82
Na_2O	0.18	0.27	0.11
K_2O	0.05	0.09	0.04
MnO	0.12	0.17	0.11
TiO_2	0.07	0.13	0.06
P_2O_5	0.02	0.06	—
NiO	0.27	0.31	0.26
Cr_2O_3	0.44	0.44	0.32

Maaløe and Aoki (1977).

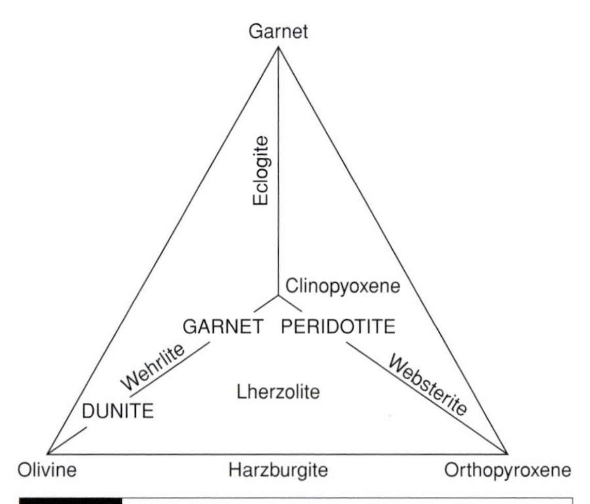

Fig. 15.1 Nomenclature tetrahedron for assemblages of olivine, clinopyroxene, orthopyroxene and garnet. Dunites and garnet peridotites lie within the tetrahedron.

to be the residue after melt extractions. Some peridotites are shallow cumulates deposited from cooling basalts and are therefore not direct samples of the mantle. Alumina in peridotites is distributed among the pyroxenes and accessory minerals such as plagioclase, spinel and garnet. At higher pressure most of the $A1_2O_3$ would be in garnet. Garnet-rich peridotite, or pyrolite, is the commonly assumed parent of mantle basalts. This variety is fertile peridotite since it can provide basalt by partial melting. Most peridotites, however, have relatively low Al_2O_3 and can be termed barren. These are commonly thought to be residual after melt extraction. Al_2O_3-poor peridotites are less dense than the fertile variety and should concentrate in the shallow mantle. Given sufficient water at crustal and shallow mantle temperatures, peridotite may be converted to serpentinite with a large reduction in density and seismic velocity. Hydrated upper mantle may therefore be seismically indistinguishable from lower crustal minerals. Similiarly, basaltic crust at depths greater than some 50-km depth converts to eclogite and this is similar to UMR assemblages in physical properties.

Table 15.2 Mineralogy of Lherzolites

Mineral	Spinel Lherzolite: Average (wt. pct.)	Spinel Lherzolite: Range (vol. pct.)	Garnet Lherzolite: Average (vol. pct.)	Garnet Lherzolite: Range (vol. pct.)
Olivine	66.7	65–90	62.6	60–80
Orthopyroxene	23.7	5–20	30	20–40
Clinopyroxene	7.8	3–14	2	0–5
Spinel	1.7	0.2–3	—	—
Garnet	—	—	5	3–10
Phlogopite	—	—	0.4	0–0.5

Maaløe and Aoki (1977).

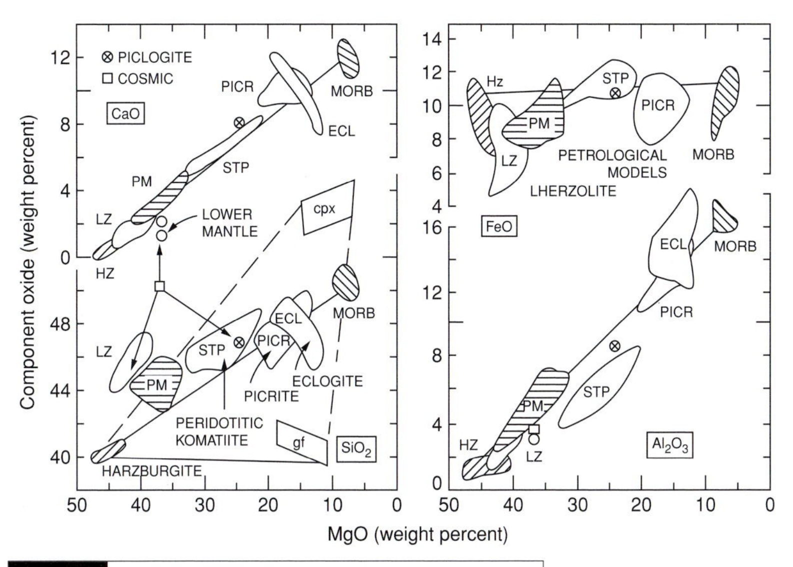

Fig. 15.2 Oxides versus MgO for igneous rocks. The basalt source region probably has a composition intermediate between basalt (MORB) and harzburgite. Most petrological models (PM) of the major-element chemistry of the source region favor a small basalt fraction. STP (spinifex textured peridotites) are high-temperature MgO-rich magmas. Picrites (PICR) are intermediate in composition between STP and MORB and may evolve to MORB by olivine separation. Picrites and eclogites (ECL) overlap in composition. Lherzolites (LZ) contain an orthopyroxene component, but the other rock types are mainly clinopyroxene + garnet and olivine. Squares represent estimates of primitive mantle composition based on a chondritic model. If the upper mantle is primarily lherzolite, basalt and harzburgite, the lower mantle (open dot) will be primarily orthopyroxene. The composition of the MORB source (piclogite model) probably falls between PICR and PM or STP.

Lherzolites typically contain 60–80% olivine, 20–40% orthopyroxene, less than 14% clinopyroxene and 1–10% of an aluminous phase such as spinel or garnet. Spinel lherzolites, the lower-pressure assemblages, dredged from the ocean bottom are similar in composition to those found in alkali basalts and kimberlites on oceanic islands and continents. Garnet lherzolites are denser than spinel lherzolites only when they contain appreciable garnet. They would become less dense at higher temperature, lower pressure or if partially molten.

The major-element chemistries of lherzolites vary in a systematic fashion. Most of the oxides vary linearly with MgO content. These trends are generally consistent with variable amounts of a basaltic component. However, the basaltic component is not tholeiitic or MORB. If lherzolites represent olivine–orthopyroxene-rich rocks with variable amounts of melt extraction or addition, this melt component is andesitic in major elements.

The major-element trends of lherzolites may also be controlled by melt-crystal equilibration at various depths in the mantle. Lherzolites, and most other ultramafic rocks, are generally thought to be the refractory residue complementary to melts presently being extracted from the mantle. They differ, however, from primitive mantle compositions. In particular they contain more olivine and less orthopyroxene than would be appropriate for a chondritic or 'cosmic' mantle. Upper-mantle lherzolites and basalts may be complementary to the lower mantle, representing melts from the original, accretional differentiation of the mantle. The MgO content of melts increases with temperature and with depth of melting. At great depth (>200 km) relatively low-MgO phases, such as orthopyroxene and garnet–majorite may remain behind, giving olivine-rich melts. The major-element trends in lherzolites may therefore represent trends in high-pressure melts. Unserpentinized peridotites have seismic velocities and anisotropies appropriate for the shallow mantle. This situation is often generalized to the whole mantle, but seismic data for depths greater than 400 km are

Table 15.3 Compositions of peridotites and pyroxenites

	Lherzolites							
	Spinel		Garnet	Dunite	Pyroxenite	Peridotites		
Oxide	(1)	(2)	(3)	(4)	(5)	(6)	(7)	(8)
SiO_2	44.15	44.40	44.90	41.20	48.60	44.14	46.36	42.1
Al_2O_3	1.96	2.38	1.40	1.31	4.30	1.57	0.98	
FeO	8.28	8.31	7.89	11.0	10.0	8.31	6.56	7.10
MgO	42.25	42.06	42.60	43.44	19.10	43.87	44.58	48.3
CaO	2.08	1.34	0.82	0.80	13.60	1.40	0.92	
Na_2O	0.18	0.27	0.11	0.08	0.71	0.15	0.11	
K_2O	0.05	0.09	0.04	0.016	0.28	—	—	
MnO	0.12	0.17	0.11	0.15	0.18	0.11	0.11	
TiO_2	0.07	0.13	0.06	0.06	0.83	0.13	0.05	
P_2O_5	0.02	0.06	—	0.10	0.10	—	—	
NiO	0.27	0.31	0.26	—	—	—	—	
Cr_2O_3	0.44	0.44	0.32	—	—	0.34	0.33	
H_2O	—	—	—	0.50	0.90	—	—	

(1) Average of 301 continental spinel lherzolites (Maaløe and Aoki, 1977).
(2) Average of 83 oceanic spinel lherzolites (Maaløe and Aoki, 1977).
(3) Average garnet lherzolite (Maaløe and Aoki, 1977).
(4) Dunite (Beus, 1976).
(5) Pyroxenite (Beus, 1976).
(6) High-T peridotites, South Africa (Boyd, 1987).
(7) Low-T peridotites, South Africa (Boyd, 1987).
(8) Extrapolated lherzolite trend (~0 percent Al_2O_3, CaO, Na_2O, etc.).

not in agreement with that hypothesis. It is not even clear that peridotite has the proper seismic properties for the lower lithosphere. In the depth interval 200–400 km both eclogite and peridotite can satisfy the seismic data.

Garnet pyroxenites and eclogites are also found among the rocks brought up from the mantle as xenoliths, and they have physical properties that overlap those of the ultramafic rocks. Some garnet-rich pyroxenites and eclogites are denser than some peridotites but densities overlap. The extrapolation of the properties of peridotites to the deep upper mantle, much less the whole mantle, should be done with caution. Not only do other rock types emerge from the mantle, but there is reason to believe that peridotites will be concentrated in the shallow mantle and to be over-represented in our rock collections (except that eclogites are exotic-looking rocks and are preferred by some).

If picrites are the parent for tholeiitic basalts, then roughly 30% melting is implied for generation from a shallow peridotitic parent. If the parent is eclogitic, then similar temperatures would cause more extensive melting. Generation of basaltic magmas from an eclogitic parent does not require extensive melting. Melts of basaltic composition are provided over a large range of partial melting, and basalts and eclogites come in a variety of flavors. The sources of basalts may not even be rocks, as conventionally defined. The fertile components may be km-size eclogitic blobs separated by tens of km in a refractory mantle. Melting, particularly at midocean ridges, takes place over large regions; the various components do not have to be distributed over grain-scale or hand-specimen size domains.

Basalts and peridotites are two of the results of mantle differentiation. They both occur near

Table 15.4 | Representative compositions of pyrolites and peridotites (wt.%)

	Pyrolite				Garnet Peridotite	
Oxide	(1)	(2)	(3)	(4)	(5)	(6)
SiO_2	45.1	42.7	46.1	45.0	42.5	46.8
TiO_2	0.2	0.5	0.2	0.2	0.1	0.0
Al_2O_3	3.3	3.3	4.3	4.4	0.8	1.5
Cr_2O_3	0.4	0.5	—	0.5	—	—
MgO	38.1	41.4	37.6	38.8	44.4	42.0
FeO	8.0	6.5	8.2	7.6	3.8	4.3
MnO	0.15	—	—	0.11	0.10	0.11
CaO	3.1	2.1	3.1	3.4	0.5	0.7
Na_2O	0.4	0.5	0.4	0.4	0.1	0.1
K_2O	0.03	0.18	0.03	0.003	0.22	0.02

(1) Ringwood (1979), p. 7.
(2) Green and Ringwood (1963).
(3) Ringwood (1975).
(4) Green and others (1979).
(5) Boyd and Mertzman (1987).
(6) Boyd and Mertzman (1987).

the surface of the Earth and may not represent the whole story. They are also not necessarily the result of a single-stage differentiation process.

Source rocks

Pyrolite

Pyrolite (pyroxene–olivine-rock) is a hypothetical primitive mantle material that on fractional melting yields a typical basaltic magma and leaves behind a residual refractory dunite-peridotite. It is approximately one part basalt and 3–4 parts dunite, assuming that 20–40% melting is necessary before liquid segregates and begins an independent existence. Garnet pyrolite is essentially identical with garnet peridotite but is more fertile than most natural samples. Pyrolite compositions have been based on three parts dunite plus one part of the averages of tholeiitic and alkali olivine basalt and a three-to-one mix of Alpine-type peridotite and a Hawaiian olivine–tholeiite. Table 15.4 gives compositions of some of these pyrolite models. Pyrolite compositions are arbitrary and are based entirely on major elements and on several arbitrary assumptions regarding allowable amounts of basalt and melting in the source region. They do not satisfy trace-element or isotopic data and they violate chondritic abundances and evidence for mantle heterogeneity. On the other hand, mantle compositions based on isotopic constraints alone are equally arbitrary and do not satisfy elementary petrological considerations.

Eclogites

The most abundant material coming out of the mantle is basalt and eclogites are the high-pressure forms of basalts. The term 'eclogite' refers to rocks composed of omphacite (diopside plus jadeite) and garnet, occasionally accompanied by kyanite, zoisite, amphibole, quartz and pyrrhotite. Natural eclogites have a variety of associations, chemistries, mineralogies and origins, and many names have been introduced to categorize these subtleties. Garnet pyroxenites are essentially eclogites that have less omphacite, or sodium.

'Eclogite' implies different things to different workers. To some eclogites mean metamorphic crustal rocks, and to others the term implies bimineralic kimberlite xenoliths. The chemical similarity of some eclogites to basalts prompted early investigators to consider eclogite as

Table 15.5 | Typical trace element concentrations (ppm)

	Eclogites		Synthetic	Peridotites		
	Omphacite	Garnet	Eclogite	Diopside	Garnet	MORB
K	1164	337	820	615	296	700
Rb	0.565	1.14	0.7	2.1	1.45	0.4
Sr	249	8.25	95	337	5.50	110
Na	52 244	74	12 700	1332	2420	17 300
Ti	4856	899	2500	659	959	5500
Zn	106	15	55	28	69	80
Rb/Sr	0.002	0.14	—	0.006	0.26	0.004
Sm/Nd	0.206	0.522	—	0.211	0.590	0.335
Rb/K	4.9*	33.8*	8.5*	34.1*	49.0*	5.7*

Basu and Tatsumoto (1982), Wedepohl and Muramatsu (1979).
* $\times 10^{-4}$.

Table 15.6 | Comparison of eclogites and other mafic rocks

Oxide	Eclogite		Picrite	MORB	Ocean Crust
(percent)	(1)	(2)	(3)	(4)	(5)
SiO_2	45.2	47.2	44.4	47.2	47.8
TiO_2	0.5	0.6	1.18	0.7	0.6
Al_2O_3	17.8	13.9	10.2	15.0	12.1
Cr_2O_3	0.4	—	0.22	—	—
Fe_2O_3	—	—	—	3.4	—
FeO	11.2	11.0	10.92	6.6	9.1
MgO	13.1	14.3	18.6	10.5	17.8
MnO	0.3	—	0.17	0.1	—
CaO	9.6	10.1	9.7	11.4	11.2
Na_2O	1.6	1.6	1.37	2.3	1.3
K_2O	0.03	0.8	0.13	0.1	0.03

(1) Bobbejaan eclogite (Smyth and Caporuscio, 1984).
(2) Roberts Victor eclogite (Smyth and Caporuscio, 1984).
(3) Picrite, Svartenhuk (Clarke, 1970).
(4) Oceanic tholeiite (MORB).
(5) Average oceanic crust (Elthon, 1979).

a possible source of basalts but more recently has been taken as evidence that these eclogites are simply subducted oceanic crust or basaltic melts that have crystallized at high pressure. Some eclogites are demonstrably metamorphosed basalts, while others appear to be igneous rocks ranging from melts to cumulates. Eclogite cumulates may accompany the formation of island arcs and batholiths and delamination of thick arc crust is another way to introduce eclogite into the mantle. The trend in petrology has been toward the splitters rather than the lumpers, and toward explanations that emphasize the derivative and secondary nature of

Table 15.7 | Heat-producing elements (ppm)

Rocks	Uranium	Thorium	Potassium
Eclogites			
Roberts Victor	0.04–0.8	0.29–1.25	83–167
	0.04–0.05	1.3	1833
Zagadochnaya	0.024		
Jagersfontein	0.07	0.17	1000
Bultfontein	0.07	0.31	1300
MORB	0.014	0.035	660
Garnet peridotite	0.22	0.97	663
Ultramafic rock	0.26	0.094	35
Primitive Mantle	0.02	0.077	151

Note: Measured values of $^{206}Pb/^{204}Pb$ for eclogites are 15.6–19.1; for primitive mantle it is ~17.5.

eclogite rather than the possible importance of eclogite as a source rock for basaltic magmas. Recently, however, eclogites have been considered as important source rocks.

Pyroxene-garnet rocks with jadeite-bearing pyroxenes are found as inclusions in alkali basalt flows as layers in ultramafic intrusions, as inclusions in kimberlite pipes, as tectonic inclusions in metamorphic terranes associated with gneiss and schist, and as inclusions in glaucophane schist areas. Jadeite-poor garnet clinopyroxenites are abundant in Salt Lake Crater, Hawaii. The eclogites from kimberlite pipes, and perhaps Hawaii, from great depth. Some contain diamonds. Peridotites and lherzolites predominate over all other rock types as inclusions in diamond-bearing kimberlites.

The presence of diamond indicates origin depths of at least 130 km, and other petrological indicators suggest depths even greater. In a few kimberlite pipes eclogites form the majority of inclusions and eclogite inclusions are common in diamonds. The overwhelming majority of peridotites and lherzolites are infertile, that is, very low in aluminum, sodium and calcium. If the distribution of rock types in kimberlite inclusions is representative of the source region, the majority of basaltic components in the upper mantle reside in dispersed eclogites rather than in an olivine-rich rock.

Table 15.5 shows typical clinopyroxene and garnet compositions of eclogites and peridotites, a synthetic two-mineral eclogite and for comparison, a typical MORB. Note that, in general, diopside plus garnet from peridotite does not approximate the composition of MORB. In most cases, however, omphacite and garnet from eclogite bracket MORB compositions, and therefore eclogite is a more appropriate source rock. Table 15.6 gives comparisons of the bulk chemistry of some South African eclogites and MORB and an estimate of the average composition of the oceanic crust. There is a close correspondence between the composition of kimberlite eclogites and the material in the oceanic crust.

Many kimberlite eclogites show signs of garnet exsolution from clinopyroxene, implying that there can be substantial changes in the density of eclogites as a function of temperature in the subsolidus region. Clinopyroxenes in eclogites have exsolved 20% or more garnet, implying a substantial increase in pressure or decrease in temperature. A representative eclogite can increase in density by 2.5% by cooling from 1350 °C to 950 °C at 30–50 kbar. The reverse process can happen as garnet plus clinopyroxene is elevated into a lower pressure regime along an adiabat. For example, the density of an eclogite can decrease by about 3% simply by rising 50 km. Thus, garnet exsolution caused by pressure release can accomplish more than a 1000 °C rise in temperature, by thermal expansion, all in the subsolidus regime. This plus the low melting point of eclogite means that rising eclogite-rich convection currents obtain considerable buoyancy. The MORB source region has been depleted by removal of either a small melt fraction or a late-stage intercumulus fluid. The abundances of uranium, thorium and potassium in various mantle samples (given in Table 15.7) show that eclogite xenoliths found in kimberlite are often less rich in these elements than peridotites found in the same pipe, although both may have been contaminated by the kimberlite matrix. These elements are, nevertheless, higher than in primitive mantle.

An eclogite cumulate layer is one possible product of a differentiated chondritic Earth and a possible source for basaltic magmas. Recycled

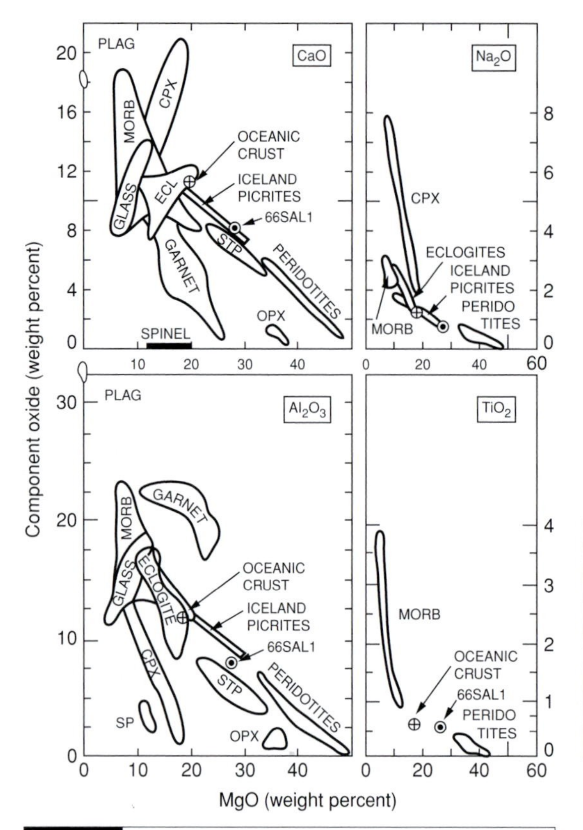

Fig. 15.3 Major-element oxides of igneous rocks and rock-forming minerals. If peridotites represent residues after basalt extraction and MORB and picrites represent melts, and if these are genetically related, the MgO content of the basalt source is probably between about 20 and 30%, much less than most lherzolites or peridotites. 66 SAL-1 is a garnet pyroxenite from Salt Lake Crater, Hawaii, and falls in the composition gap between basalts and peridotites. It is a piclogite.

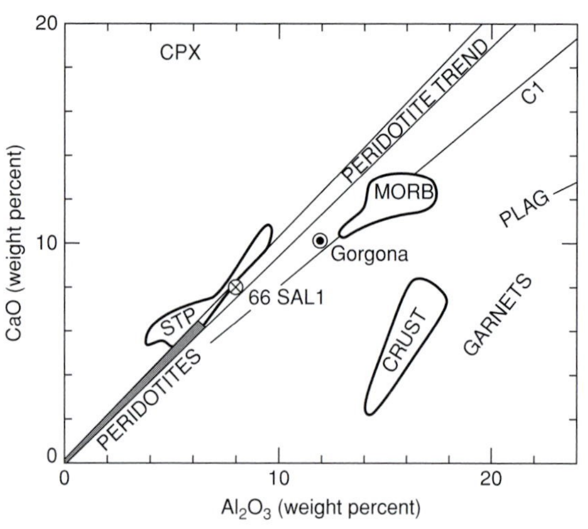

Fig. 15.4 CaO versus Al_2O_3 for igneous rocks and minerals. Note that peridotites and STP are CaO-rich compared to CI chondrites, and meteorites in general, while the continental crust in Al_2O_3-rich. MORB falls on the CI trend, possibly suggesting that it is an extensive melt of primitive mantle. Gorgona is an island in the Pacific that has komatiitic magmas. The solar CaO/Al_2O_3 ratio is slightly higher than peridotites (Chapter 1).

oceanic crust and delaminated continental crust place large blobs of eclogite, of variable density, into the mantle; 10–20% melting of eclogite can cause it to become buoyant and this happens well below the solidus temperature of mantle peridotites. I refer to a basalt- or eclogite-rich assemblage as a piclogite. Large eclogite blobs surrounded by peridotite can be a piclogite, even if hand-specimen-size samples from the region are either eclogite or peridotite.

Piclogite

Peridotites and pyrolite are rock types composed primarily of the refractory crystals olivine and orthopyroxene. Olivine is generally more than 60% by volume, and both minerals together typically constitute more than 80% of the rock. Clinopyroxene and aluminous phases such as spinel, garnet and jadeite – the basalt assemblage – are minor constituents. At the other extreme are rocks such as pyroxenites, clinopyroxenites and eclogites, which are low in orthopyroxene and olivine. Pyroxenites, by definition, contain <40% olivine. Intermediate are rocks such as picrites, olivine eclogites and komatiites. Rocks that have less than 50% olivine plus orthopyroxene have been given the general name piclogites. In major elements they fall between dunites and basalts and between lherzolites and picrites. They contain a larger basaltic fraction than peridotites, although they contain the same minerals: olivine, orthopyroxene, diopside-jadeite and garnet. At high pressure they are denser than lherzolite, but at high temperature they can become less dense. Piclogites can represent garnet-rich cumulates or frozen high-pressure melts and can generate basaltic melts over a wide range of temperatures. They

Table 15.8 Comparisons of picrites, komatiites and garnet pyroxenite

	Group II[1]		Komatiite		
Oxide	Svartenhuk	Baffin Is.	S. Africa[2]	Gorgona[3]	66SAL-1[4]
SiO_2	44.4	45.1	45.7	45.3	44.8
TiO_2	1.18	0.76	0.91	0.60	0.52
Al_2O_3	10.2	10.8	7.74	10.6	8.21
Cr_2O_3	0.22	0.27	—	—	—
FeO	10.92	10.26	—	10.9	9.77
MnO	0.17	0.18	0.23	0.18	
MgO	18.6	19.7	19.43	21.9	26.53
CaO	9.7	9.2	8.04	9.25	8.12
Na_2O	1.37	1.04	0.61	1.04	0.89
K_2O	0.13	0.08	—	0.02	0.03
P_2O_5	0.14	0.09	0.12	—	0.04
Fe_2O_3	—	—	15.44	—	—
NiO	0.13	0.12	—	—	—
H_2O	2.33	2.14	3.46	—	—

[1]Clarke (1970).
[2]Jahn and others (1980).
[3]Echeverria (1980).
[4]Frey (1980).

are not compositionally equivalent to basalts and do not necessarily require large degrees of partial melting to generate basaltic magmas. However, if piclogites are an important component in the midocean-ridge-basalt source rock, the garnet must be mainly eliminated in order to satisfy the HREE data and in order to decrease the density of the material so that it can rise into the shallow mantle. Therefore, clinopyroxene probably exceeds garnet in any component with clinopyroxene rather than garnet as a near-liquidus phase. Piclogites can be regarded as *assemblages* that are sampled by large volume melts; they do not have to exist on the hand specimen scale.

Peridotitic komatiites (STP, for spinifex-texture peridotites) have major-element chemistries about midway between harzburgites and tholeiites and occupy the compositional gap between lherzolites and picrites (Figures 15.4 and 15.4). The present rarity of komatiitic magmas may be partially due to their high density and the thick lithosphere. Komatiites evolve to picrites and then to tholeiites as they crystallize and lose olivine, becoming less dense all the while.

Compositions of picrites, komatiites and a garnet pyroxenite (66 SAL-1) from Salt Lake Center, Hawaii are given in Table 15.8. These are all examples of rocks that fall between basalts and peridotites in their major-element chemistry. In terms of mineralogy at modest pressure they would be olivine eclogites, garnet pyroxenites or piclogites. These and lower crustal cumulates are all suitable source rocks for basalts.

Chapter 16

Noble gas isotopes

^{3}He is not here, for ^{3}He is risen!

Anon.

Overview

The group of elements known as the rare, inert or noble gases possess unique properties that make them important as geodynamic tracers. The daughter isotopes fractionate readily from their parents, they are inert, and they differ from other geochemical tracers in being gases and diffusing relatively rapidly, at least at low pressure and high temperature. Nevertheless, they can be trapped in crystals for long periods of time. They give information about the degassing history of the mantle and magmas, the formation of the atmosphere, and about mixing relationships between different mantle components. Isotopes made during the Big Bang are called the 'primordial' isotopes. Some of the noble gas isotopes are cosmogenic, nucleogenic or radiogenic and these were made at later times. The presence of ^{3}He in a rock or magma has often been taken as evidence for the existence of a *primordial reservoir* in the mantle, one that had not previously been melted or degassed. A *primordial reservoir* is different from a *primordial component*. Primordial and solar components are still raining down on Earth. The most primordial materials on Earth, in terms of noble gases, are on mountain tops, in the stratosphere and in deep-ocean sediments.

Evidence for recent additions of noble gases to the Earth comes from deep-sea sediments where high concentrations of He (written [He]), high ^{3}He/^{4}He ratios (written R) and 'solar' neon are found. Mantle-derived basalts and xenoliths having similar *primordial* or *solar* characteristics are generally attributed to plumes from the *undegassed primitive* lower mantle reservoir. This is based on the assumption that high ^{3}He/^{4}He and ^{36}Ar/^{40}Ar ratios in basalts require high ^{3}He and ^{36}Ar contents in their sources, rather than low ^{4}He and ^{40}Ar, and that the whole upper mantle is characterized by low ratios and low abundances.

The critical issues are when and how the 'primordial' noble gases entered the Earth, where they are stored and whether transport is always outwards and immediate upon melting, as often assumed. It is therefore important to understand the noble-gas budget of the Earth, both as the Earth formed and as it aged. The major part of terrestrial formation probably occurred at high temperature from dry, volatile-poor, degassed particles. During the early stages of planetary formation, core differentiation and massive bolide impacts heated the Earth to temperatures that may have been high enough for a substantial part of it to have melted and vaporized. This is suggested by the fact that a large fraction (30–60%) of the incompatible elements (e.g. U, Th, Ba, K, Rb) are in the Earth's crust, which is consistent with an extensively differentiated mantle.

Some of the components now in the mantle fell to Earth in a *late veneer*, mostly around 3.8 Ga – the period of late bombardment – that added material to the Earth after completion of core formation. Late low-energy accretion probably introduced most of the volatiles and trace siderophiles (*iron-lovers*, such as Os and Ir) that are in the upper mantle today. However, what fraction of the noble gases in mantle basalts is truly *primordial* vs. a *late veneer* - or even later additions of cosmic dust - or recent contamination is unknown.

The hot-accretion model of Earth origin and evolution contrasts with the 'standard model' of noble-gas geochemistry that assumes that the bulk of the mantle accreted in a primordial undegassed state with chondritic, or cosmic, abundances of the noble-gas elements. Such material is widely assumed to make up the deep mantle today and to be accessible to surface volcanoes. `Helium fundamentals`, the `distribution of the lighter noble gases — helium, neon and argon` — and the `role of noble gases in geochemistry and cosmochemistry` are extensively discussed in monographs and review articles and on web sites `[mantleplumes, helium paradoxes]`.

Components and reservoirs

The noble gases do not necessarily enter the mantle in the same way or reside in the same components or reservoirs as other geochemical tracers. Noble gases reside in bubbles, along crystalline defects, in deep-sea sediments, and in rocks exposed to cosmic rays. Olivine-rich cumulates in the mantle may trap CO_2 and helium, but little else. U, Th and K have quite different fates and reside in quite different places than He and Ar. The noble gases are fractionated from their parents and from each other by melting, transport and degassing processes. Therefore they cannot be treated as normal LIL-elements. Solubility in magmas is an additional issue for gases.

Isotope ratios

In common with much of isotope geochemistry, the noble gases are discussed in terms of ratios. The absolute abundances of the noble gases in a rock, magma or the mantle are seldom known or believed. This complicates matters since normal statistics and mixing relations cannot be applied to ratios. The most useful light-noble-gas isotopic ratios, in petrology are shown below (* signifies radiogenic or nucleogenic; there may have been some ^{4}He and ^{40}Ar in primordial matter but most of it in the Earth was created later, and this is designated $^{4}He^{*}$ and $^{40}Ar^{*}$ to distinguish them from primordial isotopes, if the distinction is necessary):

$^{3}He/^{4}He^{*}$ $^{3}He/^{22}Ne$ $^{3}He/^{20}Ne$ $^{3}He/^{36}Ar$
$^{4}He^{*}/^{21}Ne^{*}$ $^{4}He^{*}/^{40}Ar^{*}$
$^{20}Ne/^{22}Ne$
$^{21}Ne^{*}/^{22}Ne$
$^{40}Ar^{*}/^{36}Ar$

For convenience, these will sometimes be abbreviated as 3/4, 3/22, 21/22 and so on, and sometimes written as $^{3}He/^{4}He$, or R, $^{20}Ne/^{22}Ne$ and so on. Absolute concentrations will be written [He], [^{3}He] and so on.

Helium

Helium is very rare in the Earth and it continuously escapes from the atmosphere. Outgassing of ^{4}He from the mantle does not occur at a high enough rate to correspond to the outflow of heat generated by U + Th decay. This is known as the `helium heat-flow paradox`. Since it is probably more difficult to degas the mantle now than during the early, high-temperature period of Earth history, it is likely that the stable noble gas isotopes - those that have not increased over time - are more degassed than the radiogenic and nucleogenic ones e.g. $^{4}He^{*}$ and $^{40}Ar^{*}$.

There are few constraints on how much helium may be in the Earth or when it arrived. It is readily lost from the atmosphere via escape into space. It is therefore not very useful for studying the long-term outgassing history of the Earth. Neon and argon are potentially more useful in this regard since they do not escape as readily; they accumulate in the atmosphere. We know, for example, that most (~70%) of the argon produced in the Earth is now in the atmosphere, even though much of it was not in the mantle during the earliest high-temperature phases of accretion, degassing and evolution. This argues against a large undegassed reservoir in the Earth. On the other hand, the imbalance between heat

Table 16.1 Statistics of helium in MORB [statistical tests helium reservoirs]

	Mean	Standard deviation	n
All ridge	9.14	3.59	503
Atlantic	9.58	2.94	236
Pacific	8.13	0.98	245
Indian	8.49	1.62	177
OIB	7.67	3.68	23

flow and ^{4}He flux from the mantle implies that He is trapped in the upper mantle; it is not escaping as fast as it is produced. Apparently, ^{4}He is accumulating in the mantle. Helium is more soluble in magma than the heavier noble gases and may be trapped in residual melts and cumulates. Helium trapped in residual melts or fertile mantle will evolve to low R because of the presence of U and Th. Helium trapped in olivine, olivine-rich cumulates or depleted restites will maintain nearly its original isotopic ratio. Thus, high ^{3}He/^{4}He samples may retain an ancient *frozen-in* ratio rather than a current *primordial* ratio. Such material is probably intrinsic to the upper mantle and upper mantle processes, such as degassing of ascending magmas.

The reference isotope for helium is ^{3}He and the ratio ^{3}He/(^{4}He+^{4}He*) is usually written ^{3}He/^{4}He or R and this in turn is referenced to the current atmospheric ratio Ra (or R_A), and written R/Ra (or R/R_A). The atmospheric value involves degassing from the mantle and crust - which have quite different values and time scales - a contribution from interplanetary dust particles (IDP), and the rate of escape from the atmosphere. These all change with time and the atmospheric ratio is also expected to change with time. There is no guarantee that ancient magmas, corrected for radioactive decay, had the same relation to the atmospheric values at the time as they have to present-day atmospheric ratios. It is always assumed in mantle geochemistry calculations that the atmospheric ratio is invariant with time and that the atmosphere is a well-stirred homogenous reservoir. It would be useful to be able to measure the time variability, just as it is useful to determine the variation of ^{87}Sr/^{86}Sr in seawater with time.

The helium isotopic ratio in MORB is usually quoted as 8 ± 1 R_A but this represents strongly filtered data, filtered to 'remove any plume influence'. Unfiltered data gives 9.1 ± 3.6 R_A (Table 16.2). Spreading ridges average the isotopic ratios from large volumes of the mantle and it is fairly straightforward to compile meaningful statistics. Unfortunately, the helium data on and near spreading ridges is highly selected and *filtered* prior to statistical analysis. Table 16.2 gives both filtered and unfiltered estimates. Ocean island basalt data is harder to analyze because the samples are not collected in systematic or random ways and anomalous regions are over-sampled (the reverse of the situation for MORB and ridges). Nevertheless, one can compile averages of all samples for each island and average these to get global estimates of OIB statistics.

The statistics and the distribution of helium in mantle magmas show that there is no statistical difference in ^{3}He/^{4}He (R) ratios between the available data for midoceanic ridge basalts (MORBs) and ocean island basalts (OIBs). It is usually assumed, however, that these two classes of basalts require the existence of at least two distinct mantle reservoirs that have been preserved over long periods of the Earth's history. The distribution in ocean ridge basalts (Figure 16.1) is more Gaussian than OIB. The variance is relatively low but, importantly, is a large fraction of the mean. OIBs have a non-Gaussian distribution, with a very large variance. The largest R values in the OIB population are often considered to be diagnostic of the OIB reservoir and large values in MORB are considered to be plume contamination. There is no obvious cut-off between MORB values and OIB values. Model calculations suggest that the separation between the low R ratio components and the high ratios found in some OIB-type components is relatively recent (Anderson, 1998a, b; Seta *et al.*, 2001). The variance differences, and differences in extreme values, between various basalt populations is a consequence of the *averaging during the sampling process*. The extreme ^{3}He/^{4}He ratios are averaged out upon melting and averaging (SUMA), as at the global

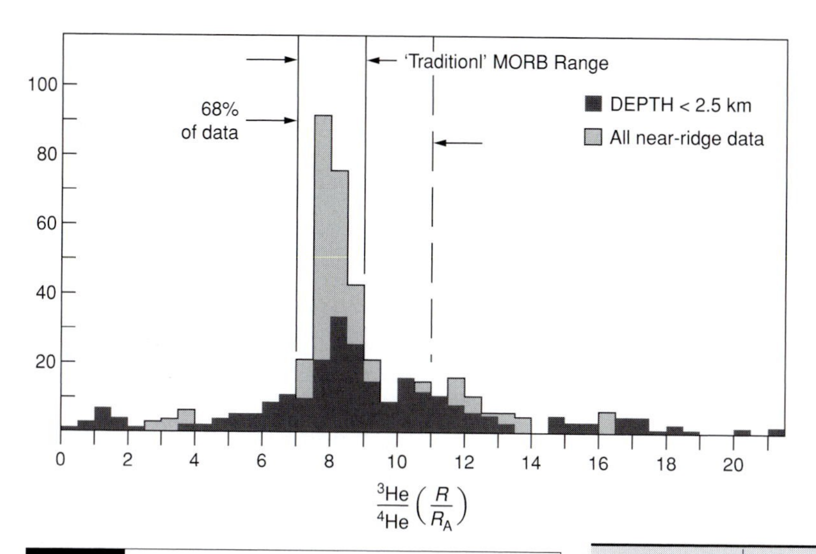

Fig. 16.1 Histogram of helium isotopes measured in samples along the global spreading ridge system, including new ridges, back-arc basins and near-ridge seamounts. In most compilations much of the right-hand side of the distribution is missing since values higher than about 9 *R*/*R*a are filtered out. High values are attributed to plume contamination. Anomalous sections of the global ridge system are also avoided. The traditional MORB range, after filtering the data, is from 7–9 R/R_A. High *R* values are found in depleted components of basalts that have been identified as the most common mantle component, possibly peridotite. The high *R*/*R*a carriers may be peridotites or olivines with low U/He ratios. The helium may be acquired from wall-rocks, by ascending magmas.

spreading ridge system, and can only be sampled by smaller scale processes, such as individual lava flows on islands. Estimates of R/R_A for OIB and for filtered ridge samples are given in Table 16.2.

Ratios that are sensitive to degassing of magma, and atmospheric contamination, such as He/Ne and He/Ar, show a clear distinction between MORB and OIB. The distribution of the lighter noble gases — helium, argon and neon — in mantle magmas suggests that degassing of magmas near the Earth's surface, trapping of volatiles, and possibly more recent atmospheric contamination, may be involved in the differences [mantleplumes]. The heavier noble gases in mantle magmas may entirely be due to atmospheric and seawater contamination.

Table 16.2 Helium isotope data, as compiled over the years. The year of the compilation is also given

R/R_A	S.D.	Year
All ridges		
9.06	3.26	
'Filtered' data		
8.4	0.36	1982
8	2	1986
8.2	0.2	1991
8.3	0.3	1996
8.58	1.81	2000
8.67	1.88	2000
8.67	1.88	2000
8.75	2.14	2002
Back-arc basin basalts		
8.4	3.2	2002
Ocean island basalts		
7.67	3.68	1995
9	3.9	2001
Seamounts		
6.58	1.7	
9.77	1.4	

S.D., standard deviation.

Most workers assume a lower-mantle plume source for high-*R* magmas - and a very high [^{3}He] - which must be isolated and preserved against convective homogenization with the more radiogenic upper mantle. A 'primordial' or 'relatively undegassed' reservoir, however, probably does not exist. The mantle is degassed, but some parts (e.g. 'the MORB - reservoir') are more enriched in radiogenic helium due to high-[U+Th] concentrations combined with great age. Since MORBs are LIL-depleted and have low R, it has been assumed that the upper mantle is degassed. MORB, however, has high [^{3}He] compared with other basalts; this is inconsistent with the standard model. Primordial He may be present in the mantle but a large, coherent, ancient, primordial, undegassed region is unlikely. The highest R/R_A basalts are similar to MORB, or the MORB source, in their heavy isotopes.

The standard model of mantle noble gas geochemistry divides the mantle into a depleted degassed upper mantle (DUM) homogenized by convection, and an undegassed primordial lower mantle (PM) that is not well stirred. This model is based on a string of assumptions.

(1) *R* values higher than the MORB average reflect high [^{3}He].
(2) High $^3He/^4He$ imply a primordial undegassed reservoir, or a reservoir more primitive and less outgassed than the mantle source for MORB, assumed to be the upper mantle.
(3) This primordial reservoir must be isolated from the upper mantle.
(4) It is therefore deep and *is* the lower mantle.
(5) This deep isolated reservoir can be tapped by oceanic islands.

The first two items are logical fallacies. There are many other ratios that can be formed with ^{3}He, e.g. $^3He/^{22}Ne$, $^3He/^{20}Ne$, $^3He/^{36}Ar$ . . . and these are all lower in OIB than in MORB except for $^3He/^{238}U$. The latter is consistent with a low-U source, e.g. peridotite, for high-*R* OIB.

Complex models have been devised to explain the apparent coexistence of a `depleted degassed well-stirred upper mantle reservoir` and a heterogenous undegassed primordial - and accessible - lower mantle reservoir, assuming that OIB and MORB represent distinct isolated reservoirs, that the upper mantle can only provide homogenous MORB-like materials, and that low-*R* implies low-[^{3}He]. The rather restricted range of helium isotope ratios in MORB is more plausibly interpreted as a result

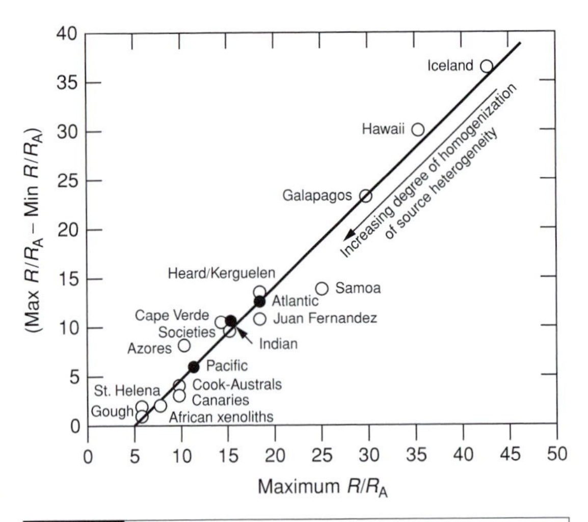

Fig. 16.2 Maximum value of *R* vs. the spread in values showing that high-*R* islands also have the largest range in values (figure courtesy of A. Meibom, `www.mantleplumes.org`).

of the vast amount of averaging that midocean ridges perform on the mantle that they sample. Both high-*R* and low-*R* domains may coexist in the upper mantle, as long as they are larger than diffusion distances; these can be commingled during the melting process or sampled separately if sampling is done at a small scale.

The so-called *high-He* hotspots exhibit very large variance (Figure 16.2). This is suggestive of the predictions of the central limit theorem and implies that mid-ocean ridges sample larger volumes of a heterogenous mantle than do oceanic islands.

The decay of uranium and thorium produces both radiogenic helium (alpha particles) and heat. The amount of uranium required to generate most of the Earth's oceanic helium flux only produces about 5–10% of the oceanic heat flow; this is the `helium heat-flow paradox`. Helium, along with CO_2, from degassing magmas, may be trapped in the shallow mantle. Trapped CO_2 and ^{4}He along with high U+Th, contributes to the low $^3He/^4He$ and high [He] of some upper mantle components, and ubiquitous carbonatitic metasomatism.

Helium apparently is not as mobile an element as generally thought, opening up the possibility that helium in various upper mantle components with differing He/U ratios, can diverge *in situ* in their isotopic ratios, removing the requirement for ancient or large gas-rich reservoirs. The component that carries the high *R* signature may be a peridotite, or gas bubbles in a depleted olivine-rich cumulate. There are no consistent global correlations between He-isotopes and other isotopes. The helium in ocean island basalts may be contained in recycled lithosphere or may be acquired during ascent through a U–Th depleted cumulate [`large scale melting and averaging SUMA`].

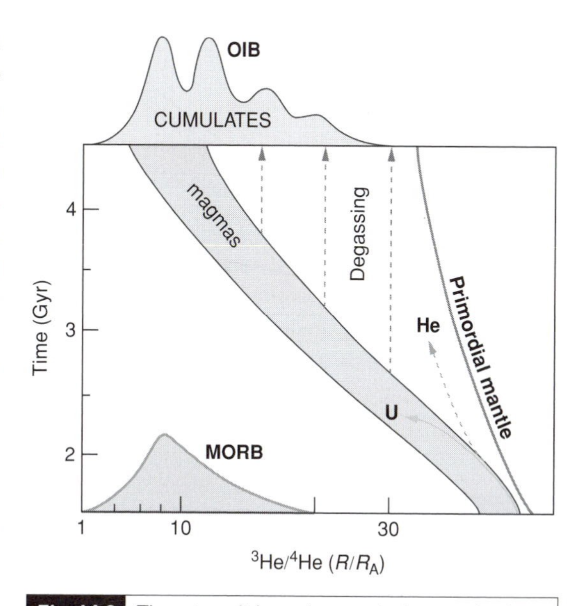

Fig. 16.3 The primordial mantle curve is the growth of $^3He/^4He$ in a hypothetical primordial reservoir from Big Bang values to present (4.5Gyr). This is what the lower mantle should provide in the standard model. At various times, including during accretion, magma is extracted from accreting material; this evolves faster because the magma gets the U and Th and gas is lost (and some is retained by residual mantle). Details aside, the idea is that differentiation and degassing changes the $^3He/(U,Th)$ ratio of the products and field labelled MAGMAS reflects this partitioning, containing higher U/He ratios than primordial mantle. The magmas lose gas as they rise to shallow depths, decreasing the He/U ratio and increasing the He/Ar and He/Ne ratios of the residual magma. The gas ends up in cumulates and gas inclusions in peridotites, giving them high He/U ratios and $^3He/^4He$ ratios of the magmas. Ancient $^3He/^4He$ ratios get frozen in, in whatever material the degassed gas finds itself, while the residual degassed magma continues to evolve to low values. The trajectories of He (and CO_2) and of U (and Th) upon melting and degassing events are shown schematically. Each melting/degassing event gives this kind of fractionation. The dashed lines are simply an indication that ancient $^3He/^4He$ ratios are frozen in as He is removed from U and Th and stored in olivine-rich mantle, while magmas go elsewhere. The MORB and OIB histograms are plotted at arbitrary positions on the time axis; they represent current values, and involve mixtures of partially degassed magmas, previously degassed gases and air, of a variety of ages. MORB is a cross section across the magma field, today. The OIB field is a mix of magmas and gases of various ages, in cumulates etc. . . . and is closer to the actual mantle distribution because it is sampled in smaller batches than MORB. By convolving OIB with itself we have a more MORB-like distribution. Note that the low values of MORB do not have to evolve from high mantle values or high OIB values.

Helium history

Because of radioactive decay and nucleogenic processes the isotopic ratios of the noble gases in the mantle change with time. If the helium content of part of the mantle is very high, the isotopic ratios change slowly with time. If $^3He/^4He$ is also high, it will remain high. This is the reasoning behind the standard undegassed mantle hypothesis of noble gas geochemistry. Fertile parts of the mantle have high U and Th concentrations and these generate large amounts of ^4He over time. Today's MORB has much higher $^3He/^4He$ ratios than ancient MORB. Ancient MORB, ascending and degassing, generated gases with high $^3He/^4He$ ratios and some of this may have been trapped in the shallow mantle. The present upper mantle contains materials that melted and degassed at various times. The helium-heatflow paradox and the amount of ^{40}Ar in the atmosphere suggests that noble gases can be trapped in the mantle, probably the upper mantle. Figure 16.3 illustrates the various stages in the helium history of the mantle, starting at the lower right.

Neon

Atmospheric Ne is believed to have been significantly depleted in the atmosphere by intense solar irradiation. Ne *3-isotope plots* allow atmospheric contamination of basalts to be monitored for this isotope. On the other hand, the heavy rare gases (argon, krypton and xenon) have accumulated in the atmosphere over Earth history, and air, or seawater, contamination is a serious problem, even for mantle samples.

Neon isotope ratios for mantle rocks are typically displayed on a *3-isotope plot* of $^{20}Ne/^{22}Ne$ vs.

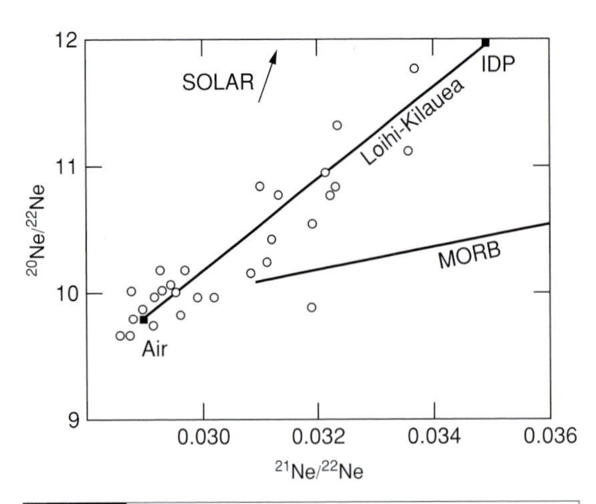

Fig. 16.4 'Three-isotope plot.' Some ocean islands appear to be mixtures of air and solar or interplanetary dust particles (IDP). MORB has relatively more ^{21}Ne suggesting that its source is older or has higher U than the OIB sources.

$^{21}Ne/^{22}Ne$ (Figure 16.4). Neon isotope ratios are more susceptible to atmospheric contamination than helium isotope ratios because Ne is more abundant in the atmosphere. Most basalts plot on what appear to be mixing lines between atmospheric values (~10), and solar wind - SW or interplanetary dust particle - IDP values (13.7) of $^{20}Ne/^{22}Ne$. The corresponding values for the $^{21}Ne/^{22}Ne$ ratios are 0.029 (air), >0.04 (OIB) and >0.06 (MORB).

The mixing line for MORB extends to greater values of $^{21}Ne/^{22}Ne$ than the mixing line for OIB, indicating more ^{21}Ne in-growth in MORB than in OIB. This is analogous to the higher $^{4}He/^{3}He$ ratios found in some MORB samples compared with the extremes found in some OIB samples. The $^{20}Ne/^{22}Ne$ ratios in mantle-derived rocks, both MORB and OIB, extend from atmospheric to the solar or IDP ratio. The identification of solar-like ('primordial' or IDP) Ne isotopic ratios in some OIB and MORB samples implies that solar neon trapped within the Earth has remained virtually unchanged over the past 4.5 Gyr (the standard 'primordial mantle' model) or, alternatively, that noble gases have been added to the mantle since ~2 Ga, perhaps from noble-gas-rich sediments or cosmic rays. Some basalts and some diamonds essentially have pure 'solar' (SW or IDP) $^{20}Ne/^{22}Ne$ ratios. The fact that the $^{20}Ne/^{22}Ne$ ratio of rocks or magmas thought to come from the mantle is greater than the atmospheric ratio is another paradox for standard models of mantle degassing and atmospheric evolution because ^{20}Ne and ^{22}Ne are essentially primordial in the sense that ^{20}Ne is stable and only very small amounts of ^{22}Ne are produced in the Earth. Their ratio is thus expected to be uniform throughout the Earth, if the solar system reservoir is uniform and invariant with time.

Non-atmospheric neon found in MORB, OIB, volcanic gases and diamonds is enriched in ^{20}Ne and nucleogenic ^{21}Ne, relative to ^{22}Ne. This have been attributed to primordial components, possibly solar neon. Basalt glasses from Hawaii have a range of Ne isotope ratios, stretching from atmospheric to ^{20}Ne-enriched compositions that approaches the solar wind composition. These signatures are similar to those in cosmic dust particles accumulated in deep-sea sediments. These dust particles become impregnated with Ne from the solar wind during their exposure in space. If these particles can survive the `noble gas subduction barrier` and deliver cosmic (solar) Ne to the mantle this could explain the ratios of submarine glasses without having to invoke ancient primordial Ne in the Earth, or an undegassed lower mantle reservoir.

Inter-element ratios

Both melting and degassing fractionate the noble gases, and separate the parent from the daughter isotopes. Helium and the light noble gases are more readily retained in magma than the heavy noble gases during degassing. Thus, the He/Ne and He/Ar ratios of different basalt types contain information about their degree of degassing. The He/Ne and He/Ar ratios of MORB are generally higher than OIB suggesting that MORB is more degassed than OIB. Nevertheless, [He] is higher in MORB than in OIB, which suggests that OIB initially contained less [He] and possibly less [Ne] than MORB, prior to MORB degassing. Some of the more noble-gas-rich igneous rocks are the 'popping rocks' found along the mid-Atlantic ridge (e.g. Sarda and Graham, 1990; Javoy and Pineau, 1991; Staudacher *et al.*, 1989). Ordinary MORB appear to be degassed versions of

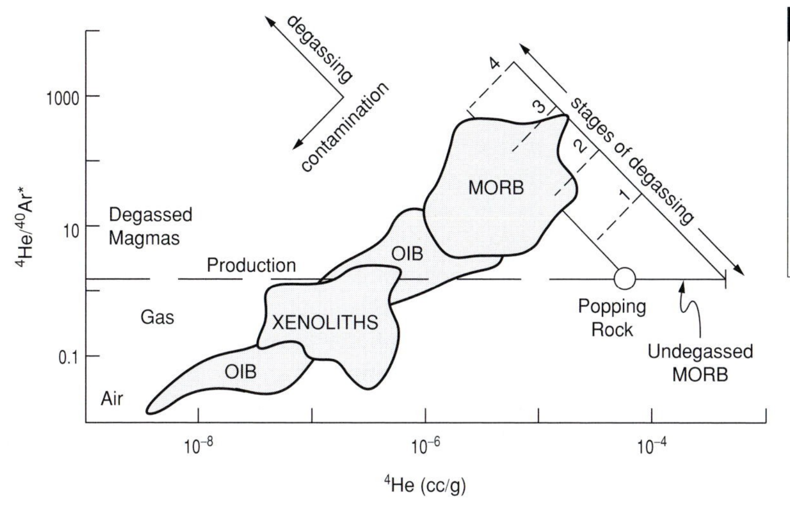

Fig. 16.5 Plot showing relation between helium and argon isotopes in MORB and OIB and the effects of degassing and 'contamination' with air and older CO_2-rich vesicles. The relative rates of production of ^4He and ^{40}Ar are shown by the horizontal dashed line (after Anderson, 2000a,b).

these popping rocks, and even some of these gas-rich rocks appear to have lost some gas.

Histogram of $^3He/^{22}Ne$ and $^4He/^{21}Ne$ in MORB and OIB show that MORB and OIB are fractionated in opposite directions relative to the solar ratio and the production ratios, respectively (mantleplumes noble gases).

A plot of $^4He/^{40}Ar$ vs. [^4He] and related inter-element plots (e.g. $^3He/^{21}Ne$ vs. [He]) shows a continuous transition from MORB to OIB to air (e.g. Figure 16.5). The MORB to OIB trends could be interpreted as air-contamination of noble gases degassed from MORB and trapped in vesicles or cumulates.

So-called primitive ratios of noble gas isotopes may be preserved by ancient separation of He and Ne from U+Th and storage of the gas in low-U+Th environments such as depleted lithosphere or olivine cumulates. In this way, the radiogenic isotopes ^4He and ^{21}Ne are not added to the gas and old isotopic ratios can be 'frozen in.' This contrasts with the situation in more fertile materials such as MORBs or undepleted peridotite, where U+Th contents are relatively high and continued radiogenic and nucleogenic ingrowth occurs. Usually, high *R* are interpreted as high-^3He but they could, with equal validity, be interpreted as low ^4He or from a low-U source.

The He/Ne ratios observed in mantle xenoliths and basaltic glasses vary by orders of magnitude and define a linear correlation with a slope of unity which passes through the point defined by the primordial $^3He/^{22}Ne$ ratio ($\sim$7.7) and the radiogenic ^4He to nucleogenic $^{21}Ne^*$ production ratio (2.2×10^7). The linear correlation implies that the elemental fractionation event (perhaps degassing of magma), which enriched MORB glasses in [He] with respect to [Ne], is recent, otherwise ingrowth of radiogenic ^4He and nucleogenic ^{21}Ne would have systematically shifted the data points from the correlation line. Basalts exhibit positive correlations between $^3He/^{22}Ne$ and [^3He], and between $^4He/^{21}Ne^*$ and [^4He]. The $^3He/^{22}Ne$, $^4He/^{21}Ne^*$ and $^4He/^{40}Ar^*$ ratios in MORB glasses are systematically higher than the primordial $^3He/^{22}Ne$ ratio or the radiogenic $^4He^*$/nucleogenic $^{21}Ne^*$ and radiogenic $^4He^*$/radiogenic $^{40}Ar^*$ production ratios.

The production rate of ^{21}Ne parallels that of ^4He since they both result from the decay of U+Th (Table 16.2). The 3-isotope plot showing OIB (Loihi) as forming one trend from atmosphere to solar and MORB as forming another (Figure 16.3) can be interpreted as a rotation of the OIB trend by nucleogenic ingrowth, i.e. MORB source is aged OIB source. In this simple model MORB = OIB + time. OIB and MORB could have evolved from a common parent at some time in the past. Old MORB gases (from ancient degassed MORB) stored in a depleted refractory host (e.g. olivine crystals, U+Th-poor lithosphere) will have high $^3He/^4He$ and low He/Ne compared with current MORB magmas. One of the mantle 'reservoirs' for noble gases may be isolated gas-filled inclusions or vugs in a peridotite. This 'reservoir' would contribute nothing besides He and CO2 and perhaps some Ne and Os, thereby decoupling He from other isotopic tracers.

Argon

The main isotopes of argon found on Earth are ^{40}Ar, ^{36}Ar, and ^{38}Ar. Naturally occurring ^{40}K with a half-life of 1.25×10^9 years, decays to stable ^{40}Ar. We have estimates of the amount of potassium in the Earth so we can estimate the efficiency of degassing of the crust and mantle from the argon content of air.

The initial amount of ^{40}Ar contained in the Earth is negligible compared with the amount that was produced subsequently. This makes it a useful tool with which to study mantle degassing. The amount of K in the crust and mantle may be estimated by calculating the mix of mantle and crustal components that satisfies the cosmic ratios of the refractory elements. The amount of K in the silicate Earth (the mantle and crust) so determined is 151 ppm of which 46% is in the crust. The amount of ^{40}Ar in the atmosphere represents 77% of that produced by the decay of this amount of potassium over the age of the Earth. This implies that either 23% of the Earth is still undegassed or that the degassing process is not 100% efficient, i.e. that there is a delay between the production of ^{40}Ar and its release into the atmosphere. `Ar may be compatible at depths >~150 km` and may therefore not be easily extracted from the deeper Earth by volcanism. However, most of the K is probably in the crust and shallow mantle. The values of ^{40}Ar/^{36}Ar (or ^{40}Ar*/^{36}Ar) measured, or estimated, in various materials are:

Air	~300
OIB	atmospheric to ~13 000
MORB	atmospheric to ~44 000

In the standard model, which assumes a lower-mantle source for OIB, the high ratios for OIB and MORB, compared with air, have been taken to indicate that both the upper (MORB) and the lower mantles (OIB) have been degassed such that little ^{36}Ar remains; but large quantities of ^{40}Ar have been produced by radiogenic decay and are retained. The higher ^{40}Ar/^{36}Ar in mantle rocks than in the atmosphere contrasts with the situation for He, where ^{4}He/^{3}He is usually much higher in the atmosphere than in either OIB or MORB as a result of the preferential escape of the lighter ^{3}He atom. Much of the He in the atmosphere entered it in the last million years and is therefore dominated by ^{4}He, older atmospheric and crustal helium having already escaped to space.

The concept of a relatively undegassed mantle reservoir has been strongly contested. There is strong evidence from the heavy noble gases that OIB are contaminated by air or ocean sources; low ^{40}Ar/^{36}Ar can be a result of atmospheric or seawater contamination. Seawater has 2–4 orders of magnitude more ^{36}Ar than mantle basalts and two orders of magnitude less ^{3}He. Hence, Ar isotope ratios in magmas can be significantly changed by seawater without affecting their He signature. Measured [^{36}Ar] covers the same range for MORB and OIB, and [^{3}He] is higher for MORB than for OIB. This also argues against OIB arising from a reservoir much less degassed than the MORB reservoir. This model for Ar is at odds with the standard model for He that considers the lower mantle to be undegassed, little degassed or less degassed than the MORB source. The lower ratios for OIB can be explained by more extreme atmospheric contamination, or a potassium-poor or younger source. Usually, however, unradiogenic ratios are attributed to high abundances of the primordial isotope, i.e. to undegassed or primordial reservoirs. The ^{20}Ne/^{22}Ne ratios for mantle rocks that are significantly higher than atmospheric indicate that the source of noble gas in both OIB and MORB cannot be mainly present-day atmosphere.

The 'two-reservoir' model

A two-reservoir model for the mantle was originally proposed to explain argon and helium isotope systematics in basalts and xenoliths but the samples were subsequently shown to be contaminated with atmospheric argon and cosmogenic helium; they had nothing to do with the mantle. Some samples of ocean-island basalts, most notably from Hawaii, Iceland and the Galapagos have elevated ^{3}He/^{4}He values compared with most MORB samples, after the latter are 'filtered for plume influence.' These islands also have an enormous spread of isotopic ratios, overlapping the values for MORB samples. Many OIB samples have $^3He/^4He$ ratios much lower than MORB samples but there is no obvious dividing point; the data for MORB and OIB are a continuum. If MORB samples larger volumes of the mantle than OIB then they will have a more constant mean, a smaller standard deviation, and fewer outliers; these are simple consequences of the central limit theorem. The outliers will be eliminated by the averaging. Nevertheless, the two-reservoir model for mantle noble gases is widely accepted and is the cornerstone of the standard model of mantle geochemistry. The two reservoirs in the standard model are a homogenous degassed upper mantle with low $^3He/^4He$ ratios, and an undegassed lower mantle with high $^3He/^4He$ ratios; $^3He/^4He$ is used as a proxy for $[^3He]$.

The He/Ne ratios of MORB and OIB, in contrast to $^3He/^4He$, define two distinct fields, but the implications are the opposite from the standard models. It is MORB that appears to have been degassed and OIB that has inherited secondary gases from a degassed magma and from the atmosphere.

A challenge for the two-reservoir model is to explain the respective concentrations of helium and other rare gases in the reservoirs. If OIBs come from an undegassed source, they would be expected to contain more helium than MORB glasses from the degassed upper mantle. However, OIB glasses typically have ten times less ^3He than MORB. This has been explained as degassing, but that cannot explain the higher He/Ne in MORB than in OIB. This observation is one of the `helium paradoxes`, and there are several. The dynamics of mantle melting and melt segregation under ridges must be different from under oceanic islands and seamounts. Ridge magmas probably collect helium from a greater volume of mantle and blend magmas from various depths, during the melting and eruption process. This would eliminate the extremes in MORB basalts. Most workers favor the two-reservoir model for the noble gases, although the evidence is far from definitive, and the assumption of an undegassed lower mantle causes more problems that it solves.

`A shallow origin for the 'primitive' He signature` in ocean-island basalts reconciles the paradoxical juxtaposition of crustal, seawater, and atmospheric signatures with inferred 'primitive' characteristics. High 238U/204Pb components in ocean-island basalts are generally attributed to recycled altered oceanic crust. The low $^{238}U/^3He$ component may be in the associated depleted refractory mantle. High $^3He/^4He$ ratios are due to low ^4He, not excess ^3He, and do not imply or require a deep or primordial or undegassed reservoir. ^{40}Ar in the atmosphere also argues against such models. `[www.pnas.org/cgi/content/abstract/95/9/4822]`

Sampling and recycling

A useful way to look at the noble gas data from mantle samples is the following. There are mantle and recycled *components* with a large range of helium isotopic ratios - because of the distribution in ages and $^3He/U$ ratios - rather than two *reservoirs* with distinctive compositions. High 20/22, 'solar' neon and high $^3He/^4He$ components have several possible sources including IDP, cosmic rays and 'old' gas trapped in depleted U+Th rocks, minerals and diamonds. Low 3/4 and high 21/22 materials are plausibly from high-U and high-Th environments, and older environments. Air and seawater contamination lower the 4/40, 20/22 and 21/22 ratios.

Several sources of high R have been suggested; frozen in ancient values from normal mantle processes, and subduction of IDP and other late veneer material. IDP accumulate in ocean-floor sediments and raise the Os, Ir, He and Ne contents of the sediments (Figure 16.6). Cosmic dust has very high $^3He/^4He$ ratios and $[^3He]$ and some of it falls to Earth without burning up in the atmosphere. In this way ocean-floor sediments develop a 'primordial' helium isotope signature, with high R and high $[^3He]$.

Basalts degas as they rise and crystallize. If this gas is trapped in olivine crystals, or cumulates or the depleted lithosphere, away from U and Th, it will have, in time, a high R compared with the parent magma and LIL-rich sources. This component has high R and high [He].

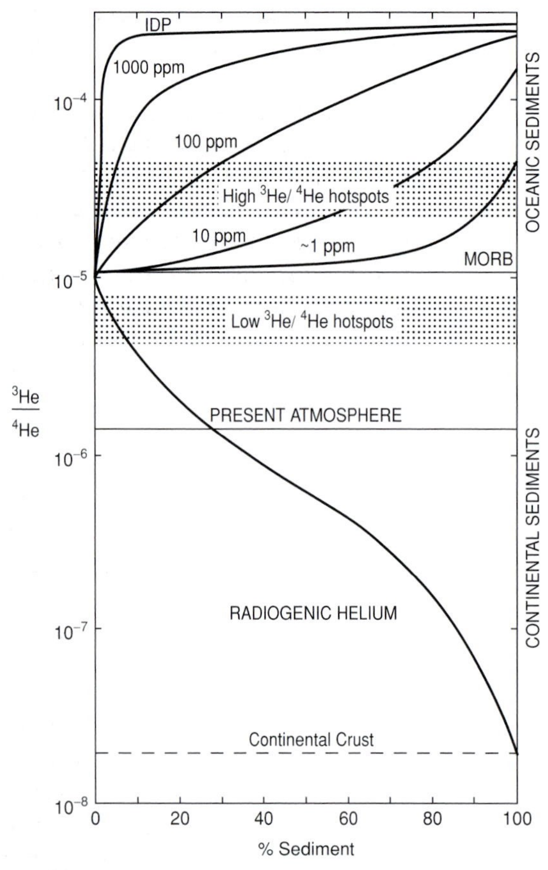

Fig. 16.6 Materials found at the surface of the Earth exhibit a range of 4 orders of magnitude in ^{3}He/^{4}He ratios. Sediments on the seafloor collect interplanetary dust particles (IDP) and these raise the helium content and helium ratio of deep sea sediments in proportion to their abundances of IDP. Continental sediments have very low ratios because of high U and Th and age. Subduction of these materials contribute to the isotopic heterogeneity of the mantle. High ^{3}He/^{4}He ratios in mantle materials can also be due to trapping of ancient gases – in U-poor environs – from degassing of rising magmas.

Many basalts with so-called primitive, primordial or solar noble gas signatures show strong evidence for seawater or atmospheric contamination. This combination of surface characteristics with ones conventionally attributed to the deep mantle is another paradox in the standard model for noble gases.

If cosmic dust and oceanic sediments survive the *subduction barrier* they have the potential to deliver He and Ne with primordial signatures into the mantle, probably its shallow levels. The ^{3}He in OIB is a small fraction of the total that is degassed from the mantle and in volume is comparable to the small amount brought onto the seafloor by IDP. However, much of the ^{3}He in the mantle may have come in as part of the late veneer; it is subduction, or the Archean equivalent, at some 1–2 Ga that is relevant, not today's flux. The ^{3}He/^{20}Ne ratio in cosmic dust is one to two orders of magnitude lower than that in MORB. Helium has a much greater diffusivity than Ne, which would promote its preferential degassing from cosmic dust grains during subduction, although this does not mean that it immediately escapes the mantle. It appears that subduction of cosmic dust *having the properties of today's dust* cannot contribute more than a small fraction of the mantle ^{3}He budget without causing excessive enrichment of ^{20}Ne in submarine glasses. The preferred explanation for the high *R* components in both OIB and MORB mantle is, therefore, evolution in low-U+Th environments such as olivine cumulates or depleted lithosphere (Anderson, 1998a, b). High ^{3}He/^{4}He and solar Ne ratios, may have been isolated in [He]-poor, U+Th-poor environments rather than in [He]-rich ones. [mantle helium neon and argon]

While some samples from Loihi, Iceland, Yellowstone and the Galapagos have high ^{3}He/^{4}He components – usually called the most primordial He compositions – some ocean islands such as Tristan, Gough, islands in the SW Pacific and the Azores, and some components (basalts, inclusions, gases) in Yellowstone, Iceland and the Galapagos, have ^{3}He/^{4}He ratios lower (more radiogenic) than MORB. If these are viewed as samples from a global population, then this is the expected result of the CLT. The low ^{3}He/^{4}He ratios require a component of radiogenic He from a long-lived U+Th-rich source such as recycled crust or sediments in the mantle, as inferred from lithophile isotope data. Parallel recycling of refractory peridotite or olivine-rich cumulates could be responsible for high ^{3}He/^{4}He ratios and these components will not have enriched signatures for other isotopes. None of these components need to be recycled very deep into the mantle. Both high- and low-^{3}He/^{4}He components can arise from recycling, but the extreme values – high and low – are averaged out at ridges by large-volume mantle sampling.

Statistics

There are as many low-R hotspots as there are high-R hotspots. The statistics (means, medians) of many hotspot datasets of helium isotopic ratios are identical or similar to midocean ridge statistics. The hypothesis that OIB and MORB, as global datasets, are drawn from the same population cannot be rejected by standard statistical tests, although selected subsets of these populations can differ. The differences in variances of the datasets are consistent with midocean ridges sampling a larger volume of a heterogenous mantle than do oceanic islands. Statistically, the MORB and OIB datasets could be drawn from the same population, at least for helium isotopes. This does not appear to be true for other isotope systems. The issue is complicated because mixing relations and averages of ratios involve also the absolute abundances, not just the ratios. If the absolute abundance of helium in the various components differ a lot, then mixtures will be dominated by the high-[He] members. On the other hand, if the Sr and Nd contents are similar, then all components contribute to the observed average ratios.

If two datasets have the same mean and different variances they may be different size samples from the same distribution. A consequence of the central limit theorem is that the ratio of the variances of a large number of samples from a heterogenous population is inversely proportional to the ratios of the number of samples contributing to the average. The larger volume sampled by ridges includes the effect of larger degrees of partial melting as well as larger physical volumes.

A common hypothesis is that helium isotopes in mantle-derived materials represent two distinct populations, the midocean-ridge (MORB) reservoir and the oceanic-island (OIB) reservoir. This based on the observation that some OIB samples have higher ratios than the mean of the MORB distribution. This is not a valid statistical argument. The alternative hypothesis is that the mantle is heterogenous on various scales and that midocean ridge and oceanic-island volcanoes sample this mantle in different ways and to different extents. Superimposed on this sampling difference, is the possibility of real lateral and vertical heterogeneity having scales that cannot be averaged out.

Basalts and hydrothermal fluids having R greater than about 8.5–9.0 R_A are commonly believed to be 'plume-type' or 'lower-mantle' helium. R can be significantly higher than the normal MORB range and these high values are, by definition 'plume-type' and, by convention, are attributed to the lower mantle. There is some circular reasoning here; although high values are rare along the global spreading ridge system – except at Iceland and the Red Sea – they are excluded, when found, and attributed to plumes. 'Hotspots' or elevated regions of ridges

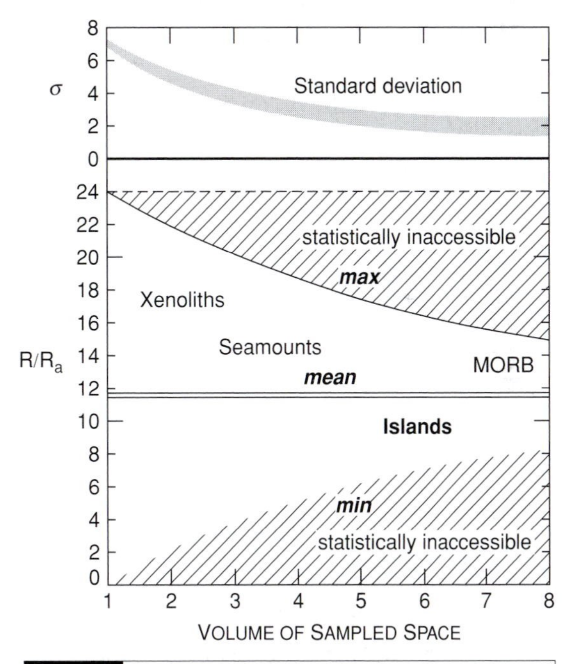

Fig. 16.7 An illustration of the central limit theorem. Midocean ridges sample very large volumes whereas seamounts and xenoliths are small-scale small-volume samples that more faithfully represent the true heterogeneity of the mantle. In this example, the range of R/Ra in the mantle is from 0 to 24. This complete range will only be evident by taking many small-scale small-volume samples. Large islands and spreading ridges average over large volumes and cannot access the whole range. The presence of high values in OIB samples does not imply that these samples are from a different part of the mantle.

are also avoided in order to avoid plume influence. We can ask the question 'Can we disprove the null hypothesis that OIB and MORB samples are drawn from the same population distribution function?' The answer is no, meaning that the datasets could be drawn from the same population. Previous studies have reached the opposite conclusion by comparing extreme values in one population with the mean value, or range, of the other population after the extreme values are removed. The inter-element isotopic ratios of MORB and OIB, however, are from distinct populations. This can be attributed to degassing and contamination processes rather than to source characteristics.

Table 16.2 compares the results for MORB (filtered and unfiltered datsets), BABB, OIB and near-ridge seamounts. The dates of the compilations of MORB are given. The means have changed little with time, in spite of the large expansion of the data. This is due to the exclusion of large ratios as being due to plume influence.

OIB samples show great diversity and exhibit values both higher and lower than found along most ridges but the median, a robust measure of central tendency, of the two populations is the same, about 8.5–9.0 R_A. It is true that more extreme values are found in some OIB samples, but this is predicted by the CLT if spreading ridges sample a larger volume of the mantle than do oceanic islands. The means of various MORB and OIB datasets are usually within one standard deviation of each other. [`statistics of mantle helium`]

The central limit theorem is illustrated in Figure 16.7.

Chapter 17

The other isotopes

Earth has a spirit of growth.

Leonardo da Vinci

Background

The various chemical elements have different properties and can therefore be readily separated from each other by igneous processes. The various isotopes of a given element are not so easily separated. The abundances of the radioactive isotopes in the crust and mantle, and their decay products, are not constant in time. Elemental compositions of magmas and residual mantle are complementary; isotopic compositions are identical, but they diverge with time. Therefore, the information conveyed by the study of isotopes is different in kind than that provided by the elements. Each isotopic system contains unique information, and the radioactive isotopes allow dating of processes in a planet's history. The unstable isotopes most useful in geochemistry have a wide range of decay constants, or half-lives, and can be used to infer processes occurring over the entire age of the Earth (Table 17.1). In addition, isotopes can be used as tracers and in this regard they complement the major- and trace-element chemistry of rocks and magmas. Isotopes in magmas and gases, however, cannot be used to infer the depth or location of the source.

Studies of isotope ratios have played an important role in constraining mantle and crustal evolution, mixing and the long-time isolation of mantle components or reservoirs. Isotope studies derive their power from the existence of suitable pairs of isotopes of a given element, one a 'primordial' isotope present in the Earth since its formation, the other a radiogenic daughter isotope produced by radioactive decay at a known rate throughout geological time. The isotopic composition of these isotope pairs in different terrestrial reservoirs – for example, the atmosphere, the ocean, and the different parts of the crust and mantle – are a function of the transport and mixing of parent and daughter elements between the reservoirs. In some cases the parent and daughter have similar geochemical characteristics and are difficult to separate in geological processes. In other cases the parent and daughter have quite different properties, and isotopic ratios contain information that is no longer available from studies of the elements themselves. For example Sr-isotope ratios give information about the time-integrated Rb/Sr ratio of the rock or its source. Since rubidium is a volatile element and separates from strontium both in pre-accretional and magmatic processes, the isotope ratios of strontium in the products of mantle differentiation, combined with mass-balance calculations, are our best guide to the rubidium content, and volatile content, of the Earth. Lead isotopes can be similarly used to constrain the U/Pb ratio, a refractory/(volatile, chalcophile) pair. The ^{40}Ar content of the atmosphere helps constrain the ^{40}K content of the Earth; both Ar and K are considered to be volatile elements in cosmochemistry. In other cases, such as the neodymium-samarium pair, the elements in question are both

Table 17.1 | Radioactive nuclides and their decay products

Radioactive Parent	Decay Product	Half-life (billion years)
^{238}U	^{206}Pb	4.468
^{232}Th	^{208}Pb	14.01
^{176}Lu	^{176}Hf	35.7
^{147}Sm	^{143}Nd	106.0
^{87}Rb	^{87}Sr	48.8
^{235}U	^{207}Pb	0.7038
^{40}K	^{40}Ar, ^{40}Ca	1.250
^{129}I	^{129}Xe	0.016
^{26}Al	^{26}Mg	8.8×10^{-4}

refractory, have similar geochemical characteristics and are probably in the Earth in chondritic ratios, or at least, in their original ratios. The neodymium isotopes can therefore be used to infer ages of mantle components or reservoirs and to discuss whether these are enriched or depleted, in terms of Nd/Sm, relative to chondritic or undifferentiated material. The Rb/Sr and Nd/Sm ratios are changed when melt is removed or added or if sediment, crust or seawater is added. With time, the isotope ratios of such components diverge.

The isotope ratios of the crust and different magmas show that mantle differentiation is ancient and that remixing and homogenization is secondary in importance to separation and isolation, at least until the magma chamber and eruption stages. Magma mixing is an efficient way to obtain uniform isotopic ratios, such as occur in MORB. Although isotopes cannot tell us where the components are, or their bulk chemistry, their long-term isolation and lack of homogenization plus the temporal and spatial proximity of their products suggests that, on average, they evolved at different depths or in large blobs that differ in lithology. This suggests that the different components differ in intrinsic density and melting point and therefore in bulk chemistry and mineralogy. Melts, and partially molten blobs, however, can be buoyant relative to the shallow mantle even if the parent blob is dense or neutrally buoyant.

The crust is extremely enriched in many of the so-called incompatible elements, particularly the *large ionic-radius lithophile* (LIL) or *high-field strength* (HFS) elements that do not readily fit into the lattices of the major mantle minerals, olivine (ol) and orthopyroxene (opx). These are also called the *crustal elements*, and they distinguish *enriched magmas* from *depleted magmas*. The crust is not particularly enriched in elements of moderate charge having ionic radii between the radii of Ca and Al ions. This suggests that the mantle has retained elements that can be accommodated in the garnet (gt) and clinopyroxene (cpx) structures. In other words, some of the so-called LIL elements are actually compatible in gt and cpx. The crust is also not excessively enriched in lithium, sodium, lead, bismuth and helium.

Isotopes as fingerprints

Box models

Radiogenic isotopes are useful for understanding the chemical evolution of planetary bodies. They can also be used to fingerprint different sources of magma. In addition, they can constrain timing of events. Isotopes are less useful in constraining the locations or depths of mantle components or reservoirs. Just about every radiogenic, nucleogenic or cosmogenic isotope has been used at one time or another to argue for a deep mantle or lower mantle source, or even a core source, for ocean island and continental flood basalts and carbonatites, but isotopes cannot be used in this way. Isotope ratios have also been used to argue that some basalts are derived from unfractionated or undegassed reservoirs, and that reservoir boundaries coincide with seismological boundaries (implying that major elements and physical properties correlate with isotopes).

Some mantle rocks and magmas have high concentrations of incompatible elements and have isotope ratios that reflect long-term enrichment of an appropriate incompatible-element parent. The crust may somehow be involved in the evolution of these magmas, either by crustal contamination prior to or during eruption, by recycling of continent-derived sediments or by delamination of the lower continental crust. Stable isotopes can be used to test these hypotheses. Early models of mantle geochemistry assumed that all potential crust-forming material had not been removed from the mantle or that the crust formation process was not 100% efficient in removing the incompatible elements from the mantle. Recycling was ignored. Later models assumed that crustal elements were very efficiently removed from the upper mantle – and, importantly, only the upper mantle – leaving it depleted. Vigorous convection then homogenized the source of midocean-ridge basalts, which was assumed to extend to the major mantle discontinuity near 650 km depth. A parallel geochemical hypothesis at the time was that some magmas represented melts from a 'primitive' mantle reservoir that had survived from the accretion of the Earth without any degassing, melting or melt extraction. The assumption underlying this model was that the part of the mantle that provided the present crust did so with 100% efficiency, and the rest of the mantle was isolated, albeit leaky. In this scenario, 'depleted' magmas were derived from a homogenized reservoir, complementary to the continental crust that had experienced a multi-stage history (stage one involved an ancient removal of a small melt fraction, the crust; stage two involved vigorous convection and mixing of the upper mantle; stage three involved a recent extensive melting process, which generated MORB). Non-MORB magmas (also called 'primitive,' 'less depleted,' 'hotspot' or 'plume' magmas) were assumed to be single-stage melts from a 'primitive' reservoir.

There is no room in these models for ancient enriched mantle components. These early 'box models' contained three boxes: the present continental crust, the 'depleted mantle' (which is equated to the upper mantle or MORB reservoir) and 'primitive mantle' (which is equated to the lower mantle) with the constraint that primitive mantle is the sum of continental crust and depleted mantle. With these simple rules many games were played with crustal recycling rates and mean age of the crust. When contradictions appeared they were, and are, traditionally explained by hiding material in the lower crust, the continental lithosphere or the core, or by storing material somewhere in the mantle for long periods of time. The products of mantle differentiation are viewed as readily and efficiently separable but, at the same time, storable for long periods of time in a hot, convecting mantle and accessible when needed.

A large body of isotope and trace-element analyses of midocean-ridge basalts demonstrates that the upper mantle is not homogenous; it contains several distinct geochemical domains on a variety of length scales. However, the physical properties of these domains, including their exact location, size, temperature and dynamics, remain largely unconstrained. Seismic data indicate that the upper mantle is heterogenous in physical properties. Plate tectonic processes create and remove heterogeneities in the mantle, and create thermal anomalies.

Global tomography and the use of long-lived isotopes are very broad brushes with which to paint the story of Earth structure, origin and evolution. Simple models such as the one- and two-reservoir models, undegassed undifferentiated lower-mantle models, and whole-mantle convection models are the results of these broad-brush paintings, as are ideas about delayed and continuous formation of the crust and core. Short-lived isotopes and high-resolution and quantitative seismic techniques paint a completely different story.

Isotopes as chronometers

Some of the great scientists, carefully ciphering the evidences furnished by geology, have arrived at the conviction that our world is prodigiously old, and they may be right, but Lord Kelvin is not of their opinion.

Mark Twain

Earliest history of the Earth

The current best estimate for the age of the Earth Moon meteorite system is 4.51 to 4.55 billion years (Dalrymple, 2001). The solar nebula cooled to the point at which solid matter could condense by ~4.566 billion years, after which the Earth grew through accretion of these solid particles; the Earth's outer core and the Moon were in place by ~4.51 billion years.

Clair Patterson of Caltech determined the age of the Earth to be 4.550 billion years (or 4.55 Gyr) ±70 million years from long-lived Pb isotopes (Patterson, 1956). This age was based on isotopic dating of meteorites and samples of modern Earth lead, all of which plot along a linear isochron on a plot of $^{207}Pb/^{204}Pb$ versus $^{206}Pb/^{204}Pb$.

Numerous other isotopic systems are used to determine the ages of solar system materials and ages of significant events, such as Moon formation. The isotopes include both those with long half lives, such as rubidium-87 – written 87Rb or ^{87}Rb (half life of 48.8 billion years), which decays to strontium-87 – written 87Sr or ^{87}Sr – and those that have half lives that are so short that the radioactive isotope no longer exists in measurable quantities. Meteorites contain evidence for decay of short-lived extinct natural radioactivities that were present when solids condensed from the primitive solar nebula. Three such short-lived radioactivities, ^{53}Mn, ^{182}Hf and ^{146}Sm, have half-lives of 3.7, 9 and 103 million years, respectively. The evidence indicates rapid accretion of solid bodies in the solar nebula, and early chemical differentiation. A hot origin of the Earth is indicated. The energetics of terrestrial accretion imply that the Earth was extensively molten in its early history; giant impacts would have raised temperatures in the Earth to about 5000–10 000 K.

The rates and timing of the early processes of Earth accretion and differentiation are studied using isotopes such as ^{129}I–^{129}Xe, ^{182}Hf– ^{182}W, ^{146}Sm–^{142}Nd, $^{235/238}U$–$^{207/206}Pb$ and ^{244}Pu–^{136}Xe and simple assumptions about how parent and daughter isotopes distribute themselves between components or geochemical reservoirs. The Hf–W and U–Pb chronometers are thought to yield the time of formation of the core, assuming that it is a unique event. The parent elements (Hf and U) are assumed to be retained in silicates during accretion and the daughters (W and Pb) to be partitioned into the core. The partitioning of W and Pb between metals and silicates – mantle and core – also depends on the oxygen fugacity and the sulfur content of the metal and the mantle. Under some conditions, W is a siderophile element while Pb is a chalcophile partitioning only slightly into the metal phase. Currently, the upper mantle is oxidized and the main 'metallic' phases are Fe-Ni sulfides. ^{146}Sm–^{142}Nd and ^{182}Hf–^{182}W chronometry indicate that core formation and mantle differentiation took place during accretion, producing a chemically differentiated and depleted mantle. The decay of ^{182}Hf into ^{182}W occurs in the silicate mantle and crust. Tungsten is then partitioned into the metal. The hafnium–tungsten pair shows that most of Earth formed within ~10 million years after the formation of the first solid grains in the solar nebula.

A plausible model for the origin of the Moon is that a Mars-sized object collided with the Earth at the end of its accretion, generating the observed angular momentum and an Fe-depleted Moon from the resulting debris disc. This may have occurred 40–50 Myr after the beginning of the solar system. The Moon-forming impact [Google images] contributed the final 10% of the Earth's mass, causing complete melting and major degassing. Core formation occurred before, during and after the giant Moon-forming impact, within tens of millions of years after the formation of the solar system. Some of the terrestrial core was probably from the impactor. Giant impacts melt a large fraction of the Earth and reset or partially reset isotopic clocks.

Mass balance calculations show that >70% of the mantle was processed in order to form the crust and upper mantle. Parts of the upper mantle are enriched but most of the mantle is either depleted and fertile (the MORB reservoir) or depleted and infertile or barren. Enriched regions (crust) or components (kimberlites, carbonatites . . .) typically are so enriched that a small volume can balance the depleted regions. Short-lived radioactivities can, in principle, determine when this fractionation occurred. Some was contemporaneous with accretion and some may have happened during Moon formation.

Elementary isotopology

Pb

Isotopes are usually expressed as ratios involving a parent and a daughter, or a decay product and a stable isotope. Absolute abundances are often not known or not treated. This introduces a hazard into statistical treatments and mixing calculations involving ratios. Ratios cannot be treated as pure numbers or absolute concentrations. The means and variances of isotopic ratios are meaningless unless all the samples have the same concentrations of the appropriate elements.

Lead has a unique position among the radioactive nuclides. Two isotopes, lead-206 and lead-207, are produced from radioactive parent isotopes of the same element, uranium-238 and uranium-235. The simultaneous use of coupled parent–daughter systems allows one to avoid some of the assumptions and ambiguities associated with evolution of a single parent–daughter system. Lead-208 is the daughter of 232-Th (half-life 14 Gyr); U and Th are geochemically similar but are separable by processes that occur near the surface of the Earth.

In discussing the uranium–lead system, it is convenient to normalize all isotopic abundances to that of lead-204, a stable nonradiogenic lead isotope. The total amount of lead-204 in the Earth has been constant since the Earth was formed; the uranium parents have been decreasing by radioactive decay while lead-206 and lead-207 have been increasing. The U/Pb ratio in various parts of the Earth changes by chemical fractionation and by radioactive decay. The $^{238}U/^{204}Pb$ ratio, calculated as of the present, can be used to remove the decay effect in order to study the chemical fractionation of various reservoirs. If no chemical separation of uranium from lead occurs, the ratio for the system remains constant. This ratio is called μ (mu). Some components of the mantle have high Pb-isotope ratios and are called HIMU.

Most lead-isotopic results can be interpreted as growth in a primitive *reservoir* for a certain period of time and then growth in reservoir with a different μ-value from that time to the present. By measuring the isotopic ratios of lead and uranium in a rock, the time at which the lead ratios were the same as inferred for the primitive reservoir can be determined, thus giving the lead-lead age of the rock. This dates the age of the uranium-lead fractionation event, assuming a two-stage growth model. In some cases multistage or continuous differentiation models are used, and similar models explain other isotope systems, e.g. fractionation of Rb/Sr, Sm/Nd, He/U etc.

A melt removed from the primitive reservoir at t_0, will crystallize to a rock composed of minerals with different μ values. If these minerals can be treated as closed systems, then they will have distinctive lead ratios that plot as a straight line on a $^{207}Pb/^{204}Pb$–$^{206}Pb/^{204}Pb$ plot (Figure 17.1). This line is an *isochron* because it is the locus of points that experienced fractionation at the same time to form minerals with differing U/Pb ratios. The residual rock will also plot on this line, on the other side of the *geochron*. The time at which the rock was fractionated can be calculated from the slope of the isochron. Mixing lines between genetically unrelated magmas will also be straight lines, in which case the age will be spurious unless both magmas formed at the same time.

In the uranium–lead decay system, the curve representing the growth of radiogenic lead in a closed system has marked curvature. This is because uranium-238 has a half-life (4.47 Gyr) comparable to the age of the Earth, whereas uranium-235 has a much shorter half-life (0.704 Gyr). In early Earth history lead-207, the daughter of uranium-235, is formed at a higher rate than lead-206. For a late fractionation event $^{207}Pb/^{204}Pb$ changes slowly with time.

For isotopic systems with very long half-lives, such as samarium-142 (106 Gyr) and rubidium-87 (48.8 Gyr), the analogous closed-system geochrons will be nearly straight lines. Isochrons and mixing lines, in general, are not straight lines. They are straight in the uranium–lead system because $^{238}U/^{204}Pb$ and $^{235}U/^{204}Pb$ have identical fractionation factors, and mixing lines for ratios are linear if the ratios have the same denominator.

The initial lead-isotopic composition in iron meteorites can be obtained since these bodies are essentially free of uranium, one of the parents. Galenas are also high in lead and low in uranium and therefore nearly preserve the lead-isotopic ratios of their parent at the time of their birth. Galenas of various ages fall close to a unique single-stage growth curve. The small departures can be interpreted as further fractionation

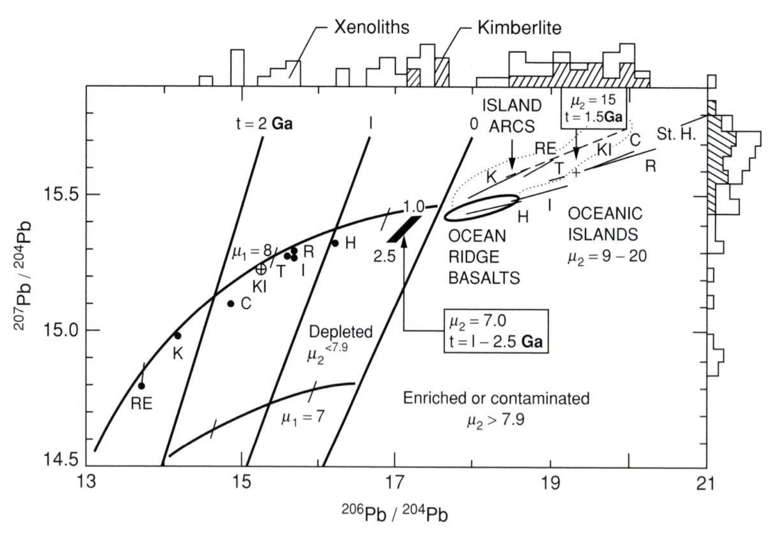

Fig. 17.1 Lead isotope diagram. Age of Earth is taken as 4.57 Ga. Straight lines labeled with letters are values for oceanic islands. Black dots are the inferred primary isotopic ratios if the island data are interpreted as secondary isochrons. Growth curves for μ values of 7.0 and 8.0 and primary isochrons at 1 Ga intervals are shown. The primary mantle reservoir appears to have a μ of 7.9. Oceanic-island basalts appear to have evolved in enriched reservoirs ranging in age from 1 to 2.5 Ga with the second-stage μ values ranging from 9 to 20. A point is shown for a two-stage model with $\mu = 7.9$ before 1.5 Ga and $= 15$ subsequently. The black bar represents the range of values for depleted reservoirs with $\mu = 7.0$ and a range of depletion ages from 1 to 2.5 Ga. The range for midocean-ridge basalts could be due to growth in an enriched reservoir or due to contamination by enriched magmas. Isotopic ratios for xenoliths and kimberlites are shown along the axes. Xenoliths are primarily from the shallow mantle and many are enriched. Kl is kimberlite. Diagram is modified from Chase (1981).

events. A few other systems also involve efficient separation of parents and daughters and ancient isotopic ratios can be *frozen-in*. These include U–He and Re–Os. U and He are fractionated and separated both by melting and degassing, so ancient high $^3He/^4He$ ratios can be frozen into gas-filled inclusions in peridotites, for example.

The μ values for basaltic magmas are usually quite high, 15–45, compared to primitive mantle. Their lead-isotopic ratios will therefore grow more rapidly with time than the primitive mantle, and the $^{206}Pb/^{204}Pb$ and $^{207}Pb/^{204}Pb$ ratios of such magmas are high. Some oceanic islands have such high lead-isotopic ratios that they must have come from ancient enriched reservoirs or contain, as a component, ancient enriched material. MORBs are thought to come from an ancient depleted reservoir but they also have ratios in excess of the geochron. This suggests that either the mantle (or upper mantle) has lost lead, relative to uranium, or that ocean-ridge basalts have been contaminated by material with high isotopic ratios, prior to eruption.

In a cooling, crystallizing mantle the μ of the residual melt will increase with time, assuming that solid silicates and sulfides retain lead more effectively than they retain uranium. Pb isotopes show that most of the mantle had solidified prior to 3.8 Ga, close to the ages of the oldest known rocks, measured by a variety of techniques. Basalts from oceanic islands have apparently experienced secondary growth in reservoirs with μ from about 10 to 20, after a long period of growth in a more 'primitive' reservoir ($\mu \sim 7.9$).

Leads from basaltic suites in many oceanic islands form linear areas on $^{206}Pb/^{204}Pb$ vs. $^{207}Pb/^{204}Pb$ diagrams (Figure 17.1). These could represent either mixing lines or secondary isochrons. Two-stage histories indicate that the leads from each island were derived from a common primary reservoir ($\mu = 7.9$) at different times from 2.5 to 1.0 Ga. Alternatively, the magmas from each island could represent mixtures between *enriched, less-enriched* or *depleted* components. In either case basalts involve a source region with ancient U/Pb enrichment, or are contaminated by a U/Pb-rich component. One mechanism for such enrichment is removal of a melt from a primitive reservoir to another part of the mantle that subsequently provides melts to the oceanic islands or contaminates MORB. Another mechanism is subduction of continental sediments. The logical default storage place is the shallow mantle, although this possibility is usually ignored in favor of deep-mantle storage.

Delamination of lower continental crust is another mechanism for *contaminating* or *polluting* the shallow mantle. In the standard model of mantle geochemistry the shallow mantle is assumed to be homogenous and any non-MORB magmas are considered to be from regions of the mantle that are `polluted` by `plumes from the deep mantle`.

To explain the various trends of the individual islands by mixing, the enriched end-members must come from parts of the crust or mantle that were enriched at different times or that have different time-integrated U/Pb ratios. In a crystallizing cumulate or magma ocean, the U/Pb ratio of the remaining fluid increases with time, and regions of the mantle that were enriched by this melt would have variable μ depending on when and how often they were enriched. If the enriched reservoir is global, as indicated by the global distribution of enriched magmas, it is plausible that different parts of it were enriched, or contaminated, at different times.

Pb-isotopes have painted a completely different story for mantle evolution and recycling than the early models based on Sr and Nd isotopes and the very incompatible elements. The major mantle reservoirs are enriched (high and increasing time-integrated U/Pb ratios) and there is no place for an accessible primordial unfractionated reservoir or evidence for a depleted reservoir, relative to primitive mantle. There are multiple stages of enrichment evident in the Pb-isotope record. Oxygen and osmium isotopes are not LIL elements and they imply recycling and a heterogeneous mantle. The importance of plate tectonics and recycling are discussed in early – but still accessible and rewarding – papers on `plumbotectonics` and the `persistent myth of continental growth`.

Os

Os is one of the platinum group elements (PGE) or metals. It is also a siderophile and a compatible element. It should be mainly in the core but there is enough in the mantle to make it a useful tracer, particularly of recycled oceanic and continental crust. It is also a useful tracer of cosmic dust, particularly in deep-sea sediments. The PGE in the crust and mantle may primarily be due to a late veneer. Because Os is *compatible* (jargon for silicate-crystal-loving rather than melt-loving) and occurs in sulfides, it tells us things that most of the other geochemically useful isotopes cannot. Peridotites, ultramafic massifs and ophiolites provide the bulk of the data. This means that the true isotope heterogeneity of the mantle can be sampled, rather than the gross averages provided by magmas. Large isotopic heterogeneities in Nd, Sr, Pb and Os have been documented in peridotite massifs related to spreading ridges on a variety of length scales, ranging from centimeters to tens of kilometers (Reisberg and Zindler, 1986), indicating a high degree of geochemical heterogeneity in the upper mantle, despite the observed homogeneity of MORB.

During partial melting, Re is mildly incompatible. This results in high Re/Os in basalts, and low Re/Os in the refractory, depleted solid residue left behind in the mantle. Thus, $^{187}Os/^{188}Os$ ratios in basalts and the residue rapidly diverge after melting and separation. Radiogenic and unradiogenic (depleted mantle residue) end-members are constantly produced by partial melting events.

Variations in osmium isotopic composition result from the alpha decay of ^{190}Pt to ^{186}Os and the beta decay of ^{187}Re to ^{187}Os. Suprachondritic $^{187}Os/^{188}Os$ ratios in intraplate volcanic rocks have been used to support models for generation of this type of volcanism from recycled oceanic crust. Suprachondritic $^{186}Os/^{188}Os$ ratios in intraplate volcanic rocks have been interpreted as indicating incorporation of outer core material into plumes. Such signatures may, however, also be generated in pyroxenites precipitated from MgO-rich melts by the preferential incorporation of Pt relative to Os in pyroxenes. High Pt/Os and Re/Os ratios, which should lead to generation of suprachondritic $^{186}Os/^{188}Os$-$^{187}Os/^{188}Os$, are properties of pyroxenites (Smith, 2003). The isotope signatures do not give any indication of depth of origin, and are consistent with shallow-source models for intraplate volcanism. `Pt-Os isotope systematics do not prove an ultra-deep origin for intraplate volcanism`.

A characteristic of the Re–Os system is the large Re/Os ratio acquired by magmas and the complementary low Re/Os ratio retained by the peridotite residues. Re/Os ratios correlate with rock type, a characteristic not shared by LIL-isotope systems. With sufficient time, differences in Re/Os ratio translate into large differences in the $^{187}Os/^{188}Os$ ratio. This makes the $^{187}Os/^{188}Os$ system a tracer of the addition of mafic crust or melt to the mantle, and Os isotopes have consequently been used to trace the involvement of recycled mafic crust in the sources of OIB (Shirey and Walker 1998).

Initial (corrected for age) $^{187}Os/^{188}Os$ values for abyssal peridotites worldwide lie between 0.117 and 0.167. Modern MORB samples have $^{187}Os/^{188}Os$ ratios clustering between 0.127 and 0.131, with some samples as high as 0.15 (Roy-Barman and Allègre, 1994).

OIB have variable and elevated $^{187}Os/^{188}Os$ ratios between 0.13 and 0.15, somewhat higher than primitive mantle values of 0.126 to 0.130 estimated from meteorites. Osmium and oxygen isotope correlate, indicating mixing of peridotitic and crustal components. Enriched mantle components (EM1, EM2, HIMU) are characterized by elevated $^{187}Os/^{188}Os$. Often, these enriched components have $\delta\ ^{18}O$ values that are different from MORB or abyssal peridotites that have been defined as upper mantle values (Eiler *et al.* 1997), indicating several source for recycled mafic material in the mantle.

Histograms of isotope ratios commonly display nearly Gaussian distributions. Such distributions can result from shallow mantle processes involving the mixing of different proportions of recycled, radiogenic and unradiogenic materials. Considering the large volume of mantle that is sampled by ocean ridge, ocean island and continental basalts, it is unlikely that pure end-members will be sampled; all basalts are blends to some extent. Small seamounts and xenoliths are more likely to have compositions close to an end-member or a pure component.

Meibom *et al.* (2002) attributed the Gaussian $^{187}Os/^{188}Os$ distribution of peridotite samples to melt-rock reactions during partial melting events in the upper mantle. During adiabatic upwelling of an upper-mantle assemblage (e.g. at a mid-ocean ridge) domains with relatively low solidus temperature and radiogenic Os isotopic compositions will melt first at depth. These melts mix with other melts and react with solid mantle material at shallower depths. Ancient unradiogenic Os is released from sulfide nuggets encapsulated in silicate and chromite host phases and mix with the more radiogenic Os in the melt. The Gaussian distribution represents random mixing between unradiogenic and radiogenic Os isotopic components of variable age.

Sr and Nd

Isotopes of Sr and Nd were used to set up the `two-reservoir model of mantle geochemistry`, a model that is inconsistent with much of petrology and geophysics, including Pb-isotopes and most other isotope systems. The idea of a two-reservoir mantle, a primordial reservoir and slow extraction of the crust from it (`the persistent myth of continental growth`) to form a depleted upper mantle were based on these systems. This so-called `standard model of mantle geochemistry and convection` conflicts with Pb-isotope data and with geophysical data and theoretical models of planetary accretion and differentiation. It also conflicts with a broader base of geochemical data that suggests very early and rapid differentiation of the silicate Earth. Unfortunately, current `hybrid mantle convection models` of mantle chemistry and dynamics are complex but retain the essence of the original model, including ready access to primitive or enriched material in the deepest mantle, and a well-stirred homogenous upper mantle. Most current models have many paradoxes associated with them [see `mantleplumes`], suggesting that the underlying assumptions are wrong.

The use of Sr and Nd isotopes usually relies on natural radioactive decay of two very long-lived radionuclides, ^{87}Rb and ^{147}Sm, which decay to ^{87}Sr and ^{143}Nd with half-lives of 49 and 106 billion years, respectively. These half-lives are significantly longer than the age of the solar system or any igneous rocks in it and therefore have little control on early differentiation processes. Sm actually decays to Nd via two radioactive decay schemes: ^{146}Sm–^{142}Nd, half-life about 10^6 years (My) and ^{147}Sm–^{143}Nd, half-life about 10^9 years (Gy). Sr, Sm and Nd are refractory lithophile (prefer silicates over metal) elements, whose relative abundances should not be affected by either volatile loss or core formation. The long-lived ^{147}Sm–^{143}Nd system has been widely used to trace planetary-scale processes such as the evolution of the crust and mantle. Because of the lack of samples from Earth's first 500 My of existence, the early epic of Earth differentiation is investigated with short-lived chronometers, such as ^{146}Sm–^{142}Nd.

Midocean-ridge basalts have ^{87}Sr/^{86}Sr less than 0.703, and 'pure' MORB may have values of 0.702 or less. Ocean-island, island-arc and continental flood basalts are generally much higher than 0.703, commonly higher than 0.71, and are more obviously mixtures. Attributing the properties of MORB to 'normal mantle' and, more recently, to the whole upper mantle, leaves crustal contamination, recycling or lower mantle sources as the only alternatives to explain ocean-island and other so-called 'plume' or 'hotspot' basalts. The standard model of mantle geochemistry originally ignored recycling and favored continuous continental growth from a primordial mantle reservoir, leaving behind a depleted homogenous upper mantle (called the 'convecting mantle'). The upper mantle, however, is inhomogenous in composition and isotopes. There is no evidence that any magma comes from a primordial or undegassed reservoir or from the lower mantle.

^{143}Nd/^{144}Nd ratios are expressed in terms of deviations, in parts per 10^4, from the value in a reservoir that has had chondritic ratios of Sm/Nd for all time. This deviation is expressed as ε_{Nd}. A chondritic unfractionated reservoir has $\varepsilon_{Nd} = 0$ at all times. Samarium and neodymium are both refractory rare-earth elements and should be in the Earth in chondritic ratios unless the Earth was assembled from some other kind of material. However, Sm and Nd are separated by magmatic process and thus record the magmatic or fractionation history of the Earth. Samarium has a higher crystal-melt partition coefficient than neodymium, and thus the Sm/Nd ratio is smaller in melts than in the original rock. The ^{143}Nd/^{144}Nd ratio, normalized as above, will therefore be positive in reservoirs from which a melt has been extracted and negative in the melts or regions of the mantle that have been infiltrated by melts. The Sm/Nd ratio depends on the extent of melting and the nature of the residual phases, and ε_{Nd} depends on the Sm/Nd ratio and the age of the fractionation event.

In spite of their geochemical differences, there is generally good correlation between neodymium and strontium isotopes. Positive ε_{Nd} correlates with low ^{87}Sr/^{86} Sr and vice versa. Midocean-ridge basalts have isotopic ratios indicating time-integrated depletions of Nd/Sm and Rb/Sr. The isotopic ratios are so extreme that the depletion must have occurred in the MORB reservoir more than 1 Ga ago. The original depletion may have occurred at the time the continental crust or the proto-crust started to form but more likely occurred throughout the accretional process of the Earth. The measured Sm/Nd and Rb/Sr ratios in MORB generally do not support such ancient ages, but the depletion may have been progressive, MORB may be mixtures of depleted and enriched materials, and other melt addition and extraction events may have occurred.

Incompatible-element ratios such as Rb/Sr and Nd/Sm are high in small-degree partial melts. However, for large fractions of partial melting the ratios are similar to the original rock. Since elements with partition coefficients much less than unity (such as Rb, Sr, Nd and Sm) are not retained effectively by residual crystals, it is difficult to change their ratio in melts, but the residual crystals, although low in these elements, have highly fractionated ratios. Partial melts representing large degrees of partial melting from primitive mantle will also have nearly primitive ratios, as will regions of the mantle invaded by these melts. If the melt cools and crystallizes, with refractory crystals being removed and isolated, the Sm/Nd ratio changes. Thus, it is dangerous to infer that a melt came from a

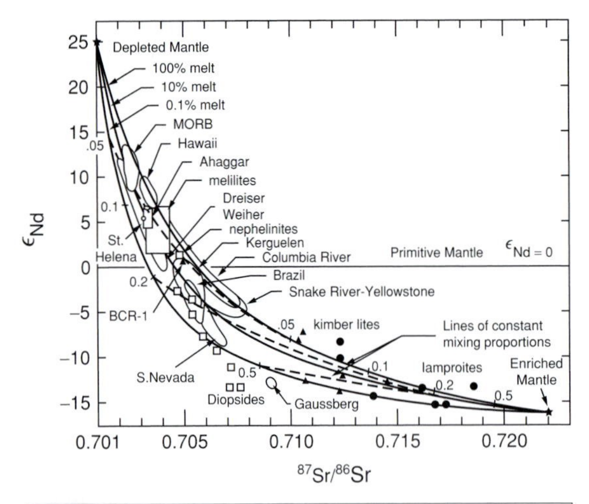

Fig. 17.2 ε_{Nd} versus $^{87}Sr/^{86}Sr$ for mixtures involving a depleted magma or residual fluids from such a magma after crystal fractionation, and an enriched component (EM). Plots such as this are known as *mantle arrays* or *multiple isotope plots.*

primitive reservoir simply because the isotopic ratios appear primitive. Similarly, magmas with ε_{Nd} near 0 can result from mixtures of melts, with positive and negative ε_{Nd}. Early claims of evidence for a primitive unfractionated reservoir based on Nd isotopes overlooked these effects. They also ignored the Pb-isotope evidence. Figure 17.2 shows isotope correlation for a variety of materials and theoretical mixing curves for fractionating melts.

Table 17.2 Parameters adopted for uncontaminated ocean ridge basalts (1) and contaminant (2)

Parameter	Pure MORB (1)	Enriched Contaminant (2)
Pb	0.08 ppm	2 ppm
U	0.0085 ppm	0.9 ppm
U/Pb	0.10	0.45
$^{238}U/^{204}Pb$	7.0	30.0
$^{206}Pb/^{204}Pb$	17.2	19.3, 21
Rb	0.15 ppm	28 ppm
Sr	50 ppm	350 ppm
$^{87}Sr/^{86}Sr$	0.7020	0.7060
Sm	2 ppm	7.2 ppm
Nd	5 ppm	30 ppm
$^{143}Nd/^{144}Nd$	0.5134	0.5124

Enrichment Factors*

Pb	24.6	U/Pb	4.5
Sr	7.0	Rb/Sr	26.7
Nd	6.0	Sm/Nd	0.60

(1) Assumed composition of uncontaminated midocean-ridge basalts.
(2) Assumed composition of contaminant. This is usually near the extreme end of the range of oceanic-island basalts.
*Ratio of concentration in two end-members.

The lead paradox

There are a large number of paradoxes involving U and its products (He, Pb and heat). Paradoxes are not intrinsic to data; they exist in relation to a model or a paradigm.

Both uranium and lead are incompatible elements in silicates but uranium enters the melt more readily than lead. The U/Pb ratio should therefore be high in melts and low in the solid residue, relative to the starting material. One would expect, therefore, that the MORB reservoir should be depleted in U/Pb as well as Rb/Sr and Nd/Sm. A time-average depletion would give Pb-isotope ratios that fall to the left of the primary geochron and below the mantle growth curve. Figure 17.1 shows, however, that both MORB and ocean-island tholeiites appear enriched relative to the primary growth curve. This implies that MORB has been contaminated by high-uranium or high-U/Pb material before being sampled, or that lead has been lost from the MORB reservoir. Early lead loss to the core, in sulfides, is possible, but the isotopic results, if interpreted in terms of lead removal, also require lead extraction over an extended period.

Contamination may have affected MORB or MORB source rocks. In order to test if contamination or magma mixing is a viable explanation for the location of the field of MORB on lead-lead isotopic diagrams, we need to estimate the lead content of uncontaminated depleted magmas and the lead and lead-isotopic ratios of possible contaminants. Table 17.2 lists some plausible parameters.

The results of mixing calculations are shown in Figure 17.3. The differences between the lead and other systems is striking. A small amount

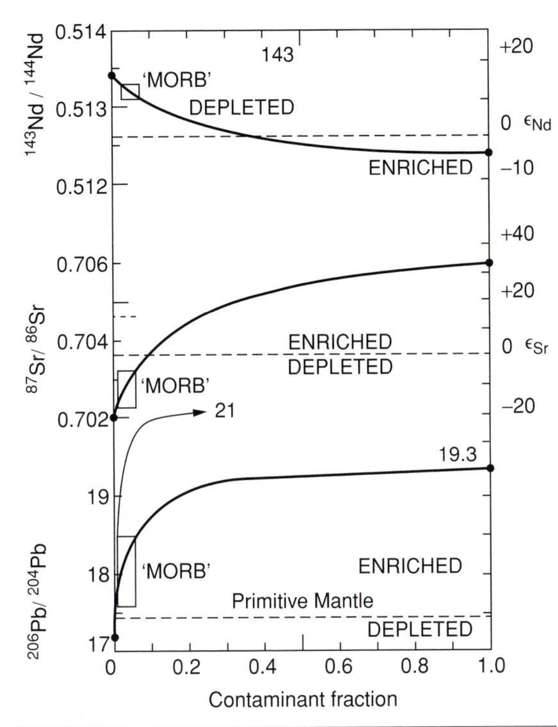

Fig. 17.3 Isotopic ratios versus contamination. Note that a small amount of contamination has a large effect on the lead system. Enriched magmas and slightly contaminated depleted magmas will both fall in the 'enriched' field relative to primitive mantle and will both give 'future' ages on a single-stage Pb–Pb evolution diagram. Slight contamination has less effect on Nd and Sr isotopes, and contaminated MORB will still appear depleted. The correlation line cannot be used to estimate the primitive value for $^{87}Sr/^{86}Sr$ if basalts are mixtures (e.g. Figure 17.2).

of contamination, less than 0.5%, pushes MORB compositions into the enriched field for lead but not for neodymium or strontium. In terms of single-stage 'evolution,' both observed (contaminated) MORB and oceanic-island basalt will appear to have *future ages* on a lead–lead geochron diagram. The neodymium and strontium isotopic ratios are not affected as much, and contaminated MORB will appear to come from depleted reservoirs.

Oxygen isotopes

The $^{18}O/^{16}O$ ratio is a powerful geochemical diagnostic because of the large difference between crustal and mantle rocks. Basalts also exhibit differences in this ratio. The relative abundances of oxygen isotopes are dominated by low-temperature mass-dependent fractionation processes, such as water–rock interactions. Rocks that react with the atmosphere or hydrosphere become richer in oxygen-18. Crustal contamination and flux of crustal material into the mantle serve to increase the $^{18}O/^{16}O$ ratio.

Oxygen is not a trace element; it comprises about 50 wt.% of most common rocks. There are three stable isotopes of oxygen, mass 16, 17 and 18, which occur roughly in the proportions of 99.76%, 0.038% and 0.20%. Variations in measured $^{18}O/^{16}O$ are expressed as ratios relative to a standard;

$$\delta^{18}O\,(0/00)$$
$$= 1000 \times [(^{18}O/^{16}O)m - (^{18}O/^{16}O)smow]/(^{18}O/^{16}O)smow$$

where *smow* is Standard Mean Ocean Water.

Oxygen isotopic variations in rocks are chiefly the result of fractionation between water and minerals; the fractionation between minerals is small. $\delta^{18}O$ values of minerals decreases in the order quartz–feldspar–pyroxene–garnet–ilmenite, with olivine and spinel showing large variations. Metasomatism increases $\delta^{18}O$ and the LIL elements. At high-temperature clinopyroxene and garnet have lower $\delta^{18}O$ values than olivine.

Oxygen isotope ratios constrain the origins of chemical and isotopic heterogeneity in the sources of basalts and the role of subducted material. $\delta^{18}O$ of oceanic basalts correlate with trace element abundances and radiogenic isotope tracers such as Sr, Nd, Pb and Os isotopes (Eiler *et al.* 1997). The oxygen isotope variations directly trace those parts of the mantle that have interacted with water at or near the Earth's surface. Osmium and oxygen isotopes are useful tracers of crustal involvement in the sources of MORB and OIB.

Large variations in $^{18}O/^{16}O$ are produced in the oceanic crust and associated sediments from interaction with the hydrosphere, either becoming ^{18}O-enriched by alteration and low-temperature exchange (<300 °C) in layered dike sequences, pillow basalts and sediments at the top of the subducted crustal section, or moderately ^{18}O-depleted by high-temperature exchange with seawater in the gabbroic lower crust.

These large variations in ^{18}O in subducted material translate to measurable ^{18}O variations in ocean island basalts that sample subducted crust. Oxygen isotopes can be used to quantify crustal contributions to mantle sources. Unlike other isotope tracers the concentration of oxygen is similar in all rocks and mass balance calculations can be attempted; oxygen isotopes are complementary to the LIL as tracers of crustal recycling.

Typical upper-mantle olivines have oxygen isotope ratios in the range +5.0 to +5.4 per mil and most unaltered MORB has $\delta^{18}O$ near +5.7 per mil. Mantle peridotites and fresh basaltic rocks have oxygen isotope ratios primarily from +5.5‰ to +5.9‰, which is taken to be the isotopic composition of common mantle oxygen or as a typical upper mantle value. Alkali basalts and mantle-derived eclogite xenoliths depart from these values. $\delta^{18}O$ values of eclogite garnets range from +1.5‰ to +9.3‰. Such oxygen isotope signatures were initially interpreted as the products of fractional crystallization of mantle magmas, although oxygen isotope fractionation effects are now known to be small under mantle conditions. The isotope and trace-element geochemistry of mantle eclogite and pyroxenite xenoliths and the anomalous oxygen isotope ratios are now taken as evidence that these rocks are derived from subducted altered ocean-floor basalts and gabbros, and possibly from delaminated continental crust. The changes in oxygen-isotope ratios and other geochemical characteristics of oceanic lithosphere that occur through exchange with seawater and hydrothermal fluids near ocean ridges have been documented in studies of ocean-floor rocks and obducted ophiolite equivalents. Metamorphic dehydration reactions do not cause large shifts in oxygen isotope ratios.

Most ocean-ridge tholeiites have $\delta^{18}O$ between +5 and +5.7 per mil. Some pillow lavas in ophiolites have $\delta^{18}O$ values of over +12 per mil. Potassic lavas have values of +6.0 to +8.5, continental tholeiites range up to +7.0. Oceanic alkalic basalts go as high as +10.7. Kimberlites and carbonatites have values up to 26. EM2 basalts range from +5.4 to +6.1 per mil.

It is possible that some tholeiites originate from garnet–clinopyroxenite or eclogite while some involve garnet peridotite; alkalic basalts may have more olivine in their sources. In addition, olivine fractionation at low temperature increases the $\delta^{18}O$ of residual melts. $\delta^{18}O$ values of lower crustal granulites range from +4.8‰ to +13.5‰. If these values are unchanged by eclogitization then this will be the range expected for delaminated lower-crustal eclogites, a potential mantle component. In fact, this range encompasses the entire OIB range, from HIMU to EM2.

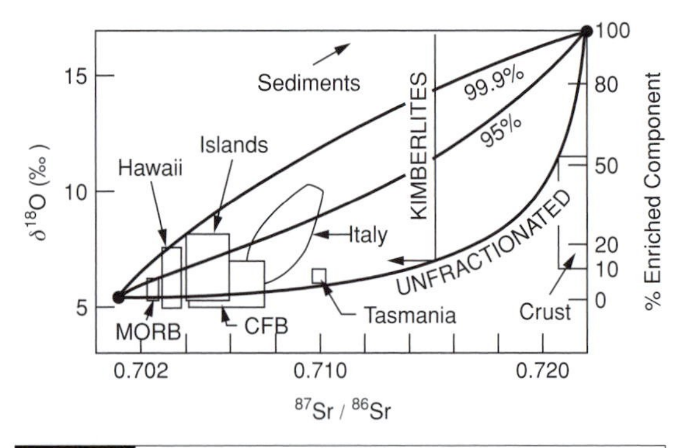

Fig. 17.4 $\delta^{18}O$ versus $^{87}Sr/^{86}Sr$ for magmas, oceanic sediments, kimberlites and continental crust. Mixing curves are shown for various degrees of crystal removal from the depleted end-member. Oxygen fractionation is ignored in this calculation.

The maximum variations in rocks known to be derived from the mantle are those recorded in eclogite, +1.5 to +9.0 per mil. The range of oxygen isotope ($\delta^{18}O \sim +6.5$ to +8.5) from Sierran pyroxenites show consistently higher values than measured in spinel+garnet and garnet peridotites (<6.4) from the same locations. Peridotites are generally +5.2 to +7.0. The range in pyroxenes are very much smaller, + 5.3 to +6.5 per mil. Garnet in peridotite is on the low end of this range.

$^{87}Sr/^{86}Sr$ is positively correlated with $\delta^{18}O$ among OIBs and other enriched magmas. Pb, Os and He isotopes also show some correlations with oxygen isotopes; HIMU lavas and other lavas with low $^{3}He/^{4}He$ are often depleted in ^{18}O, consistent with recycled or assimilated lower oceanic crust or lithosphere component. A sample mixing calculation is shown in Figure 17.4. In order to match the higher $\delta^{18}O$ values found for some oceanic islands with reasonable amounts of fractionation and contamination, a $\delta^{18}O$ of about +17 per mil is implied for the enriched or recycled component.

Contamination of mantle-derived magmas by the shallow mantle, lithosphere or crust may be caused by bulk assimilation of solid rock, by isotopic and trace-element exchange between magma and wallrock, or by magma mixing between the original melt and melts from the wallrock. Isotopic and trace-element exchange between magma and solid rock are likely to be too inefficient to be important because of the very low diffusivities. Diffusion distances, and therefore equilibration distances, are only a few centimeters per million years. Bulk assimilation or isotope exchange during partial melting are probably the most efficient means of magma contamination. This is not a simple two-component mixing process. It involves three end-members, the magma, the contaminant and a cumulate phase, which crystallizes to provide the heat required to partially melt the wallrock or dissolve the assimilated material.

Eclogitic garnets have higher $\delta^{18}O$ values than peridotite garnets. $\delta^{18}O$ values in some eclogites are equivalent to the extreme $\delta^{18}O$ values found in low-temperature altered oceanic basalts. Diamond inclusions tend to have higher $\delta^{18}O$ than eclogite xenoliths, suggesting that the latter may have exchanged with 'normal' mantle. The anomalous oxygen-isotope ratios of some eclogites might be the result of processes other than subduction and metamorphism of altered ocean-floor basalt, such as melting of delaminated lower continental crust. Low $\delta^{18}O$ values, 3‰–4‰, appear to require remelting of hydrothermally altered oceanic crust or meteoric–hydrothermal alteration after magma crystallization.

Hawaiian basalts have significantly lower $\delta^{18}O$ than typical upper mantle olivines (Eiler *et al.*, 1996a,b); low values are associated with radiogenic Pb-isotope ratios and with *depleted* Sr- and Nd-isotope ratios and relatively low $^3He/^4He$ ratios. The low-^{18}O-end member may represent Pacific ocean crust that underlies Hawaii, or a recycled upper-mantle lithosphere component. Hawaiian lavas also contain a high (i.e. >5.2 per mil) $\delta^{18}O$ end member, enriched in Sr and Nd isotope ratios and low (for Hawaii) $^3He/^4He$ ratios (the 'Koolau' end-member). This component of Hawaiian basalts was originally thought to be a pure deep-mantle plume end-member but was later attributed to sediments. $\delta^{18}O$ values similar to MORB and peridotite xenoliths rules out a large contribution of subducted sediment. With high radiogenic osmium, heavy oxygen and low $^3He/^4He$ (comparable to some values found along spreading ridges), the Koolau end-member may represent upper oceanic crust and/or mixing of EM2 and HIMU.

Some of the components in ocean-island basalts (e.g. EM1, EM2, HIMU, high- and low-$^3He/^4He$) identified with radiogenic isotopes also have distinctive oxygen-isotope ratios. There are correlations between oxygen and radiogenic isotopes. EM2 basalts are enriched in oxygen isotopes relative to MORB, consistent with the presence of subducted sediments in their sources; $^{87}Sr/^{86}Sr$ is positively correlated with $\delta^{18}O$ among all OIBs; HIMU lavas and lavas with low $^3He/^4He$ are often depleted in ^{18}O relative to normal upper mantle, consistent with the presence of recycled lower crust. Low $\delta^{18}O$ values are associated with radiogenic Pb isotope ratios and with *depleted* Sr and Nd isotope ratios and relatively low $^3He/^4He$ ratios.

The various recycled components – sediments, lower crust, upper crust, lithosphere – that have been identified in OIB by $\delta^{18}O$ variations imply a heterogenous dynamic upper mantle. The various heterogeneities introduced by plate tectonics and surface processes have different melting points and densities. These components all coexist in the shallow mantle. The isotope evidence does not imply that these components can only be returned to the shallow mantle by deep plumes.

Reservoirs

The standard model

At one time it was routinely assumed that $^3He/^4He$ and other isotopic ratios in mantle-derived materials represented two distinct populations corresponding to two distinct reservoirs.

These were the midocean-ridge basalt (MORB) reservoir and an ocean-island basalt (OIB) and continental flood basalt (CFB) reservoir. The former was postulated to fill up the whole upper mantle, and the latter was assumed to be a primordial (chondritic) lower mantle. It is now generally believed that the noble gases - actually only He and Ne - are the only reliable geochemical indicators of lower mantle involvement in surface volcanism. All other indicators have been traced to the recycling of crust and sediments, or fluids therefrom, or to delamination. High $^{3}He/^{4}He$ has traditionally been assumed to result from a high abundance of ^{3}He, and this has been used to argue that the lower mantle is undegassed in primordial volatiles. Plumes are assumed to carry the high $^{3}He/^{4}He$ signal from the deep mantle to the surface. However, the assumptions underlying this model require that the deep mantle has a high absolute abundance of He. In this model, the observed low abundance of He in OIB is a paradox, one of the `helium paradoxes`.

The MORB reservoir was originally thought to be homogenous because some isotopic ratios show less scatter in MORB than in OIB. The common explanation was that MORB were derived from a well-stirred, convecting part of the mantle while OIB were derived from a different, deeper reservoir. Alternatively, the homogeneity of MORB can be explained as a consequence of the sampling process, and the central limit theorem. MORB is more likely to result from a *process* rather than from a *reservoir*; it is an average, not a component.

The standard, two-reservoir model is reinforced by data selection and data filtering practices. Samples along ridges that are judged to be contaminated by plumes - magmas that are assumed to be from a different reservoir - are often removed from the dataset prior to statistical analysis. The definition of *plume influence* is arbitrary. For example, isotopic ratios which exceed an arbitrary cutoff may be eliminated from the dataset. In this way, the MORB dataset is forced to appear more homogenous than it really is. Despite this, various ridges still have different means and variances in their isotopic ratios, and these depend on spreading rate and ridge maturity. The variance for many ridge segments increases as spreading rate decreases and, by analogy, the observed high variance of various OIB data-sets is consistent with slow spreading, small sampled volume, or low degrees of melting or degassing.

Geochemical variations in a well-sampled system such as a midocean ridge or an oceanic island can be characterized by an average value, or mean, and a measure of dispersion such as the standard deviation or variance. When dealing with isotopic ratios the appropriate measures of central tendency are the *median* and the *geometric mean*, since these are invariant to inversion of the ratio. Likewise, when dealing with ratios, the absolute concentrations must be taken into account, in addition to the ratios. That is, the ratios must be weighted appropriately before being analysed or averaged.

Recycling of crustal materials and mantle heterogeneity

Chemical and isotopic diversity within the mantle, and fertility variations, may be related to recycling of oceanic crust, lithosphere and sediments into the mantle by plate tectonic processes. Delamination of the eclogitic portion of over-thickened continental crust is an important process for fertilizing the mantle. The partial extraction of melts at ridges and volcanic islands contributes, over time, to the isotopic diversity of the mantle.

On the basis of radiogenic Pb, Nd and Sr isotopes, mainly in ocean-island basalts, five isotopically extreme mantle components or reservoirs have been defined: (1) MORB (depleted or normal MORB); (2) HIMU ('high-μ' where μ is $^{238}U/^{204}Pb$); (3) EM1; (4) EM2; and (5) LOMU ('low-μ') with low time-integrated U/Pb. 'Enriched' refers to time-integrated Rb/Sr, Sm/Nd or (U,Th)/Pb ratios higher than primitive mantle (bulk silicate Earth). In addition, new isotopic systems have been developed using isotopes of hafnium, osmium, oxygen and neon. The end-member and reservoir description of mantle variability, which includes MORB, FOZO, HIMU, EM1 and EM2 was based on isotopes of Sr, Nd and Pb. Osmium, oxygen and helium isotope data have revealed the limitations of this five-component classification. For reviews of the rhenium–osmium and oxygen isotope systems, see Shirey and Walker (1998), Eiler (2001).

A large number of other components have been suggested: DUPAL, FOZO, C (Common), PHEM, LONU and so on. Most of these have been attributed to plumes, or the lower mantle, or the core–mantle boundary, but with little justification. Isotopes cannot constrain the locations, depths, protoliths or lithologies of the sources of these isotopic signatures. The components have been attributed to separate reservoirs, such as DMM (depleted MORB mantle), DUM (depleted upper mantle), EM (enriched mantle), PM (primitive mantle), PREMA (prevalent mantle) and so on. These hypothetical reservoirs have been equated with mantle subdivisions adopted by seismologists, e.g. crust, upper mantle, transition region, lower mantle and D″. The components, however, may be distributed throughout the upper mantle.

In addition to MORB, there are also a variety of depleted magmas including picrites and komatiites at *hotspot islands* (Hawaii, Iceland, Gorgona) and at the base of continental flood basalts. Many so-called hotspot basalts have very low ^{3}He concentrations and low ^{3}He/^{4}He isotope ratios. The crust and shallow mantle account for most of the incompatible elements in the Earth, such as U, Th, K, Ba, Rb, Sr and isotopes that are used to finger-print so-called plume influence. The origins and locations of these components are actively debated. It is conventional, in isotope geochemistry treatises, to assume that the upper mantle is the source of depleted MORB, and only of depleted MORB. Thus, anything other than depleted MORB must come from the deep mantle. The reasoning is as follows: MORB is the most abundant magma type; it erupts passively at ridges; the MORB-source must therefore be shallow; since MORB is derived from the upper mantle, nothing else can be. MORB, at one time, was thought to be a common component in magmas from the mantle; the most likely location for such a common component is the shallowest mantle. Early mass balance calculations suggested that about 30% of the mantle was depleted by melt extraction. If this was due to removal of the continental crust, then a volume equivalent to the mantle above about 650-km depth is depleted.

These arguments are not valid. Based on more complete mass-balance calculations in *Theory of the Earth*, I showed that most of the mantle must be depleted and much of it must be infertile [www.resolver.caltech.edu/CaltechBOOK:1989.001].

Most of the mantle is not basaltic or fertile (there is not enough Ca, Al, Na and so on). About 70–90% of the mantle must be depleted to explain the concentrations of some elements in the continental crust alone and the amount of ^{40}Ar in the atmosphere. Enriched basalts are the first to emerge at a new ridge, or upon continental break-up, and enriched MORB (EMORB) occurs along spreading ridges. Enriched islands and seamounts occur on and near ridges. A chemically and isotopically heterogenous shallow mantle is indicated. In the early isotope models of mantle geochemistry and crustal growth, there was no recycling; crustal growth and upper mantle depletion were one-way and continuous processes. The upper mantle was assumed to be vigorously convecting and chemically homogenous. The plume hypothesis emerged naturally from these assumptions.

Plate tectonics continuously subducts sediments, oceanic crust and lithosphere and recycles them back into the mantle. The basalts and peridotites are generally altered; subduction zone processes modify this material. Upper continental crust enters subduction zones at trenches, and lower continental crust enters the mantle by delamination and subcrustal erosion. Some of the melts and gases in the mantle are trapped and never make it to the surface. These are some of the components that one might expect to find in the mantle. They have distinctive trace element and isotope signatures.

The following are some of the isotopically distinctive components that have been defined in oceanic and continental magmas and which may correspond to some of the above materials mantle mixing [see HIMU EM1 EM2 DMM for examples of mixing trends].

HIMU may represent old recycled hydrothermally altered oceanic crust, dehydrated and converted to eclogite during subduction. It is depleted in Pb and K and enriched in U, Nb and

Ta relative to MORB and has high $^{206}Pb/^{204}Pb$ and $^{207}Pb/^{204}Pb$ ratios. Type-islands include Tubuaii, Mangaia and St. Helena.

EM1 has extremely low $^{143}Nd/^{144}Nd$, moderately high $^{87}Sr/^{86}Sr$ and very low Pb-isotope ratios. EM1 includes either recycled oceanic crust plus a few percent pelagic sediment or metasomatized continental lithosphere. Examples are Pitcairn and the Walvis Ridge.

EM2 is an intermediate Pb-isotope component with enriched Nd- and Sr-isotope signatures. Samoa and Taha (Societies) are examples, and Kerguelen trends towards EM2. EM2 may be recycled oceanic crust containing a few percent of continent-derived sediment.

LOMU is distinguished from EM1 by its unusually high $^{87}Sr/^{86}Sr$ and $^{207}Pb/^{204}Pb$ ratios, and very low $^{143}Nd/^{144}Nd$ and $^{177}Hf/^{176}Hf$ ratios, suggesting an ancient continental source. EM1, EM2, LOMU and HIMU and mixtures of these are referred to as EM components; all may include recycled materials of continental derivation.

LONU is low $^{238}U/^{3}He$ (ν), an ancient component that retains high $^{3}He/^{4}He$ ratios over time compared with high ν components such as MORB and other melts. This component is most likely U- and Th-depleted peridotites or cumulates with trapped CO_2-rich inclusions. Gases that escape from ascending magmas or crystallizing cumulates are partially trapped at shallow depths in LIL-poor surroundings, such as the lithosphere and olivine-rich cumulates and the $^{3}He/^{4}He$ ratio is frozen in. Meanwhile, the U and Th in fertile and LIL-rich mantle causes the ^{4}He content to increase with time in these components. High $^{3}He/^{4}He$ components do not need to represent undegassed mantle; they can be ^{3}He-poor.

The end-members constrain the compositional, or at least the isotopic, extremes in magmas. Some isotopic data (Sr, Nd, Pb) from ocean islands form arrays that trend toward a limited region of isotopic space that has been given various names, including C ('common' mantle component), FOZO ('focus zone') and PHEM ('primary helium mantle'). The fact that there appears to be a common component suggests that melt percolation through a shallow buoyant peridotite, such as depleted lithosphere, may be involved.

The helium carrier

The carrier of the high *R*/*R*a signature is unknown. An ancient gas-rich bubble in the interior of an olivine grain represents one extreme, but plausible, carrier. Newly formed bubbles derived from ascending MORB magmas, will have present-day MORB isotopic signatures.

Locations with the highest $^{3}He/^{4}He$ materials – Hawaii, Iceland, Galapagos – also have the highest variance and are associated with basalts with MORB-like ratios. On the basis of Pb, Sr and Nd isotopes, and LIL ratios, the highest $^{3}He/^{4}He$ magmas have depleted MORB-like signatures. They also exhibit large variance in $^{3}He/^{4}He$ and have low ^{3}He abundances. The Os isotopes suggest a peridotitic protolith, such as a depleted restite or cumulate. This would explain why the high $^{3}He/^{4}He$ component is the C- or 'common component' and why it is low in ^{3}He. Peridotites can freeze-in ancient He isotopic signatures because of their low U/He ratios.

If the mantle is composed of peridotite with blobs of recycled eclogitized oceanic or lower continental crust (HIMU), then metasomatism or mixing could occur during ascent of diapers or in ponded melts. In such a mixture, concentrations of noble gases and compatible elements (Ni, Cr, Os and heavy rare-earth elements) would come almost completely from the peridotitic mantle, whereas the highly incompatible elements are supplied by melts. Ascending magmas interact with the surrounding material – wall-rock reactions – exchanging heat, fluids and gases, and causing melting and crystallization. Such processes may explain the complexities of isotopic and trace-element arrays.

Mixing arrays

The chemical and isotopic properties of many mantle magmas can be approximated by binary mixing between two end-members [see `mantle mixing trends HIMU EM-1 EM-2 DMM`]. However, the mantle contains many different components. Some of the 'end-members' themselves appear to be mixtures. Magmas are likely to be blends of melts that represent different degrees of partial melting and crystal fractionation from several lithologies. Mixing arrays are therefore likely to be very complex.

Mixing arrays involving ratios

Data arrays in isotope space reflect mixing between distinct mantle or melt components, with different isotopic and elemental concentrations. The curvature of two-component mixing curves is related to differences in the elemental abundance ratios of the components and as such, mixing curves can, in principle, be used to estimate the relative abundances of Sr, Nd, Pb, He, and Os in various magmas. Linear mixing arrays with little scatter can indicate similar relative abundances of these elements in the mantle components. For isotope arrays of Sr, Nd, Pb and Hf, the assumption is usually made that all components have similar elemental concentrations.

In order to estimate trace-element concentrations from isotopic mixing arrays, simple assumptions have to be made about the relative abundances of Sr, Nd, Pb and so on in the mixing end-members. The same is true for mixing arrays involving trace-element ratios. Often it is assumed that the elemental abundances are the same in all end-members. Such assumptions are not valid if the mixing arrays reflect mixing of melts from sources with various proportions of these components. They are not valid if the mixing components are fractionating melts, blends of variable partial melts, or are lithologically very different (e.g. peridotite, recycled mafic crust, sediment).

Oxygen isotopes play a key role in identifying components because oxygen is not a minor or trace element; the abundance of oxygen varies little among various lithologies. Mixing relationships with oxygen isotopes therefore can constrain the absolute abundances of various elements in the end-members of mixing arrays.

Midocean-ridge basalts (MORB) are among the most depleted (low concentrations of LILs, low values of Rb/Sr, Nd/Sm, $^{87}Sr/^{86}Sr$, $^{144}Nd/^{143}Nd$, $^{206}Pb/^{204}Pb$) and the most voluminous magma type. They erupt through thin lithosphere and have experienced some crystal fractionation prior to eruption. The melting region under ridges involves sections of extensive melting and deeper sections, and areas on the wings, having much smaller degrees of melting. MORB are blends of these various magmas. The blending process erases the diversity of the components that go into MORB, but nevertheless different species of MORB have been identified (DMORB, NMORB, EMORB, TMORB and PMORB). MORBs typically contain orders of magnitude more ^{3}He than other basalts. Lead isotopes suggest that MORB has experienced contamination prior to eruption. MORB itself is a hybrid magma.

Basalts with high Rb/Sr, La/Yb and Nd/Sm generally have high $^{87}Sr/^{86}Sr$ and low $^{143}Nd/^{144}Nd$. These can be explained by binary mixing of depleted and enriched magmas or by mixtures of a depleted magma and a component representing varying degrees of melting of an enriched reservoir ore blob. The variable LIL ratios of such an enriched component generates a range of mixing hyperbolas or 'scatter' about a binary mixing curve, even if there are only two isotopically distinct end-members. Thus a model with only two isotopically distinct reservoirs can generate an infinite variety of mixing lines. In some regions, however, the inverse relationship between LIL and isotopic ratios cannot be explained by binary mixing. These regions are all in midplate or thick lithosphere environments, and sublithospheric crystal fractionation involving garnet and clinopyroxene might be expected prior to eruption.

Magma mixing can happen in many different ways. A heterogenous mantle can partially melt. A depleted diapir can cause variable degrees of melting of enriched components. An undepleted or enriched magma can ascend and partially melt a depleted layer. A fractionating (cooling and crystallizing) depleted magma can interact with the shallow mantle. To fix ideas, consider a depleted magma from a depleted-mantle reservoir to be the parent of MORB. If this magma is brought to a near-surface environment, it may crystallize olivine, plagioclase and orthopyroxene. If arrested by thick lithosphere it may precipitate garnet and clinopyroxene. Melts can represent varying degrees of crystal fractionation, or varying degrees of partial melting. These are called *evolved magmas*. A fertile blob may be a garnet- and clinopyroxene-rich region, such as a

Table 17.3 | Parameters of end-members

Parameter	Depleted Mantle (1)	Enriched Mantle (2)	Enrichment Ratio (2)/(1)	
$^{87}Sr/^{86}Sr$	0.701–0.702	0.722	Sr_2/Sr_1	13.8
ϵ^{Nd}	24.6	−16.4	Nd_2/Nd_1	73.8
$^{206}Pb/^{204}Pb$	16.5–17.0	26.5	Pb_2/Pb_1	62.5
$^3He/^4He$ (R_a)	6.5	150	He_2/He_1	2.0
$\delta^{18}O$ (permil)	5.4	17	O_2/O_1	1.0
La/Ce	0.265	0.50	Ce_2/Ce_1	165.0
Sm/Nd	1.50–0.375	0.09–0.39		

Partition Coefficients

Rb	0.02	Sr	0.04	La	0.012	Ce	0.03
Sm	0.02	Pb	0, 0.002	He	0.50	Nd	0.09

garnet–pyroxenite cumulate, delaminated lower continental crust or an eclogite slab.

There are two situations that have particular relevance to mantle-derived magmas. Consider a 'normal' depleted mantle magma. If it can rise unimpeded from its source to the surface, such as at a rapidly spreading ridge, it yields a relatively unfractionated, uncontaminated melt, but can nevertheless be a blend of melts. Suppose now that the magma rises in a midplate environment, and its ascent is impeded by thick lithosphere. The magma will cool and crystallize – evolve – simultaneously partially melting or reacting with the surrounding mantle. Thus, crystal fractionation and mixing occur together, and the composition of the hybrid melt changes with time and with the extent of fractionation. Fractionation of garnet and clinopyroxene from a tholeiitic or picritic magma at sublithospheric depths (>50 km) can generate alkalic magmas with enriched and fractionated LIL patterns. For purposes of illustration, let us investigate the effects of combined eclogite fractionation (equal parts of garnet and clinopyroxene) and 'contamination' on melts from depleted mantle. 'Contamination' is modeled by mixing an enriched component with the fractionating depleted magma. This component is viewed as a partial melt generated by the latent heat associated with the crystal fractionation. The assumed geochemical properties of the end-members, the enrichment factors of the elements in question and the partition coefficients, *D*, assumed in the modeling are given in Table 17.3. The figures that follow show various ratios for mixes of a fractionating depleted melt and an enriched component. We assume equilibrium crystal fractionation, as appropriate for a turbulent or permeable magma body, and constant *D*. La/Ce and Sm/Nd are used in the following examples but the discussion is general and the results are typical of other LIL-pairs.

La/Ce versus $^{87}Sr/^{86}Sr$

La/Ce is high in melts relative to crystalline residues containing garnet and clinopyroxene. It is low in melt-depleted reservoirs and high in enriched reservoirs. Low and high values of $^{87}Sr/^{86}Sr$ are characteristics of time-integrated depleted and enriched magmas, respectively. Since high $^{87}Sr/^{86}Sr$ implies a time-integrated enrichment of Rb/Sr, there is generally a positive correlation of Rb/Sr and La/Ce with $^{87}Sr/^{86}Sr$. Some magmas, however, exhibit high La/Ce and low $^{87}Sr/^{86}Sr$. This cannot be explained by binary mixing of two homogenous magmas, but can be explained by mixing of magmas from a cooling partially molten mantle.

On a theoretical La/Ce vs. $^{87}Sr/^{86}Sr$ plot (Figure 17.5), the mixing lines between the crystallizing MORB and enriched mantle components reverse slope when MORB has experienced slightly more than 99% crystal fractionation. The relationships for equilibrium partial melting

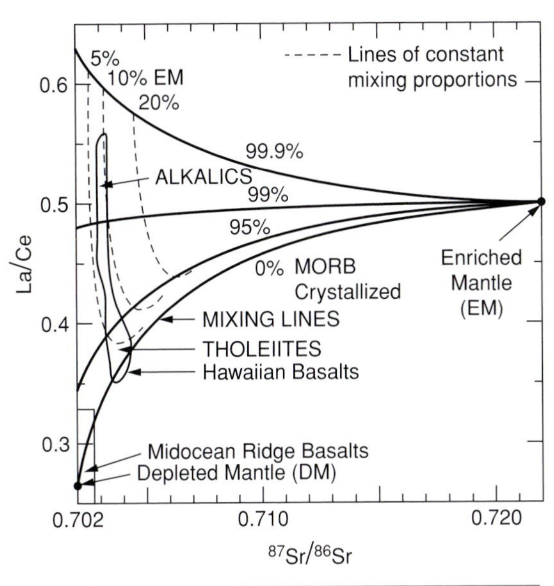

Fig. 17.5 La/Ce versus Sr-isotopes for the fractionation–contamination model, compared with Hawaiian basalts. Unfractionated MORB (DM) has La/Ce = 0.265. La/Ce of the depleted end-member increases as crystal fractionation proceeds. The enriched end-member (EM) has La/Ce = 0.5, in the range of kimberlitic magmas. Hawaiian tholeiites can be modeled as mixes ranging from pure MORB plus 2–7% of an enriched component to melts representing residuals after 95% clinopyroxene-plus-garnet crystal fractionation and 5–8% enriched component. Alkali basalts involve more crystal fractionation, or smaller-degree partial melts, and more contamination. In this and subsequent figures solid curves are mixing lines between EM and melts representing fractionating depleted magmas. Dashed curves are trajectories of constant mixing proportions. Note that some basalts are not intermediate in La/Ce to the end-members. Similar 'discrepencies' in other geochemical ratios are often called *paradoxes*.

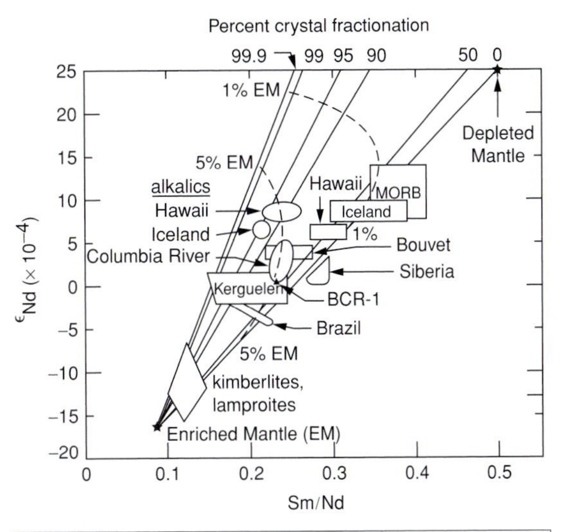

Fig. 17.6 $^{143}Nd/^{144}Nd$ (relative to primitive mantle) versus Sm/Nd for mixtures of enriched mantle (EM) and residual melts resulting from high-pressure crystal fractionation of MORB-like depleted magma.

and equilibrium, or batch, crystallization are the same. Therefore, large degrees of crystallization of a MORB-like melt or small amounts of partial melting of a depleted source are implied by an inverse relationship between La/Ce (or Rb/Sr, La/Yb, Nd/Sm, and so on) and $^{87}Sr/^{86}Sr$ (or $^{143}Nd/^{144}Nd$) such as observed at many mid-plate environments. The apparently contradictory behavior of magmas with evidence for current enrichment and long-term depletion is often used as evidence for 'recent mantle metasomatism.' Figure 17.5 illustrates an alternative explanation. Note that, with the parameters chosen, Hawaiian alkalics have up to 10% contamination by an enriched component.

Neodymium isotopes versus $^{87}Sr/^{86}Sr$

Isotopic ratios for ocean-island and continental basalts are compared with mixing curves (Figure 17.2). These basalts can be interpreted as mixes between a fractionating depleted magma and an enriched component. The value for primitive mantle is also shown. The primitive mantle value of $^{87}Sr/^{86}Sr$ is unknown and cannot be inferred from basalts that are themselves mixtures.

Neodymium isotopes versus Sm/Nd

The mixing–fractionation curves for Nd isotopes versus Sm/Nd are shown in Figure 17.6. High-Sm/Nd basalts from Iceland, Hawaii, Siberia, Kerguelen, and Brazil all fall near the curve for unfractionated MORB with slight, 1–5%, contamination. Alkalics from large oceanic islands with thick crust (Hawaii, Iceland and Kerguelen) are consistent with large amounts of crystal fractionation and moderate (5–10%) amounts of contamination. The interpretation is that the more voluminous tholeiites are slightly fractionated and contaminated, while the alkalics have experienced sublithospheric crystal fractionation and contamination or magma mixing prior to eruption. MORB itself has about 1% contamination, similar to that required to explain lead isotopes.

The Hawaiian tholeiites can be modeled with variable degrees of deep crystal fractionation (0–95%), mixed with 1–5% of an enriched component. Alkali basalts represent greater extents of crystal fractionation and contamination. All of this is consistent with magma evolution beneath thick crust or lithosphere, the main tectonic differences between midocean ridges and oceanic islands.

Lead and Sr isotopes

Rubidium, strontium and the light REEs are classic incompatible elements, and the effects of partial melting, fractionation and mixing can be explored with some confidence. Relations between these elements and their isotopes should be fairly coherent. Some enriched magmas also have high $^{3}He/^{4}He$, $\delta^{18}O$ and $^{206}Pb/^{204}Pb$. These isotopes provide important constraints on mantle evolution, but they may be decoupled from the LIL variations and are usually assumed to be so. $^{3}He/^{4}He$ depends on the uranium and thorium content and age of the enriched component. $^{206}Pb/^{204}Pb$ and $^{3}He/^{4}He$ are sensitive to the age of various events affecting a reservoir because of the short half-life of uranium. The U/Pb ratio may be controlled by sulfides and metals as well as silicates.

A minimum of three components is required to satisfy the combined strontium and lead isotopic systems when only simple mixing is considered. A mixing-fractionation curve for Sr and Pb isotopes (Figure 17.7) shows that the data can be explained with only two isotopically distinct components. Enriched components can have variable isotope ratios since these are sensitive to the U/Pb ratio and the age of enrichment events. The various contributions to mantle variability (subduction, trapped magmas, trapped bubbles) make it likely that the OIB 'reservoir' is in the shallow mantle and is laterally inhomogenous. In this sense the mantle contains multiple 'reservoirs'; their dimensions are likely to be of the order of tens of kilometers.

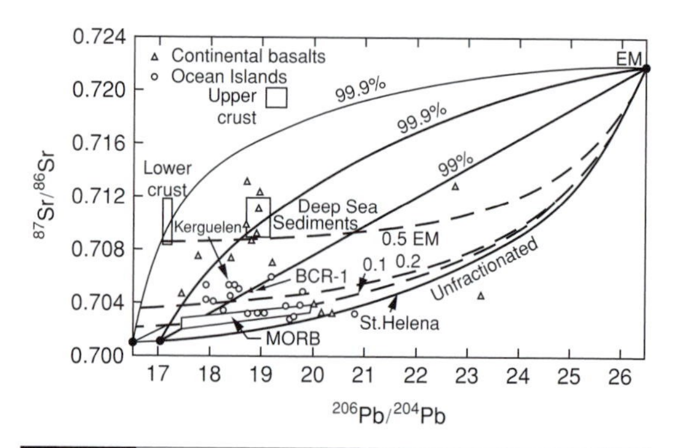

Fig. 17.7 Sr versus Pb isotopes; trajectories for mixtures of a fractionating depleted MORB-like magma and an enriched component. Mixing hyperbolas are solid lines. Hawaiian basalts fall in the region of 50–99% crystal fractionation (garnet plus clinopyroxene) and 10–30% contamination by EM (dashed lines). The enriched component may be recycled mafic crust or cumulates.

Part V

Mineral physics

Chapter 18

Elasticity and solid-state geophysics

Mark well the various kinds of minerals, note their properties and their mode of origin.

Petrus Severinus (1571)

The seismic properties of a material depend on composition, crystal structure, temperature, pressure and in some cases defect and impurity concentrations. Most of the Earth is made up of crystals. The elastic properties of crystals depend on orientation and frequency. Thus, the interpretation of seismic data, or the extrapolation of laboratory data, requires knowledge of crystal or mineral physics, elasticity and thermodynamics. But one cannot directly infer composition or temperature from mantle tomography and a table of elastic constants derived from laboratory experiments. Seismic velocities are not unique functions of temperature and pressure alone nor are they linear functions of them. Tomographic cross-sections are not maps of temperature. The mantle is not an ideal linearly elastic body or 'an ideal harmonic solid'. Elastic 'constants' are frequency dependent and have intrinsic, extrinsic, anharmonic and anelastic contributions to the pressure and temperature dependencies.

Elastic constants of isotropic solids

The elastic behavior of an isotropic, ideally elastic, solid – at infinite frequency – is completely characterized by the density ρ and two elastic constants. These are usually the bulk modulus K and the rigidity G or μ Young's modulus E and Poisson's ratio σ are also commonly used. There are, correspondingly, two types of elastic waves; the compressional, or primary (P), and the shear, or secondary (S), having velocities derivable from

$$\rho V_P^2 = K + 4G/3 = K + 4\mu/3$$
$$\rho V_S^2 = G = \mu$$

The interrelations between the elastic constants and wave velocities are given in Table 18.1. In an anisotropic solid there are two shear waves and one compressional wave in any given direction.

Only gases, fluids, well-annealed glasses and similar noncrystalline materials are strictly isotropic. Crystalline material with random orientations of grains can approach isotropy, but rocks are generally anisotropic.

Laboratory measurements of mineral elastic properties, and their temperature and pressure derivatives, are an essential complement to seismic data. The high-frequency elastic-wave velocities are known for hundreds of crystals. Compilations of elastic properties of rocks and minerals complement those tabulated here. In addition, ab initio calculations of elastic properties of high-pressure minerals can be used to supplement the measurements.

Elastic properties depend on both crystal structure and composition, and the understanding of these effects, including the role of temperature and pressure, is a responsibility of a discipline called *mineral physics* or *solid-state geophysics*. Most measurements are made under

Table 18.1 | Connecting identities for elastic constants of isotropic bodies

K	$G = \mu$	λ	σ
$\lambda + 2\mu/3$	$3(K - \lambda)/2$	$K - 2\mu/3$	$\dfrac{\lambda}{2(\lambda + \mu)}$
$\mu\dfrac{2(1+\sigma)}{3(1-2\sigma)}$	$\lambda\dfrac{1-2\sigma}{2\sigma}$	$\mu\dfrac{2\sigma}{1-2\sigma}$	$\dfrac{\lambda}{3K - \lambda}$
$\lambda\dfrac{1+\sigma}{3\sigma}$	$3K\dfrac{1-2\sigma}{2+2\sigma}$	$3K\dfrac{\sigma}{1+\sigma}$	$\dfrac{3K - 2\mu}{2(3K + \mu)}$
$\rho(V_p^2 - 4V_s^2/3)$	ρV_s^2	$\rho(V_p^2 - 2V_s^2)$	—
—	—	—	$\dfrac{1}{2}\dfrac{(V_p/V_s)^2 - 2}{(V_p/V_s)^2 - 1}$

λ, μ = Lamé constants
G = Rigidity or shear modulus $= \rho V_s^2 = \mu$
K = Bulk modulus
σ = Poisson's ratio
E = Young's modulus
ρ = density
V_p, V_s = compressional and shear velocities

$$\rho V_p^2 = \lambda + 2\mu = 3K - 2\lambda = K + 4\mu/3 = \mu\frac{4\mu - E}{3\mu - E}$$
$$= 3K\frac{3k + E}{9K - E} = \lambda\frac{1-\sigma}{\sigma} = \mu\frac{2 - 2\sigma}{1 - 2\sigma} = 3K\frac{1-\sigma}{1+\sigma}$$

conditions far from the pressure and temperature conditions in the deep crust or mantle. The frequency of laboratory waves is usually far from the frequency content of seismic waves. The measurements themselves, therefore, are just the first step in any program to predict or interpret seismic velocities.

Some information is now available on the high-frequency elastic properties of all major rock-forming minerals in the mantle. On the other hand, there are insufficient data on any mineral to make assumption-free comparisons with seismic data below some 100 km depth. It is essential to have a good theoretical understanding of the effects of frequency, temperature, composition and pressure on the elastic and thermal properties of minerals so that laboratory measurements can be extrapolated to mantle conditions. Laboratory results are generally given in terms of a linear dependence of the elastic moduli on temperature and pressure. The actual variation of the moduli with temperature and pressure is more complex. It is often not justified to assume that all derivatives are linear and independent of temperature and pressure; it is necessary to use physically based equations of state. Unfortunately, many discussions of upper-mantle mineralogy and interpretations of tomography ignore the most elementary considerations of solid-state and atomic physics.

The functional form of $\alpha(T, P)$, the coefficient of thermal expansion, is closely related to the specific-heat function, and the necessary theory was developed long ago by Debye, Grüneisen and Einstein. Yet $\alpha(T, P)$ is sometimes assumed to be independent of pressure and temperature, or linearly dependent on temperature. Likewise, interatomic-potential theory shows that the pressure derivative of the bulk modulus dK/dP must decrease with compression, yet the moduli are often assumed to increase linearly with pressure. There are also various thermodynamic relationships that must be satisfied by any self-consistent equation of state, and certain inequalities regarding the strain dependence of anharmonic properties. Processes within the Earth are not expected to give random orientations of the constituent anisotropic minerals. On the other hand the full elastic tensor is difficult to determine from seismic data. Seismic data usually provide some sort of average of the velocities in a given region and, in some cases, estimates of the anisotropy. The best-quality laboratory data are obtained from high-quality single crystals. The full elastic tensor can be obtained in these cases, and methods are available for computing average properties from these data.

It is simpler to tabulate and discuss average properties, as I do in this section. It should be kept in mind, however, that mantle minerals are anisotropic and they tend to be readily oriented by mantle processes. Certain seismic observations in subducting slabs, for example, are best interpreted in terms of oriented crystals and a resulting seismic anisotropy. If all seismic observations are interpreted in terms of isotropy, it is possible to arrive at erroneous conclusions. The debates about the thickness of the lithosphere, the deep structure of continents, the depth of slab penetration, and the scale of mantle convection are, to some extent, debates about the anisotropy and mineral physics of the mantle and the interpretation of seismic data. Although it is important to understand the effects of temperature and pressure on physical properties, it is also important to realize that changes in crystal structure (solid-solid phase changes) and orientation have large effects on the seismic velocities. Tomographic images are often interpreted in terms of a single variable, temperature. Thus *blue regions* on tomographic cross-sections are often called *cold slabs* and *red regions* are often called *hot plumes*. Many of the current controversies in mantle dynamics and geochemistry, such as deep slab penetration, whole mantle convection and the presence of plumes can be traced to over-simplified or erroneous scaling relations between seismic velocities and density, temperature and physical state.

Table 18.2 is a compilation of the elastic properties, measured or estimated, of most of the important mantle minerals, plus pressure and temperature derivatives.

Anharmonicity

There are various routes whereby temperature affects the elastic moduli and seismic velocity. The main ones are anelasticity and anharmonicity. The first one does not depend, to first order, on volume or density and therefore many geodynamic scaling relations are invalid if anelasticity is important. Elastic moduli also depend on parameters other than temperature, such as composition. The visual interpretations of tomographic color cross-sections assume a one-to-one correspondence between seismic velocity, density and temperature.

The thermal oscillation of atoms in their (asymmetric) potential well is anharmonic or nonsinusoidal. Thermal oscillation of an atom causes the mean position to be displaced, and thermal expansion results. (In a symmetric, or parabolic, potential well the mean positions are unchanged, atomic vibrations are harmonic, and no thermal expansion results.) `Anharmonicity` causes atoms to take up new average positions of equilibrium, dependent on the amplitude of the vibrations and hence on the temperature, but the new positions of dynamic equilibrium remain nearly harmonic. At any given volume the harmonic approximation can be made so that the characteristic temperature and frequency are not explicit functions of temperature. This is called the `quasi-harmonic approximation`. If it is assumed that the frequency of each normal mode of vibration is changed in simple proportion as the volume is changed. There is a close relationship between lattice thermal conductivity, thermal expansion and other properties that depend intrinsically on anharmonicity of the interatomic potential. The atoms in a crystal vibrate about equilibrium positions, but the normal modes are not independent except in the idealized case of a harmonic solid. The vibrations of a crystal lattice can be resolved into interacting traveling waves that interchange energy due to anharmonic, nonlinear coupling.

In a harmonic solid:

there is no thermal expansion;
adiabatic and isothermal elastic constants are equal;

Table 18.2 Elastic properties of mantle minerals (Duffy and Anderson, 1988)

Formula (structure)	Density (g/cm^3)	K_s (GPa)	G (GPa)	K'_s	G'	$-\dot{K}_s$ (GPa/K)	$-\dot{G}$ (GPa/K)
$(Mg,Fe)_2SiO_4$ (olivine)	3.222+ 1.182x_{Fe}	129	82– 31x_{Fe}	5.1	1.8	0.016	0.013
$(Mg,Fe)_2SiO_4$ (β-spinel)	3.472+ 1.24x_{Fe}	174	114– 41x_{Fe}	4.9	1.8	0.018	0.014
$(Mg,Fe)_2SiO_4$ (γ-spinel)	3.548+ 1.30x_{Fe}	184	119– 41x_{Fe}	4.8	1.8	0.017	0.014
$(Mg,Fe)SiO_3$ (orthopyroxene)	3.204+ 0.799x_{Fe}	104	77– 24x_{Fe}	5.0	2.0	0.012	0.011
$CaMgSi_2O_6$ (clinopyroxene)	3.277	113	67	4.5	1.7	0.013	0.010
$NaAlSi_2O_6$ (clinopyroxene)	3.32	143	84	4.5	1.7	0.016	0.013
$(Mg,Fe)O$ (magnesiowustite)	3.583+ 2.28x_{Fe}	163– 8x_{Fe}	131 77x_{Fe}	4.2	2.5	0.016	0.024
Al_2O_3 (corundum)	3.988	251	162	4.3	1.8	0.014	0.019
SiO_2 (stishovite)	4.289	316	220	4.0	1.8	0.027	0.018
$(Mg,Fe)_3Al_2Si_3O_{12}$ (garnet)	3.562+ 0.758x_{Fe}	175+ 1x_{Fe}	90+ 8x_{Fe}	4.9	1.4	0.021	0.010
$Ca_3(Al,Fe)_2Si_3O_{12}$ (garnet)	3.595+ 0.265x_{Fe}	169– 11x_{Fe}	140– 14x_{Fe}	4.9	1.6	0.016	0.015
$(Mg,Fe)SiO_3$ (ilmenite)	3.810+ 1.10x_{Fe}	212	132– 41x_{Fe}	4.3	1.7	0.017	0.017
$(Mg,Fe)SiO_3$ (perovskite)	4.104+ 1.07x_{Fe}	266	153	3.9	2.0	0.031	0.028
$CaSiO_3$ (perovskite)	4.13	227	125	3.9	1.9	0.027	0.023
$(Mg,Fe)_4Si_4O_{12}$ (majorite)	3.518+ 0.973x_{Fe}	175+ 1x_{Fe}	90+ 8x_{Fe}	4.9	1.4	0.021	0.010
$Ca_2Mg_2Si_4O_{12}$ (majorite)	3.53	165	104	4.9	1.6	0.016	0.015
$Na_2Al_2Si_4O_{12}$ (majorite)	4.00	200	127	4.9	1.6	0.016	0.015

x_{Fe} is molar fraction of Fe endmember.

Alternative values for parameters in Table 18.2 and data for other minerals are given in the following references.

Anderson, O. L. and Isaak, D. G. (1995) Elastic constants of mantle minerals at high temperature, in *Mineral Physics and Crystallography: A Handbook of Physical Constants*, pp. 64–97, ed. T. J. Ahrens, American Geophysical Union, Washington, DC.

Bass, J. D. (1995) Elasticity of minerals, glasses, and melts, in *Mineral Physics and Crystallography: A Handbook of Physical Constants*, pp. 45–63, ed. T. J. Ahrens, American Geophysical Union, Washington, DC.

Duffy, T. and Anderson, D. L. (1989) Seismic velocities in mantle minerals and the mineralogy of the upper mantle. *J. Geophys. Res.*, **94**(B2), 1895–912.

Li, B. and Zhang, J. (2005) Pressure and temperature dependence of elastic wave velocity of $MgSiO_3$ perovskite and the composition of the lower mantle, *Phys. Earth Planet. Inter*, **151**, 143–54.

Mattern, E., Matas, J., Ricard, Y. and Bass, J. D. (2005) Lower mantle composition and temperature from mineral physics and thermodynamic modelling, *Geophys. J. Int.*, **160**, 973–90.

the elastic constants are independent of pressure and temperature;
the heat capacity is constant at high temperature ($T > 0$); and
the lattice thermal conductivity is infinite.

These are the result of the neglect of anharmonicity (higher than quadratic terms in the interatomic displacements in the potential energy). In a real crystal the presence of lattice vibration causes a periodic elastic strain that, through anharmonic interaction, modulates the elastic constants of a crystal. Other phonons are scattered by these modulations. This is a nonlinear process that does not occur in the absence of anharmonic terms.

The concept of a strictly harmonic crystal is highly artificial. It implies that neighboring atoms attract one another with forces proportional to the distance between them, but such a crystal would collapse. We must distinguish between a harmonic solid in which each atom executes harmonic motions about its equilibrium position and a solid in which the forces between individual atoms obey Hooke's law. In the former case, as a solid is heated up, the atomic vibrations increase in amplitude but the mean position of each atom is unchanged. In a two- or three-dimensional lattice, the net restoring force on an individual atom, when all the nearest neighbors are considered, is not Hookean. An atom oscillating on a line between two adjacent atoms will attract the atoms on perpendicular lines, thereby contracting the lattice. Such a solid is not harmonic; in fact it has negative thermal expansion.

The quasi-harmonic approximation takes into account that the equilibrium positions of atoms depend on the amplitude of vibrations, and hence temperature, but that the vibrations about the new positions of dynamic equilibrium remain closely harmonic. One can then assume that at any given volume V the harmonic approximation is adequate. In the simplest quasi-harmonic theories it is assumed that the frequencies of vibration of each normal mode of lattice vibration and, hence, the vibrational spectra, the maximum frequency and the characteristic temperatures are functions of volume alone. In this approximation γ is independent of temperature at constant volume, and α has approximately the same temperature dependence as molar specific heat.

Correcting elastic properties for temperature

The elastic properties of solids depend primarily on static lattice forces, but vibrational or thermal motions become increasingly important at high temperature. The resistance of a crystal to deformation is partially due to interionic forces and partially due to the radiation pressure of high-frequency acoustic waves, which increase in intensity as the temperature is raised. If the increase in volume associated with this radiation pressure is compensated by the application of a suitable external pressure, there still remains an intrinsic temperature effect. Thus, these equations provide a convenient way to estimate the properties of the static lattice, that is, $K(V, 0)$ and $G(V, 0)$, and to correct measured values to different temperatures at constant volume. The static lattice values should be used when searching for velocity-density or modulus-volume systematics or when attempting to estimate the properties of unmeasured phases.

The first step in forward modeling of the seismic properties of the mantle is to compile a table of the ambient or zero-temperature properties, including temperature and pressure derivatives, of all relevant minerals. The fully normalized extrinsic and intrinsic derivatives are then formed and, in the absence of contrary information, are assumed to be independent of temperature. The coefficient of thermal expansion can be used to correct the density to the temperature of interest at zero pressure. It is important to take the temperature dependence of $\alpha(T)$ into account properly since it increases rapidly from room temperature but levels out at high T. The use of the ambient α will underestimate the effect of temperature; the use of α plus the initial slope will overestimate the volume change at high temperature. Fortunately, the shape of $\alpha(T)$ is well known theoretically (a Debye function) and has been measured for many mantle minerals (see Table 18.5). The moduli can then be corrected for the volume change using equations given. The normalized parameters can then be used in an equation of state to calculate $M(P, T)$, for example, from finite strain. The normalized form of the pressure derivatives can be assumed to be either independent of temperature or functions of $V(T)$. Temperature is less effective in causing variations in density and elastic properties in the deep mantle than in the shallow mantle and the relationships between these variations are different from those observed in the laboratory. Anharmonic effects also are predicted to decrease rapidly with compression. Anharmonic contributions are particularly important at high temperature and low pressure. Intrinsic or anharmonic effects probably remain important in the lower mantle, even if they decrease with pressure. Anelastic thermal effects on moduli are also probably more important in the upper mantle than at depth. Large velocity variations in the deep mantle are probably due to phase changes, composition and melting, or the presence of a liquid phase.

Temperature and pressure derivatives of elastic moduli

The pressure derivatives of the adiabatic bulk modulus for halides and oxides generally fall in the range 4.0 to 5.5. Rutiles are generally somewhat higher, 5 to 7. Oxides and silicates having ions most pertinent to major mantle minerals have a much smaller range, usually between 4.3 and 5.4. MgO has an unusually low value, 3.85. The density derivative of the bulk modulus,

$$(\partial \ln K_S/\partial \ln \rho)_T = (K_T/K_S)K_S'$$

for mantle oxides and silicates that have been measured usually fall between 4.3 and 5.4 with MgO, 3.8, again being low.

The rigidity, G, has a much weaker volume or density dependence. The parameter

$$(\partial \ln G/\partial \ln \rho)_T = (K_T/G)G'$$

generally falls between about 2.5 and 2.7. The above dimensionless derivatives can be written $\{K_S\}_T$ and $\{G\}_T$ respectively, and this notation will be used later.

For many minerals, and the lower mantle

$$(\partial \ln K_S/\partial \ln G)_T = (K_S/G)$$

is a very good approximation.

Intrinsic and extrinsic temperature effects

Temperature has several effects on elastic moduli; intrinsic, extrinsic, anharmonic and anelastic. The effects of temperature on the properties of the mantle must be known for various geophysical calculations. Because the lower mantle is under simultaneous high pressure and high temperature, it is not clear that the simplifications that can be made in the classical high-temperature limit are necessarily valid. For example, the coefficient of thermal expansion, which controls many of the anharmonic properties, increases with temperature but decreases with pressure. At high temperature, the elastic properties depend mainly on the volume through the thermal expansion. At high pressure, on the other hand, the intrinsic effects of temperature may become relatively more important.

The temperature derivatives of the elastic moduli can be decomposed into extrinsic and intrinsic components:

$$(\partial \ln K_S/\partial \ln \rho)_T = (\partial \ln K_S/\partial \ln \rho)_T - \alpha^{-1}(\partial \ln K_S/\partial T)_V$$
$$\text{or } \{K_S\}_P = \{K_S\}_T - \{K_S\}_V$$

for the adiabatic bulk modulus, K_s, and a similar expression for the rigidity, G. Extrinsic and intrinsic effects are sometimes called 'volumetric' or 'quasi-harmonic' and 'anharmonic.' The anelastic effect on moduli, treated later, is also important.

The intrinsic contribution is

$$(d \ln M/dT)_V = \alpha[(\delta \ln M/\delta \ln \rho)_T - (\delta \ln M/\delta \ln \rho)_P]$$

where M is K_s or G. There would be no intrinsic or anharmonic effect if $\alpha = 0$ or if $M(V, T) = M(V)$. The various terms in this equation require

Table 18.3 | Extrinsic and intrinsic of derivations

Substance	Extrinsic $\left(\frac{\partial \ln K_S}{\partial \ln \rho}\right)_P$	Extrinsic $\left(\frac{\partial \ln K_S}{\partial \ln \rho}\right)_T$	Intrinsic $\frac{1}{\alpha}\left(\frac{\partial \ln K_S}{\partial T}\right)_V$	Intrinsic $\left(\frac{\partial \ln K_S}{\partial T}\right)_V$ (10^{-5}/deg)
MgO	3.0	3.8	0.8	2.3
CaO	3.9	4.8	0.9	2.6
SrO	4.7	5.1	0.4	1.6
TiO_2	10.5	6.8	−3.7	−8.7
GeO_2	10.2	6.1	−4.1	−5.6
Al_2O_3	4.3	4.3	0.0	0.0
Mg_2SiO_4	4.7	5.4	0.7	0.03
$MgAl_2O_4$	4.0	4.2	0.2	0.4
Garnet	5.2	5.5	0.3	0.5
$SrTiO_3$	7.9	5.7	−2.2	−6.2
Substance	$\left(\frac{\partial \ln G}{\partial \ln \rho}\right)_P$	$\left(\frac{\partial \ln G}{\partial \ln \rho}\right)_T$	$\frac{1}{\alpha}\left(\frac{\partial \ln G}{\partial T}\right)_V$	$\left(\frac{\partial \ln G}{\partial T}\right)_V$ (10^{-5}/deg)
MgO	5.8	3.0	−2.8	−8.6
CaO	6.3	2.4	−3.9	−11.2
SrO	4.9	2.2	−2.7	−11.3
TiO_2	8.4	1.2	−7.2	−16.9
GeO_2	5.8	2.1	−3.7	−5.1
Al_2O_3	7.5	2.6	−4.9	−16.5
Mg_2SiO_4	6.5	2.8	−3.7	−9.2
$MgAl_2O_4$	5.9	1.3	−4.6	−7.4
Garnet	5.4	2.6	−2.8	−6.1
$SrTiO_3$	7.9	3.1	−4.8	−13.4

Sumino and Anderson (1984).

measurements as a function of both temperature and pressure. These are tabulated in Table 18.3 for a variety of oxides and silicates.

Generally, the intrinsic temperature effect on rigidity is greater than on the bulk modulus; G is a weaker function of volume at constant T than at constant P; and G is a weaker function of volume, at constant T, than is the bulk modulus. The extrinsic terms are functions of pressure through the moduli and their pressure derivatives.

The coefficient of thermal expansion, at constant temperature, decreases by about 80% from $P = 0$ to the base of the mantle. It is hard to create buoyancy at the base of the mantle. This also means that all temperature derivatives of the moduli and the seismic velocities are low in the deep mantle. The inferred intrinsic temperature effect on the bulk modulus is generally small and can be either negative or positive (Table 18.3). The bulk modulus and its derivatives must be computed from differences of the directly measured moduli and therefore they have much larger errors than the shear moduli. The intrinsic component of the temperature derivative of the rigidity is often larger than the extrinsic, or volume-dependent, component and is invariably negative; an increase in temperature causes

Table 18.4 | Normalized pressure and temperature derivatives

Substance	$\left(\frac{\partial \ln K_S}{\partial \ln \rho}\right)_T$	$\left(\frac{\partial \ln G}{\partial \ln \rho}\right)_T$	$\left(\frac{\partial \ln K_S}{\partial \ln G}\right)_T$	$\frac{K_S}{G}$	$\left(\frac{\partial \ln K_S}{\partial \ln \rho}\right)_P$	$\left(\frac{\partial \ln G}{\partial \ln \rho}\right)_P$	$\left(\frac{\partial \ln K_S}{\alpha \partial T}\right)_V$	$\left(\frac{\partial \ln G}{\alpha \partial T}\right)_V$
MgO	3.80	3.01	1.26	1.24	3.04	5.81	0.76	−2.81
Al_2O_3	4.34	2.71	1.61	1.54	4.31	7.45	0.03	−4.74
Olivine	5.09	2.90	1.75	1.64	4.89	6.67	0.2	−3.76
Garnet	4.71	2.70	1.74	1.85	6.84	4.89	−2.1	−2.19
$MgAl_2O_4$	4.85	0.92	5.26	1.82	3.84	4.19	1.01	−3.26
$SrTiO_3$	5.67	3.92	1.45	1.49	8.77	8.70	−3.1	−4.78

G to decrease due to the decrease in density, but a large part of this decrease would occur at constant volume. Increasing pressure decreases the total temperature effect because of the decrease of the extrinsic component and the coefficient of thermal expansion. The net effect is a reduction of the temperature derivatives, and a larger role for rigidity in controlling the temperature variation of seismic velocities in the lower mantle. This is consistent with seismic data for the lower mantle.

The total effects of temperature on the bulk modulus and on the rigidity are comparable under laboratory conditions (Table 18.3). Therefore the compressional and shear velocities have similar temperature dependencies. On the other hand, the thermal effect on bulk modulus is largely extrinsic, that is, it depends mainly on the change in volume due to thermal expansion. The shear modulus is affected both by the volume change and a purely thermal effect at constant volume.

Although the data in Table 18.3 are not in the classical high-temperature regime it is still possible to separate the temperature derivatives into volume-dependent and volume-independent parts. Measurements must be made at much higher temperatures in order to test the various assumptions involved in quasi-harmonic approximations. One of the main results I have shown here is that, in general, the relative roles of intrinsic and extrinsic contributions and the relative temperature variations in bulk and shear moduli will not mimic those found in the restricted range of temperature and pressure presently available in the laboratory. The Earth can be used as a natural laboratory to extend conventional laboratory results.

It is convenient to treat thermodynamic parameters, including elastic moduli, in terms of volume-dependent and temperature-dependent parts, as in the Mie–Grüneisen equation of state. This is facilitated by the introduction of dimensionless anharmonic (DA) parameters. The Gruneisen ratio is such a parameter. The pressure derivatives elastic moduli are also dimensionless anharmonic parameters, but it is useful to replace pressure, and temperature, by volume. This is done by forming logarithmic derivatives with respect to volume or density, giving dimensionless logarithmic anharmonic (DLA) parameters. They are formed as follows:

$$(\partial \ln M/\partial \ln \rho)_T = \frac{K_T}{M}\left(\frac{\partial M}{\partial P}\right) = \{M\}_T$$

$$(\partial \ln M/\partial \ln \rho)_P = (\alpha M)^{-1}\left(\frac{\partial M}{\partial T}\right)_T = \{M\}_P$$

$$(\partial \ln M/\alpha \partial T)_V = \{M\}_T - \{M\} = \{M\}_V$$

where we use braces $\{\}$ to denote DLA parameters and the subscripts T, P, V and S denote isothermal, isobaric, isovolume and adiabatic conditions, respectively. The $\{\}_V$ terms are known as *intrinsic derivatives*, giving the effect of temperature or pressure at constant volume. Derivatives for common mantle minerals are listed in Table 18.4. Elastic, thermal and anharmonic parameters are relatively independent of temperature at constant volume, particularly at high temperature. This simplifies temperature corrections for the elastic moduli. I use density rather than volume in order to make most of the parameters positive.

The DLA parameters relate the variation of the moduli to volume, or density, rather than to temperature and pressure. This is useful since the variations of density with temperature, pressure, composition and phase are fairly well understood. Furthermore, anharmonic properties tend to be independent of temperature and pressure at constant volume. The anharmonic parameter known as the thermal Grüneisen parameter γ is relatively constant from material to material as well as relatively independent of temperature.

Anelastic effects

Solids are not ideally elastic; the moduli depend on frequency. This is known as 'dispersion' and it introduces an additional temperature dependency on the seismic-wave velocities. This can be written

$$\mathrm{d}\ln v/\mathrm{d}\ln T = (1/2\pi)\, Q^{-1}_{\max}[2E^*/RT]$$

where v is a seismic velocity, T is absolute temperature, $Q^{-1}_{\max}$ is the peak value of the absorption band and E^* is the activation energy. The temperature dependence arises from the relaxation time, which is an activated parameter that depends exponentially on temperature. Many other examples are given in the chapter on dissipation. The above example applies inside the seismic absorption band.

Seismic constraints on thermodynamics of the lower mantle

For most solids at normal conditions, the effect of temperature on the elastic properties is controlled mainly by the variation of volume. Volume-dependent extrinsic effects dominate at low pressure and high temperature. Under these conditions one expects that the relative changes in shear velocity, due to lateral temperature gradients in the mantle, should be similar to changes in compressional velocity. However, at high pressure, this contribution is suppressed, particularly for the bulk modulus, and variations of seismic velocities are due primarily to changes in the rigidity.

Intrinsic effects are more important for the rigidity than for the bulk modulus. Geophysical results on the radial and lateral variations of velocity and density provide constraints on high-pressure–high-temperature equations of state. Many of the thermodynamic properties of the lower mantle, required for equation-of-state modeling, can be determined directly from the seismic data. The effect of pressure on the coefficient of thermal expansion, the Grüneisen parameters, the lattice conductivity and the temperature derivatives of seismic-wave velocities should be taken into account in the interpretation of seismic data and in convection and geoid calculations.

The lateral variation of seismic velocity is very large in the upper 200 km of the mantle but decreases rapidly below this depth. Velocity itself generally increases with depth below about 200 km. This suggests that temperature variations are more important in the shallow mantle than at greater depth. Most of the mantle is above the Debye temperature and therefore thermodynamic properties may approach their classical high-temperature limits. What is needed is precise data on variations of properties with temperature at high pressure and theoretical treatments of properties of solids at simultaneous high pressure and temperature.

I use the following relations and notation:

$$(\partial \ln K_{\mathrm{T}}/\partial \ln \rho)_{\mathrm{T}} = (\partial K_{\mathrm{T}}/\partial P)_{\mathrm{T}} = K'_{\mathrm{T}} = \{K_{\mathrm{T}}\}_{\mathrm{T}}$$
$$(\partial \ln K_{\mathrm{S}}/\partial \ln \rho)_{\mathrm{T}} = (K_{\mathrm{T}}/K_{\mathrm{S}})K'_{\mathrm{S}} = \{K_{\mathrm{S}}\}_{\mathrm{T}}$$
$$(\partial \ln K_{\mathrm{S}}/\partial \ln \rho)_{\mathrm{P}} = -(\alpha K_{\mathrm{S}})^{-1}(\partial K_{\mathrm{S}}/\partial T)_{\mathrm{P}} = \delta_{\mathrm{S}} = \{K_{\mathrm{S}}\}_{\mathrm{P}}$$
$$(\partial \ln G/\partial \ln \rho)_{\mathrm{T}} = (K_{\mathrm{T}}/G)(\partial G/\partial P)_{\mathrm{T}} = (K_{\mathrm{T}}/G)G' = G^* = \{G\}_{\mathrm{T}}$$
$$(\partial \ln G/\partial \ln \rho)_{\mathrm{P}} = -(\alpha G)^{-1}(\partial G/\partial T)_{\mathrm{P}} = g = \{G\}_{\mathrm{P}}$$
$$-(\partial \ln \alpha/\partial \ln \rho)_{\mathrm{T}} \simeq \delta_{\mathrm{S}} + \gamma = -\{\alpha\}_{\mathrm{T}}$$
$$(\partial \ln K_{\mathrm{T}}/\partial T)_{\mathrm{V}} = \alpha[(\partial \ln K_{\mathrm{T}}/\partial \ln \rho)_{\mathrm{T}} - (\partial \ln K_{\mathrm{T}}/\partial \ln \rho)_{\mathrm{P}}] = \alpha(K'_{\mathrm{T}} - \delta_{\mathrm{T}})$$

Table 18.5 Dimensionlss logarithmic anharmonic parameters

Substance	α (10^{-6}/K)	K_S (kbar)	G (kbar)	$\{K_S\}_T$	$\{G\}_T$	$\{K_T\}_P$	$\{K_S\}_P$	$\{G\}_P$	$\{K_T\}_V$	$\{K_S\}_V$	$\{G\}_V$	$\{K-G\}_V$	K_S/G	γ thermal	γ BR
LiF	98	723	485	4.90	3.97	4.69	2.42	6.35	0.5	2.5	−2.4	4.9	1.49	1.66	1.92
NaF	98	483	314	4.96	2.62	5.80	3.75	5.06	−0.6	1.2	−2.4	3.6	1.54	1.51	1.37
KF	99	318	164	4.81	1.97	5.05	3.18	6.17	0.0	1.6	−4.2	5.8	1.94	1.50	1.12
RbF	95	280	127	5.35	1.63	4.77	2.97	5.95	0.8	2.4	−4.3	6.7	2.20	1.43	1.05
LiCl	134	318	193	4.65	4.45	5.40	3.32	6.84	−0.4	1.3	−2.4	3.7	1.65	1.82	2.11
NaCl	118	252	148	5.10	3.00	5.45	3.74	5.07	−0.1	1.4	−2.1	3.4	1.71	1.51	1.55
Kcl	105	181	93	5.10	2.01	7.48	5.67	5.54	−2.0	−0.6	−3.5	3.0	1.95	1.39	1.17
RbCl	119	163	78	5.09	1.79	5.81	4.11	5.30	−0.4	1.0	−3.5	4.5	2.10	1.44	1.09
AgCl	93	440	81	6.21	2.83	10.6	7.94	11.0	−3.8	−1.7	−8.2	6.5	5.44	2.08	1.77
NaBr	135	207	114	4.63	3.15	7.34	5.28	4.24	−2.2	0.7	−1.1	0.4	1.81	1.72	1.59
Kbr	116	150	79	5.12	2.01	5.64	3.94	4.68	−0.2	1.2	−2.7	3.8	1.90	1.45	1.16
RbBr	113	137	65	5.05	1.81	6.27	4.54	5.96	−0.9	0.5	−4.2	4.7	2.09	1.47	1.09
NaI	138	161	91	5.11	3.22	4.79	2.75	5.41	0.6	2.4	−2.2	4.5	1.76	1.74	1.65
KI	126	122	60	4.82	2.29	5.95	4.05	4.89	−0.8	0.8	−2.6	3.4	2.02	1.41	1.26
RbI	119	111	50	5.14	1.97	6.17	4.49	6.05	−0.7	0.6	−4.1	4.7	2.21	1.51	1.17
CsCl	140	182	101	5.20	4.75	6.28	3.82	6.01	−0.6	1.4	−1.3	2.6	1.80	2.04	2.30
TlCl	158	240	92	6.00	4.45	7.78	5.18	7.35	−1.0	0.8	−2.9	3.7	2.60	2.46	2.31
CsBr	138	156	88	4.97	4.53	6.33	3.86	5.90	−0.9	1.1	−1.4	2.5	1.76	1.98	2.18
TlBr	170	224	88	5.80	5.39	7.51	4.75	6.36	−0.8	1.1	−1.0	2.0	2.55	2.76	2.66
CsI	138	126	72	5.06	4.51	6.40	3.86	6.09	−0.9	1.2	−1.6	2.8	1.73	1.94	2.18
MgO	31	1628	1308	3.80	3.01	5.48	3.04	5.81	−1.6	0.8	−2.8	3.6	1.24	1.52	1.41
CaO	29	1125	810	4.78	2.42	5.45	3.92	6.29	−0.6	0.9	−3.9	4.7	1.39	1.27	1.25
SrO	42	912	587	5.07	2.25	6.99	4.68	4.98	−1.8	0.4	−2.7	3.1	1.55	1.74	1.22
BaO	38	720	367	5.42	1.95	9.46	7.34	7.88	−3.9	−1.9	−5.9	4.0	1.96	1.56	1.17
Al_2O_3	16	2512	1634	4.34	2.71	6.84	4.31	7.45	−2.5	0.0	−4.7	4.8	1.54	1.27	1.33
Ti_2O_3	17	2076	945	4.10	2.23	7.78	6.66	12.9	−3.6	−2.6	−11	8.2	2.20	1.13	1.15
Fe_2O_3	33	2066	910	4.44	1.63	5.70	3.68	3.34	−1.2	0.8	−1.7	2.5	2.27	1.99	0.95
TiO_2	24	2140	1120	6.83	1.09	12.7	10.5	8.37	−5.7	−3.7	−7.3	3.6	1.91	1.72	0.96
GeO_2	14	2589	1509	6.12	2.10	11.9	10.2	5.83	−5.7	−4.1	−3.7	−0.3	1.72	1.17	1.27
SnO_2	10	2123	1017	5.09	1.25	9.90	8.64	6.21	−4.8	−3.6	−5.0	1.4	2.09	0.88	0.85

MgF_2	38	1019	547	5.01	1.34	5.36	4.17	3.83	−0.3	0.8	−2.5	3.3	1.86	1.22	0.87
NiF_2	23	1207	459	4.98	−1.4	8.80	7.47	1.96	−3.8	−2.5	−3.4	0.9	2.63	0.88	−0.2
ZnF_2	29	1052	394	4.51	0.13	9.49	8.23	4.73	−4.9	−3.7	−4.6	0.9	2.67	0.97	0.39
CaF_2	61	845	427	4.55	2.26	5.85	3.86	4.75	−1.1	0.7	−2.5	3.2	1.98	1.83	1.21
SrF2	47	714	350	4.67	1.66	6.02	4.75	4.94	−1.2	−0.1	−3.3	3.2	2.04	1.30	0.98
CdF_2	66	1054	330	5.77	4.05	8.36	6.04	7.11	−2.2	−0.3	−3.1	2.8	3.19	2.45	2.10
BaF_2	61	581	255	4.89	0.88	6.52	4.54	3.83	−1.4	0.3	−3.0	3.3	2.28	1.80	0.71
Opx	48	1035	747	9.26	3.18	7.26	5.43	3.34	2.2	3.8	−0.2	4.0	1.39	1.87	1.95
Olivine	25	1294	791	5.09	2.90	6.70	4.89	6.67	−1.6	0.2	−3.8	4.0	1.64	1.16	1.49
Olivine	27	1292	812	4.83	2.85	6.61	4.67	6.27	−1.7	0.2	−3.4	3.6	1.59	1.25	1.44
Olivine	25	1286	811	5.32	2.83	6.55	4.73	6.50	−1.2	0.6	−3.7	4.3	1.59	1.16	1.48
Garnets															
Fe_{16}	19	1713	927	4.71	2.70	7.89	6.84	4.89	−3.1	−2.1	−2.2	0.1	1.85	1.05	1.38
Fe_{36}	19	1682	922	4.71	2.67	7.42	5.98	5.05	−2.7	−1.3	−2.4	1.1	1.82	1.01	1.37
Fe_{54}	24	1736	955	5.38	2.52	6.79	5.52	4.81	−1.3	−0.1	−2.3	2.2	1.82	1.28	1.37
$MgAl_2O_4$	21	1969	1080	4.85	0.92	5.47	3.84	4.19	−0.6	1.0	−3.3	4.3	1.82	1.40	0.67
$SrTiO_3$	25	1741	1168	5.67	3.92	11.2	8.77	8.70	−5.3	−3.1	−4.8	1.7	1.49	1.63	1.06
$KMgF_3$	60	751	488	4.87	2.98	5.02	3.46	4.72	−0.0	1.4	−1.7	3.1	1.54	1.60	1.50
$RbMnF_3$	57	675	341	4.80	3.69	4.48	3.01	5.04	0.4	1.8	−1.4	3.1	1.98	1.49	1.80
$RbCdF_3$	40	614	257	1.09	2.15	4.12	3.06	3.27	3.0	2.0	−1.1	0.8	2.39	1.06	1.80
$TlCdF_3$	49	609	228	7.43	0.95	5.09	3.86	3.14	2.4	3.6	−2.2	5.8	2.68	1.24	1.03
ZnO	15	1394	442	4.76	−2.2	9.60	6.22	3.02	−4.8	−1.5	−5.2	3.7	3.15	0.81	−0.4
BeO	18	2201	1618	5.48	1.19	5.17	3.08	4.19	0.3	2.4	−3.0	5.4	1.36	1.27	0.79
SiO_2	35	378	455	6.37	0.35	3.28	2.43	−0.3	3.1	3.9	0.7	3.3	0.85	0.67	0.40
$CaCO_3$	17	747	318	5.36	−3.5	24.00	23.1	18.5	−19	−18	−22	4.2	2.35	0.56	−1.0

Table 18.6 Dimensionless logarithmic anharmonic derivatives

Substance	$\{K_S\}_T$	$\{G\}_T$	$\{K_S\}_P$	$\{G\}_P$	$\{K_S\}_V$	$\{G\}_V$
Averages						
Halides	5.1	2.6	4.1	5.9	0.9	−3.3
Perovskites*	4.8	2.7	4.4	5.0	0.3	−2.2
Garnets*	4.9	2.6	6.1	4.9	−1.2	−2.3
Fluorites*	5.0	2.2	4.8	5.2	0.2	−2.9
Oxides	5.3	2.0	5.7	5.8	−0.4	−3.7
Silicates	5.6	2.8	5.4	5.4	0.2	−2.6
Grand average	5.1	2.5	5.0	5.7	0.1	−3.2
	(±1.0)	(±1.3)	(±1.9)	(±1.9)	(±1.9)	(±1.9)
Olivine	5.1	2.9	4.9	6.7	0.2	−3.8
Olivine	4.8	2.9	4.7	6.3	0.2	−3.4
$MgAl_2O_4$-spinel	4.9	0.9	3.8	4.2	−0.6	+1.0

$\{M\} = \partial \ln M/\partial \ln \rho$

* Structures.

Table 18.7 Anharmonic parameters for oxides and silicates and predicated values for some high-pressure phases

Mineral	$\{K_T\}_T$	$\{K_S\}_T$	$\{G\}_T$	$\{K_T\}_P$	$\{K_S\}_P$	$\{G\}_P$
α-olivine	5.0	4.8	2.9	6.6	4.7	6.5
β-spinel*	4.9	4.8	3.0	6.6	4.7	6.4
γ-spinel*	5.1	5.0	3.1	6.8	4.9	6.5
Garnet	4.8	4.7	2.7	7.9	6.8	4.9
$MgSiO_3$ (majorite)*	4.9	4.7	2.6	7.9	6.0	4.5
Al_2O_3-ilmenite	4.4	4.3	2.7	6.8	4.3	7.5
$MgSiO_3$-ilmenite*	4.7	4.5	2.7	7.0	4.3	6.0
$MgSiO_3$-perovskite*	4.5	4.5	3.5	7.5	5.5	6.5
*	4.2	4.1	2.8	6.2	4.0	5.7
SiO_2-(stishovite)*	4.5	4.4	2.4	7.5	6.4	5.0

* Predicted.

$$(\partial \ln G/\partial T)_V = \alpha[(\partial \ln G/\partial \ln \rho)_T - (\partial \ln G/\partial \ln \rho)_P]$$

$$K_S = K_T(1+\alpha\gamma T)$$

$$\delta_T \approx \delta_S + \gamma$$

This notation stresses the volume, or density, dependence of the thermodynamic variables and is particularly useful in geophysical discussions.

Values of most of the dimensionless logarithmic parameters are listed in Table 18.5 for many halides, oxides and minerals. Average values for chemical and structural classes are extracted in Table 18.6, and parameters for mineral phases of the lower mantle are presented in Table 18.7.

$$(\partial \ln V_P/\partial \ln \rho) = \frac{1}{2}\left[\frac{3}{5}(\partial \ln K_S/\partial \ln \rho) + \frac{2}{5}(\partial \ln G/\partial \ln \rho) - 1\right]$$

where I have used $K_s = 2G$, a value appropriate for the lower mantle.

Theoretically, the effects of temperature decrease with compression, and this is borne out by the seismic data. In the lower mantle the lateral variation of seismic velocities is dominated by variations in the rigidity. This is similar to the situation in the upper mantle where it is caused by nonelastic processes, such as partial melting and dislocation relaxation, phenomena accompanied by increased attenuation, and phase changes. In the lower mantle the effect is caused by anharmonic phenomena and intrinsic temperature effects that are more important in shear than in compression. Iron-partitioning and phase changes, including spin-pairing and melting, may be important in the deep mantle. One cannot assume that physical properties are a function of volume alone, or that classical high-temperature behavior prevails, or that shear and compressional modes exhibit similar variations. On a much more basic level, one cannot simply adopt laboratory values of temperature derivatives to estimate the effect of temperature on density and elastic-wave velocities in slabs and in the deep mantle.

If one ignores the possibility of phase changes and chemical stratification, one can estimate lower-mantle properties. From the seismic data for the lower mantle, we can obtain the following estimates of *in-situ* values:

$$(\partial \ln G/\partial \ln \rho)_P \quad 5.8 \text{ to } 7.0$$
$$(\partial \ln G/\partial \ln \rho)_T \quad 2.6 \text{ to } 2.9$$
$$(\partial \ln K_S/\partial \ln \rho)_T \quad 1.8 \text{ to } 1.0$$
$$(\partial \ln K_S/\partial \ln \rho)_T \quad 2.8 \text{ to } 3.6$$
$$(\partial \ln \gamma/\partial \ln \rho)_T \quad -1+\varepsilon$$
$$-(\partial \ln \alpha/\partial \ln \rho)_T \quad 3 \text{ to } 2$$

For the intrinsic temperature terms we obtain:

$$\frac{1}{\alpha}\left[\frac{\partial \ln G}{\partial T}\right]_V \approx -3.2 \text{ to } -4.1$$

$$\frac{1}{\alpha}\left[\frac{\partial \ln K_S}{\partial T}\right]_V \approx +1 \text{ to } +2.6$$

An interesting implication of the seismic data is that the bulk modulus and rigidity are similar functions of volume at constant temperature. On the other hand, G is a stronger function of volume and K_s a much weaker function of volume at constant pressure than they are at constant temperature.

Extrapolation of lower mantle values to the surface, ignoring chemical and phase changes, gives:

$$\rho_o = 3.97-4.00 \text{ g/cm}^3$$
$$K_o = 2.12-2.23 \text{ kbar}$$
$$G_o = 1.30-1.35 \text{ kbar}$$
$$(K_o')_S = 3.8-4.1$$
$$(G_o')_S = 1.5-1.8$$

Chapter 19

Dissipation

As when the massy substance of the Earth quivers.

Marlowe

Real materials are not perfectly elastic. Solids creep when a high stress is applied, and the strain is a function of time. These phenomena are manifestations of anelasticity. The attenuation of seismic waves with distance and of normal modes with time are other examples of anelastic behavior. Generally, the response of a solid to a stress can be split into an elastic or instantaneous part and an anelastic or time-dependent part. The anelastic part contains information about temperature, stress and the defect nature of the solid. In principle, the attenuation of seismic waves can tell us about such things as small-scale heterogeneity, melt content, dislocation density and defect mobility. These, in turn, are controlled by temperature, pressure, stress, history and the nature of the defects. If these parameters can be estimated from seismology, they can be used to estimate other anelastic properties such as viscosity.

For example, the dislocation density of a crystalline solid is a function of the non-hydrostatic stress. These dislocations respond to an applied oscillatory stress, such as a seismic wave, but they are out of phase because of the finite diffusion time of the atoms around the dislocation. The dependence of attenuation on frequency can yield information about the dislocations. The longer-term motions of these same dislocations in response to a higher tectonic stress gives rise to a solid-state viscosity. Seismic waves also attenuate due to macroscopic phenomena, such as scattering and interactions between fluid, or molten, parts of the interior and the solid matrix. Anelasticty causes the elastic moduli to vary with frequency; *elastic constants* are not constant. The equations in this section can be used to correct seismic velocities for temperature effects due to *anelasticity*. These are different from the *anharmonic effects* discussed in other chapters.

Seismic-wave attenuation

The travel time or velocity of a seismic wave provides an incomplete description of the material it has propagated through. The amplitude and frequency of the wave provide some more information. Seismic waves attenuate or decay as they propagate. The rate of attenuation contains information about the anelastic properties of the propagation medium.

A propagating wave can be written

$$A = A_0 \exp\ i(\omega t - \kappa x)$$

where A is the amplitude, ω the frequency, κ the wave number, t the travel time, x the distance and $c = \omega/\kappa$ the phase velocity. If spatial attenuation occurs, then κ is complex. The imaginary part of $\boldsymbol{\kappa}$, κ^* is called the spatial attenuation coefficient.

The elastic moduli, M, are now also complex:

$$\boldsymbol{M} = \mathrm{M} + \mathrm{i}\mathrm{M}^*$$

The specific quality factor, a convenient dimensionless measure of dissipation, is

$$Q^{-1} = M^*/M$$

This is related to the energy dissipated per cycle. Since the phase velocity

$$c = \frac{\omega}{k} = \sqrt{M/\rho}$$

it follows that

$$Q^{-1} = 2\frac{k^*}{k} = \frac{M^*}{M} \quad \text{for } Q \gg 1$$

In general, all the elastic moduli are complex, and each wave type has its own Q and velocity, both frequency dependent. For an isotropic solid the imaginary parts of the bulk modulus and rigidity are denoted as K^* and G^*. Most mechanisms of seismic-wave absorption affect the rigidity more than the bulk modulus, and shear waves more than compressional waves.

Frequency dependence of attenuation

In a perfectly elastic homogenous body, the elastic wave velocities are independent of frequency. Variations with temperature, and pressure, or volume, are controlled by anharmonicity. In an imperfectly elastic, or anelastic, body the velocities are dispersive; they depend on frequency, and this introduces another mechanism for changing moduli, and seismic velocities with temperature. This is important when comparing seismic data taken at different frequencies or when comparing seismic and laboratory data. When long-period seismic waves started to be used in seismology, it was noted that the free oscillation, or normal mode, models, differed from the classical body wave-models, which were based on short-period seismic waves. This is the `body-wave-normal-mode discrepancy`. The discrepancy was resolved when it was realized that in a real solid, as opposed to an ideally elastic one, the elastic moduli were functions of frequency. One has to allow for this when using body waves, surface waves and normal modes in the inversion for velocity vs. depth. The absorption, or dissipation, of energy, and the frequency dependence of seismic velocity, can be due to *intrinsic anelasticity*, or due to *scattering*. [`seismic wave scattering Monte Carlo`]

A variety of physical processes contribute to attenuation in a crystalline material: motions of point defects, dislocations, grain boundaries and so on. These processes all involve a high-frequency, or unrelaxed, modulus and a low-frequency, or relaxed, modulus. At sufficiently high frequencies, or low temperatures, the defects, which are characterized by a time constant, do not have time to contribute, and the body behaves as a perfectly elastic body. Attenuation is low and Q, the seismic quality factor, is high in the high-frequency limit. At very low frequencies, or high temperature, the defects have plenty of time to respond to the applied force and they contribute an additional strain. Because the stress cycle time is long compared with the response time of the defect, stress and strain are in phase and again Q is high. Because of the additional relaxed strain, however, the modulus is low and the relaxed seismic velocity is low. When the frequency is comparable to the characteristic time of the defect, attenuation reaches a maximum, and the wave velocity changes rapidly with frequency. Similar effects are seen in porous or partially molten solids; the elastic moduli depend on frequency.

These characteristics are embodied in the `standard linear solid`, which is composed of an elastic spring and a dashpot (or viscous element) arranged in a parallel circuit, which is then attached to another spring. At high frequencies the second, or series, spring responds to the load, and this spring constant is the effective modulus that controls the total extension. At low frequencies the other spring and dashpot both extend, with a time constant characteristic of the dashpot, the total extension is greater, and the effective modulus is therefore lower. This system is sometimes described as a *viscoelastic solid*. The temperature dependence of the spring constant, or modulus, represents the anharmonic contribution to the temperature dependence of the overall modulus of the system. The temperature dependence of the viscosity of the dashpot introduces another term – the *anelastic term* – in

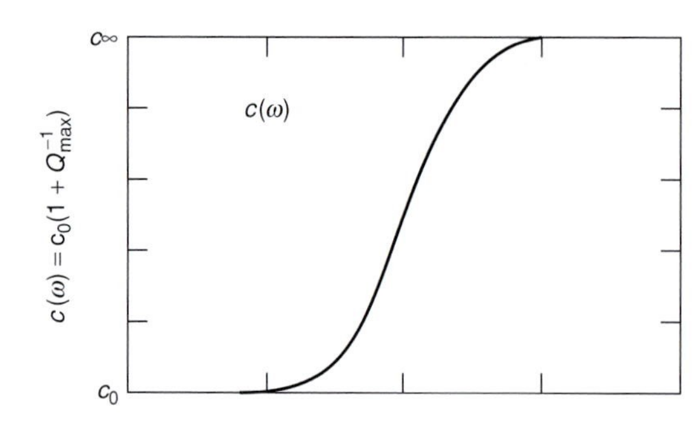

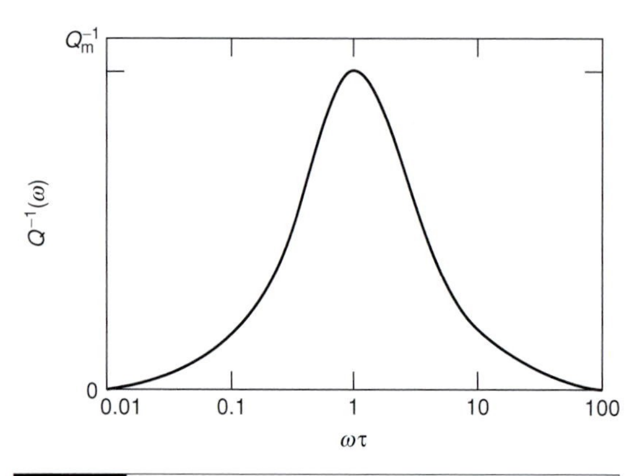

Fig. 19.1 Q^{-1} and phase velocity as a function of frequency for a standard linear solid with a single relation time.

the temperature dependence of moduli or seismic velocities.

The Q^{-1} of such a system is

$$Q^{-1}(\omega) = \frac{k_2}{k_t}\frac{\omega\tau}{1+(\omega\tau)^2}$$

where k_2 and k_1 are, respectively, the spring constants (or moduli) of the series spring and the parallel spring and τ is the relaxation time.

Clearly, Q^{-1}, the dimensionless attenuation, is a maximum at $\omega\tau = 1$, and

$$Q^{-1}(\omega) = 2Q^{-1}_{\max}\frac{\omega\tau}{1+(\omega\tau)^2} \tag{1}$$

The resulting absorption peak is shown in Figure 19.1. This can be considered a plot of attenuation and velocity vs. either frequency, or temperature, since τ is a function of temperature, an exponential function for thermally activated processes.

The phase velocity is approximately given by

$$c(\omega) = c_0\left(1 + Q^{-1}_{\max}\frac{(\omega\tau)^2}{1+(\omega\tau)^2}\right) \tag{2}$$

where c_0 is the zero-frequency velocity. The high-frequency or elastic velocity is

$$c_\infty = c_0\left(1 + Q^{-1}_{\max}\right) \tag{3}$$

Far away from the absorption peak, the velocity can be written

$$c(\omega) \approx c_0\left(1 + \frac{1}{2}\frac{k_1}{k_2}Q^{-2}\right) \quad \text{for } \omega\tau \ll 1$$

$$= c_\infty\left(1 - \frac{k_1^2}{(2k_1 + k_2)k_2}Q^{-2}\right) \quad \text{for } \omega\tau \gg 1$$

and the Q effect is only second order. In these limits, velocity is nearly independent of frequency, but Q is not; Q and c cannot both be independent of frequency. Velocity depends on the attenuation. When Q is constant, or nearly so, the fractional change in phase velocity becomes a first-order effect.

For activated processes,

$$\tau = \tau_0 \exp E^*/RT \tag{4}$$

where E^* is an activation energy. This is where the temperature dependence of seismic velocities comes in, in anelastic processes. Velocity is not a simple linear function of temperature.

For activated processes, then,

$$Q^{-1}(\omega) = 2Q^{-1}_{\max}\{\omega\tau_0 \exp E^*/RT\}/\{1 + (\omega\tau_0)^2 \times \exp 2E^*/RT\} \tag{5}$$

The relaxation peak can be defined either by changing ω or changing T.

At high temperatures, or low frequencies,

$$Q^{-1}(\omega) = 2Q^{-1}_{\max}\omega\tau_0 \exp E^*/RT \tag{6}$$

This is contrary to the general intuition that attenuation must increase with temperature. However, if τ differs greatly from seismic periods, it is possible that we may be on the low-temperature or high-frequency portion of the absorption peak, and

$$Q^{-1}(\omega) = 2Q^{-1}_{\max}/(\omega\tau_0 \exp E^*/RT), \quad \omega\tau \gg 1 \tag{7}$$

In that case Q does decrease with an increase in T, and in that regime Q increases with frequency. This appears to be the case for short-period waves in the mantle. It is also generally observed that low-Q and low-velocity regions of the upper mantle are in tectonically active and high heat-flow areas. Thus, seismic frequencies appear to be near the high-frequency, low-temperature side of the absorption peak in the Earth's upper mantle. This may not be true in the lower mantle; the absorption band shifts with frequency, as in the absorption band model for mantle attenuation.

The characteristic relaxation time also changes with pressure,

$$\tau = \tau_0 \exp(E^* + PV^*)/RT \tag{8}$$

where V^*, the activation volume, controls the effect of pressure on τ and Q, and seismic velocity.

Most mechanisms of attenuation at seismic frequencies and mantle temperatures can be described as activated relaxation effects. Increasing temperature drives the absorption peak to higher frequencies (characteristic frequencies increase with temperature). Increasing pressure drives the peak to lower frequencies.

Absorption in a medium with a single characteristic frequency gives rise to a bell-shaped Debye peak centered at a frequency $\omega\tau = 1$, as shown in Figure 19.1. The specific dissipation function and phase velocity satisfy the differential equation for the standard linear solid and can be written

$$c^2(\omega) = c_0^2\left(1+\omega^2\tau^2 c_\infty^2/c_0^2\right) / \left[(1+\omega^2\tau^2)^2 + 2\omega^2\tau^2 Q_{\max}^{-1}\right]^{1/2}$$

The high-frequency (c) and low-frequency velocities are related by

$$\frac{c_\infty^2 - c_0^2}{c_0 c_\infty} = 2Q_{\max}^{-1}$$

so that the total dispersion depends on the magnitude of the peak dissipation. For a Q of 200, a typical value for the upper mantle, the total velocity dispersion is 2%.

Solids in general, and mantle silicates in particular, are not characterized by a single relaxation time and a single Debye peak. A distribution of relaxation times broadens the peak

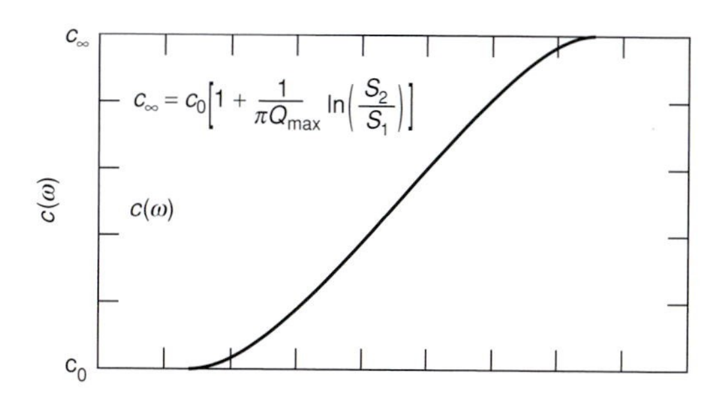

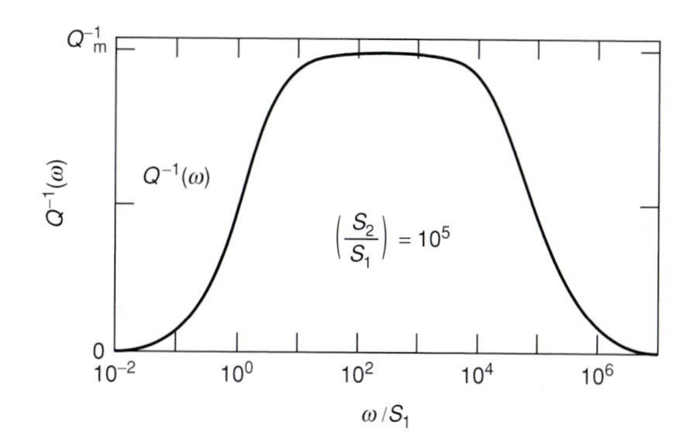

Fig. 19.2 Absorption band with a distribution of relaxation times (Kanamori and Anderson, 1977).

and gives rise to an absorption band (Figure 19.2). Q can be weakly dependent on frequency in such a band. Seismic Q values are nearly constant with frequency over much of the seismic band. A nearly constant Q can be explained by involving a spectrum of relaxation times and a superposition of elementary relaxation peaks, giving

$$Q^{-1}(\omega) = \left(2Q_{\max}^{-1}/\pi\right)\tan^{-1}[\omega(\tau_1-\tau_2)/(1+\omega^2\tau_1\tau_2)]$$

and

$$c(\omega) = c_0\left(1 + \left(Q_{\max}^{-1}/2\pi\right)\ln\left[\left(1+\omega^2\tau_1^2\right)/\left(1+\omega^2\tau_2^2\right)\right]\right)$$

For $\tau_1 \ll \omega^{-1} \ll \tau_2$ the value of Q^{-1} is constant and equal to $Q_{\max}^{-1}$. The total dispersion in this case is

$$\frac{c_\infty - c_0}{c_0} = \left(Q_{\max}^{-1}/\pi\right)\ln(\tau_1/\tau_2)$$

Which depends on the ration τ_1/τ_2, which is the width of the absorption band. The spread in τ can be due to a distribution of τ_0 or of E^*.

Attenuation mechanisms

The actual physical mechanism of attenuation in the mantle is uncertain, but it is likely to be a relaxation process involving a distribution of relaxation times, or a scattering process involving a distribution of scatterers. Many of the attenuation mechanisms that have been identified in solids occur at relatively low temperatures and high frequencies and can therefore be eliminated from consideration. These include point-defect and dislocation resonance mechanisms, which typically give absorption peaks at kilohertz and megahertz frequencies at temperatures below about half the melting point. The so-called grain-boundary and cold-work peak and the 'high-temperature background' occur at lower frequencies and higher temperatures. These mechanisms involve the stress-induced diffusion of dislocations. The Bordoni peak occurs at relatively low temperature in cold-worked metals but may be a higher-temperature peak in silicates. It is apparently due to the motion of dislocations since it disappears upon annealing. Because of the large wavelengths of seismic waves, it is not required that the dissipation mechanism be microscopic, or grain-scale.

Spectrum of relaxation times

Relaxation mechanisms lead to an internal friction peak of the form

$$Q^{-1}(\omega) = \Delta \int_{-\infty}^{\infty} D(\tau)\left[\omega\tau/(1+\omega^2\tau^2)\right] d\tau$$

where $D(\tau)$ is called the retardation spectrum and Δ is the modulus defect, the relative difference between the high-frequency, unrelaxed shear modulus and the low-frequency, relaxed modulus. The modulus defect is also a measure of the total reduction in modulus that is obtained in going from low temperature to high temperature.

Convenient forms of the retardation spectrum are given in Minster and Anderson (1981) and discussed at greater length in Theory of the Earth. These equations are now widely used in correcting seismic velocities in the mantle to standard temperatures and frequencies, the so-called anelastic correction. For simple dislocation and grain boundary networks the difference between the relaxed and unrelaxed moduli is about 8%. Anharmonic and anelastic corrections to seismic velocities are now routinely applied in comparisons with laboratory values. These are more complex, but more physically based, than the simple linear scalings between velocity and temperature or density that were used in the past.

The relationship between retardation spectra and transient creep and the Jeffreys-Lomnitz creep law are given in Theory of the Earth, Chapter 14. Jeffreys used this law as an argument against continental drift.

Partial melting

Seismic studies indicate that increased absorption, particularly of S-waves, occurs below volcanic zones and is therefore presumably related to partial melting. Regional variations in seismic absorption are a powerful tool in mapping the thermal state of the crust and upper mantle. It has also been suggested that partial melting is the most probable cause of the low-velocity layer in the upper mantle of the Earth. Thus the role of partial melting in the attenuation of seismic waves may be a critical one, at least in certain regions of the Earth. Studies of the melting of polycrystalline solids have shown that melting begins at grain boundaries, often at temperatures far below the melting point of the main constituents of the grains. This effect is caused by impurities that have collected at the grain boundaries during the initial solidification.

In principle, a large anelastic contribution can cause a large decrease in seismic velocity without partial melting but many of the low-Q regions are volcanic. In high heat-flow areas it is difficult to design a geotherm that does not imply upper mantle melting. Partial melting is a possible cause of seismic attenuation, and low velocity, particularly at very low frequencies.

Figure 19.3 shows the Q in an ice–brine–NaCl system for a concentration of 2% NaCl. Note the abrupt drop in Q as partial melting is initiated at the eutectic temperature. There is a corresponding, but much less pronounced, drop in velocity at the same temperature.

Bulk attenuation

Most of the mechanisms of seismic-wave attenuation operate in shear. Shear losses, generally, are much larger than compressional losses, and therefore shear waves attenuate more rapidly than compressional waves. Bulk or volume attenuation can become important in certain circumstances. One class of such mechanisms is due to thermoelastic relaxation. An applied stress changes the temperature of a sample relative to its surrounding, or of one part of a specimen relative to another. The heat flow that then occurs in order to equalize the temperature difference gives rise to energy dissipation and to anelastic behavior. The difference between the *unrelaxed* and *relaxed* moduli is the difference between the *adiabatic* and *isothermal* moduli. Under laboratory conditions this is typically 1% for oxides and silicates.

Absorption-band Q model for the Earth

Attenuation in solids and liquids, as measured by the quality factor, or specific dissipation function, Q, is frequency dependent. In seismology, however, Q is often assumed to be independent of frequency. The success of this assumption is a reflection of the limited precision, resolving power and bandwidth of seismic data and the trade-off between frequency and depth effects rather than a statement about the physics of the Earth's interior.

Frequency-independent Q models provide an adequate fit to most seismic data including the normal-mode data. There is evidence, however, that short-period body waves may require higher Q values. Some geophysical applications require estimates of the elastic properties of the Earth outside the seismic band. These include calculations of tidal Love numbers, Chandler periods and high-frequency moduli for comparison with ultrasonic data. The anelastic temperature derivatives of the seismic velocities, required for the interpretation of tomographic images, requires a theoretically sound dissipation fuction. The constant-Q models cannot be used for these purposes. For these reasons it is important to have a good attenuation model for the Earth.

The theory of seismic attenuation has been worked out in some detail and discussed more fully than here in Theory of the Earth, Chapter 14.

For a solid characterized by a single relaxation time Q^{-1} is a Debye function with a single narrow peak. For a solid with a spectrum of relaxation times, the band is broadened and the maximum attenuation is reduced. For a polycrystalline solid with a variety of grain sizes, orientations and activation energies, the absorption band can be appreciably more than several decades wide. If, as seems likely, the attenuation mechanism in the mantle is an activated process, the relaxation times should be a strong function of temperature and pressure. The location of the absorption band, therefore, changes with depth. Q can be a weak function of frequency only over a limited bandwidth. If the material has a finite elastic modulus at high frequency and a nonzero modulus at low frequency, there must be high- and low-frequency cutoffs in the relaxation spectrum. Physically this means that relaxation times cannot take on arbitrarily high and low values. The anelastic temperature correction, likewise, depends on temperature and frequency. In some published studies the anelastic correction for temperature implies an infinite or semi-infinite absorption band.

The relationship between Q and bandwidth indicates that a finite Q requires a finite bandwidth of relaxation times and therefore an absorption band of finite width. Q can be a weak function of frequency, and velocity can be a strong function of temperature, only in this band.

We can approximate the seismic absorption band in the mantle and core in the following

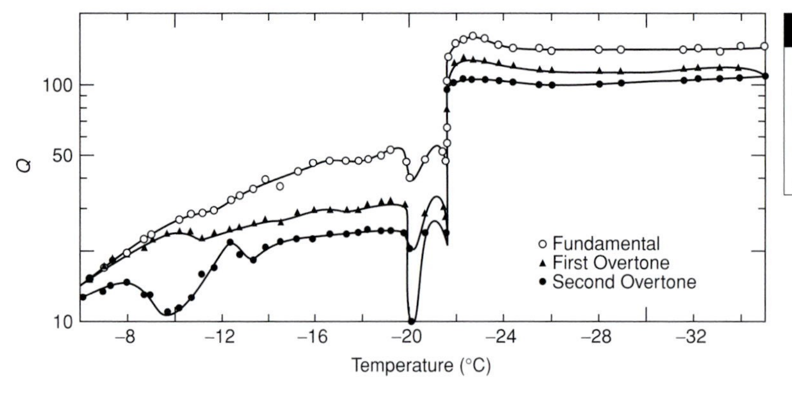

Fig. 19.3 Q of ice containing 2% NaCl. At low temperatures this is a solid solution. At temperatures higher than the eutectic the system is an ice–brine mixture (Spetzler and Anderson, 1968).

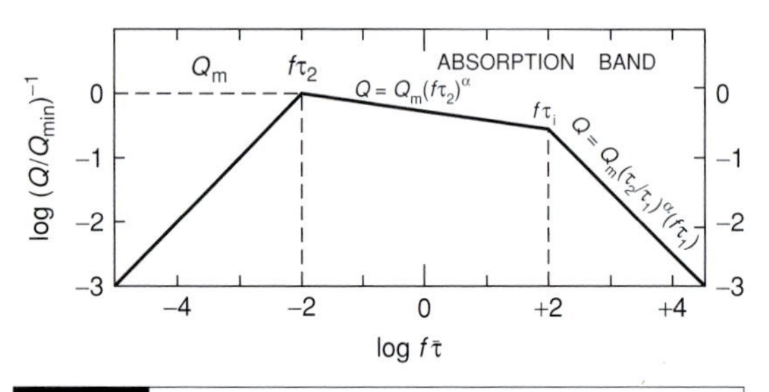

Fig. 19.4 Schematic illustration of an absorption band.

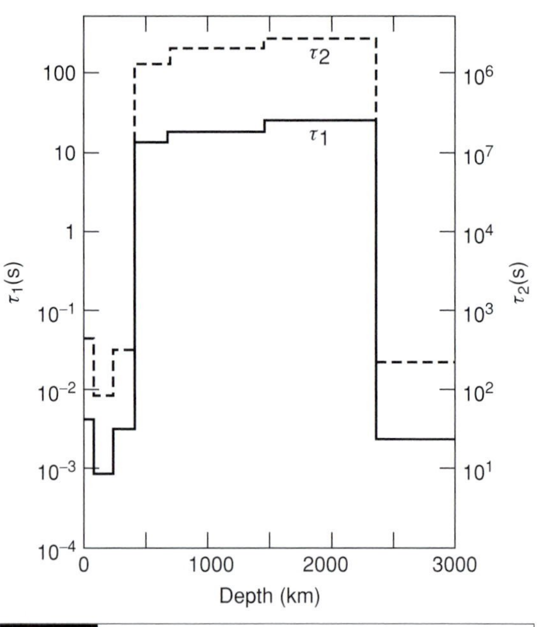

Fig. 19.5 Relaxation times as a function of depth in the mantle for the absorption band model (ABM) (Anderson and Given, 1982).

way;

$$Q = Q_{\min}(f\tau_2)^{-1}, \quad f < 1/\tau_2$$
$$Q = Q_{\min}(f\tau_2)^{\alpha}, \quad 1/\tau_2 < f < 1/\tau_1$$
$$Q = Q_{\min}(\tau_2/\tau_1)^{\alpha}(f\tau_1), \quad f > 1/\tau_1$$

where f is the frequency of the wave, τ_1 is the short-period cutoff, τ_2 is the long-period cutoff and $Q_{\min}$ is the minimum Q. These parameters are shown in Figure 19.4.

The relaxation time for an activated process depends exponentially on temperature and pressure. Characteristic lengths, such as dislocation or grain size, are a function of tectonic stress, which is a function of depth. The location of the band, therefore, depends on tectonic stress, temperature and pressure, at least for microscopic mechanisms. The width of the band is controlled by the distribution of relaxation times, which in turn depends on the distribution of grain sizes, dislocation lengths and so on.

The effect of pressure dominates over temperature for most of the upper mantle, and tectonic stress decreases with depth and away from shear boundary layers. The absorption band is expected to move to longer periods with increasing depth. A reversal of this trend may be caused by steep stress or temperature gradients across boundary layers, or by enhanced diffusion due to the presence of fluids, changes in crystal structure or in the nature of the point defects. If we assume that the parameters of the absorption band are constant throughout the mantle, we can use the seismic data to determine the location of the band as a function of depth. This assumption is equivalent to assuming that the activation energy, E*, and activation volume, V*,

Table 19.1 Absorption band parameters for Q_S

Radius (km)	Depth (km)	τ_1 (s)	τ_2/τ_1	Q_s (min) (τ_1)	Q_s (100 s)	α
1230	5141	0.14	2.43	35	1000	0.15
3484	2887	—	—	—	—	—
4049	2322	0.0025	10^5	80	92	0.15
4832	1539	25.2	10^5	80	366	0.15
5700	671	12.6	10^5	80	353	0.15
5950	421	0.0031	10^5	80	330	0.15
6121	250	0.0009	10^5	80	95	0.15
6360	11	0.0044	10^5	80	90	0.15
6371	0	0	∞	500	500	0

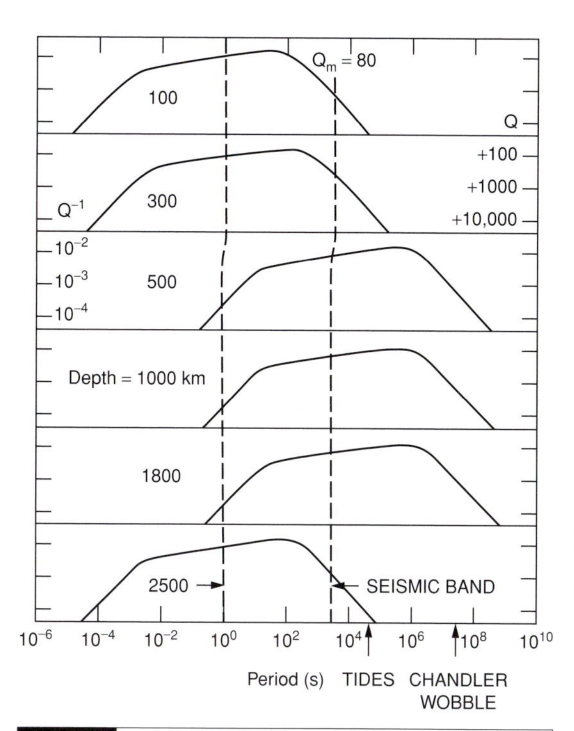

Fig. 19.6 The location of the absorption band for Q_S as a function of depth in the mantle (Anderson and Given, 1982).

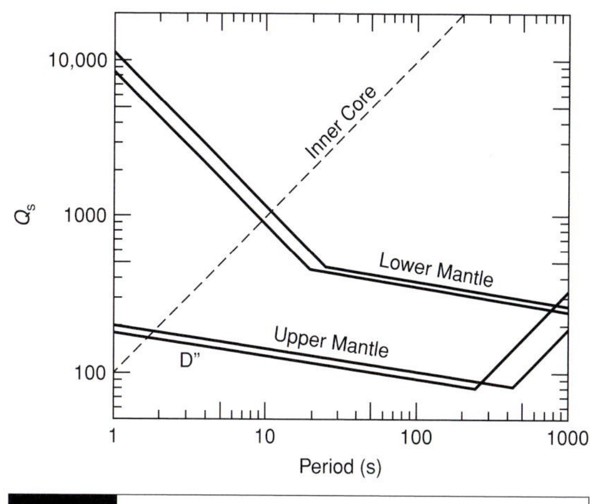

Fig. 19.7 Q_S as a function of period for the mantle and inner core for ABM. Note the similarity of the upper mantle and the lowermost mantle (D). These may be thermal (high-temperature gradient) and mechanical (high-stress) boundary layers associated with mantle convection (Anderson and Given, 1982).

are fixed. By assuming that the characteristics of the absorption band are invariant with depth, we are assuming that the width of the band is controlled by a distribution of characteristic relaxation times rather than a distribution of activation energies. Although this assumption can be defended, to some extent it has been introduced to reduce the number of model parameters. If a range of activation energies is assumed, the shape of the band (its width and height) varies with temperature and pressure. The parameters of the absorption bands are given in Table 19.1. The locations of the bands as a function of depth are shown in Figures 19.5 through 19.7. I refer to the absorption-band model as ABM.

The variation of *the characteristic times* with depth in the mantle is shown in Figure 19.5. Note that both decrease with depth in the uppermost mantle. This is expected in regions of steep thermal gradient. They increase slightly below 250 km and abruptly at 400 km. No abrupt change occurs at 670 km. Apparently, a steep

Table 19.2 Q_S and Q_P as function of depth and period for model ABM

Depth (km)	Q_S 1s	10s	100s	1000s	Q_P 1s	10s	100s	1000s
5142	100	1000	10,000	10^5	511	454	3322	3.3×10^4
4044	—	—	—	—	4518	493	1000	10^4
2887	—	—	—	—	7530	753	600	6000
2843	184	130	92	315	427	345	247	846
2400	184	130	92	315	427	345	247	846
2200	11 350	1135	366	259	2938	2687	942	668
671	8919	892	353	250	2921	2060	866	615
421	5691	569	330	234	741	840	819	603
421	190	134	95	254	302	296	244	659
200	157	111	90	900	270	256	237	2365
11	200	141	100	181	287	262	207	377
11	500	500	500	500	487	767	1168	1232

Table 19.3 Average mantle Q values as function of period. Model ABM

Region	0.1s	1s	4s	10s	100s	1000s
			Q_S			
Upper mantle	379	267	210	173	127	295
Lower mantle	1068	721	520	382	211	266
Whole mantle	691	477	360	280	176	274
			Q_P			
Upper mantle	513	362	315	354	311	727
Lower mantle	586	1228	1262	979	550	671
Whole mantle	562	713	662	639	446	687
			$Q({}_0S_2)$			
		3232 s		12 h		14 mos
		596		514		463

Table 19.4 Relaxation times and Q, for bulk attenuation at various periods

Region	τ_1(s)	τ_2(s)	Q_K 1s	10s	100s	1000s
Upper mantle	0	3.33	479	1200	1.2×10^4	1.2×10^5
Lower mantle	0	0.20	2000	2×10^4	2×10^5	2×10^6
Outer core	15.1	66.7	7530	753	600	6000
	9.04	40.0	4518	493	1000	10 000
Inner core	3.01	13.3	1506	418	3000	3×10^4

Q_K (min) $= 400$, $\alpha = 0.15$.

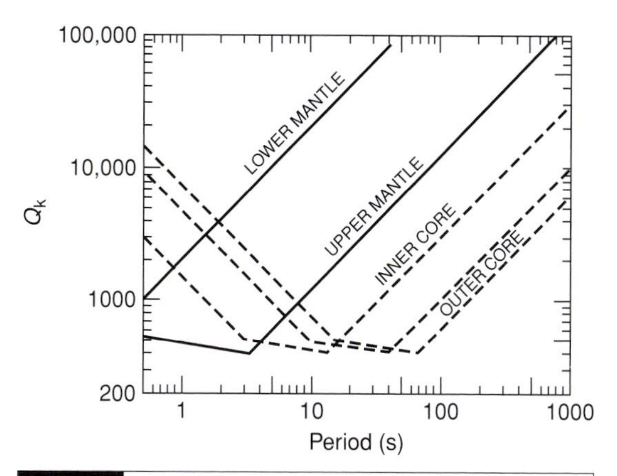

Fig. 19.8 Q_K as a function of frequency for various regions in the Earth (Anderson and Given, 1982).

temperature gradient and high tectonic stresses can keep the absorption band at high frequencies throughout most of the upper mantle, but these effects are overridden below 400 km where most mantle minerals are in the cubic structure. A phase change, along with high pressure and low stress, may contribute to the lengthening of the relaxation times. Relaxation times change only slightly through most of the lower mantle.

Most of the lower mantle therefore has high Q, for body waves and low Q, for free oscillations (Tables 19.2 and 19.3). The relationship between the P, S and Rayleigh wave Q is;

$$Q_P^{-1} = L\,Q_s^{-1} + (1 - L)\,Q_K^{-1}$$

where $L = (4/3)(\beta/\alpha)^2$ and β and α are the shear and compressional velocities.

For a Poisson solid with $Q_K^{-1} = 0$,

$$Q_P = (9/4)Q_s$$

Chapter 20

Fabric of the mantle

Anisotropy

And perpendicular now and now transverse, Pierce the dark soil and as they pierce and pass Make bare the secrets of the Earth's deep heart.

Shelley, Prometheus Unbound

Anisotropy is responsible for large variations in seismic velocities; changes in the orientation of mantle minerals, or in the direction of seismic waves, cause larger changes in velocity than can be accounted for by changes in temperature, composition or mineralogy. Plate-tectonic processes, and gravity, create a fabric in the mantle. Anisotropy can be microscopic – orientation of crystals – or macroscopic – large-scale laminations or oriented slabs and dikes. Discussions of velocity gradients, both radial and lateral, and chemistry and mineralogy of the mantle must allow for the presence of anisotropy. Anisotropy is not a second-order effect. Seismic data that are interpreted in terms of isotropic theory can lead to models that are not even approximately correct. Slab anisotropy can cause artifacts in tomographic models. A wealth of new information regarding mantle structure, history, mineralogy and flow is becoming available as the anisotropy of the mantle is becoming better understood.

Introduction

In the first edition of `Theory of the Earth` there were sections that are largely missing in this edition. There was a very large section on anisotropy since it was a relatively new concept to seismologists. There was also a large section devoted to the then-novel thesis that seismic velocities were not independent of frequency, and that anelasticity had to be allowed for in estimates of mantle temperatures. Earth scientists no longer need to be convinced that anisotropy and anelasticity are essential elements in Earth physics, but there may still be artifacts in tomographic models or in estimates of errors that are caused by anisotropy. At the time of the first edition of this book – 1989 – the Earth was usually assumed to be perfectly elastic and isotropic to the propagation of seismic waves. These assumptions were made for mathematical and operational convenience. The fact that a large body of seismic data can be satisfactorily modeled with these assumptions does not prove that the Earth is isotropic or perfectly elastic. There is often a direct trade-off between anisotropy and heterogeneity, and between frequency dependence and depth dependence of seismic velocities. An anisotropic structure can have characteristics, such as travel times, normal-mode frequencies and dispersion curves, that are identical, or similar, to a different isotropic structure. A layered solid, for example, composed of isotropic layers that are thin compared to a seismic wavelength will behave as an anisotropic solid – the velocity of propagation depends on direction. The effective long-wavelength elastic constants depend on the thicknesses and elastic properties of the individual layers. The reverse is also true; an anisotropic solid with these same elastic constants can be modeled exactly as a stack of isotropic layers. The same holds true for an isotropic solid permeated by oriented cracks or aligned inclusions. This serves to illustrate the trade-off between heterogeneity and anisotropy. Not all anisotropic structures, however, can be modeled by laminated solids.

The crystals of the mantle are anisotropic, and rocks from the mantle show that these crystals exhibit a high degree of alignment. There is also evidence that crystal alignment is uniform over large areas of the upper mantle. At mantle temperatures, crystals tend to be easily recrystallized and aligned by the prevailing stress and flow fields. But there may also be large-scale fabric in the upper mantle, caused by orientation of subducted slabs or dikes, for example.

The effects of anisotropy are often subtle and, if unrecognized, are usually modeled as inhomogeneities, for example, as layering or gradients. The most obvious manifestations of anisotropy are:

(1) [shear-wave splitting] or birefringence – the two polarizations of S-waves arrive at different times;
(2) [azimuthal anisotropy] – the arrival times, or apparent velocities of seismic waves at a given distance from an event, depend on azimuth; and
(3) an apparent discrepancy between Love waves and Rayleigh waves [Love Rayleigh discrepancy].

Even these are not completely unambiguous indicators of anisotropy. Effects such as P to S conversion, dipping interfaces, attenuation, and density variations must be properly taken into account. There is now growing acceptance that much of the upper mantle may be anisotropic to the propagation of seismic waves.

It has been known for some time that the discrepancy between mantle Ray- leigh and Love waves could be explained if the vertical P and S velocities in the upper mantle were 7–8% less than the horizontal velocities. The Love–Rayleigh discrepancy has survived to the present, and average Earth models have SV in the upper mantle less than SH by about 3%. Some early models were based on separate isotropic inversions of Love and Rayleigh waves (pseudo-isotropic inversions). This is not a valid procedure. There is a trade-off between anisotropy and structure. In particular, the very low upper-mantle shear velocities, 4.0–4.2 km/s, found by many isotropic and pseudo-isotropic inversions, are not a characteristic of models resulting from full anisotropic inversion. The P-wave anisotropy makes a significant contribution to Rayleigh wave dispersion. This must also be allowed for in tomographic surface wave inversions, but seldom is.

Since intrinsic anisotropy requires both anisotropic crystals and preferred orientation, the anisotropy of the mantle contains information about the mineralogy and stress gradients. For example, olivine, the most abundant upper-mantle mineral, is extremely anisotropic for both P-wave and S-wave propagation. It is readily oriented by recrystallization in the ambient stress field. Olivine-rich outcrops show a consistent preferred orientation over large areas. In general, the seismically fast axes of olivine are in the plane of the flow, with the *a*-axis, the fastest direction, pointing in the direction of flow. The *b*-axis, the minimum velocity direction, is generally normal to the flow plane, or vertical. Pyroxenes are also very anisotropic. The magnitude of the anisotropy in the mantle is comparable to that found in ultramafic rocks (Figure 20.1). Soft layers or oriented fluid-filled cracks also give an apparent anisotropy. Much seismic data that are used in upper-mantle modeling are averages over several tectonic provinces or over many azimuths. Azimuthal anisotropy may therefore be averaged out, but differences between vertical and horizontal velocities are not.

Origin of mantle anisotropy

Nicholas and Christensen (1987) elucidated the reason for strong preferred crystal orientation in deformed rocks. First, they noted that in homogenous deformation of a specimen composed of minerals with a dominant slip system, the preferred orientations of slip planes and slip directions coincide respectively with the orientations of the flow plane and the flow line.

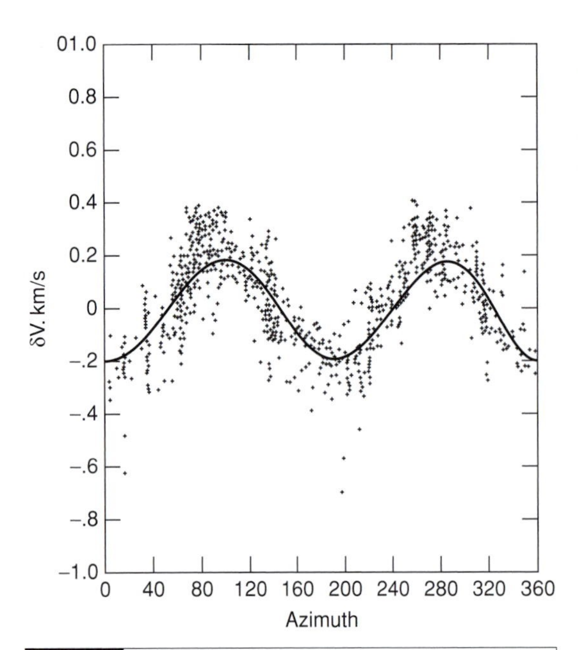

Fig. 20.1 Azimuthal anisotropy of Pn waves in the Pacific upper mantle. The unique anisotropy of the Pacific upper mantle has also been mapped with surface waves (after Morris *et al.*, 1969).

Simple shear in a crystal rotates all the lines attached to the crystal except those in the slip plane. This results in a bulk rotation of crystals so that the slip planes are aligned, as required to maintain contact between crystals. The crystal reorientations are not a direct result of the applied stress but are a geometrical requirement. Bulk anisotropy due to crystal orientation is therefore induced by plastic strain and is only indirectly related to stress. The result, of course, is also a strong anisotropy of the viscosity of the rock, and presumably attenuation, as well as elastic properties. This means that seismic techniques can be used to infer flow in the mantle. It also means that mantle viscosity inferred from postglacial rebound is not necessarily the same as that involved in plate tectonics and mantle convection.

Peridotites from the upper mantle display a strong preferred orientation of the dominant minerals, olivine and orthopyroxene. They exhibit a pronounced acoustic-wave anisotropy that is consistent with the anisotropy of the constituent minerals and their orientation. In igneous rocks preferred orientation can be caused by grain rotation, recrystallization in a nonhydrostatic stress field or in the presence of a thermal gradient, crystal setting in magma chambers, flow orientation and dislocation-controlled slip. Macroscopic fabrics caused by banding, cracking, sill and dike injection can also cause anisotropy. Eclogites and basalts are much less anisotropic; anisotropy can therefore be used as a petrological tool.

Plastic flow induces preferred orientations in rock-forming minerals. The relative roles of deviatoric stresses and plastic strain have been long debated. In order to assure continuity of a deforming crystal with its neighbors, five independent degrees of motion are required (the Von Mises criterion). This can be achieved in a crystal with the activation of five independent slip systems or with a combination of fewer slip systems and other modes of deformation. In silicates only one or two slip systems are activated under a given set of conditions involving a given temperature, pressure and deviatoric stresses. The homogenous deformation of a dominant slip system and the orientation of slip planes and slip directions tend to coincide with the flow plane and the flow direction.

Mantle peridotites typically contain more than 65% olivine and 20% orthopyroxene. The high-P wave direction in olivine (Figure 20.2) is along the *a*-axis [1001], which is also the dominant slip direction at high temperature. The lowest velocity crystallographic direction is [0101], the *b*-direction, which is normal to a common slip plane. Thus, the pattern in olivine aggregates is related to slip orientations. There is no such simple relationship with shear waves and, in fact, the S-wave anisotropy of peridotites is small. Orthopyroxenes also have large P-wave anisotropies and relatively small S-wave anisotropies. The high-Vp direction coincides with the [1001] pole of the unique slip plane and the intermediate Vp crystallographic direction coincides with the unique [0011] slip line (Figure 20.2). In natural peridotites the preferred orientation of olivine is more pronounced than the other minerals. Olivine is apparently the most ductile and easily oriented upper-mantle mineral, and therefore controls the seismic anisotropy of the

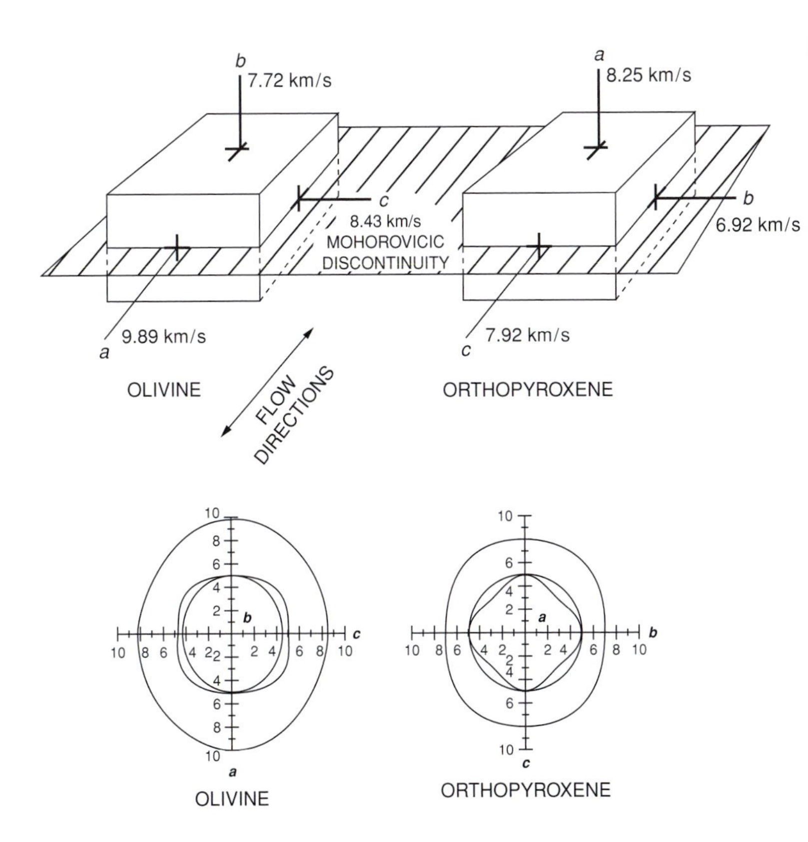

Fig. 20.2 Possible olivine and orthopyroxene orientations within the upper mantle showing compressional velocities for the three crystallographic axes, and compressional and shear velocities in the olivine *a*–*c* plane and orthopyroxene *b*–*c* plane (after Christensen and Lundquist, 1982).

upper mantle. The anisotropy of β-spinel, a high-pressure form of olivine that is expected to be a major mantle component below 400 km, is also high. The γ-spinel form of olivine, stable below about 500 km, is much less anisotropic. Recrystallization of olivine to spinel forms can be expected to yield aggregates with preferred orientation but with perhaps less pronounced P-wave anisotropy. β-spinel has a strong S-wave anisotropy (24% variation with direction, 16% maximum difference between polarizations). The fast shear directions are parallel to the slow P-wave directions, whereas in olivine the fast S-directions correspond to intermediate P-wave velocity directions. Orthopyroxene transforms to a cubic garnet-like structure that is stable over much of the transition region part of the upper mantle. This mineral, majorite, is expected to be relatively isotropic. Therefore, most of the mantle between 400 and 650 km depth is expected to have relatively low anisotropy, with the anisotropy decreasing as olivine transforms to the spinel structures. At low temperatures, as in subduction zones, the stable form of pyroxene is an ilmenite-type structure that is extremely anisotropic. Thus, the deep part of slabs may exhibit pronounced anisotropy, a property that could be mistaken for deep slab penetration in certain seismic experiments.

Petrofabric studies combined with field studies on ophiolite harzburgites give the following relationships.

(1) Olivine *c* axes and orthopyroxene *b* axes lie approximately parallel to the inferred ridge axis in a plane parallel to the Moho discontinuity.
(2) The olivine *a* axes and the orthopyroxene *c* axes align subparallel to the inferred speading direction.
(3) The olivine *b* axes and the orthopyroxene *a* axes are approximately perpendicular to the Moho.

These results indicate that the compressional velocity in the vertical direction increases with the orthopyroxene content, whereas horizontal velocities and anisotropy decrease with increasing orthopyroxerie content.

The maximum compressional wave velocity in orthopyroxene (along the *a* axis) parallels the minimum (*b* axis) velocity of olivine. For olivine *b* axis vertical regions of the mantle, as in ophiolite peridotites, the vertical P-velocity increases with orthopyroxene content. The reverse is true for other directions and for average properties. Appreciable shear-wave birefringence is expected in all directions even if the individual shear velocities do not depend much on azimuth. The total P-wave variation with azimuth in olivine- and orthopyroxene-rich aggregates is about 4–6%, while the S-waves only vary by 1 to 2% (Figure 20.3). The difference between the two shear-wave polarizations, however, is 4–6%. The azimuthal variation of S-waves can be expected to be hard to measure because the maximum velocity difference occurs over a small angular difference and because of the long-wavelength nature of shear waves.

The shear-wave anisotropy in the ilmenite structure of pyroxene, expected to be important in the deeper parts of subducted slabs, is quite pronounced and bears a different relationship to the P-wave anisotropy than that in peridotites. One possible manifestation of slab anisotropy is the variation of travel times with take-off angle from intermediate- and deep-focus earthquakes. Fast in-plane velocities, as expected for oriented olivine, and probably spinel and ilmenite, may easily be misinterpreted as evidence for deep slab penetration. The mineral assemblages in cold slabs are also different from the stable phases in normal and hot mantle. The colder phases are generally denser and seismically fast. Anisotropy and isobaric phase changes in the source region have been ignored in most studies purporting to show deep slab penetration into the lower mantle. There is a trade-off between the length of a high-velocity slab and its velocity contrast and anisotropy and structure at the source.

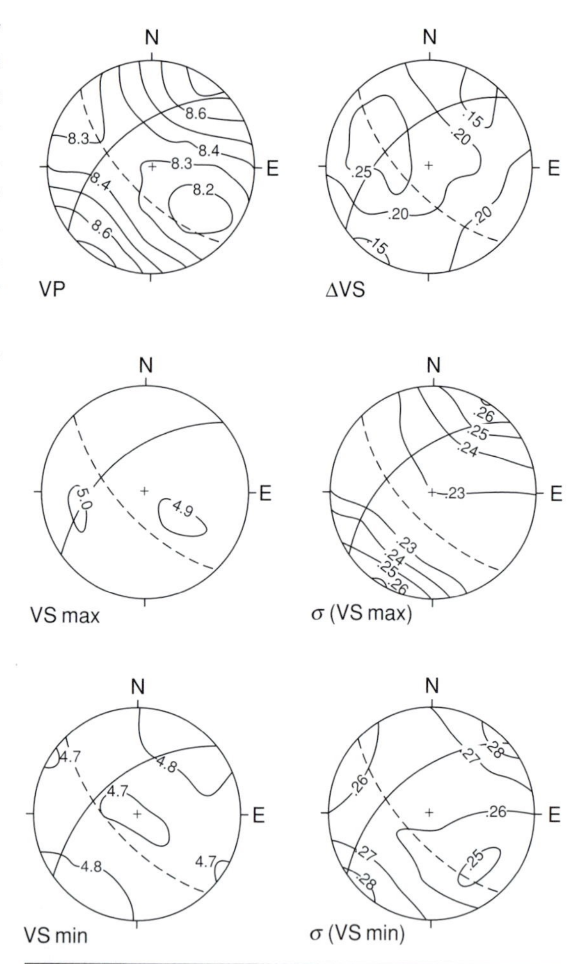

Fig. 20.3 Equal area projection of the acoustic velocities measured on samples of peridotite. Dashed line is vertical direction, solid great circle is the horizontal (after Christensen and Salisbury, 1979).

Anisotropy of crystals

Because of the simplicity and availability of the microscope, the optical properties of minerals receive more attention than the acoustic properties. It is the acoustic or ultrasonic properties, however, that are most relevant to the interpretation of seismic data. Being crystals, minerals exhibit both optical and acoustic anisotropy. Aggregates of crystals, rocks, are also anisotropic and display fabrics that can be analyzed in the same terms used to describe crystal symmetry. Tables 20.1, 20.2 and 20.3 summarize the acoustic anisotropy of some important rock-forming minerals. Pyroxenes and olivine are unique in having a greater P-wave anisotropy than S-wave anisotropy. Spinel and garnet, cubic crystals,

Table 20.1 | Anisotropic properties of rock forming minerals

Mineral	Symmetry	P direction Max.	P direction Min.	Max S Directionl Polarization	Anisotropy† (percent) P	Anisotropy† (percent) S
Olivine	Orthorhombic	[100]	[010]	45°/135°*	25	22
Garnet	Cubic	[001]	~50°*	[110] / [110]	0.6	1
Orthopyroxene	Orthorhombic	[100]	[010]	[010] / [001]	16	16
Clinopyroxene	Monoclinic	[001]	[101]	[011] / $[0\bar{1}1]$	21	20
Muscovite	Monoclinic	[110]	[001]	[010] / [100]	58	85
Orthoclase	Monoclinic	[010]	[101]	[011] / $[01\bar{1}]$	46	63
Anorthite	Triclinic	[010]	[101]	[011] / $[01\bar{1}]$	36	52
Rutile	Tetragonal	[001]	[100]	[100] / [010]	28	68
Nepheline	Hexagonal	[001]	[100]	[011] / $[0\bar{1}1]$	24	32
Spinel	Cubic	[101]	[001]	[100] / [010]	12	68
β-Mg_2SiO_4	Orthorhombic	[010]	[001]	—	16	14

Babuska (1981), Sawamoto and others (1984).
*Relative to [001].
†$[(V_{max} - V_{min}) / V_{mean}] \times 100$.

have low P-wave anisotropy. Hexagonal crystals, and the closely related class of trigonal crystals, have high shear-wave anisotropies. This is pertinent to the deeper part of cold subducted slabs in which the trigonal ilmenite form of pyroxene may be stable. Deep-focus earthquakes exhibit a pronounced angular variation in both S- and P-wave velocities and strong shear-wave birefringence. Cubic crystals do not necessarily have low shear-wave anisotropy. The major minerals of the shallow mantle are all extremely anisotropic. β-spinel and clinopyroxene are stable below 400 km, and these are also fairly anisotropic. Below 400 km the major mantle minerals at high temperature, γ-spinel and garnet-majorite are less anisotropic. At the temperatures prevailing in subduction zones, the cold high-pressure forms of orthopyroxene, clinopyroxene and garnet are expected to give high velocities and anisotropies. If these are lined up, by stress or flow or recrystallization, then the slab itself will be anisotropic. This effect will be hard to distinguish from a long isotropic slab, if only sources in the slab are used. The mantle minerals outside of a sinking slab are also likely to be aligned.

The elastic properties of simple oxides and silicates are predominantly controlled by the oxygen anion framework, especially for hexagonally close-packed and cubic close-packed structures, but also by the nature of the cations occurring within the oxygen interstices. Corundum (Al_2O_3) consists of a hexagonal close-packed array of large oxygen ions into which the smaller aluminum ions are inserted in interstitial positions. Forsterite (Mg_2SiO_4) consists of a framework of approximately hexagonal close-packed oxygen ions with the $Mg^{2\sim}$ cations occupying one-half of the available octahedral sites (sites surrounded by six oxygen ions) and the $Si^{4\sim}$ cations occupying one-eighth of the available tetrahedral sites (sites surrounded by four oxygens). The packing of the oxygens depends on the nature of the cations.

Theory of anisotropy

In a transversely isotropic solid with a vertical axis of symmetry, we can define four elastic constants in terms of P- and S-waves propagating perpendicular and parallel to the axis of symmetry. The fifth elastic constant, F, requires information from another direction of propagation. PH, SH are waves propagating and polarized in the horizontal direction and PV, SV are waves propagating in the vertical direction. In the vertical direction SH = SV; the two shear waves travel with the same velocity, and this velocity is the same as SV waves traveling in the horizontal direction. There is no azimuthal variation of velocity in the horizontal, or symmetry, plane.

Love waves are composed of SH motions, and Rayleigh waves are a combination of P and SV motions. In isotropic material Love waves and Rayleigh waves require only two elastic constants to describe their velocity. In general, more than two elastic constants at each depth are required to satisfy seismic surface-wave data, even when the azimuthal variation is averaged out, and complex vertical variations are allowed.

The upper mantle exhibits what is known as polarization anisotropy. In general, four elastic constants are required to describe Rayleigh-wave propagation in a homogenous transversely or equivalent transversely isotropic mantle.

Transverse isotropy, although a special case of anisotropy, has quite general applicability in geophysical problems. This kind of anisotropy is exhibited by laminated or layered solids, solids containing oriented cracks or melt zones, peridotite massifs, harzburgite bodies, the oceanic upper mantle and floating ice sheets. A mantle containing small-scale layering, sills or randomly oriented dikes will also appear to be macroscopically transversely isotropic. Since seismic waves have wavelengths of tens to hundreds of kilometers, the scale of the layering can actually be quite large. If flow in the upper mantle is mainly horizontal, then the evidence from fabrics of peridotite nodules and massifs suggests that the average vertical velocity is less than the average horizontal velocity, and horizontally propagating SH-waves will travel faster than SV-waves. In regions of upwelling and subduction, the slow direction may not be vertical, but if these regions are randomly oriented, the average Earth will still display the spherical equivalent of transverse isotropy. Since the upper mantle is composed primarily of the very anisotropic crystals olivine and pyroxene, and since these crystals tend to align themselves in response to flow and non-hydrostatic stresses, it is likely that the upper mantle is anisotropic to the propagation of elastic waves. Although the preferred orientation in the horizontal plane can be averaged out by determining the velocity in many directions or over many plates with different motion vectors, the vertical still remains a unique direction. It can be shown that if the azimuthally varying elastic velocities are replaced by the horizontal averages, then many problems in seismic wave propagation in more general anisotropic media can be reduced to the problem of transverse isotropy.

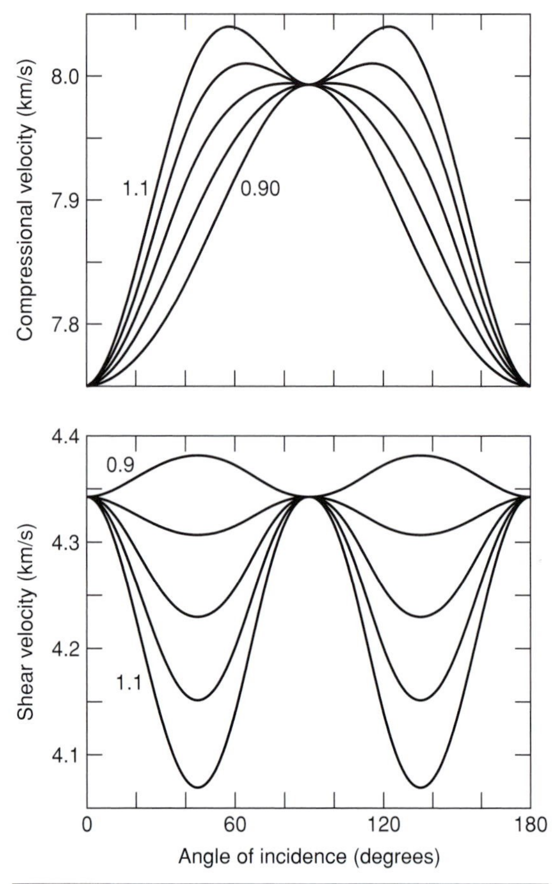

Fig. 20.4 P and S velocities as a function of angle of incidence relative to the symmetry plane and an anisotropic parameter, which varies from 0.9 to 1.1 at intervals of 0.05. Parameters are $V_{PV} = 7.752$, $V_{PH} = 7.994$, $V_{SV} = 4.343$, all in km/s (after Dziewonski and Anderson, 1981).

The inconsistency between Love- and Rayleigh-wave data, first noted for global data, has now been found in regional data sets and it appears that anisotropy is an intrinsic and widespread property of the uppermost mantle. The crust and exposed sections of the upper mantle exhibit layering on scales ranging from meters to kilometers. Such layering in the deeper mantle would be beyond the resolution of seismic waves and would show up as an apparent anisotropy. This, plus the preponderance of aligned olivine in mantle samples, means that at least five elastic constants are probablyrequired to properly describe the elastic response of the upper mantle. It is clear that inversion of P-wave data, for example, or even of P and SV data cannot provide all of these constants. Even more serious, inversion of a limited data set, with the assumption of isotropy, does not necessarily yield the proper structure. The variation of velocities with angle of incidence, or ray parameter, will be interpreted as a variation of velocity with depth. In principle, simultaneous inversion of Love-wave and Rayleigh-wave data along with shear-wave splitting data can help resolve the ambiguity.

Because seismic waves have such long wavelengths, a mantle composed of oriented subducted slabs with partially molten basaltic crusts could be misinterpreted as a partially molten aggregate with grain boundary films and oriented crystals. Global tomography can only resolve large-scale features. Anisotropy and anelasticity, however, contain information about the smaller scale fabric of the mantle. Both anisotropy and anelasticity may be the result of kilometer, or tens of kilometer, scale features or the result of grain-scale phenomena.

Transverse isotropy of layered media

A material composed of isotropic layers appears to be transversely isotropic for waves that are long compared to the layer thicknesses. The symmetry axis is obviously perpendicular to the layers. All transversely isotropic material, however, cannot be approximated by a laminated solid. For example, in layered media, the velocities parallel to the layers are greater than in the perpendicular direction. This is not generally true for all materials exhibiting transverse or hexagonal symmetry. Backus (1962) derived other inequalities which must be satisfied among the five elastic constants characterizing long-wave anisotropy of layered media.

The effect of oriented cracks on seismic velocities

The velocities in a solid containing flat-oriented cracks, or magma-filled sills, depend on the elastic properties of the matrix, porosity or melt content, aspect ratio of the cracks, the bulk modulus of the pore fluid, and the direction of propagation. Substantial velocity reductions, compared with those of the uncracked solid, occur in the direction normal *to* the plane of the cracks. Shear-wave birefringence also occurs in rocks with oriented cracks.

Figure 20.5 gives the intersection of the velocity surface with a plane containing the unique axis for a rock with ellipsoidal cracks. The short-dashed curves are the velocity surfaces, spheres, in the crack-free matrix. The long-dashed curves are for a solid containing 1% by volume of aligned spheroids with $\alpha = 0.5$ and a pore–fluid bulk modulus of 100 kbar. The solid curves are for the same parameters as above but for a relatively compressible fluid in the pores with modulus of 0.1 kbar. The shear-velocity surfaces do not depend on the pore-fluid bulk modulus. Note the large compressional-wave anisotropy for the solid containing the more compressible fluid. The ratio of compressional velocity to shear velocity is strongly dependent on direction and the nature of the fluid phase.

There is always the question in seismic interpretations whether a measured anisotropy is due to intrinsic anisotropy or to heterogeneity, such as layers, sills or dikes. The magnitude of the anisotropy often can be used to rule out an apparent anisotropy due to layers if the required velocity contrast between layers is unrealistically large. The velocities along the layers, in the

Table 20.2 | Elastic coefficients of equivalent transversely isotropic models of the upper mantle.

	Olivine Model		Petrofabric		PREM	
	(1)	(2)	(3)	(4)	(5)	(6)
			Mbar			
A	2.416	2.052	2.290	2.208	2.251	2.176
C	2.265	3.141	2.202	2.365	2.151	2.044
F	0.752	0.696	0.721	0.724	0.860	0.831
L	0.659	0.723	0.770	0.790	0.655	0.663
N	0.824	0.623	0.784	0.746	0.708	0.661
			km/s			
V_{PH}	8.559	7.888	8.324	8.174	8.165	8.049
V_{PV}	8.285	9.757	8.166	8.460	7.982	7.800
V_{SH}	5.000	4.347	4.871	4.752	4.580	4.436
V_{SV}	4.470	4.684	4.828	4.889	4.404	4.441
			Mbar/Mbar			
ξ	1.250	0.861	1.018	0.944	1.081	0.997
ϕ	0.937	1.531	0.961	1.071	0.956	0.939
η	0.686	1.151	0.963	1.153	0.914	0.977
			g/cm^3			
ρ	3.298	3.298	3.305	3.305	3.377	3.360

(1) Olivine based model; *a* horizontal, *b* horizontal, *c* vertical (Nataf *et al.*, 1986).
(2) *a*-vertical, *b*-horizontal, *c*-horizontal.
(3) Petrofabric model; horizontal flow (Montagner and Nataf, 1986).
(4) Petrofabric model; vertical flow.
(5) PREM, 60 km depth (Dziewonski and Anderson, 1981).
(6) PREM, 220 km depth.

symmetry plane, are faster than velocities perpendicular to the layers. No such restrictions apply to the general case of crystals exhibiting hexagonal symmetry or to aggregates composed of crystals having preferred orientations. In a laminated medium, with a vertical axis of symmetry, the P and SH velocities decrease monotonically from the horizontal to the vertical. A cracked solid with flat aligned cracks, or an asthenosphere with sills, behaves as a transversely isotropic solid. If the oceanic lithosphere is permeated with aligned dikes, it will behave a hexagonal crystal, with a horizontal symmetry axis.

Oriented cracks are important in crustal seismic studies and indicate the direction of the prevailing stress or a paleostress field. The orientation of the shear waves and their velocities will be controlled by the orientation of the cracks. The magnitude of the velocity difference will be controlled by the crack density and the nature of the pore fluid. Cracks may form and open up as a result of tectonic stresses.

Global maps of transverse isotropy as a function of depth

There are numerous studies of upper mantle shear wave velocity and anisotropy. Azimuthal anisotropy can reach 10% in the shallowest mantle. Polarization anisotropy up

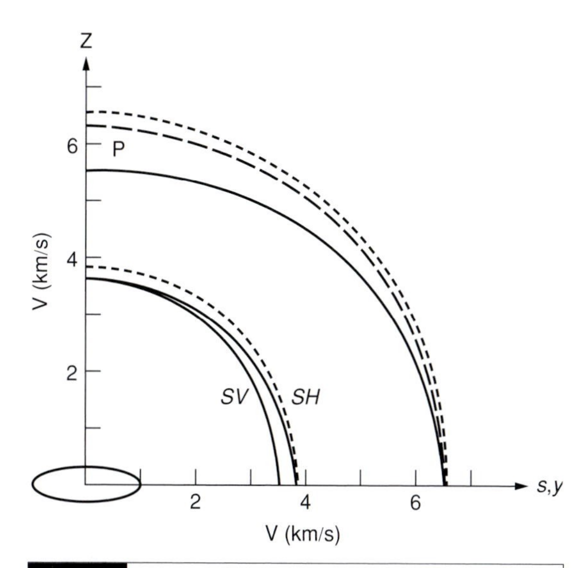

Fig. 20.5 Velocities as a function of angle and fluid properties in granite containing aligned ellipsoidal cracks (orientation shown at origin) with porosity = 0.01 and aspect ratio = 0.05. The short dashed curves are for the isotropic uncracked solid, the long dashes for liquid-filled cracks (K_L = 100 kbar) and the solid curves for gas-filled cracks (K_L = 0.1 kbar) (after Anderson *et al.*, 1974).

to 5% has been inferred from surface-wave studies in order to fit Love waves and Rayleigh waves. Azimuthal anisotropy can be averaged out. We are then left with only polarization anisotropy and can use a transversely isotropic parameterization.

Shear-velocity maps and cross-sections are shown in Figures 20.6 through 20.8. Shear velocity shows a strong correlation with surface tectonics down to about 200 km. Deeper in the mantle the correlation vanishes, and some long-wavelength anomalies appear. At 50–100 km depth heterogeneities are closely related to surface tectonics. All major shields show up as fast regions (Canada, Africa, Antarctica, West Australia, South America). All major ridges show up as slow regions (East Pacific, triple junctions in the Indian Ocean and in the Atlantic, East African rift). The effect of the fast mantle beneath the shields is partially offset by the thick crust. Old oceans also appear to be fast, but not as fast as shields. A few regions seem to be anomalous, considering their tectonic setting: a slow region around French Polynesia in average age ocean, a fast region centered southeast of South America. At 100 km depth the overall variations are smaller than at shallower depth. Triple junctions are slow. Below 200 km depth, the correlation with surface tectonics starts to break down. Shields are fast, in general, but ridges do not show up systematically. The East African region, centered on the Afar, is slow. The south-central Pacific is faster than most shields. An interesting feature is the belt of slow mantle at the Pacific subduction zones, a manifestation of the volcanism and marginal sea formation induced by the sinking ocean slab. Below 340 km, the same belt shows up as fast mantle; the effect of cold subducted material that was formerly part of the surface thermal boundary layers. Many ridge segments are now fast. At larger depths the resolution becomes poor, but these trends seem to persist.

At intermediate depths, regions of uprising (ridges) or downwelling (subduction zones) have an SV > SH anisotropy. Shallow depths (50 km) show very large anisotropy variations (± 10%). From observed Pn anisotropy and measured anisotropy of olivine, such values are not unreasonable. At 100 km the amplitude of the variations is much smaller (±5%), but the pattern is similar. The Mid-Atlantic Ridge has SV > SH, whereas the other ridges show no clear-cut trend. Under the Pacific there appear to be some parallel bands trending northwest-southeast with a dominant SH > SV anomaly. This is the expected anisotropy for horizontal flow of olivine-rich aggregates. At 340 km, most ridges have SH < SV (vertical flow). Antarctica and South America have a strong SV < SH anomaly (horizontal flow). North America and Siberia are almost isotropic at this depth. They exhibit, however, azimuthal anisotropy. The central Pacific and the eastern Indian Ocean have the characteristics of vertical flow. These regions have faster than average velocities at shallow depths and may represent sinkers.

Age-dependent transverse isotropy

A wide band of anomalous bathymetry and magmatism extends across the Pacific plate from

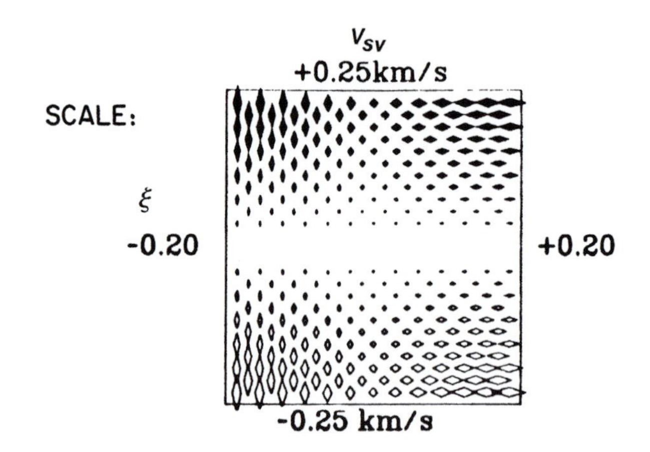

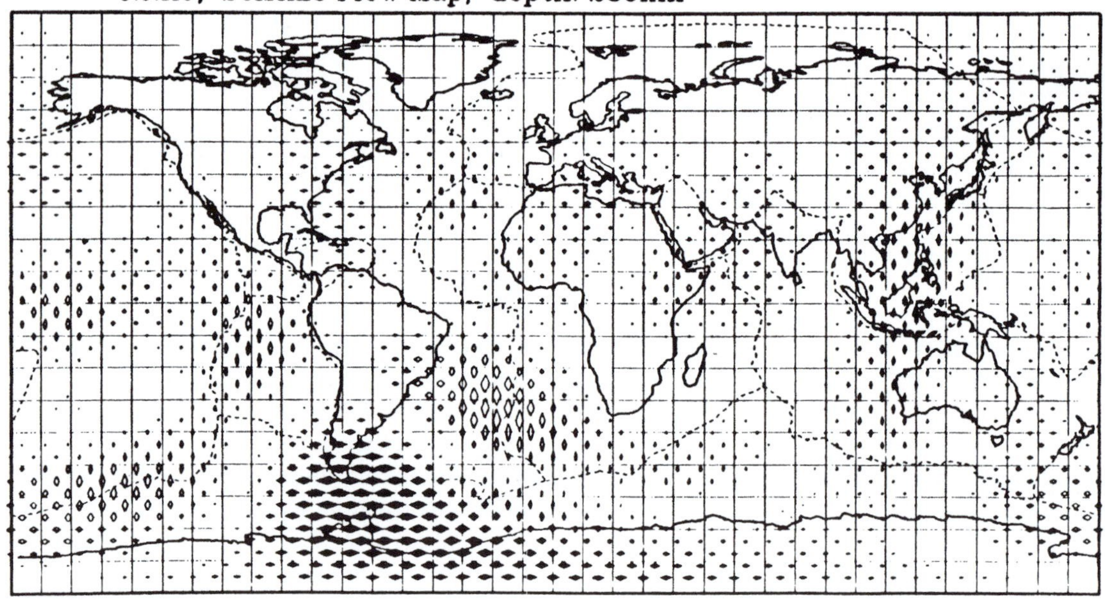

Fig. 20.6 Seismic flow map at 280 km depth. This combines information about shear velocity and polarization anisotropy. Open symbols are slow, solid symbols are fast. Vertical diamonds are SV > SH. Horizontal diamonds are SH > SV. Slow velocities are at least partially due to high temperatures and, possibly, partial melting.

the Samoa–Tonga elbow to the Juan Fernandez microplate on the East Pacific plate (Montagner, 2002). This band is also reflected in seismic velocities and anisotropy. This band continues across the Nazca plate, the Challenger fracture zone and the Juan Fernandez volcanic chain to the south end of the volcanic gap in Chile. To first order, the azimuthal anisotropy below the Pacific reflects the plate motion as a whole. It is in good agreement with models of plate motions. This anomalous band may represent a future plate plate boundary between the North and South Pacific.

If the mantle is anisotropic, the use of Rayleigh waves or P-waves alone is of limited usefulness in the determination of mantle structure because of the trade-off between anisotropy and structure. Love waves and shear waves provide additional constraints. If it is assumed that available surface-wave data are an azimuthal average, we can treat the upper mantle as a transversely isotropic solid with five elastic constants. The azimuthal variation of long-period surface waves is small.

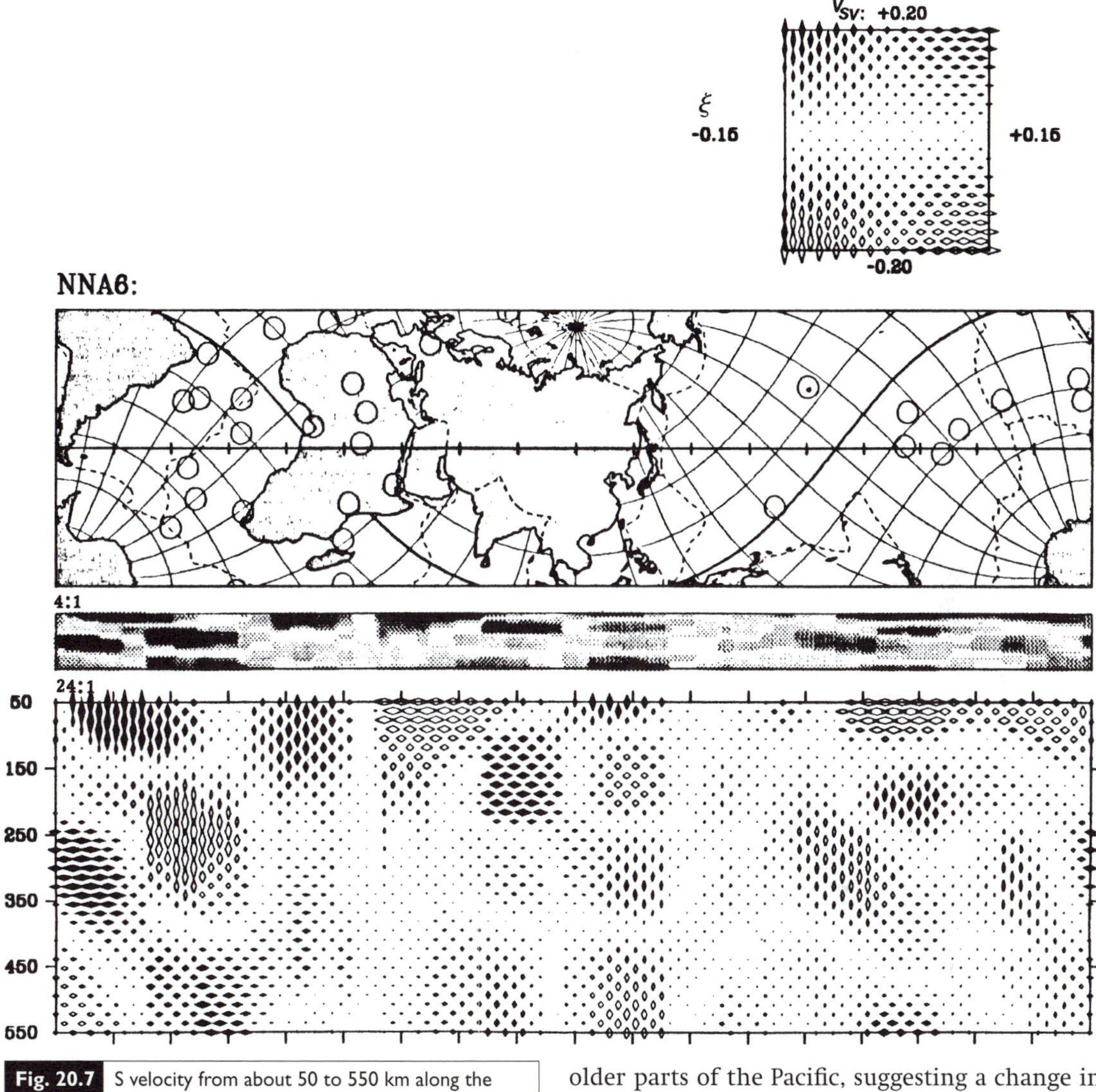

Fig. 20.7 S velocity from about 50 to 550 km along the great-circle path shown. Cross-sections are shown with two vertical exaggerations. Velocity variations are much more extreme at depths less than 250 km than at greater depths. The circles on the map represent hotspots.

Mapping mantle flow

The combined inversion of Rayleigh waves and Love waves across the Pacific has led to models that have age-dependent properties; LID thicknesses, seismic velocities and anisotropies. In general, the seismic lithosphere increases in thickness with age and $V_{SH} > V_{SV}$ for most of the Pacific. However, $V_{SH} < V_{SV}$ for the younger and older parts of the Pacific, suggesting a change in the flow regime or the mantle fabric.

The variation of velocities and anisotropy with age suggests that stress- or flow-aligned olivine may be present. The velocities depend on temperature, pressure and crystal orientation. An interpretation based on flow gives the velocity depth relations illustrated in Figure 20.9. The upper-left diagram illustrates a convection cell with material rising at the midocean range (R) and flowing down at the trench (T).

The lower left of Figure 20.9 shows the schematic temperature profile for such a cell. The seismic velocities decrease with temperature and increase with pressure. Combining the effects of temperature and pressure, one obtains a

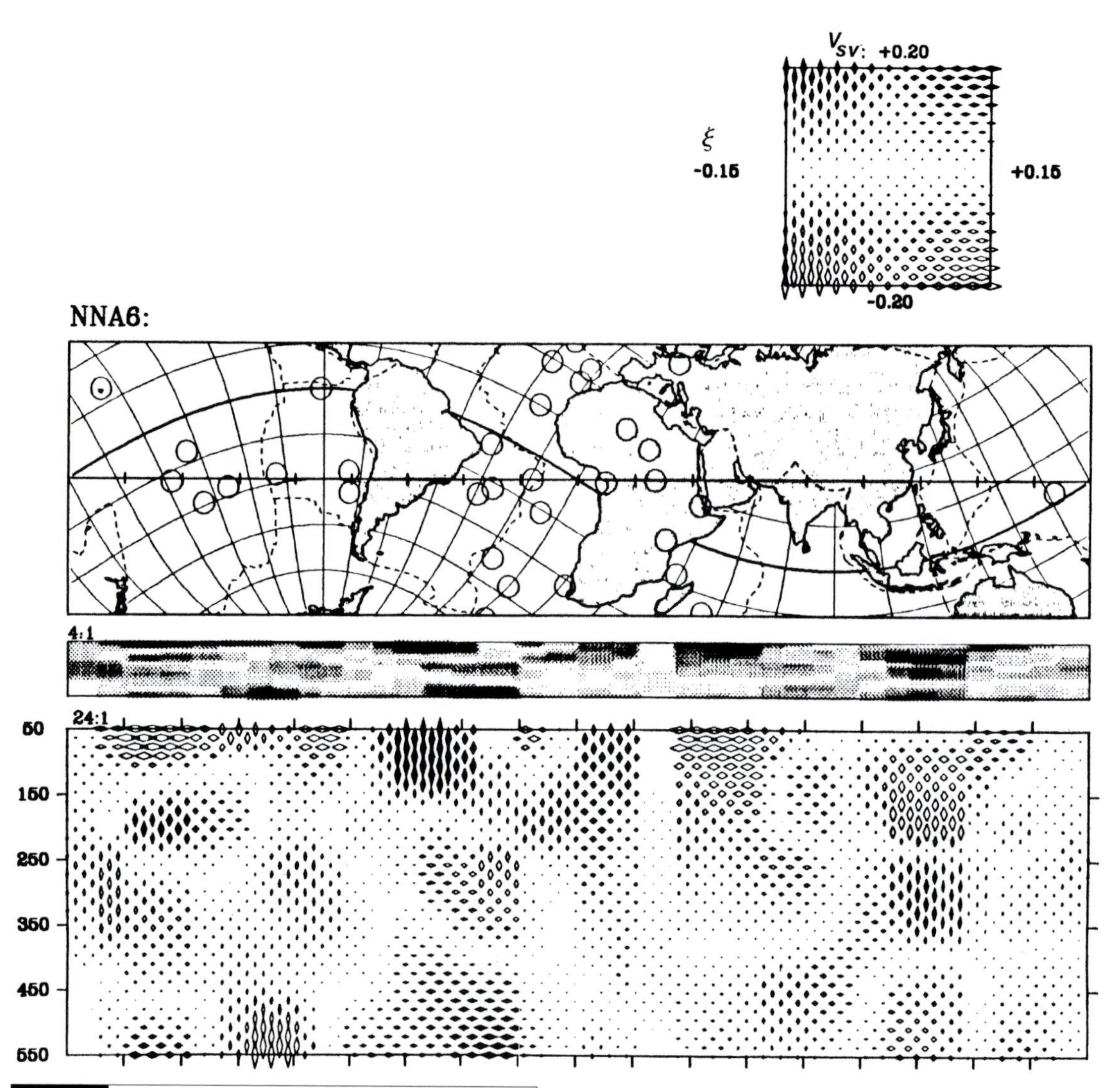

Fig. 20.8 S velocity in the upper mantle along the cross-section shown. Note low velocities at shallow depth under the western Pacific, replaced by high velocities at greater depth. The eastern Pacific is slow at all depths. The Atlantic is fast below 400 km.

relation between velocity and depth. At the ridge the temperature increases very rapidly with depth near the surface; thus, the effects of temperature dominate over those of pressure, and velocities decrease. Deeper levels under the ridge are almost isothermal; thus, the effect of pressure dominates, and the velocities increase. At the trench the temperature gradient is large near the base of the cell and nearly isothermal at shallower depths. Therefore, the velocity response is a mirror image of that at the ridge. Midway between the ridge and the trench (M), the temperature increases rapidly near the top and bottom of the cell. Thus, the velocities decrease rapidly in these regions.

The crystal orientation, if alignment with flow is assumed, is with the shortest and slowest axis (*b* axis) perpendicular to the flow. Thus, at the ridge and the trench where the flow is near vertical, the *b* axis is horizontal, and midway between where the flow is horizontal, the *b* axis is vertical. VPH and VSH between the upward and downward flowing edges of the convection cell are controlled by the velocities along the *a* axis and *c* axis. Thus, at midpoint, and wherever flow is horizontal, SH > SV and PH > PV.

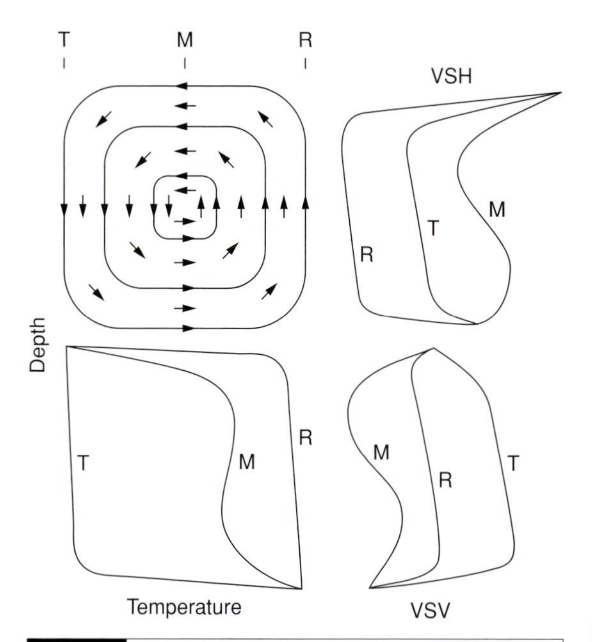

Fig. 20.9 Schematic representation of seismic velocities due to temperature, pressure, and crystal orientation assuming a flow-aligned olivine model. The upper left diagram shows a convection cell with arrows indicating flow direction. The trench is indicated by T, the ridge by R, and the midpoint by M. The lower left diagram shows temperature depth profiles for the trench, ridge and midpoint. The upper and lower right diagrams show the nature of the velocity depth structure of *VSH* and *VSV*, respectively due to pressure, temperature and crystal orientation.

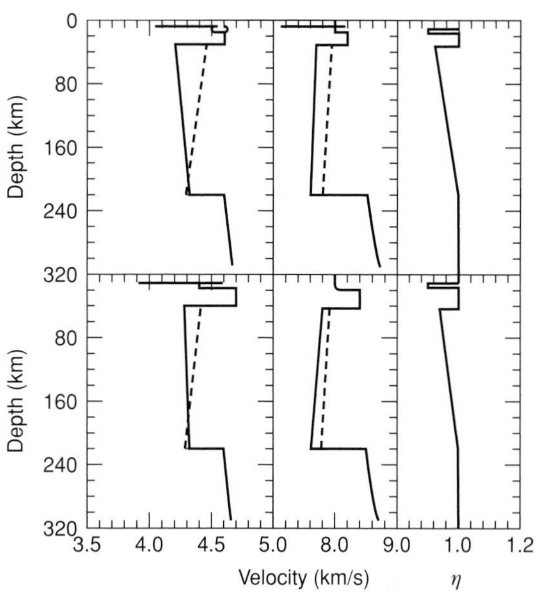

Fig. 20.10 Velocity depth profiles for the 0–20 Ma (upper set) and the 20–50 Ma (lower set) old oceanic-age provinces. From left: PV, SV, and ETA; dashes are PH and SH.

At the ridge and at the trench the flow is vertical, rapidly changing to horizontal at the top and bottom of the cell. For vertical flow the horizontal velocity is controlled by the *b*-axis and *c*-axis velocities, so SH < SV and PH < PV. The values at the top and bottom of the cell rapidly change to the horizontal flow values. Between the midpoint and the trench or ridge, the transition from horizontal to vertical flow velocities becomes sharper, and the depth extent of constant vertical velocities increases.

For the 100–200 km depth range for the youngest regions, SH > SV. The vertical flow expected in the ridge-crest environment would exhibit this behavior. The temperature gradients implied are 5–8 °C per kilometer for older ocean. The young ocean results are consistent with reorientation of olivine along with a small temperature gradient. With these temperature and flow models, the velocity of Love waves along ridges is expected to be extremely slow. The velocity of Rayleigh waves is predicted to be high along subduction zones. For midplate locations, Love-wave velocities are higher and Rayleigh-wave velocities are lower than at plate boundaries.

The average Earth model (Table 20.9) takes into account much shorter period data than used in the construction of PREM. Note that a high-velocity LID is required by this shorter wavelength information. This is the seismic lithosphere. It is highly variable in thickness, and an average Earth value has little meaning. The seismic LID is about the same thickness as the strong lithosphere, and much thinner than the thermal boundary layer and, probably, the plate.

Azimuthal anisotropy

Maps of global azimuthal and polarization anisotropy are now readily available. Anisotropy of the upper mantle may originate from preferred orientation of olivine – and other – crystals or from a larger-scale fabric perhaps related to ancient slabs in the mantle. The

Table 20.3 Upper mantle velocities for the average Earth model.

	Thickness (km)	V_{PV} (km/s)	V_{PH} (km/s)	V_{SV} (km/s)	V_{SH} (km/s)	η	Q_μ
Water	3.00	1.45	1.45	0.00	0.00	1.00	∞
Crust1	12.00	5.80	5.80	3.20	3.20	1.00	600
Crust2	3.40	6.80	6.80	3.90	3.90	1.00	600
LID	28.42	8.02	8.19	4.40	4.61	.90	600
LVZ top		7.90	8.00	4.36	4.58	.80	80
LVZ bottom		7.95	8.05	4.43	4.44	.98	80
220 km		8.56	8.56	4.64	4.64	1.00	1.43
400 km		8.91	8.91	4.77	4.77	1.00	143

Regan and Anderson (1984).

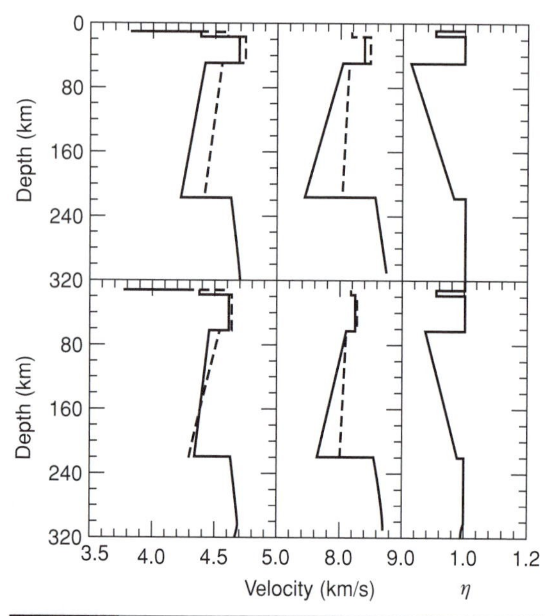

Fig. 20.11 Velocity depth profiles for the 50–100 Ma (upper set) and the >100 Ma (lower set) old oceanic regions.

a-axes of olivine-rich aggregates tend to cluster around the flow direction, the a and c axes concentrate in the flow plane, and the b axes align perpendicular to the flow plane [Nicolas & Poirier anisotropy]. For P-waves the a, b, and c axes are, respectively, the fast, slow, and intermediate velocity directions. If the flow plane is horizontal, the azimuthal P-wave velocity

There is good correlation of fast Rayleigh wave directions with the upper-mantle return flow models derived from kinematic considerations (Hager and O'Connell, 1979). This is consistent with the fast (a-axis) of olivine being aligned in the flow direction. The main differences between the kinematic return-flow models and the Rayleigh-wave azimuthal variation maps occur in the vicinity of hotspots. A large part of the return flow associated with plate tectonics appears to occur in the upper mantle, and this in turn requires a low-viscosity channel. Figure 20.12 is a map of the azimuthal results for 200-s Rayleigh waves. The lines are oriented in the maximum velocity direction, and the length of the lines is proportional to the anisotropy. The azimuthal variation is low under North America and the central Atlantic, between Borneo and Japan, and in East Antarctica. Maximum velocities are oriented northeast-southwest under Australia, the eastern Indian Ocean and northern South America and east-west under the central Indian Ocean; they vary under the Pacific Ocean from north-south in the southern central region to more northwest-southeast in the northwest part. The fast direction is generally perpendicular to plate boundaries.

Hager and O'Connell (1979) calculated flow in the upper mantle by taking into account the drag of the plates and the return flow from subduction zones to spreading centers. Flow lines for a model that includes a low-viscosity channel in the upper mantle are shown in Figure 20.12. Flow under the large fast-moving plates is roughly antiparallel to the plate motions. Thermal buoyancy is ignored in these calculations, and there is no lateral variation in viscosity. An interesting feature

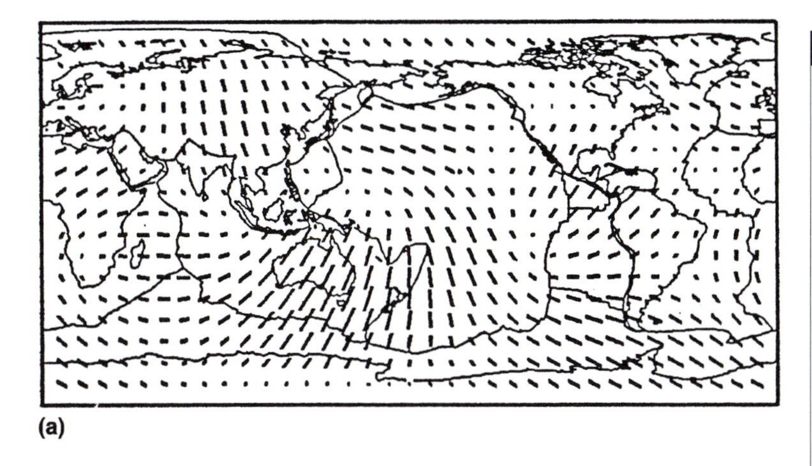

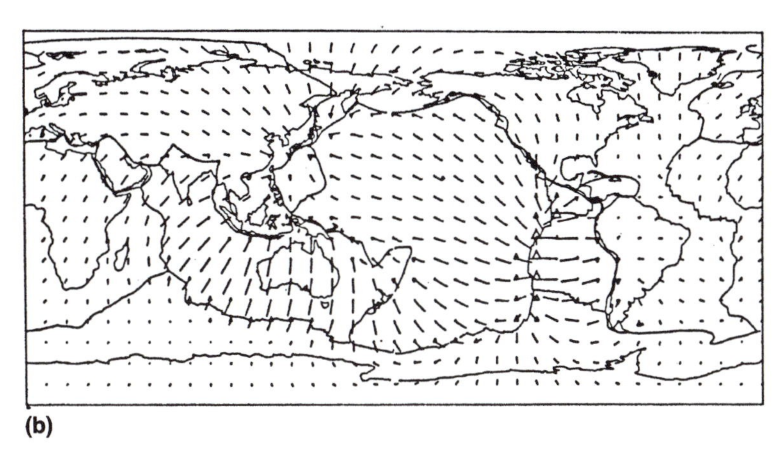

Fig. 20.12 (a) Azimuthal anisotropy of 200-s Rayleigh waves. The lines indicate the fast phase velocity direction. The length of the lines is proportional to the anisotropy (Tanimoto and Anderson, 1984). (b) Flow lines at 260 km depth for the upper-mantle kinematic flow model of Hager and O'Connell (1979). The model includes a low-viscosity channel in the upper mantle. The flow lines mimic hotspot tracks, and the relative motions of hotspots, suggesting that entrained asthenospheric heterogeneities might explain melting anomalies on plates.

of this map is that the vectors are parallel to so-called hotspot tracks; if fertile heterogeneities occur in the asthenosphere and are responsible for hotspots then these hotspots will appear to be motionless on a given plate, but will move with respect to hotspots on other plates. For example, the hotspot tracks on the Nazca plate will appear to move relative to those on the Pacific plate, and hotspots in the Atlantic, Antarctic and Africa will appear to be stationary.

In the kinematic flow model the flow is nearly due south under Australia, shifting to southwest under the eastern Indian Ocean, or directly from the subduction zones to the nearest ridge. In the anisotropic map the inferred flow is more southwestward under Australia, nearly parallel to the plate motion, shifting to east-west in the eastern Indian Ocean. The southeastern Indian Ridge is fast at depth, suggesting that this ridge segment is shallow. The Mid-Indian Ridge, the Indian Ocean triple junction, and the Tasman Sea regions are slow, suggesting deep hot anomalies in these regions. These deep anomalies are offset from those implicit in the kinematic model and apparently are affecting the direction of the return flow.

Similarly, the flow under the northern part of the Nazca plate is diverted to the southwest relative to that predicted, consistent with the velocity anomaly observed near the southern part of the Nazca-Pacific ridge. The flow lines in the mantle under the Nazca plate are parallel to the hotspot tracks. The anisotropy due north of India indicates north-south flow, perpendicular to the plate motion of Eurasia and the theoretical return-flow direction. One interpretation is that the Indian plate has subducted beneath the Tibetan plateau and extends far into the continental interior.

Note that anisotropy and predicted flow vectors are subparallel under the Pacific plate but are different for the Nazca plate and the Indo-Australian plate. If these are indeed flow vectors then fertile heterogeneities in the astheneosphere will show little relative motion under a given plate. In fact, the vectors are similar in relative motions to hotspot tracks, suggesting that hotspots may have a shallow origin. The mantle is likely to be heterogenous in its melting point and ability to produce basalt (fertility). If `fertile blobs` are embedded in the upper mantle return flow channel the above map will give their relative directions and velocities. For example, fertile blobs under the Pacific plate will trace out parallel paths and move at about the same velocity with respect to one another. The blobs under the African and Antarctic plates will be almost motionless. The blobs under the Indian plate will move north and those under the Nazca plate will move east-west. These are similar to the motions of hotspot tracks and to the relative motions of hotspots. If the return flow channel is 3 to 4 times thicker than the plates, then the velocities will be 3 to 4 times slower than plate velocities. This is an explanation for the near fixity of hotspots relative to one another.

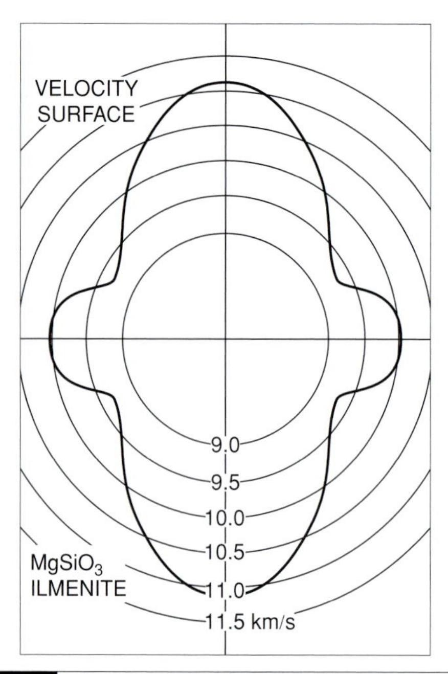

Fig. 20.13 Variation of compressional velocity with direction in the ilmenite form of $MgSiO_3$. This is a stable mineral below about 500 km in cold slabs. $MgSiO_3$-ilmenite is a platy mineral and may be oriented by stress, flow and recrystallization in the slab. Ice (in glaciers) and calcite (in marble) have similar crystal structures and are easily oriented by flow, giving anisotropic properties to ice and marble masses. The deep slab may also be anisotropic.

Shear-wave splitting and slab anisotropy

In an anisotropic solid there are two shear waves, having mutually orthogonal polarizations, and they travel with different velocities. This is known as `shear-wave splitting or birefringence`. Since shear waves are secondary arrivals and generally of long period, it requires special studies to separate the two polarizations from each other and from other later arrivals. Deep-focus events are the most suitable for this purpose; many studies have clearly demonstrated the existence of splitting.

Ando *et al.* (1983), in an early pioneering study, analyzed nearly vertically incident shear waves from intermediate and deep-focus events beneath the Japanese arc. The time delay between the two nearly horizontal polarizations of the shear waves was as much as 1 s. The polarization of the maximum-velocity shear waves changed

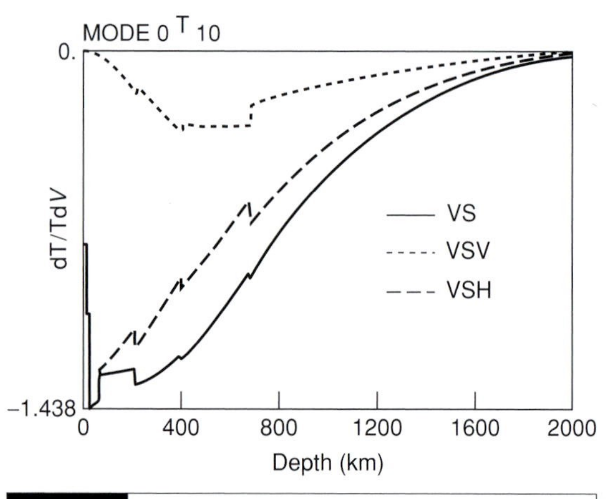

Fig. 20.14 Partial derivatives for a relative change in period of toroidal mode (Love wave) $_0T_{10}$ due to a change in shear velocity as a function of depth. The solid line gives the isotropic partial derivative. The dashed lines give the effect of perturbations in two components of the velocity. Period is 720 s (after Anderson and Dziewonski, 1982).

from roughly north-south in the northern part of the arc to roughly east-west further south. The anisotropic regions were of the order of 100 km in extent and implied a 4% difference in shear-wave velocities in the mantle wedge above the slab.

In another early paper, Fukao (1984) studied ScS splitting from a deep-focus event in the Kuriles recorded in Japan. The uniformity in polarization across the Japanese arc is remarkable. The faster ScS phase had a consistent polarization of north-northwest – south-southeast and an average time advance of 0.8 ± 0.4s over the slower ScS wave. The splitting could occur anywhere along the wave path, but the consistency of the results over the arc and the difference from the direct S results, from events beneath Japan, suggests that the splitting occurs in the vicinity of the source. The fast polarization direction is nearly parallel to the dip direction of the Kurile slab and the fast P-wave direction of the Pacific plate in the vicinity of the Kurile Trench. The stations are approximately along the strike direction of the deep Kurile slab. All of this suggests that the splitting occurs in the slab beneath the earthquake. This earthquake has been given various depths ranging from 515 to 544 km, the uncertainty possibly resulting from deep-slab anisotropy. If the slab extends to 100 km beneath the event, the observed splitting could be explained by 5% anisotropy. This event shows a strong S-wave residual pattern with the fast directions along the strike direction. The residuals vary by about 6 s. The waves showing the earliest arrival times spend more time in the slab than the nearly vertical ScS waves. They also travel in different azimuths. If the fast shear-velocity directions are in the plane of the slab, this will add to the effect caused by low temperatures in the slab. Thus, a large azimuthal effect can accumulate along a relatively short travel distance in the slab. If the slab is 5% faster due to temperature and 5% anisotropic, then rays travelling 300 km along the strike direction will arrive 6 s earlier than waves that exit the slab earlier. Actually, the anisotropy implied by the vertical ScS waves just gives the difference in shear-wave velocities in that (arbitrary) direction and is not a measure of the total azimuthal S-wave velocity variation, which can be much larger. The presence of near-source anisotropy can give results similar to those caused by a long cold isotropic slab and can cause artifacts in global tomographic models, e.g. fast blue bands in the lower mantle on strike with the deep focus earthquakes used in the inversion.

The mineral assemblage in the deeper parts of the slab are different from those responsible for the anisotropy in the plate and in the shallower parts of the slab. The orientation of high-pressure phases is possibly controlled both by the ambient stress field and the orientation of the 'seed' low-pressure phases. Results to date are consistent with the fast crystallographic axes being in the plane of the slab. The most anisotropic minerals at various depths are olivine ($<$ 400 km), modified-spinel (400–500 km) and $MgSiO_3$-ilmenite ($>$ 500 km). The last phase is not expected to be stable at the higher temperatures in normal mantle, being replaced by the more isotropic garnet-like phase majorite. Thus, the deep slab cannot be modeled as simply a colder version of normal mantle. It differs in mineralogy and therefore in intrinsic velocity and anisotropy. Ilmenite is one of the most anisotropic of mantle minerals (Figure 20.13), especially for shear waves. If it behaves in aggregate as do ice and calcite, which are similar structures, then a cold slab can be expected to be extremely anisotropic. Seismic waves which travel along the South American slab for large distances on their way to North American stations can be expected to arrive very early, especially at shield stations. Some of this effect may be mapped as fast bands in the deep mantle under North America and the Atlantic and will be interpreted in terms of `deep slab penetration`.

Chapter 21

Nonelastic and transport properties

Shall not every rock be removed out of his place?

Job 18:4

Most of the Earth is solid, and much of it is at temperatures and pressures that are difficult to achieve in the laboratory. The Earth deforms anelastically at small stresses and, over geological time, this results in large deformations. Most laboratory measurements are made at high stresses, high strain rates and low total strain. Laboratory data must therefore be extrapolated in order to be compared with geophysical data, and this requires an understanding of solid-state physics. In this chapter I discuss processes that are related to rates or time. Some of these are more dependent on temperature than those treated in previous chapters. These properties give to geology the 'arrow of time' and an irreversible nature.

Thermal conductivity

There are three mechanisms contributing to thermal conductivity in the crust and mantle. The lattice part is produced by diffusion of thermal vibrations in a crystalline lattice and is also called the phonon contribution. The radiative part is due to the transfer of heat by infrared electromagnetic waves, if the mantle is sufficiently transparent. The exciton part, is due to the transport of energy by quasiparticles composed of electrons and positive holes; this becomes dominant in intrinsic semiconductors as the temperature is raised. Thus, thermal conduction in solids arises partly from electronic and partly from atomic motion and, at high temperature, from radiation passing through the solid.

Debye theory regards a solid as a system of coupled oscillators transmitting thermoelastic waves. For an ideal lattice with simple harmonic motion of the atoms, the conductivity would be infinite. In a real lattice, anharmonic motion couples the vibrations, reducing the mean free path and the lattice conductivity. Thermal conductivity is related to higher-order terms in the potential and should be correlated with thermal expansion. Lattice conductivity can be viewed as the exchange of energy between high-frequency lattice vibrations – elastic waves. An approximate theory for the lattice conductivity, consistent with the Grüneisen approximation, gives

$$K_{\mathrm{L}} = a/3\gamma^2 T K_{\mathrm{T}}^{3/2} \rho^{1/2}$$

where a is the lattice parameter, γ is the Grüneisen parameter, T is temperature, K_{T} is the isothermal bulk modulus and ρ is density. This is valid at high temperatures, relative to the Debye temperature. This relation predicts that the thermal conductivity decreases with depth in the top part of the upper mantle.

The thermal conductivities of various rock-forming minerals are given in Table 21.1. Note that the crust-forming minerals have about one-half to one-third of the conductivity of mantle minerals. This plus the cracks present at low crustal pressures means that a much higher thermal gradient is maintained in the crust,

Table 21.1 Thermal conductivity of minerals

Mineral	Thermal Conductivity 10^{-3}cal/cm s °C
Albite	4.71
Anorthite	3.67
Microcline	5.90
Serpentine	7.05
Diopside	11.79
Forsterite	13.32
Bronzite	9.99
Jadeite	15.92
Grossularite	13.49
Olivine	6.7–13.6
Orthopyroxene	8.16–15.3

Horai (1971), Kobayzshigy (1974).

relative to the mantle, to sustain the same conducted heat flux. The gradient can be higher still in sediments.

The thermal gradient decreases with depth in the Earth. If the crustal radioactivity and mantle heat flow are constant and the effects of temperature are ignored, regions of thick crust should have relatively high upper-mantle temperatures.

Thermal conductivity is strongly anisotropic, varying by about a factor of 2 in olivine and orthopyroxene as a function of direction. The highly conducting axes are [100] for olivine and [001] for orthopyroxene. The most conductive axis for olivine is also the direction of maximum P-velocity and one of the faster S-wave directions, whereas the most conductive axis for orthopyroxene is an intermediate axis for P-velocity and a fast axis for S-waves. In mantle rocks the fast P-axis of olivine tends to line up with the intermediate P-axis of orthopyroxene. These axes, in turn, tend to line up in the flow direction, which is in the horizontal plane in ophiolite sections. The vertical conductivity in such situations is much less than the average conductivity computed for mineral aggregates. Conductivity decreases with temperature and may be only half this value at the base of the lithosphere. The implications of this anisotropy in thermal conductivity and the lower than average vertical conductivity have not been investigated. Two obvious implications are that the lithosphere can support a higher thermal gradient than generally supposed, giving higher upper-mantle temperatures, and that the thermal lithosphere grows less rapidly than previously calculated. For example, the thermal lithosphere at 80 Ma can be 100 km thick for $K = 0.01$ cal/cm s °C and only 30 km thick for 0.3 cal/cm s °C. The low lattice conductivity of the oceanic crust is usually also ignored in these calculations, but this may be counterbalanced by water circulation in the crust.

The lattice (phonon) contribution to the thermal conductivity decreases with temperature, but at high temperature radiative transfer of heat may become significant, depending on the opacities of mantle rocks, which depend on grain size and iron content.

Convection is probably the dominant mode of heat transport in the Earth's deep interior, but conduction is not irrelevant to the thermal state and history of the mantle as heat must be transported across thermal boundary layers by conduction. Thermal boundary layers exist at the surface of the Earth, at the core–mantle boundary and, possibly, at chemical interfaces internal to the mantle. Conduction is also the mechanism by which subducting slabs cool the mantle, and become heated up. The thicknesses and thermal time constants of boundary layers are controlled by the thermal conductivity, and these regulate the rate at which the mantle cools and the rate at which the thermal lithosphere grows. The importance of radiative conductivity in the deep mantle is essentially unconstrained.

The thermal conductivity goes through a minimum at about 100 km depth in the upper mantle. Higher thermal gradients are then needed to conduct the same amount of heat out, and this results in a further lowering of the lattice conductivity. The conductivity probably increases by at least a factor of 3 to 4 from 100 km depth to the core–mantle boundary (CMB).

The parameters that enter into a theory of lattice conductivity are fairly obvious; temperature, specific heat and the coefficient of thermal expansion, some measure of anharmonicity, a measure of a mean free path or a mean

Table 21.2 Estimates of lattice thermal diffusivity in the mantle

Depth (km)	κ (cm^2/s)
50	5.9×10^{-3}
150	3.0×10^{-3}
300	2.9×10^{-3}
400	4.7×10^{-3}
650	7.5×10^{-3}
1200	7.7×10^{-3}
2400	8.1×10^{-3}
2900	8.4×10^{-3}

Horai and Simmons (1970).

collision time or a measure of the strength and distribution of scatterers, velocities of sound waves and the interatomic distances.

Both thermal conductivity and thermal expansion depend on the anharmonicity of the interatomic potential and therefore on dimensionless measures of anharmonicity such as γ or $\alpha\gamma T$. The lattice or phonon conductivity is

$$K_{\mathrm{L}} = C_{\mathrm{V}} V l / 3$$

Where V is the mean sound speed, l is the mean free path, which depends on the interatomic distances and the isothermal bulk modulus, K_{T}. This gives

$$[\delta \ln K_{\mathrm{L}}/\delta \ln \rho] = [\delta \ln K_{\mathrm{T}}/\delta \ln \rho] - 2[\delta \ln \gamma/\delta \ln \rho] + \gamma - 1/3$$

For lower-mantle properties this expression is dominated by the bulk modulus term and the variation of K_{L} with density is expected to be similar to the variation of K_{T}.

The lattice conductivity decreases approximately linearly with temperature, a well-known result, but increases rapidly with density. The temperature effect dominates in the shallow mantle, but pressure dominates in the lower mantle. This has important implications regarding the properties of thermal boundary layers, the ability of the lower mantle to conduct heat from the core, and the convective mode of the lower mantle. The spin-pairing and post-perovskite transitions in the deep mantle may cause a large increase in lattice conductivity. Other pressure effects on physical properties – viscosity, thermal expansion – all go in the direction of suppressing thermal instabilities – narrow plumes – at the core-mantle boundary.

The ratio α/K_{L} decreases rapidly with depth in the mantle, thereby decreasing the Rayleigh number. Pressure also increases the viscosity, an effect that further decreases the Rayleigh number of the lower mantle. The net effect of these pressure-induced changes in physical properties is to make convection sluggish in the lower mantle, to decrease thermally induced buoyancy, to increase the likelihood of chemical stratification, and to increase the thickness of the thermal boundary layer in D″, above the core–mantle boundary.

The mechanism for transfer of thermal energy is generally well understood in terms of lattice vibrations, or high-frequency sound waves. This is not enough, however, since thermal conductivity would be infinite in an ideal harmonic crystal. We must understand, in addition, the mechanisms for scattering thermal energy and for redistributing the energy among the modes and frequencies in a crystal so that thermal equilibrium can prevail. An understanding of thermal 'resistivity,' therefore, requires an understanding of higher order effects, including anharmonicity.

Debye theory explains the thermal conductivity of dielectric or insulating solids in the following way. The lattice vibrations can be resolved into traveling waves that carry heat. Because of anharmonicities the thermal fluctuations in density lead to local fluctuations in the velocity of lattice waves, which are therefore scattered. Simple lattice theory provides estimates of specific heat and sound velocity and how they vary with temperature and volume. The theory of attenuation of lattice waves involves an understanding of how thermal equilibrium is attained and how momentum is transferred among lattice vibrations.

The thermal resistance is the result of interchange of energy between lattice waves, that is, scattering. Scattering can be caused by static imperfections and anharmonicity. Static imperfections include grain boundaries, vacancies, interstitials and dislocations and their associated strain fields, which considerably broadens the defects cross section. These 'static' mechanisms generally become less important at high temperature. Elastic strains in the crystal scatter because of the strain dependence of the elastic properties, a nonlinear or anharmonic effect.

Diffusion and viscosity

Diffusion and viscosity are activated processes and depend more strongly on temperature and pressure than the properties discussed up to now. The diffusion of atoms, the mobility of defects, the creep of the mantle and seismic wave attenuation are all controlled by the diffusivity.

$$D(P, T) = \zeta a^2 v \exp[-G^*(P, T)/RT]$$

where G^* is the Gibbs free energy of activation, ζ is a geometric factor and v is the attempt frequency (an atomic vibrational frequency). The Gibbs free energy is

$$G^* = E^* + PV^* - TS^*$$

where E^*, V^* and S^* are activation energy, volume and entropy, respectively. The diffusivity can therefore be written

$$D = D_o \exp -(E^* + PV^*)/RT$$
$$D_o = \zeta a^2 v \exp S^*/RT$$

Typical D_o values are in Table 21.4. The theory for the volume dependence of D_o is similar to that for thermal diffusivity, $\kappa = K_L/\rho C_v$. It increases with depth but the variation is small, perhaps an order of magnitude, compared to the effect of the exponential term. The product of K_L times viscosity is involved in the Rayleigh number, and the above considerations show that the temperature and pressure dependence of this product depend mainly on the exponential terms.

The activation parameters are related to the derivative of the rigidity (Keyes, 1963):

$$V^*/G^* = (l/K_T)\left[\left(\frac{\partial \ln G}{\partial \ln \rho}\right)_T - 1\right]$$

The effect of pressure on D can be written

$$-\frac{RT}{K_T}\left(\frac{\partial \ln G}{\partial \ln \rho}\right)_T = V^*$$
$$= \frac{1}{K_T}\left[\left(\frac{\partial \ln G}{\partial \ln \rho}\right)_T - 1\right]G^*$$

or

$$\left(\frac{\partial \ln D}{\partial \ln \rho}\right)_T = -\frac{G^*}{RT}\left[\left(\frac{\partial \ln G}{\partial \ln \rho}\right)_T - 1\right]$$

For a typical value of 30 for G^*/RT we have V^* decreasing from 4.3 to 2.3 cm^3/mole with depth in the lower mantle. This gives a decrease in diffusivity, and an increase in viscosity, due to compression, across the lower mantle. Phase changes, chemical changes, and temperature also affect these parameters.

Viscosity

There are also effects on viscosity that depend on composition, defects, stress and grain size. Viscosities will tend to decrease across mantle chemical discontinuities, because of the high thermal gradient, unless the activation energies are low for the dense phases. Even if the viscosities of two materials are the same at surface conditions, the viscosity contrast at the boundary depends on the integrated effects of T and P, and the activation energies and volumes.

The combination of physical parameters that enters into the Rayleigh number decreases rapidly with compression. The decrease through the mantle due to pressure, and the possibility of layered convection, may be of the order of 10^6 to 10^7. The increase due to temperature mostly offsets this; there is a delicate balance between temperature, pressure and stable stratification. All things considered, the Rayleigh number of deep mantle layers may be as low as 10^4. The local Rayleigh number in thermal boundary layers increases because of the dominance of the thermal gradient over the pressure gradient. Most convection calculations, and convective mixing calculations, use Rayleigh numbers appropriate for whole-mantle convection and no pressure effects. These can be high by many orders of magnitude. The mantle is unlikely to be vigorously convecting or well-stirred by convection.

Diffusion

Diffusion of atoms is important in a large number of geochemical and geophysical problems: metamorphism, element partitioning, creep, attenuation of seismic waves, electrical conductivity and viscosity of the mantle. Diffusion means a local non-convective flux of matter under the action of a chemical or electrochemical potential gradient.

The net flux J of atoms of one species in a solid is related to the gradient of the concentration, N, of this species

$$J = -D \text{ grad } N$$

where D is the diffusion constant or diffusivity and has the same dimensions as the thermal diffusivity. This is known as Fick's law and is analogous to the heat conduction equation.

Usually the diffusion process requires that an atom, in changing position, surmount a potential energy barrier. If the barrier is the height G^*, the atom will have sufficient energy to pass over the barrier only a fraction $\exp(-G^*/RT)$ of the time. The frequency of successes is therefore

$$\nu = \nu_o \exp(-G^*/RT)$$

where ν_o is the attempt frequency, usually taken as the atomic vibration, or Debye, frequency, which is of the order of 10^{14} Hz. The diffusivity can then be written

$$D = \zeta \nu a^2$$

where ζ is a geometric factor that depends on crystal structure or coordination and that gives the jump probability in the desired direction and a is the jump distance or interatomic spacing.

Regions of lattice imperfections in a solid are regions of increased mobility. Dislocations are therefore high-mobility paths for diffusing species. The rate of diffusion in these regions can exceed the rate of volume or lattice diffusion. In general, the activation energy for volume diffusion is higher than for other diffusion mechanisms. At high temperature, therefore, volume diffusion can be important. In and near grain boundaries and surfaces, the jump frequencies and diffusivities are also high. The activation energy for surface diffusion is related to the enthalpy of vaporization.

Table 21.3 | Diffusion in silicate minerals

Mineral	Diffusing Species	T (K)	D (m^2/s)
Forsterite	Mg	298	2×10^{-18}
	Si	298	10^{-19}–10^{-21}
	O	1273	2×10^{-20}
Zn_2SiO_4	Zn	1582	3.6×10^{-15}
Zircon	O	1553	1.4×10^{-19}
Enstatite	Mg	298	10^{-20}–10^{-21}
	O	1553	6×10^{-16}
	Si	298	6.3×10^{-22}
Diopside	Al	1513	6×10^{-16}
	Ca	1573	1.5×10^{-15}
	O	1553	2.4×10^{-16}
Albite	Ca	523	10^{-14}
	Na	868	8×10^{-17}
Orthoclase	Na	1123	5×10^{-15}
	O	~1000	10^{-20}

Freer (1981).

The effect of pressure on diffusion is given by the activation volume, V^*:

$$V^* = RT(\partial \ln D/\partial P)_T - RT\left(\frac{\partial \ln \zeta a^2 \nu}{\partial P}\right)_T$$

The second term can be estimated from lattice dynamics and pressure dependence of the lattice constant and elastic moduli. This term is generally small. V^* is usually of the order of the atomic volume of the diffusing species. The activation volume is also made up of two parts, the formational part, and the migrational part.

For a vacancy mechanism the V^* of formation is simply the atomic volume since a vacancy is formed by removing an atom. This holds if there is no relaxation of the crystal about the vacancy. Inevitably there must be some relaxation of neighboring atoms inward about a vacancy and outward about an interstitial, but these effects are small. In order to move, an atom must squeeze through the lattice, and the migrational V^* can also be expected to about an atomic volume.

Table 21.4 Diffusion parameters in silicate minerals

Mineral	Diffusing Species	D_o (m^2/s)	Q ($kJ\ mol^{-1}$)
Olivine	Mg	4.1×10^{-4}	373
	Fe	4.2×10^{-10}	162
	O	5.9×10^{-8}	378
	Si	7.0×10^{-13}	173
	Fe-Mg	6.3×10^{-7}	239
Garnet	Sm	2.6×10^{-12}	140
	Fe-Mg	6.1×10^{-4}	344
Ca_2SiO_4	Ca	2.0×10^{-6}	230
$CaSiO_3$	Ca	7	468
Albite	Na	1.2×10^{-7}	149
	O	1.1×10^{-9}	140
Orthoclase	Na	8.9×10^{-4}	220
	O	4.5×10^{-12}	107
Nepheline	Na	1.2×10^{-6}	142
Glass			
Albite	Ca	3.1×10^{-5}	193
Orthoclase	Ca	2.6×10^{-6}	179
Basalt	Ca	4.0×10^{-5}	209
	Na	5×10^{-10}	41.8

Freer (1981).

V^* is about 8–12 cm^3/mole for oxygen self-diffusion in olivine and about 2–5 cm^3/mole in the lower mantle, decreasing with depth. The effect of pressure on ionic volumes leads one to expect that V^* will decrease with depth and, therefore, that activated processes became less sensitive to pressure at high pressure. Indeed, both viscosity and seismic factor Q do not appear to increase rapidly with depth in the lower mantle. Pressure also suppresses the role of temperature in the deep mantle. Another way to look at this is to consider that solids become more incompressible with depth, and therefore density, or volume, variations are less, and many lattice properties vary with volume.

Homologous temperature

In many processes a scaled temperature is more useful than an absolute temperature. The Debye temperature and the homologous temperature are two of these scaled, or dimensionless, temperatures. The ratio E^*/T_m is nearly constant for a variety of materials, though there is some dependence on valency and crystal structure. Thus, the factor E^*/T in the exponent for activated processes can be written $\lambda T_m/T$ where λ is roughly a constant and T_m is the melting temperature. If this relation is assumed to hold at high pressure, then the effect of pressure on G^*, that is, the activation volume V^*, can be estimated from the effect of pressure on the melting point:

$$D(P, T) = D_o \exp[-\lambda T_m(P)/RT]$$

and

$$V^* = E^* \frac{dT_m}{dP} \Big/ T_m$$

which, invoking the Lindemann law, becomes

$$V^* = 2E^*(\gamma - 1/3)/K_T$$

which is similar to expressions given above. The temperature T, normalized by the 'melting temperature,' T_m is known as the `homologous temperature`. It is often assumed that activated properties depend only on T_m/T and that the effect of pressure on these properties can be estimated from $T_m(P)$. Experimentally determined diffusion parameters are given in Tables 21.3 and 21.4.

The melting point of a solid is related to the equilibrium between the solid and its melt and not to the properties of the solid alone. Various theories of melting have been proposed that involve lattice instabilities, critical vacancy concentrations or dislocation densities, or amplitudes of atomic motions. These are not true theories of melting since they ignore the properties of the melt phase, which must be in equilibrium with the solid at the melting point.

Dislocations

Dislocations are extended imperfections in the crystal lattice and occur in most natural crystals. They can result from the crystal growth process itself or by deformation of the crystal. They can be partially removed by annealing. Although dislocations occur in many complex forms, all can be obtained by the superposition of two basic types: the edge dislocation and the screw dislocation. These can be visualized by imagining a cut made along the axis of a cylinder, extending from the edge to the center and then shearing the cylinder so the material on the cut slides radially (edge dislocation) or longitudinally (screw dislocation). In the latter case the cylinder is subjected to a torque. Dislocation theory has been applied to the creep and melting of solids, and to the attenuation of seismic waves.

Melting and origin of magmas

There are several ways to generate melts in the mantle; raise the absolute temperature, lower the pressure, change the composition or raise the homologous temperature. Melting can occur *in situ* if the temperature is increasing with depth and eventually exceeds the solidus. The main source of heating in the mantle is the slow process of radioactive decay. Heating also leads to buoyancy and convection, a relatively rapid process that serves to bring heated material toward the surface where it cools. The rapid ascent of warm material leads to decompression, another mechanism for melting due to the relative slopes of the adiabat and the melting curve. Extensive *in situ* melting, without adiabatic ascent, is unlikely except in layers or blobs that are intrinsically denser than the overlying mantle. In this case melting can progress to a point where the intrinsic density contrast is overcome by the elimination of a dense phase such as garnet. Below about 100 km the effect of pressure on the melting point is much greater than the adiabatic gradient or the geothermal gradient in homogenous regions of the mantle. The melting curve of peridotite levels off at pressures greater than about 100 kb. This combined with chemical boundary layers at depth makes it possible to envisage the onset of melting at depths between about 300 and 400 km or deeper. Because of the high temperature gradient at the interface between chemically distinct layers, melting is most likely to initiate in chemical boundary layers. Whether melting is most extensive above or below the interface depends on the mineralogy and the amount of the lower-melting-point phase. The Earth is slowly cooling with time and therefore melting was more extensive in the past and probably extended, on average, to both shallower and greater depths than at present. Eclogitic portions of the mantle can also have long melting columns and low melting temperatures.

The dependence of melting point T_m on pressure is governed by the Clausius–Clapeyron equation

$$dT_m/dP = \Delta V/\Delta S = T\,\Delta V/L$$

where ΔV and ΔS are the changes in volume and entropy due to melting and L is the latent heat of melting. ΔS is always positive, but ΔV can be either positive or negative. Therefore the slope of the melting curve can be either negative or positive but is generally positive. Although this equation is thermodynamically rigorous, it does not provide us with much physical insight and is not suitable for extrapolation since the parameters all depend on pressure. Because of the high compressibility of liquids, ΔV decreases rapidly with pressure, and $\Delta V/V$ and ΔS probably approach limiting values at high pressure.

The Lindemann melting equation states that melting occurs when the thermal oscillation of atoms reaches a critical amplitude,

$$T_m = A\,m\Theta^2 V^{2/3}$$

where m is the mass of the atoms, V is the volume, Θ is the Debye temperature and A is a constant. Gilvarry (1956) rewrote this in terms of the bulk modulus and volume of the solid at the melting point,

$$T_m/T_o = (V_o/V_m)^{2(\gamma-1/3)}$$

and obtained

$$c = (6\gamma+1)/(6\gamma-2)$$

Other theories of melting assume that some critical density of dislocations or vacancies causes the crystal to melt or that a crystal becomes unstable when one of the shear moduli vanishes. All of the above theories can be criticized because they do not involve the properties of the melt or considerations of solid–melt equilibrium. They correspond rather to an absolute stability limit of a crystal, which may differ from the crystal-liquid transition.

The `Stacey Irvine melting relation` resembles Lindemann's equation, and was derived from a simple adaptation of the Mie–Grüneisen equation without involving the vibration amplitude assumption,

$$\mathrm{d}T_\mathrm{m}/T_\mathrm{m}\ \mathrm{d}P = 2(\gamma - 2\gamma^2 \alpha T_\mathrm{m})/K_\mathrm{T}$$

The `Lindemann law` itself can be written

$$\mathrm{d}T_\mathrm{m}/T_\mathrm{m}\ \mathrm{d}P = 2(\gamma - 1/3)/K_\mathrm{T}$$

which gives almost identical numerical values. These can be compared with the expression for the adiabatic gradient

$$\mathrm{d}\ln T/\mathrm{d}P = \gamma/K_\mathrm{T}$$

Since γ is generally about 1, the melting point gradient is steeper than the adiabatic gradient. For $\gamma < 2/3$ the reverse is true.

If the above relations apply to the mantle, the adiabat and the melting curve diverge with depth. This means that melting is a shallow-mantle phenomenon and that deep melting will only occur in thermal boundary layers. In thermal boundary layers the thermal gradient is controlled by the conduction gradient, which is typically 10–20 °C/km compared to the adiabatic gradient of 0.3 °C/km. Both the melting gradient and the adiabatic gradient decrease with depth in the mantle.

The Lindemann law was motivated by the observation that the product of the coefficient of thermal expansion and the melting temperature T_m was very nearly a constant for a variety of materials. This implies that

$$\mathrm{d}T_\mathrm{m}/T_\mathrm{m}/\mathrm{d}P = 2(\gamma - 1/3)/K_\mathrm{T}$$

which can be written

$$(\mathrm{d}T_\mathrm{m}/T_\mathrm{m})/\mathrm{d}P = -(\delta \ln K_\mathrm{T}/\delta \ln V)_\mathrm{p}/K_\mathrm{T}$$

Thus, the increase of melting temperature with pressure can be estimated from the thermal and elastic properties of the solid. Typical values for silicates give

$$\mathrm{d}T_\mathrm{m}/\mathrm{d}P \sim 6\ ^\circ\mathrm{C/kbar}$$

Typical observed values are 5 to 13 °C/kbar.

Chapter 22

Squeezing: phase changes and mantle mineralogy

It is my opinion that the Earth is very noble and admirable . . . and if it had contained an immense globe of crystal, wherein nothing had ever changed, I should have esteemed it a wretched lump of no benefit to the Universe.

Galileo

Overview

Before one can infer the composition of the mantle from physical properties, one must deal with the mineralogy of the mantle, and the role of partial melting and solid–solid phase changes. These issues straddle the disciplines of petrology and mineral physics. Pressure-induced phase changes and chemical variations are important in understanding the radial structure of the Earth. The advent of tomography has made it important to understand lateral changes in physical properties. Phase changes and compositional changes are probably more important than temperature changes in the interpretation of tomographic images. Tomographic cross-sections are not maps of temperature.

The densities and seismic velocities of rocks are relatively weak functions of temperature, pressure and composition unless these are accompanied by a drastic change in mineralogy. The physical properties of a rock depend on the proportions and compositions of the various phases or minerals – the mineralogy. These, in turn, depend on temperature, pressure and composition. In general, one cannot assume that the mineralogy is constant as one varies temperature and pressure. Lateral and radial variations of physical properties in the Earth are primarily due to changes in mineralogy. Changes of composition with depth in the mantle are subtle and there can be chemical discontinuities with little jump in seismic velocity. The mineralogy of the mantle changes at constant pressure, if the temperature or composition changes. Tomography maps the lateral changes in seismic velocity. These changes are due to changes in mineralogy and composition, and changes in crystal orientation, or fabric.

Spherical ions and crystal structure

It is useful to think of a crystal as a packing of different-size spheres (ions), the small spheres – usually cations – occupying interstices in a framework of larger ones – usually oxygen. In ionic crystals each ion can be treated as a ball with certain radius and charge. The arrangement of these balls, the crystal structure, follows certain simple rules. The crystal must contain ions in ratios such that the crystal is electrically neutral. Maximum stability is associated with regular arrangements that place as many cations around anions as possible, and vice versa, without putting ions with similar charge closer together than their radii allow while bringing cations and anions as close together as possible. In other words, we

Table 22.1 | Ionic radii for major mineral forming elements

Ion	Coordination Number	Ionic Radius	Ion	Coordination Number	Ionic Radius
Al^{3+}	IV	0.39	Fe^{3+}	IV	0.49(HS)*
	V	0.48		VI	0.55(LS)
	VI	0.53		VI	0.65(HS)
Ca^{2+}	VI	1.00	Mg^{2+}	IV	0.49
	VII	1.07		VI	0.72
	VIII	1.12		VIII	0.89
	IX	1.18	Fe^{2+}	IV	0.63(HS)
	X	1.28		VI	0.61(LS)
	XII	1.35		VI	0.77(HS)
Si^{4+}	IV	0.26	Ti^{4+}	V	0.53
	VI	0.40		VI	0.61
Na^{+}	VI	1.02	K^{+}	VI	1.38
	VIII	1.16		VIII	1.51
O^{2-}	II	1.35	F^{-}	II	1.29
	III	1.36		III	1.30
	IV	1.38		IV	1.31
	VI	1.40		VI	1.33
	VIII	1.42	Cl^{-}	VI	1.81

* HS, high spin; LS, low spin.

pack the balls together as closely as possible considering their size and charge. Many crystals are based on cubic close packing or hexagonal close packing of the larger ions. The stable packing and interatomic distances change with temperature, pressure and composition. Most physical properties are strong functions of interatomic distances.

Ionic crystal structures, such as oxides and silicates, consist of relatively large ions, usually the oxygens, in a closest-pack arrangement with the smaller ions filling some of the interstices. The large ions arrange themselves so that the cations do not 'rattle' in the interstices. The 'non-rattle' requirement of tangency between ions is another way of saying that ions pack so as to minimize the potential energy of the crystal. The so-called *large-ion lithophile (LIL)* or *incompatible elements* are not essential parts of the crystal structure but are guest phases that are excluded to varying degrees upon partial melting.

In high-pressure language, mineral names such as *spinel, ilmenite, rocksalt* and *perovskite* refer to structural analogs in silicates rather than to the minerals themselves. This has become conventional in high-pressure petrology and mineral physics, but it can be confusing to those trained in conventional mineralogy with no exposure to the high-pressure world. To complicate matters further, high-pressure silicate phases have been given names; `majorite, ringwoodite, wadsleyite, akimotoite` and so on.

Interatomic distances in dense silicates

The elastic properties of minerals depend on interatomic forces and hence on bond type, bond length and packing. As minerals undergo phase changes, the ions are rearranged, increasing the length of some bonds and decreasing others. For a given coordination the cation–anion distances are relatively constant. This is the basis for ionic radius estimates. Cation–anion distances increase with coordination, as required by packing considerations. The increases of density and bulk modulus with pressure are controlled by the increase in packing efficiency of the oxygen ions. Table 22.1 gives the ionic radii for the most common mineral-forming ions.

Minerals and phases of the mantle

As far as physical properties and major elements are concerned, the most important upper-mantle minerals are olivine, orthopyroxene, clinopyroxene and aluminum-rich phases such as plagioclase, spinel and garnet. Olivine and orthopyroxene are the most refractory phases and tend to occur together, with only minor amounts of other phases, in peridotites. Clinopyroxene and garnet are the most fusible components and also tend to occur together as major phases in rocks such as *eclogites* and *garnet clinopyroxenites*.

All of the above minerals are unstable at high pressure and therefore only occur in the upper part of the mantle. Olivine and orthopyroxene start to collapse at depths near 400 km. Clinopyroxene – diopside plus jadeite – may be stable to depths as great as 500 km. Garnet is stable to much greater depths. Olivine transforms successively to β-spinel, a distorted spinel-like structure, near 400 km and to γ-spinel, a true cubic spinel, near 500 km. At high pressure it disproportionates to $(Mg,Fe)SiO_3$ in the perovskite structure plus (Mg,Fe)O, magnesiowustite, which has the rocksalt structure. FeO is strongly partitioned into the (Mg,Fe)O phase. One useful rule of thumb is that adding Fe generally decreases transition pressures, and adding Al generally increases the stability of an assemblage.

A few examples of high-pressure phases will be discussed. More detail is found in the first edition of Theory of the Earth; (http://caltechbook.library.caltech.edu/14/17/TOE16.pdf) and websites, journal articles and books on `mantle mineralogy and high-pressure phase diagrams`. See also http://www.uni-wuerzburg.de/mineralogie/links/teach/diagramteach.html, http://www.agu.org/reference/minphys.html

$(Mg, Fe)_2SiO_4$—olivine

Olivine is the name for a series between two end-members, fayalite and forsterite. Fayalite is the iron-rich member with a formula of Fe_2SiO_4. Forsterite is the magnesium-rich member. The two minerals form a series where the iron and magnesium are substituted for each other without much effect on the crystal structure.

The olivine structure is based on a nearly hexagonal closest-packing of oxygen ions:

$$^{VI}Mg_2{}^{IV}SiO_4$$

The four (IV) oxygen atoms surrounding each silicon atom are not linked to any other silicon atom. Mg is surrounded by 6 (VI) oxygen atoms. Olivine is a very anisotropic mineral and is easily aligned or recrystallized by stress and flow, making peridotites anisotropic as well. Peridotites are named after peridot, another name for olivine – which is named after olives because of its distinctive color. No one knows the origin of the word *peridot*. Olivine crystals may contain CO_2 and helium-rich fluid inclusions but are hosts for few other of the incompatible elements that form the basis for trace element and isotope models of the mantle.

$(Mg, Fe)_2SiO_4$—spinel

Natural olivines (ol) and orthopyroxenes (opx) transform to higher pressure phases at about the right temperature and pressure to explain the 410-km discontinuity (Figure 22.1). Fayalite transforms directly to the *spinel structure*; olivines having high MgO contents occur in two modifications related to the spinel structure, β-spinel and γ-spinel. Both of these have much higher elastic-wave velocities than appropriate for the mantle just below 400 km, so there must be other components in the mantle that dilute the effect of the $\alpha-\beta$ phase change and phase changes in opx. Although *normal*, or *real*, spinel is a structural analog to silicate *spinel*, it should be noted that Mg occurs in 4-coordination in $MgAl_2O_4$ and in 6-coordination in β-and γ-*spinel* with a consequent change in the Mg–O distance and the elastic properties. ^{IV}Mg is an unusual coordination for Mg, and the elastic properties and their derivatives cannot be assumed to be similar for ^{IV}MgO and ^{VI}MgO compounds. In particular, normal spinel has an unusually low pressure derivative of the rigidity, a property shared by other 4-coordinated compounds but not ^{VI}MgO-bearing compounds.

The spinel structure consists of an approximate cubic close packing of oxygen anions. Spinel

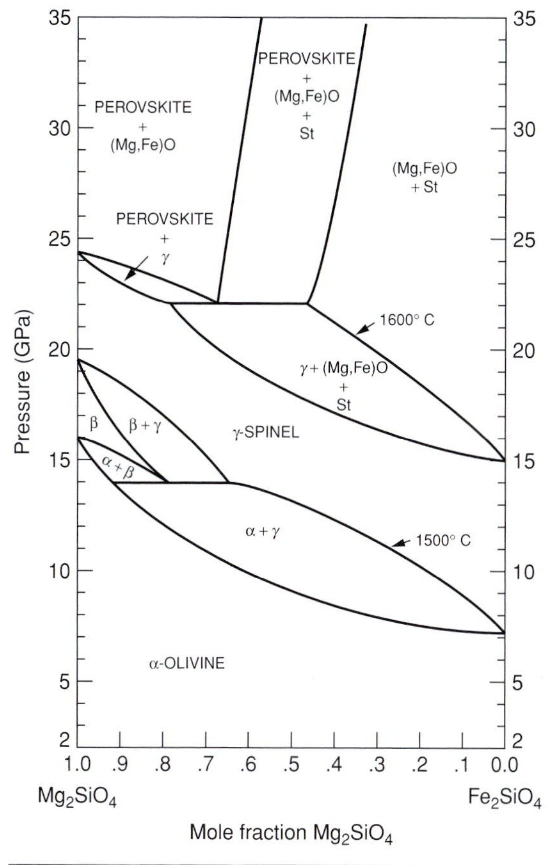

Fig. 22.1 Phase relations in the (Mg, Fe)²SiO⁴ system.

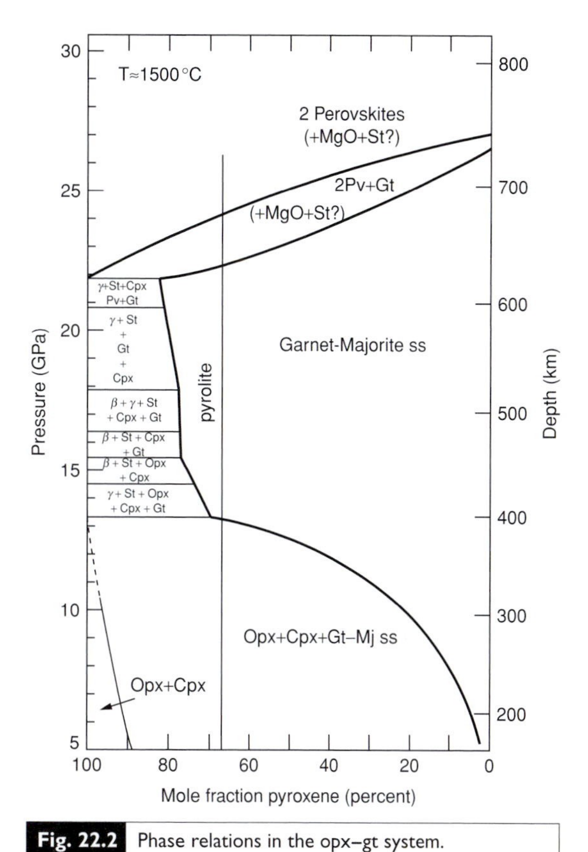

Fig. 22.2 Phase relations in the opx–gt system.

is crystallographically orthorhombic, but the oxygen atoms are in approximate cubic close packing. It is sometimes referred to as a *distorted* or *modified spinel structure*. It is approximately 7% denser than olivine. *β-spinel* is an elastically anisotropic mineral. The transformation to *γ-spinel* results in a density increase of about 3% with no overall change in coordination.

Orthopyroxene, opx

Enstatite is the magnesium end-member of the orthopyroxene series. Hypersthene is the intermediate member with about 50% iron and ferrosilite is the iron-rich end member of the series. These minerals have low seismic velocities. Orthopyroxene, (Mg,Fe)SiO_3 transforms to a distorted garnet-like phase, *majorite*, with an increase in coordination of some of the magnesium and silicon:

$$^{VIII}(Mg,Fe)_3{}^{VI}Mg^{VI}Si^{VI}Si_3O_{12}$$

where the Roman numerals signify the coordination. This can be viewed as a garnet with MgSi replacing the Al_2. This is a high-temperature transformation. Phase changes in olivine and orthoproxene both contribute to the 410-km discontinuity (Figures 22.1 and 22.2) but the seismic velocity jump at this depth is much less than predicted if these are the only two minerals in the mantle. Other mantle minerals, garnet (gt) and clinopyroxene (cpx), do not transform at 410 km (Figure 22.3). In principle, this can be used to constrain the gt + cpx, i.e. eclogitic or fertile, fraction of the mantle.

$MgSiO_3$–Majorite

Pyroxene enters the garnet structure at high pressure via the substitution

$$^{VI}[MgSi] \rightarrow Al_2$$

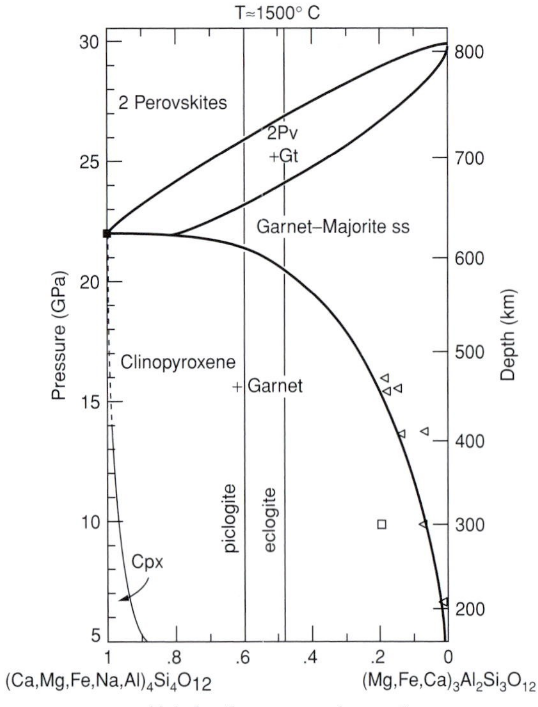

Fig. 22.3 Phase relations in the cpx–gt system.

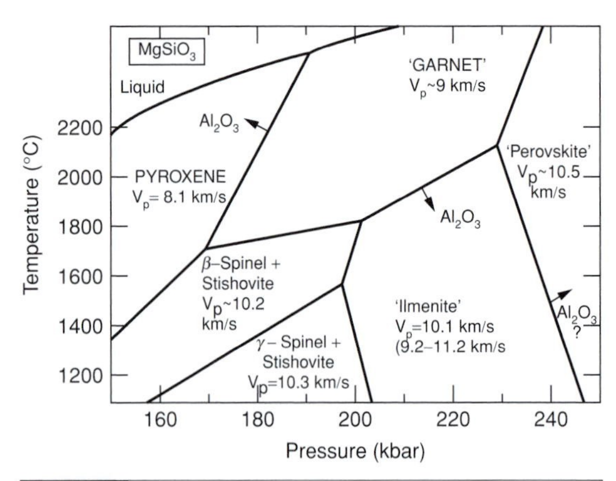

Fig. 22.4 Phase relations in $MgSiO_3$. The arrows show the direction that the phase boundaries are expected to move when corundum or garnet, is added. The approximate compressional velocities are shown for each phase. [akimotoite]

Then mineral

$$^{VIII}Mg_3{}^{VI}[MgSi]^{IV}Si_3O_{12}$$

is known as majorite, and it exhibits a wide range of solubility in garnet. Note that one-fourth of the Si atoms are in 6-coordination. However, the elastic properties of MgO plus SiO_2–stishovite are similar to Al_2O_3, so we expect the elastic properties of majorite to be similar to garnet. Majorite also has a density similar to that of garnet.

$MgSiO_3$–ilmenite

The structural formula of 'ilmenite,' the hexagonal high-pressure form of enstatite, is

$$^{VI}Mg^{VI}SiO_3$$

At low temperature the following transformations occur with pressure:

$$2MgSiO_3(opx) \rightarrow Mg_2SiO_4(\beta - sp) + SiO_2(st)$$
$$\rightarrow Mg_2SiO_4(\gamma - sp) + SiO_2(st) \rightarrow 2MgSiO_3$$
$$(ilmenite)$$
$$\rightarrow 2MgSiO_3(perovskite)$$

Mg–ilmenite is isostructural with true ilmenite and corundum. Silicon is 6-coordinated, as it is in stishovite and perovskite, dense high-elasticity phases. Because of similarity in ionic radii, we expect that extensive substitution of Al for MgSi is possible. Mg–ilmenite is a platy mineral, suggesting that it may be easily oriented in the mantle. It is also a very anisotropic mineral, rivaling olivine in its elastic anisotropy. The arrangement of oxygen atoms is based on a distorted hexagonal closest packing having a wide range of O–O distances.

The *ilmenite* form of $MgSiO_3$, also known as akimotoite, is stable at pressures between *c.* 18 and 25 GPa and temperatures from <1100 to 1900 K in the pure $MgSiO_3$ system. The presence of iron lowers the transition pressure from akimotoite to perovskite. Aluminum plays an important role in the akimotoite stability field in that pyrope makes complete solid solution with the $MgSiO_3$ end-member majorite, enlarging the garnet field at the expense of that of akimotoite both in pressure and temperature. Depending on temperature and Al content, akimotoite may be present in certain areas of the transition zone but absent in others (Figure 22.4) causing significant lateral variations in seismic velocities and density. This complicates the interpretation of seismic observations, particularly tomographic cross-sections;

seismically observed lateral variations will not merely reflect temperature derivatives even in a chemically homogenous mantle.

Akimotoite may dominate in cold parts of the transition zone, i.e. in areas near subduction zones. It has higher velocities than garnet at similar depths in hotter regions of the transition zone, producing a local velocity high that may be misinterpreted as pile-up of subducted materials.

Garnet, gt

Garnets are cubic minerals of various compositions and can incorporate Mg, Fe^{2+} or Ca for the common garnets or almost any 2+ element. Garnet exists in two compositional groups, which form intragroup but not intergroup solid solutions. The calcic group contains *uvarovite, grossular* and *andradite*. The non-calcic group contains *pyrope, almandine* and *spessartite*.

Some natural garnets have Cr^{3+} or Fe^{3+} instead of Al^{3+}. Garnets are stable over an enormous pressure range, reflecting their close packing and stable cubic structure. They are probably present over most of the upper mantle and, perhaps, below 650 km, at least when they are colder than ambient mantle. Furthermore, they dissolve pyroxene at high pressure, so their volume fraction in the mantle expands with pressure. Garnets are the densest common upper-mantle mineral, and therefore eclogites and fertile (undepleted) peridotites are denser than basalt-depleted peridotites or harzburgites. On the other hand, they are less dense than other phases that are stable at the base of the transition region. Therefore, eclogite can become less dense than the dominant mantle lithology at great depth. Most eclogites are less dense then γ-spinel and some majorites, so a perched eclogite-rich layer may form near 500–600 km depth, in the middle of the transition regions. Silica-poor eclogites may be trapped by the 400 km discontinuity. Garnet has a low melting point and is eliminated from peridotites in the upper mantle at small degrees of partial melting. The large density change associated with partial melting of a garnet-bearing rock is probably one of the most important sources of buoyancy in the mantle. Cold eclogite – subducted oceanic crust and delaminated continental crust – are stable at depth in the mantle, until they warm up to ambient mantle temperatures. Garnets and eclogites are nearly elastically isotropic.

Garnets and *perovskites* are both very accommodating of major elements and incompatible elements – they have been called *junk-box minerals*. They are expected to give diagnostic signatures to any melts that they interact with.

Clinopyroxene, cpx

Clinopyroxene consists of diopside ($CaMgSi_2O_6$), hedenbergite ($CaFeSi_2O_6$), and jadeite ($NaAlSi_2O_6$); and the orthopyroxenes, enstatite ($Mg_2Si_2O_6$) and ferrosillite ($Fe_2Si_2O_6$).

Diopside is a pyroxene mineral that forms a solid solution series with hedenbergite and augite $(Ca,Na)(Mg,Fe^{2+},Al,Ti)(Si,Al)_2O_6$. The ionic radius of calcium is much greater than aluminum, and this is expected to make the transition pressure to the garnet structure much higher than for orthopyroxene. The pyroxene garnets form solid solutions with ordinary aluminous garnets, the transition pressure decreasing with Al content.

The mineralogy in the transition region, at normal mantle temperatures, is expected to be the olivine–spinels plus garnet solid solutions. At colder temperatures, as in slabs and delaminated continental crust, the mineralogy at the base of the transition region includes *ilmenite solid-solution* (Figure 22.4). The cold parts and the warm parts of the mantle do not necessarily have the same mineralogy, even at the same depth. This can be more important that density differences associated with thermal expansion, the main buoyancy effect considered in fluid dynamic simulations of mantle convection. This can also be important in interpreting tomographic cross sections. The garnet component of the mantle is stable to very high pressure, becoming, however, less aluminous and more siliceous as it dissolves the pyroxenes. At low temperature and at pressures equivalent to those in the lower part of the transition region, the garnet as well as the pyroxenes are probably in ilmenite solid solutions.

The *ilmenite structure* of orthopyroxene can be regarded as a substitution of MgSi for 2Al in the corundum structure. The transformation of

$CaMgSi_2O_6$ clinopyroxene to ilmenite, if it occurs, is probably a higher-pressure transition. It may transform to the perovskite structure without an intervening field of ilmenite:

$$^{VIII}(Ca_{0.5})_3\,^{VI}(Ca_{0.5}Mg_{0.5})^{VI}Si^{IV}Si_3O_{12}$$
$$\text{'majorite'} \rightarrow\, ^{VIII-XII}(Ca_{0.5}Mg_{0.5})_4\,^{VI}Si_4O_{12}$$
$$\text{(perovskite)}$$

The ionic radii (in angstroms) of some of the ions involved in the above reactions are:

^{VI}Al, 0.53	^{VIII}Mg, 0.89	^{VI}Ca, 1.00
^{XII}Ca, 1.35	^{VI}Si, 0.40	^{VI}[CaSi], 0.70
^{VI}Mg, 0.72	^{XII}Mg, 1.07	^{VIII}Ca, 1.12
^{IV}Si, 0.26	^{IV}[MgSi], 0.56	

This table provides a guide as to whether substitutions are possible.

$MgSiO_3$–perovskite

The structural formula of the high-pressure phase of enstatite, or 'perovskite,' is

$$^{VIII-XII}Mg^{VI}SiO_3$$

Abbreviated Mg-pv, it appears to be stable throughout most of the lower mantle and is therefore the most abundant mineral in the mantle. Mg–perovskite is about 3% denser than the isochemical mixture stishovite plus periclase.

There are a variety of Mg–O and Si–O distances in Mg–pv. The mean Si–O distance is similar to that of stishovite. The structure is orthorhombic and represents a distortion from ideal cubic perovskite. In the ideal cubic perovskite the smaller cation has a coordination number of 6, whereas the larger cation is surrounded by 12 oxygens. In Mg–pv the 12 Mg–O distances are divided into four short distances, four fairly long distances and four intermediate distances, giving an average distance appropriate for a mean coordination number of eight.

A slightly denser, `post-perovskite phase` (ppv), has recently been discovered. The importance of this new phase lies in its appetite for iron. Anything that strongly affects the Fe-partitioning will affect the role of radiative conductivity and compositional layering in the deep mantle.

Olivine compositions also transform to perovskite-bearing assemblages Mg–pv + mw. The seismic velocity of mw is low and there cannot be too much of it in the lower mantle if seismic data is to be satisfied. This is one of the arguments for a chemically stratified mantle.

(Mg,Fe)O, magnesiowüstite

The post-*spinel* phases of mantle minerals are mixtures of magnesiowüstite and perovskite. Magnesiowüstite [(Mg,Fe)O], a cubic mineral, is the second most abundant mineral of Earth's lower mantle. Mg-rich magnesiowüstite may be stable in the rock-salt structure throughout the lower mantle. Iron-rich magnesiowüstites may decompose into two components, Fe-rich and Mg-rich magnesiowüstites, particularly if the low-spin transition in FeO takes place. Magnesiowüstite in the lowermost mantle may remove FeO from the outer core.

In the lower mantle Fe^{2+} favors (Mg,Fe)O over *perovskite*. The post-*perovskite* phase in the deepest mantle is also probably very FeO-rich. When the Fe^{2+} high-spin–low-spin transition occurs, somewhere deep in the lower mantle, solid solution between Fe^{2+} and Mg^{2+} is probably no longer possible because of the disparity in ionic radii, and a separate FeO-bearing phase is likely. At high pressure this phase may dissolve extensively in any molten iron that traverses the region on the way to the core, or to be stripped out of any mantle that comes into contact with the core in the course of mantle convection. An FeO-poor lower mantle is therefore a distinct possibility. The corollary is an iron–FeO core. If FeO is stripped out of the lower mantle, or if the FeO exists in dispersed phases or layers, the radiative conductivity and viscosity of the deep mantle may be quite different than generally assumed. The seismic properties of the lower mantle are broadly consistent with (Mg,Fe)SiO_3-*perovskite*, although other phases are certainly present, such as (Mg,Fe)O. If the mantle was efficiently differentiated during accretion, with upward removal of most melts, then the deep mantle may be deficient in Ca and Al, as well as the LIL and heat producing elements.

Low-spin Fe^{2+}

Fe and Mg have similar ionic radii at low-pressure and substitute readily for each other in upper mantle minerals. Fe is more-or-less uniformly partitioned among the major minerals. This and low temperatures suppress the role of radiative transport of heat.

Two electronic configurations, high-spin and low-spin, are possible for Fe^{2+} in the lower mantle. The high-spin (HS) state is usually stable in silicates and oxides at normal pressures. The ionic radius of the low-spin (LS) state is much smaller than the high-spin state, and a spin-pairing transition is induced by increased pressure. A large increase in density accompanies this phase transformation. For example, the volume change accompanying a phase change in FeO due to the high-spin–low-spin transition, is expected to be 11–15%. This far exceeds other phase changes in the deep mantle. Partial transformation is also possible so smaller volume changes may also occur.

The small ionic radius of Fe^{2+}(LS) probably means that Fe^{2+} will not readily substitute for Mg^{2+} under lower-mantle conditions. Additional Fe^{2+}(LS)O-bearing phases will form with high densities and bulk modulus and this means that the lower mantle could be enriched in FeO and SiO_2 relative to the upper mantle. The magnesium-rich phases of the lower mantle may be relatively iron free:

$$MgFeSiO_4 \rightarrow MgSiO_3(\text{perovskite}) + FeO(LS)$$

which would facilitate the entry of FeO into molten iron and removal to the core.

The major minerals in the deep mantle may be almost Fe-free perovskite, Mg–pv and Fe-rich magnesiowüstite [(Mg,Fe)O] and post-*perovskite* (ppv) phases. This has several important geodynamic implications. *Perovskite*, being the major phase, will control the conductivity and viscosity. Radiative conductivity and viscosity may be high in Fe-poor minerals. Both effects will tend to stabilize the mantle against convection and decrease the Rayleigh number. Over time, the dense FeO-rich phases may accumulate, irreversibly, at the base of the mantle, and, in addition, may interact with the core. The lattice conductivity of this iron-rich layer will be high and the radiative term should be low but the trade-offs are unknown. A thin layer convects sluggishly (because of the h^3 term in the Rayleigh number) but its presence slows down the cooling of the mantle and the core. The overlying FeO-poor layer may have high radiative conductivity, because of high T and transparency, and have high viscosity and low thermal expansivity, because of P effects on volume. This part of the mantle will also convect sluggishly. If it represents about one-third of the mantle (by depth) it will have a Rayleigh number about 30 times less than Rayleigh numbers based on whole mantle convection and orders of magnitude less than Ra based on $P = 0$ properties.

The LS iron in the deep mantle would behave like a different element than the HS iron at shallow depths. There may be phase separation between Fe-rich and Mg-rich phases. The melting temperature of the Fe-rich end member may be higher than the Mg-rich end members. The HS Fe^{2+} ions in the lower mantle may hinder blackbody radiation in the near-infrared, allowing more efficient radiative heat transfer.

Phase equilibria in mantle systems at high-pressure

The lateral and radial variations of seismic velocity and density in the mantle depend, to first order, on the stable mineral assemblages and, to second order, on the variation of the velocities with temperature, pressure and composition. Temperature, pressure and composition dictate the compositions and proportions of the various phases. In order to interpret observed seismic velocity profiles, or to predict the velocities for starting composition, one must know both the expected equilibrium assemblage and the properties of the phases. Olivine, orthopyroxene, clinopyroxene and an aluminous phase (feldspar, spinel, garnet) are stable in the shallow mantle. β-spinel, majorite, garnet and clinopyroxene are stable in the vicinity of 400 km, near the top of the transition region. γ-spinel, majorite or γ-spinel plus stishovite, *Ca-perovskite*, *garnet* and *ilmenite* are stable between about 500 and 650 km. *Garnet, ilmenite, Mg–perovskite, Ca–perovskite* and magnesiowustite are stable near the top of the upper mantle, and *perovskites* and magnesiowüstite are stable throughout most of the lower mantle. The details of the stable assemblages depend on composition and temperature. It is usually assumed that the radial stucture of the mantle, and mantle discontinuities, are due to equilibrium solid–solid phase changes, and that lateral changes in seismic velocity – tomography – are due to changes in temperature. This limited view has produced some strange models of mantle dynamics and chemistry.

Lateral variations in velocity due to temperature-induced phase changes can be as important as pressure-induced phase changes are in the radial direction. Phase equilibria is usually discussed in terms of the olivine system, the pyroxene system and the pyroxene–garnet system. However, at high pressures, these systems can interact with each other. Pyroxenes can tolerate a certain amount of Al_2O_3 and garnets, at high pressure, dissolve pyroxene, so pyroxenes and garnets must always be treated together. Olivine and orthopyroxene also interact at high-temperature, exchanging iron with one another.

Phase equilibria in mantle systems are summarized below in a series of figures based on available experiments and calculations. Although these are useful and informative to specialists, the main point for non-specialists – seismologists, geodynamicists, students – is that one cannot wander very far in P–T-composition-depth space without encountering large changes in density and other physical properties. Interpretations of tomographic models, and design of convection simulations must take phase changes into account in order to be realistic. On the other hand, radial or 1D Earth models may also be affected by chemical layering as well as by phase equilibria.

The main mantle systems are shown in Figures 22.1 through 22.6. More complete phase diagrams for mantle minerals are readily available:
`http://www.mpi.stonybrook.edu/ResearchResults/PhaseRelations Gasparik/figures.htm`
`http://www.springeronline.com/sgw/cda/frontpage/0,11855,1-10010-22-2155856-0,00.html`
`http://www.mantleplumes.org/Transition-Zone.html`

Pyroxene system at high pressure

Enstatite (en) and diopside (di) do not form a complete solid-solution series, but en dissolves a certain amount of di, and di contains an appreciable amount of en at moderate temperature and pressure. The amount of mutual solubility increases with temperature and decreases with pressure, and this provides a method for estimating the temperature of equilibration of mantle-derived xenoliths. Pyroxenes also react with garnet. In principle, the measurement of the compositions of coexisting pyroxenes and garnets provides information about pressures and temperatures in the mantle. At higher pressure garnet dissolves the enstatite, and this requires a change in coordination of one-fourth of the Mg and Si. Depths in excess of about 300 km are required for this change in coordination.

Natural clinopyroxenes, particularly in eclogites, are solid solutions between diopside and jadeite called omphacite. At modest pressure, equivalent to about 50–60 km depth, the solid solution series is complete (Figure 22.3). Natural clinopyroxenes from kimberlite eclogites contain up to 8 wt.% Na_2O. Clinopyroxenes from peridotites typically have much less Na and jadeite. Most garnets contain very little sodium; however, at high pressure Na_2O can enter the garnet lattice. Natural garnets associated with diamonds in kimberlite pipes contain up to 0.26% Na_2O. In the transition region, the sodium is probably contained in a complex garnet solid solution.

The other high-pressure forms of pyroxene include *ilmenite*, *spinel* plus stishovite, and *perovskite*, depending on pressure, temperature and content of calcium, aluminum and iron. The pressures at which clinopyroxene and orthopyroxene disappear are strong functions of the other variables.

The phase behavior of garnet + clinopyroxene + orthopyroxene, the peridotite assemblage, is substantially different from the behavior of garnet + clinopyroxene, the basalt–eclogite assemblage. When only clinopyroxene + garnet are present, the clinopyroxene dissolves in the

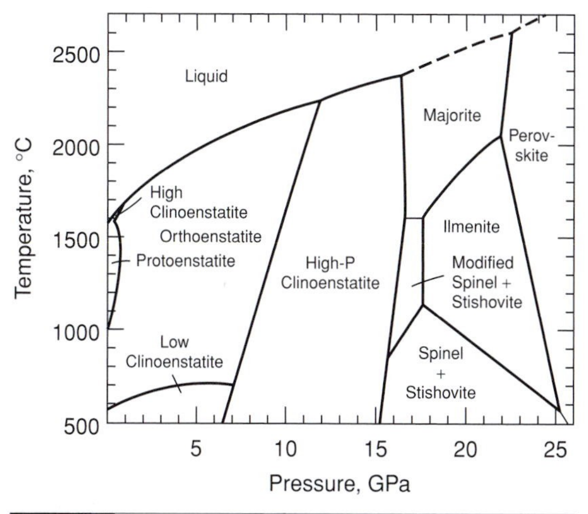

Fig. 22.5 Phase relations in $MgSiO_3$ synthesized from a large number of studies (Presnall, 1995).

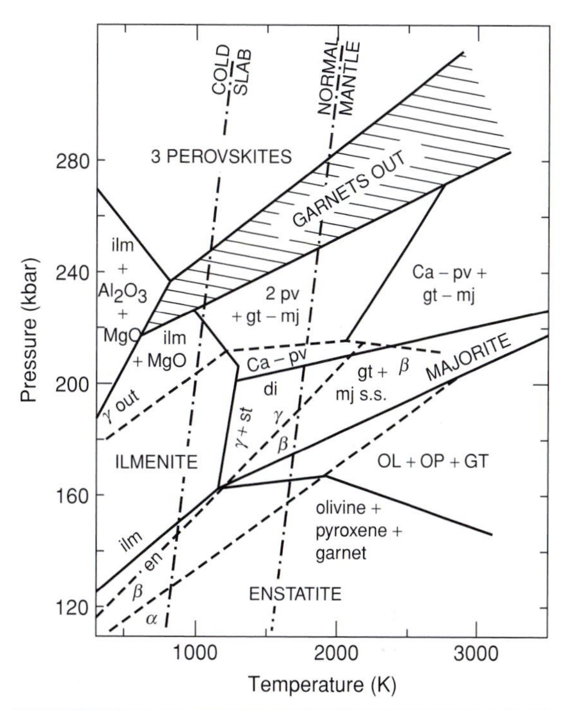

Fig. 22.6 Equilibrium phase boundaries for mantle minerals. Dashed lines are olivine system boundaries; solid lines are pyroxene–garnet system boundaries. Approximate adiabats are shown for 'normal' mantle and 'cold slab' mantle. Note the different phase assemblages, at constant pressure, for the two adiabats.

garnet with increasing pressure and eventually a homogenous garnet solid solution is formed. When orthopyroxene is also present, the garnet and clinopyroxene compositions move toward orthopyroxene with increasing pressure; that is, the orthopyroxene component dissolves in both the garnet and the clinopyroxene and eventually disappears. At this point an MgSi-rich, aluminum-deficient garnet coexists with a magnesium-rich diopside. Garnet then moves toward the clinopyroxene composition as diopside dissolves in the garnet. Therefore, in contrast to the bimineralic eclogite system, the garnet takes a detour toward $MgSiO_3$ before it heads toward $CaMgSi_2O_6$. In either case, orthopyroxene disappears at a relatively low pressure. A synthesis is given in Figure 22.5. Note that the high-temperature sequence of transitions is different from the low-temperature sequence. The resulting densities and seismic velocities are also quite different. The low-temperature minerals (*spinel* + stishovite, *ilmenite*) are 10% to 20% higher in velocity than the high-temperature minerals (pyroxene, majorite).

The CMAS system

The system $CaO–MgO–Al_2O_3–SiO_2$ (CMAS) is shown in Figure 22.6 in simplified form. High temperatures and the presence of Al_2O_3 stabilize the majorite (mj) structure, and a broad majorite-garnet solid-solution field occurs between enstatite, *ilmenite* and *perovskite*. Diopside and jadeite, components of clinopyroxene, are stable to higher pressures than enstatite. Diopside collapses to a dense calcium-rich phase, (Ca–pv), at pressures less than required to transform enstatite to *perovskite*. Garnet (gt) itself is stable throughout most of the upper mantle, although it dissolves pyroxene at high pressure. Note that the phase assemblages along the 'cold slab' adiabat are different from those along the 'normal mantle' adiabat at almost all pressures; this is true for temperature contrasts much smaller than the 800 °C chosen for purposes of illustration.

Temperature variations in a convecting mantle are expected to fluctuate by about 100 degrees on both sides of the 'normal' temperature. Temperatures in a homogenous self-compressed solid or a vigorously convecting homogenous fluid will approximate an adiabatic temperature gradient, away from thermal boundary layers. The normal-mantle adiabat is a useful reference state but it is unlikely to exist anywhere in the mantle. Different phase assemblages are encountered as one increases the temperature from the normal-mantle adiabat. The partial melt field (not shown) is encountered at low pressure. The fields of the low-pressure, low-density assemblages are expanded at high temperature, but the change in density is not symmetric about the average, or normal, temperature. The thermal expansion coefficient increases with temperature, so there is a larger decrease of density for a given increase in temperature than for the corresponding decrease. For an internally heated mantle, upwellings are broader than downwellings, so the lateral changes in physical properties are expected to be more diffuse than for the slab.

Isobaric phase changes and lateral variations of physical properties

It is important to understand the factors that influence lateral heterogeneity in density and seismic velocities. Much of the radial structure of the Earth is due to changes in mineralogy resulting from pressure-induced equilibrium phase changes or changes in composition. The phase fields depend on temperature as well as pressure so that a given mineral assemblage will occur at a different depth in colder parts of the mantle. The elevation of the olivine–*spinel* phase boundary in cold slabs is a well known example of this effect. The other important minerals of the mantle, orthopyroxene, clinopyroxene and garnet, also undergo isobaric temperature-dependent phase changes to denser phases with higher elastic moduli. These phases include majorite, *ilmenite*, *spinel* plus stishovite, and *perovskite*. The pronounced low-velocity zone (LVZ) under oceans and tectonic regions and its suppression under shields is another example of phase differences (partial melting) associated with lateral temperature gradients. Deeper LVZs can be caused by composition, e.g. eclogite vs. peridotite.

Variations in temperature, at constant pressure, can cause larger changes in the physical properties than are caused by the effect of temperature alone. In general, the sequence of phase changes that occurs with increasing pressure also occurs with decreasing temperature. There are also some mineral assemblages that do not exist under normal conditions of pressure and temperature but occur only under the extremes of temperature found in cold slabs or near the solidus in hot regions of the mantle. Generally, the cold assemblages are characterized by high density and high elastic moduli. Lithological gradients, such as from peridotite to eclogite, can result in density increases with seismic velocity decreases.

The magnitude of the horizontal temperature gradients in the mantle are unknown, but in slabs and thermal boundary layers the temperature changes about 800 °C over about 50 km. In an internally heated material the upwellings are much broader than slabs or downwellings. Tomographic results show extensive low-velocity regions associated with ridges and tectonic regions, consistent with broad high-temperature, or low-melting-point, regions. The cores of convection cells have relatively low thermal gradients. We therefore expect the role of isobaric phase changes to be most important and most concentrated in regions of subducting slabs and delaminating continental crust. The temperature drop across a downwelling is roughly equivalent to a pressure increase of 50 kbar, using typical Clapeyron slopes of upper-mantle phase transitions.

In the geophysics literature it is often assumed that lateral variations in density and seismic velocity are due to temperature alone. By contrast, it is well known that radial variations are controlled not only by temperature and pressure but also by pressure-induced phase changes. Phase changes such as partial melting, basalt-eclogite, olivine–*spinel*–*postspinel*, and pyroxene–majorite–*perovskite* dominate the radial variations in density and seismic velocity. It would be futile to attempt to explain the radial variations in the upper mantle, particularly across the 400- and 650-km discontinuities, in terms of temperature and pressure and a constant mineralogy. All of the above phase changes, plus others, also occur as the temperature is changed at constant pressure or depth.

Some of the isobaric phase changes, their approximate depth extent in 'normal' mantle and the density contrasts are the following:

50–60 km	basalt → eclogite (15%)
50–60 km	spinel peridotite → garnet peridotite (3%)
50–200 km	partial melting (10%)
400–420 km	olivine → β-spinel (7%)
300–400 km	orthopyroxene → majorite (10%)
500–580 km	→ β + st(4.5%) → γ + st(1.6%)
400–500 km	clinopyroxene → garnet (10%)
500 km	garnet–majorite s.s. (5%)
700 km	ilrnenite → perovskite (5%)

Slabs

Slabs sink into the mantle because they are cold and dense. Slabs are not simple thermal boundary layers and their properties depend on more than $\alpha\Delta T$. When phase changes are included, the average density contrast of slabs is about three times greater than would be calculated from thermal expansion alone. There are several phase changes in cold subducting material that contribute to the increase in the relative density of the slab. The basalt–eclogite transition is elevated, contributing a 15% density increase for the basaltic portion of the slab in the upper 50–60 km. The absence of melt in the slab relative to the surrounding asthenosphere increases the density and velocity anomaly of the slab in the upper 300 km of the mantle. If the slab contains volatiles, or low-melting point components, it can be both dense and low-velocity. A CO_2-rich eclogite sinker will show up as a low-velocity zone, in spite of being cold.

The olivine–*spinel* and pyroxene–majorite phase changes in cold slabs are elevated by some 100 km above the 410-km discontinuity, contributing about 10% to the density contrast in the slab. Other transitions are also elevated, adding several percent to the density of the slab between 410 and 500 km. The *ilmenite* form of pyroxene is 5% denser than garnetite, increasing the density contrast of the cold slab between 500 and 670 km, relative to normal mantle, by about a factor of 2 or 3 over that computed from thermal expansion. In addition to these effects, accumulated slabs cool off the surrounding and underlying mantle, even in a chemically stratified mantle. The formation of a detached internal thermal boundary layer can make it appear that a slab has penetrated the boundary.

The seismic anomalies associated with what are thought to be slabs in the mantle can be much greater than can be accounted for by the effect of temperature on velocity. The associated density contrast between slab and normal mantle is probably greater than between the plate and the underlying mantle as estimated from thermal expansion alone. It is also possible that what have been interpreted as slabs in the lower mantle, based on very broad bands of high velocity imaged by tomography, may not be slabs at all.

The ilmenite form of $MgSiO_3$, akimotoite, is a stable phase, at low temperature, in the lower part of the transition region. Ilmenite is about 8% denser than garnet-majorite, stable at higher temperatures, and has seismic velocities about 10% greater. Although *ilmenite* is only 4% slower than perovskite, the main lower mantle mineral, it is 7% less dense. *Ilmenite* becomes stable at slab temperatures somewhere between 450 and 600 km and is predicted to remain stable to depths greater than the *perovskite* phase boundary in higher-temperature mantle.

In most interpretations of mantle tomography it is assumed that the slab is identical in composition to the adjacent mantle and that temperature is the only variable. Yet the slab is laminated: the upper layer is basalt/eclogite, and the second layer is probably olivine–orthopyroxene–harzburgite. These undergo their own series of phase changes and, when cold, remain denser than garnet peridotite to at least 500 km. Young or thin slabs, or slabs with thick crustal portions may equilibrate in the shallow mantle, where the crustal parts – basalt, gabbro or eclogite – eventually melt. An assemblage of such slabs trapped in the upper mantle can cause the same sort of anisotropy and anelasticity that has been attributed to crystal orientation and grain boundary melting.

Cold harzburgite, a component of the slab, averages about 0.1 g/cm^3 denser than warm pyrolite between 400 and 600 km depth. At 600 km it becomes less dense. Some eclogites are denser than peridotite at the same temperature to depths of about 500–560 km and, when cold, to 680 km. Subducted oceanic crust is silica-rich and is denser than other eclogites, at depth, because of the presence of stishovite (st). Cold eclogite can be 4–5% denser than warm peridotite above 550 km depth but eclogites have a wide range of compositions, mineral proportions and densities. If the 1000-km discontinuity (Repetti Discontinuity) is a chemical boundary, it will be depressed by the integrated density excess in overlying cold mantle even if the deeper part of the slab is buoyant. Chemical interfaces in general are expected to be irregular boundaries in a chemically stratified mantle and to be much deeper under slabs. The different phase assemblages in the slab relative to warm mantle will contribute to the density and seismic velocity contrasts. An increase of intrinsic density between upper and lower mantle and a negative Clapeyron slope will inhibit slab penetration into the lower mantle (defined by Bullen to start at 1000 km depth, the top of his region D′). An increase in viscosity will also partially support the slab. Some support is required in order to explain the geoid highs associated with subduction zones. A chemical change has a similar effect in holding up the slab.

Reheated slabs and delaminates

Slabs are cold when they enter the mantle but they immediately start to thermally equilibrate. Ambient mantle warms up slabs from both sides. Slabs, in part, are composed of oceanic crust and in part of serpentinized peridotite; CO_2 and water occur in the upper parts. All of these effects serve to lower the melting point and seismic velocities of slabs compared to dry refractory peridotite. After the conversion of basalt to eclogite, the melting point is still low. Even small amounts of fluid or melt can drastically lower the seismic velocity, even if the density remains high.

The time scale for heating the slab to above the dehydration and melting points is a small fraction of the age of the plate upon subduction since the basalt and volatiles are near the top of the slab. It is even smaller for delaminated lower continental crust since this is already hot, being about midway into the thermal boundary layer. As far as the slab, or a piece of delaminated crust (*delaminate*), is concerned the surrounding mantle is an infinite heat source. The idea that low-velocity material in the mantle can be low temperature should not be overlooked when interpreting seismic images. Seismic velocity is controlled by composition, mineralogy, and volatile content as well as by temperature. If the melts and volatiles completely leave the peridotitic part of a slab then it can become a high seismic velocity anomaly, at least between depths of order 60 to 600 km. Deeper than that the eclogite in the slab can become low-density and low-velocity compared to normal mantle.

The fate of subducted and delaminated material has been controversial. Below about 50 km depth, basalts convert to eclogites with a considerable increase of density. However, eclogite is not a uniform rock type; it comes in a variety of flavors and intrinsic densities. NMORB, for example, is silica-rich and contains the dense phase `stishovite` at high pressure. Cold NMORB can probably sink to about 650 km before it is neutrally buoyant. If it is cold enough so that both stishovite and *perovskite* are stable then it can possibly breach the density barrier at 650 km, if only temporarily. Warmer slab, and SiO_2-poor eclogites can thermally equilibrate at shallower depths. The subduction depth also depends on the crustal thickness of the plate, the compositions of lower crustal cumulates, and whether the crustal part of the slab can detach from the mantle part. In addition, the 650 km phase change region is not the only plausible barrier to through-going convection or subduction.

The evidence from obduction, ophiolites, flat subduction and `exhumed ultrahigh-pressure` (UHP) crustal and slab fragments confirms the shallow, and temporary, nature of some recycling. The evidence from tomography for recumbent slabs at 650 km also suggests

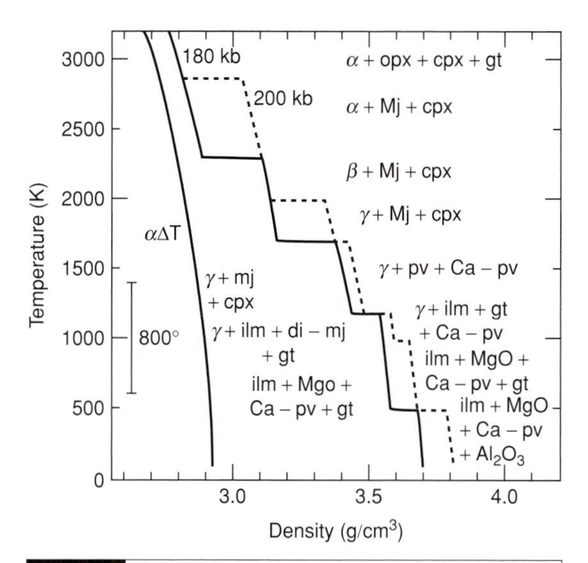

Fig. 22.7 Approximate variation of zero-pressure density with temperature taking into account thermal expansion (curve to left) and phase changes at two pressures. The effect of pressure on density is not included. The inclusion of density jumps associated with phase changes increases the average effect of temperature by a factor of 3 to 4. There are seismic velocity changes associated with these phase changes but these are usually ignored in visual interpretations of tomographic cross-sections.

an important role for upper-mantle circulation. The low-velocity zones above slabs suggest that volatile and low-melting point components leave the slabs at shallow depths. The inference that some slabs break through 650 km, even if a valid interpretation of some tomographic cross-sections, does not imply that they all do. Most of the material entering subduction zones apparently never makes it to 650 km depth, but some apparently makes it to ~1000 km. This is based on cross-correlation and plate reconstruction studies, not on visual evidence for occasional slab penetration or qualitative analysis of tomographic cross-sections.

Thermal phase changes

The approximate zero-pressure density for a peridotitic assemblage as a function of temperature, at two pressures, is shown in Figure 22.7. Temperature is plotted increasing upward to emphasize the fact that decreasing temperature has effects similar to increasing pressure – lateral temperature changes are similar to vertical pressure changes. The curve labeled $\alpha\Delta T$ is the approximate effect of thermal expansion alone on density. The bar labeled 800 °C shows the expected change in temperature across a subducted slab and is approximately half the maximum expected lateral temperature changes in the mantle. Note that a temperature change of 800 °C placed anywhere in the field of temperatures expected in the mantle will cross one, two or even three phase boundaries, each of which contributes a density change in addition to the term from thermal expansion. Changes in elastic properties are associated with these phase changes.

Figure 22.8 shows the approximate zero-pressure room-temperature density for phase relations in the CMAS system with a low-pressure mineralogy appropriate for garnet peridotite with olivine > orthopyroxene > clinopyroxene ~ garnet. Note that the low-temperature assemblages are denser than high-temperature assemblages (normal mantle) until about 600 km depth, and that the differences are particularly pronounced between about 300 and 550 km. The density change associated with a temperature change of 800 K ranges from 7–17% in the temperature interval 1000 to 2300 K at pressures near 600 km depth. This includes thermal expansion and isobaric phase changes. Thermal expansion alone gives 2–3%. Note that the density anomaly associated with a slab is far from constant with depth. Furthermore, the density anomaly of a slab with respect to the adjacent mantle is quite different from the density contrast between the surface plate and the underlying mantle, as estimated from the bathymetry-age relation for oceanic plates.

The important phase changes in the mantle mostly have Clapeyron slopes that correspond to depth variations of 30–100 km per 1000 °C. The figures in this section show that several phase boundaries are crossed in covering the normal expected range of mantle temperature, at constant pressure. At 230 kbar (23 GPa) Ca-*perovskite*, Mg-*perovskite* and magnesiowustite is the normal assemblage in peridotite; *ilmenite* replaces Mg-*perovskite* at cold temperature.

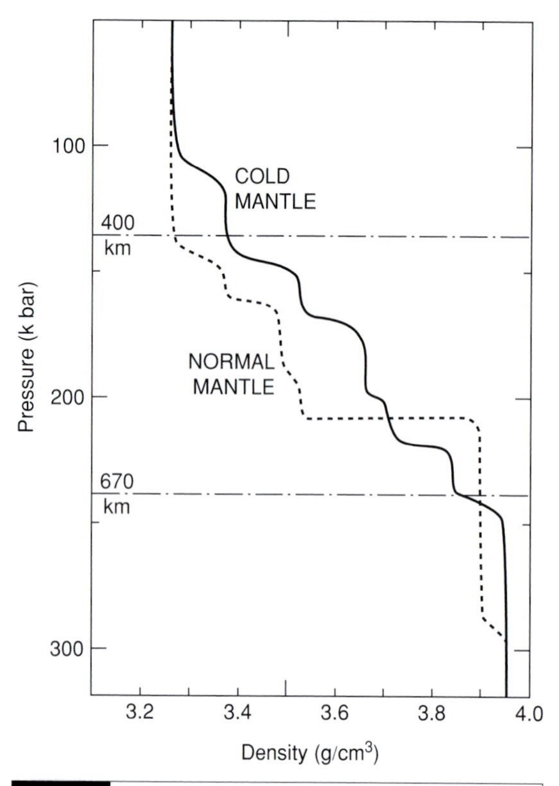

Fig. 22.8 Zero-pressure room-temperature density of FeO-free peridotite (ol> opx > cpx ~ gt) using mineral assemblages appropriate for the temperatures in warm ('normal') mantle and slab ('cold') mantle. This illustrates that lateral variations due to phase changes are as important as radial changes, and that tomographic images cannot be interpreted in terms of temperature variations operating on a constant mineralogy mantle. The existance of eclogite blobs in the mantle complicates the matter further.

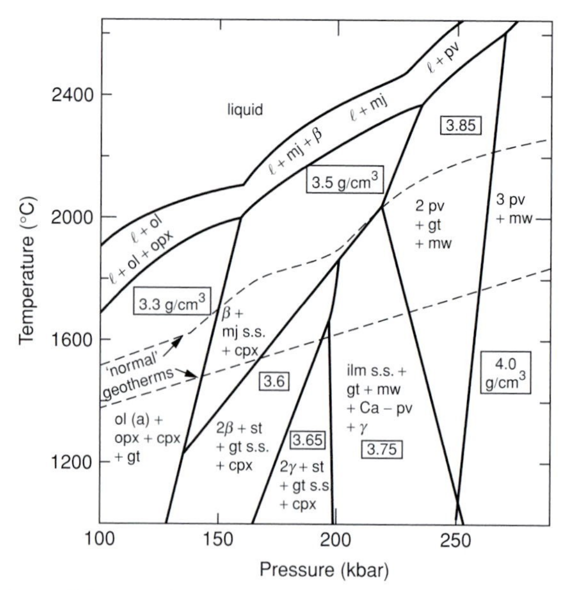

Fig. 22.9 Approximate phase relations in the mantle system with an ol > opx > cpx = gt mineralogy at low pressure. The geotherms bracket most estimates of temperatures in 'normal' or average mantle. Warmer parts of the mantle may be near the solidus; the interiors of slabs may be 800 °C colder.

Sub-solidus phase relations in MORB compositions show that, at this pressure MORB–eclogite is composed of the phases majorite + stishovite + Ca–pv. At transition zone pressures MORB contains mj + st. Si-poor eclogite, such as lower crustal cumulates or melt depleted eclogites, will not contain stishovite and are therefore buoyant at relatively moderate pressures. Garnet solid solutions, including majorite, are stable over a very large pressure and temperature range. This is important since large amounts of garnet will decrease both the radial and lateral variations in physical properties, and decrease the jumps at the 410 and 650 km phase boundaries.

In peridotite the density in the lower part of the transition region (Figures 22.9 and 22.10) is dominated by β-spinel ($\rho = 3.6$ g/cm^3) or γ-spinel (3.7 g/cm^3) and majorite (3.52 g/cm^3) or ilmenite (3.82 g/cm^3). In eclogite the mineralogy is garnet (3.6–3.8 g/cm^3), calcium-perovskite (~4.1 g/cm^3), stishovite (4.92 g/cm^3) and an aluminous phase (~4 g/cm^3). Eclogite will have a STP density somewhere between about 3.70 and 3.75 g/cm^3 at the base of the transition region, which is less than the uncompressed density of the lower mantle. Eclogites and peridotites can have the same density at transition zone pressures, but the silica-poor eclogites will have lower shear velocities. The garnet–majorite in cold quartz-rich eclogite may convert to Ca–pv + Mg–pv at the base of the TZ and this would allow the eclogite to overpower the phase change at 650 km.

The high gradient in seismic velocity between about 400 and 600 km depth implies a gradual phase change or series of phase changes occurring over this depth interval, or a change in composition. One cannot rule out an important role for eclogite in the transition region. The

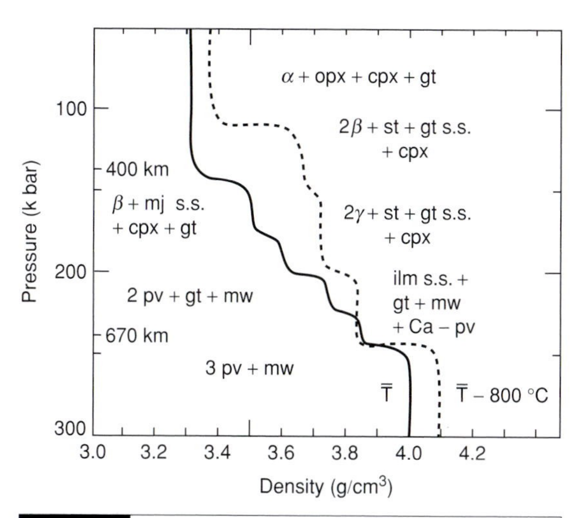

Fig. 22.10 Zero-pressure density for peridotite at two temperatures (a 'normal' geotherm and a temperature 800 °C colder). Phase changes and thermal expansion are included. The density differences would be different for an eclogitic mantle or for a slab that differs in chemistry from the surrounding mantle.

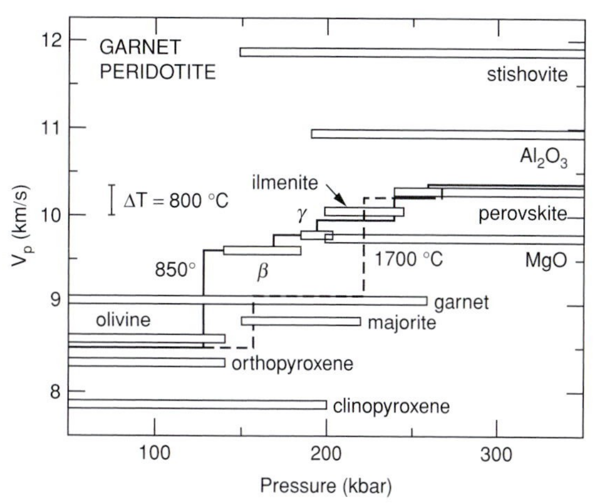

Fig. 22.11 Compressional velocities, at standard conditions, and stability fields of mantle minerals. The approximate compressional velocity of garnet peridotite, at standard (STP) conditions ($P = 0$, $T = 20$ °C) for phases stable at two temperatures (850 °C, 1700 °C) is also shown. The small bar gives the approximate change in V_P for a 800 °C change in temperature.

melting point of eclogite is so low that eclogite would likely not be a permanent resident of the TZ but would yo-yo up and down. Some eclogites may even become neutrally buoyant above 410 km.

The measured or estimated compressional velocities of the important phases of upper-mantle minerals are shown in Figure 22.11. Also shown are the estimated velocities for peridotite at two temperatures, taking into account the different stable phase assemblages. The major differences occur between about 130 and 225 kbar. The heavy lines show the approximate stability pressure range for the various phases. Garnet and clinopyroxene represent less than 20% of peridotite compositions. Since these minerals are stable to about 260 and 200 kbar, respectively, the effect of density and velocity changes associated with phase changes as both a function of temperature and pressure will be lower for an olivine and orthopyroxene-poor mantle – eclogite or piclogite.

The small bar in Figure 22.11 shows a typical change of velocity for an 800 °C temperature change, assuming no phase changes. Note that the effect of phase changes is to double or triple the effect of temperature. The largest effects occur between about 400 and 700 kilometers, the depth range of deep-focus earthquakes. Note that clinopyroxene and garnet, minor constituents of peridotite and pyrolite but major components of eclogite and piclogite have the largest stability fields of the low-pressure minerals. The presence of clinopyroxene and garnet reduces the size of the phase-change effects in the upper part of the transition region, particularly near 400 km. The smallness of the velocity jump near 400 km indicates the presence of substantial amounts of a 'neutral' component – garnet and clinopyroxene – near this depth. On the other hand, eclogite experiences major transformations and velocity increases between 200 and 260 kbar (20 and 26 GPa), which is perhaps related to deep-focus earthquakes.

Lower mantle mineralogy

Under mid-mantle conditions peridotite crystallizes into an assemblage of Mg–perovskite + Ca–perovskite + magnesiowüstite; NMORB and K-rich basalt compositions crystallize as Mg–perovskite + Ca–perovskite + stishovite + an aluminous phase with a $CaFe_2O_4$-type structure. In the bottom 1000 km of the mantle there are further phase changes in *perovskites*, stishovite and the spin-state of FeO. Below about 700 km the density of silica-rich basalt (NMORB) exceeds that of pyrolite, if it can get through the subduction barrier. The lower mantle may be enriched in both FeO and SiO_2 compared to pyrolite and the upper mantle. This has been a controversial issue, based partly on fitting equations of state to lower mantle properties, ignoring the above issues, i.e. assuming chemical and phase homogeneity, and assuming an adiabatic temperature gradient.

Part VI

Origin and evolution of the layers and blobs

Chapter 23

The upper mantle

Let's descend to that blind world below. I'll go first, and you can follow

Dante

Overview

The composition of the crust and the upper mantle are the results of a series of melting and fractionation events, including the high-temperature accretion of the planet. Early attempts to estimate upper mantle (UM) chemistry started from the assumption that UM initially was the same as `bulk silicate Earth` (BSE) and differed from it only by the extraction of the continental crust, or that the most depleted – low in LIL, U, Th, K – midocean-ridge basalts plus their depleted refractory residues – unaltered abyssal peridotites – constitute the entire upper mantle. Traditionally, geochemists have assumed that the lower mantle is *undegassed BSE* or *primitive mantle (PM)*. Geodynamicists have noted that there are heat flow problems with this model, which they fixed up by putting a large amount of U, Th and K in a `lower mantle stealth layer`.

Large-scale melting and differentiation upon accretion probably pre-enriched the upper mantle with incompatible elements, including the radioactive elements; the crust and the various enriched and depleted components sampled by current melting events were already in the upper mantle shortly after accretion and solidification.

Recycling of crust into the upper mantle is an important current process. It is possible to estimate the composition of the `fertile upper mantle` by combining known components of the upper mantle – basalts, peridotites, recycled crust and so on. The MORB source is just part of the upper mantle and it is not the only LIL-depleted part of the mantle. It is not necessarily convectively homogenized.

Attempts to establish an average composition for 'the upper mantle' focus on the most depleted MORB lavas or abyssal peridotites and involve major assumptions about melt generation, melt transport and differentiation processes that have affected these. The `depleted upper mantle`, that part of the mantle that is assumed to provide MORB by partial melting is variously called DUM, DM, DMM and the `convecting upper mantle`. Simplified mass balance calculations suggested to early workers that this depleted mantle constituted only ~30% of the mantle; the 650–670 km discontinuity was therefore adopted as the boundary between DUM and 'the primitive undepleted undegassed lower mantle.' The starting condition for the upper mantle (UM) was taken as identical to `primitive mantle` (PM) – or `bulk silicate Earth` (BSE) – and the present lower mantle (LM). Estimates of PM and BSE are based on cosmological and petrological considerations. The hypothetical `primitive upper mantle` – crust plus DUM – is labeled PUM. It is further assumed that the upper mantle – from the base of the plate to 650 km – after extraction of the crust, became well-stirred and chemically homogenous by vigorous convection. The non-MORB basalts that occur at the initiation of spreading and at various locations along the global spreading system are attributed to plumes from the chemically distinct lower mantle.

However, it can be shown that most or all of the mantle needs to be depleted and degassed to form the crust and upper mantle and to explain the amount of ^{40}Ar in the atmosphere; this was a major theme in the first edition of Theory of the Earth (Anderson, 1989). Depletion and degassing of the upper mantle alone cannot explain the observations. In addition, the MORB reservoir and the CC are not exactly complementary; another enriched component is needed. When this (Q-component) is added in, MORB + CC + Q require that most of the mantle must be processed and depleted. There must be other components and processes beyond single stage small-degree melt removal from part of the primordial mantle to form CC. There are other enriched components in the mantle, probably in the shallow mantle, and other depleted residues over and above the MORB-source. The upper mantle (UM) is still generally treated as if its composition can be uniquely determined from the properties of depleted MORB – NMORB or DMORB – and depleted peridotites, continental crust, and an undifferentiated starting condition.

The traditional hotspot and plume models of OIB and enriched magma genesis and mantle heterogeneity are unsatisfactory [mantle plumes, plume paradoxes]. Traditional models for both OIB and MORB genesis involving only peridotitic protoliths are also being re-evaluated [olivine-free mafic sources]. Although recycling has long been used as a mechanism for modifying the isotopic character of OIB, it is now becoming evident that it can also create melting anomalies. The roles of eclogite and garnet pyroxenite in petrogenesis and in the formation of melting anomalies are becoming evident (Escrig *et al.*, 2004, 2005, Gao *et al.*, 2004, Sobolev *et al.*, 2005) [delamination mantle fertility]. Midocean-ridge basalts represent large degrees of melting of a large source volume, and involve blending of magmas having different melting histories. The Central Limit Theorem explains many of the differences between MORB and other kinds of melts that sample smaller volumes of the heterogenous mantle.

Observed isotopic arrays and mixing curves of basalts, including ocean-island basalts (OIB), can be generated by various stages of melting, mixing, melt extraction, depletion and enrichment and do not require the involvement of unfractionated, primitive or lower-mantle reservoirs. However, the first stage of Earth formation – the accretional stage – does involve large degrees of melting that essentially imparts an unfractionated – but enriched – chondritic REE pattern to the upper mantle. Small-degree melts from this then serve to fractionate LIL.

What is the upper mantle?

On the basis of seismic data Bullen divided the mantle into regions labeled B, C and D. Region B is the upper mantle and C is the Transition Zone (410 to 1000 km). D, the lower mantle, starts at 1000 km depth. The upper mantle (Region B) was subsequently found to contain a high-velocity lid and a low-velocity zone (LVZ), generally associated with the asthenosphere. Region C, the mantle transition region (TR), was subsequently found to contain two abrupt seismic discontinuities near the depths of 410 and 650 km, and a region of high and variable gradient below 650 km depth. A depth of 670 km was found for the deeper discontinuity in western North America and this was adopted for the PREM model. However, the average depth of the discontinuity, globally, is 650 km, with a variation of about 30 km. Some authors have referred to the 650 (or 670) km discontinuity as the base of the upper mantle and have suggested that this represents a profound chemical and isotopic boundary between depleted upper mantle and primitive lower mantle, rather than primarily an isochemical phase change, as originally inferred (Anderson, 1967). Others have suggested that the 1000 km level is a chemical boundary and should be retained as the definition of the top of the lower mantle. Sometimes the TR is included as part of the upper mantle; sometimes it is defined as a separate region, Bullen's Region C. This confusion in terminology about what constitutes the upper mantle and the lower mantle has led to the widespread view that there are fundamental conflicts between isotope geochemistry and geophysics, and confusion about whether slabs penetrate into 'the lower mantle' or not. It appears that the upper 1000 km of the mantle – about 40% by mass of the mantle – differs from the rest of the mantle and this also appears to be the *active layer* for plate tectonics. The majority of the incompatible trace elements that are not in the crust may be confined to an even shallower depth range. Bullen's nomenclature is precise and useful and I will follow it. The terms *upper mantle* and *lower mantle* are now fuzzy concepts because of usage in the geochemical literature and the decoupling of this usage from seismological data. These terms will be used when precise depths, volumes or masses are not needed. The term *mesosphere* has also been used for the mid-mantle, Bullen's Region D′. The terms *shallow mantle* and *deep mantle* will be used to avoid the conflict between the precise seismological definitions of mantle regions and geochemical usage.

Depleted upper mantle; the DUM idea

Geochemists have ideas different from seismological conventions about what constitutes the upper mantle. They are based on compositions of *depleted midocean-ridge basalts, DMORB*, and assumptions about how these form. The definition of the upper mantle adopted by isotope geochemists is the following.

> The upper mantle is that part of the mantle that provides uniform and depleted midocean ridge basalts and that formed by removal of the continental crust; it extends from the Moho to the 670-km mantle discontinuity. It is also called 'the convecting mantle.'

The upper mantle is usually viewed by geochemists as homogenous because MORB are relatively homogenous. The assumption is that homogenous products require homogenous sources. The composition of the inferred *MORB reservoir* has been attributed to the whole upper mantle. Since MORB is depleted in LIL compared to other basalts, the above assumptions have led to the *Depleted Upper Mantle*, or *DUM*, concept. DUM is a two-component system, DMORB and a residual depleted peridotite or an unaltered abyssal peridotite. It has a simple one-stage history.

In some models of upper-mantle chemistry only the most depleted materials are used in its construction; hence Depleted Upper Mantle.

Heterogenous upper mantle

Convective stirring takes large blobs and shears and stretches them, folds them and stretches them more, repeatedly, until the dimensions are very small; for obvious reasons this is known as the Baker's transformation, a fundamental result of chaotic advection theory. This theory may not be appropriate for the mantle. It implies high Rayleigh number, stirring or folding in one direction and low-viscosity passive particles. We must therefore pay attention to the materials that enter the mantle instead of relying on averages of the magmatic products. We must keep an open mind about the possibility of large fertile blobs in the mantle, and extensive regions having *high homologous temperature*.

From a petrological point of view, the mantle can be viewed as a multi-component system. The known components of the upper mantle are recycled continental crust and other mafic components, ultramafic rocks, MORB and other basalts, depleted refractory residues and enriched components (Q, quintessence or fifth component) such as kimberlite; mixtures of these satisfy most chemical constraints on the composition of the mantle or BSE, including major and trace elements. The compositions of basalts and the compositions of continental and abyssal peridotites are available in petrological databases. These can be used to reassemble the original petrology and composition of the mantle, and with a few other assumptions, the composition of the upper mantle.

Basalt and peridotite compositions represent the culmination of melt depletion and enrichment processes over the entire history of the mantle, including the accretional process. There are a variety of basalts and peridotites. The compositions of many basalts and peridotites appear to lie along mixing lines and the end-member compositions have been interpreted as trapped melts, depleted residues and recycled and delaminated materials. On major element plots (Chapter 15) the end-members are harzburgite and MORB, or eclogite. Picrites, komatiites and primitive mantle have intermediate compositions. On LIL and REE plots, the extreme compositions are kimberlites and DMORB or abyssal peridotite.

From an isotopic point of view, oceanic basalts are also treated as multi-component systems involving mixtures of DMM, EM1, EM2 and HIMU and C or FOZO. These are shorthand names for what are thought to be the various enriched (EM) and depleted (DM) isotopic end-members of the mantle and there is a large literature on each. There is no agreement regarding the lithology or history that goes with each component. Trends of OIB and MORB isotopic compositions approach – or converge on – a hypothetical component of the mantle referred to as FOZO (focal zone) or C (common). It has been assumed that this reflects the composition of the lower mantle. It is not clear why the most common component in basalts should represent the deepest, rather than the shallowest, mantle. Melts pond beneath, and percolate through, the lithosphere, and interact with it. A lithosphere or harzburgite component may therefore be involved in most magmas. If so, ultramafic rocks (UMR) may anchor the ends of both major element and isotope mixing arrays. UMRs are certainly the most common or prevalent lithology of the shallow mantle.

The average or prevalent mantle composition has been referred to as PREMA in the isotope literature. In contrast to the end-member components, PREMA, C and FOZO are interior components – on isotope diagrams – and are therefore either mixtures or sources. The one extreme attribute of these average compositions is that basalts falling near these compositions tend to have higher variance in their $^3He/^4He$ ratios, and therefore contain some high $^3He/^4He$ samples. The most prevalent lithology of the mantle – peridotite – may be implicated in this component while EM and HIMU may reside in the fertile or mafic components.

Enrichment and depletion processes

Vigorous stirring can homogenize a fluid by a process known as chaotic advection. Diffusive and thermodynamic processes, in the absence of gravity, are homogenizers. Large-scale melting can be a homogenizer. Plate tectonic and other petrological processes, however, create heterogeneities. Removal of small-degree melts causes depletion of fertile regions (basalt sources) – or components – and refractory regions (depleted peridotites, cumulates). Small-degree melts are enriched in LIL and become the crust and the enriched components (EM) in the upper mantle, such as kimberlite and carbonatites, and in the sources of enriched magmas such as EMORB and OIB. Large-degree melting occurs at spreading centers and at thin spots of the lithosphere from upwelling mantle that has been depleted (NMORB source) or enriched (EMORB source) by the transfer of these small melt fractions. Large-degree and large-volume melts blend together large- and small-degree melts from a large volume of the mantle and give fairly uniform magmas with small variance in trace-element and isotopic ratios, even if the shallow mantle is heterogenous. This is called *melt aggregation* or *blending*.

Melt extraction from partially molten rocks or crystallizing cumulates is not 100% efficient and residual melts help explain some of the trace element and isotopic paradoxes of mantle magmatism, such as apparent contradictions between the elemental and isotopic compositions. Other sources of magmatic diversity include recycling, delamination and melting of diverse lithologies such as eclogite and peridotite, which also have experienced various levels of melt extraction and infusion. This heterogenous upper mantle or statistical upper mantle assemblage gives relatively homogenous products when it experiences large degree melting. Magmas at ridges and thin spots represent blends of melts from various depths, lithologies and extents of partial melting. The compositions of ocean island basalts, ocean ridge basalts, and residual mantle reflect upper mantle processes of melt extraction, migration and trapping – as well as recycling from the surface – and the sampling/melting process that blends the various enriched and depleted products.

During accretion, melting is extensive and there should be good separation between the elements – LIL – that enter the magmas and those that are retained by the residual solids. The LIL include U, Th and K – the heat-producing elements – the concentrations of which were much higher during the first Gyr after Earth formation. The relative buoyancy of magma, and the self-heating tendency of potentially fertile parts of the mantle combine to concentrate the radioactive elements in the outer shells of the Earth, including the crust and upper mantle. To a good approximation the large-ion and large-charge elements – LIL and HFS elements – can be considered to reside in the fertile parts of the mantle and the crust. This is not necessarily true of non-LIL elements such as He and Os, although these also are probably in the shallow mantle.

Composition of the upper mantle

An estimate of the original composition of the mantle (see Part IV) can be derived simply by mixing together all the products of mantle differentiation. For example, the mantle produces MORB, EMORB, OIB, kimberlite (KIMB), island arc basalts and so on, and has also produced the continental crust (CC). There is direct evidence that the mantle contains a variety of peridotites, pyroxenites and eclogites. One can alternatively or in addition focus on the recycled materials that are known to be entering the mantle – sediments, oceanic plates and delaminated lower continental crust. All of these can be put into the mix. If one assumes that the mixture must have chondritic ratios of the refractory elements and that the present mass of the crust is a lower bound on the CC component then one can estimate the mixing ratios of the components and the original composition of BSE, or at least that part that has provided material to the surface. This can be posed as a least-squares or geophysical inverse problem.

The components in the mantle can be alternatively subdivided into *fertile components* and *refractory* or *residual-solid* components. The fertile components are those that have relatively low melting points and which provide the main components for basalts. The incompatible LIL elements enter the melt. The refractory components have higher melting temperatures and are left behind when partial melts are extracted. There is no requirement that the fertile and refractory components are intimately mixed or remixed in the mantle or that the fertile components are veins in a refractory matrix rather than km-sized blobs. When small amounts of melting are involved the separation tendencies are quantified by partition coefficients, which give the partitioning of a given element between the magma and the residual solid. When melting is extensive, most of the incompatible and basalt forming elements in the source end up in the magma or the fertile components.

DMORB and *abyssal peridotites* are, respectively, the 'blank' slates – depleted basalt and refractory residue – upon which various enriched signatures are written. These are among the most depleted of mantle materials and form the basis for recent estimates of upper-mantle composition (e.g. Donnelly *et al.*, 2004, Salters and Stracke, 2004, Workman and Hart, 2005). These are estimates of the most depleted parts of the upper mantle, not the upper mantle as a whole.

Kimberlites and continental crust are the main enriched complements to the above. EMORB and OIB are intermediate in trace element properties – but extreme in some isotopes. Eclogite reservoirs/components are required to balance such trace elements as Re, Zr, Ti and Na but are also implied by recycling models and by mass balance using chondritic constraints. Mass-balance calculations (e.g. Part IV and Chapter 8 of Theory of the Earth) imply that the mantle contains less than about 10% of the fertile – basalt, gabbro, eclogite, garnet pyroxenite – component; this can be regarded as potential basalt for ridges, islands, CFB and LIPs; some of this material may be recycled or delaminated crust/eclogite. This is about the amount of oceanic crust recycled into the mantle over billions of years at current subduction rates. Delaminated lower continental crust also introduces fertile material – eclogite cumulates – into the upper mantle. The fertile components are generally viewed as well-mixed with the refractory infertile components – depleted peridotites, lherzolites, ultramafic residues (UMR). They are more likely to reside in large blobs.

Various combinations of materials give chondritic or BSE ratios of the refractory LIL elements. The *fertile components* can account for all the LIL and they occupy about 10% of the mantle; the average enrichment – of the fertile part of the mantle – is therefore a factor of 10 times the concentration levels of primitive mantle. About 0.5–2% melting is implied to obtain material as enriched as CC and KIMB from BSE, and to deplete DM to the extent observed; up to approximately 25% melting of this still fertile residue, at a later time, is implied in order to generate the more abundant depleted basalts. The time separation of these events can be estimated from isotopes to be of the order of 1.5 to 2.5 Gyr. This has traditionally been taken as the convective overturn time of the mantle. From a plate-tectonic or top-down point of view it is the revisitation time of a migrating spreading ridge. There is no contradiction between small-degree melting in the past, when the mantle was hotter, and large-degree melting at the present, after the mantle has cooled down. Since the mantle is close to the melting point, the formation and removal of very small-degree melts may have occurred in the thermal boundary layer at the surface of the Earth, where the current temperature rises from near 0 °C to about 1400 °C. Low-degree melting can occur in a surface or internal TBL or in recycled mafic material. High-degree melting requires special circumstances.

Kimberlites, OIB and EMORB are probably all produced from the upper mantle; there is no reason to suppose that they are not. Mass-balance constraints on the composition of the mantle can be achieved by adding EMORB, EM or KIMB to the very depleted components involved in DMORB genesis. For example, having 5% EMORB or 0.5% kimberlite plus MORB plus peridotite in the upper mantle gives approximately chondritic ratios of the refractory LIL.

Petrological and cosmochemical constraints can be satisfied if the basaltic or fertile components of the mantle are mainly DMORB but also include about 10% OIB, 10% EMORB and 0.001 – 0.002 kimberlite, or some combination. All of this material fits easily into the upper mantle, and is potentially available for incorporation into ocean-ridge basalts, ocean-island basalts, seamounts and continental basalts. Estimates of the composition of fertile mantle are given in Table 23.1.

Original unprocessed mantle may have included the equivalent of 7.5–9.5% NMORB, 1% OIB, 2% EMORB plus the present continental crust (Chapter 13). This particular combination gives chondritic ratios of the refractory LIL and accounts for almost all the LIL and Na of BSE. The crust itself accounts for about 50–80% of the original LIL material in PM. Almost all the rest is contained in the various MORB components, in kimberlites and in OIB. Other considerations suggest that the mantle may contain from 6–15% eclogite.

Melt generation

Current upper-mantle conditions allow a wide range of melt fractions, with the largest extents of melting permitted at midocean ridges and other thin-lithosphere spots where mantle near the melting point can increase its melt content by adiabatic ascent to shallow levels, a mechanism not available under thick plates. Large degrees of melting can also occur in the more fertile – or eclogite-rich – regions of the mantle. Small-degree melts occur on the wings of midocean-ridge melting zones, and at greater depths, and under thick plates, and in the mantle wedge above subducted plates. As slabs warm up to ambient mantle temperature they can also experience small-degree melting. Thus, there are many opportunities for generating and removing enriched small-melt fractions, and for enriching and depleting various regions of the upper mantle. On the present Earth, and probably throughout Earth history, small-degree and large-degree melting conditions can both operate.

The simplest melt-generation scenerio is one in which the source rock is melted by a certain amount and the melt is then completely extracted. This single-stage process can be modeled, and the melt and residue compositions

Table 23.1 Estimates of 'fertile mantle' compositions derived by mixing components (continental crust (CC), ultramafic rocks (UMR), MORB, OIB, kimberlite (Q)) in various proportions in order to match chondritic ratios of refractory elements. BSE is bulk silicate Earth (CC, upper mantle, and lower mantle). Several estimates are given for some components

%	BSE	BSE	CC	NMORB	EMORB	OIB	KIMB	UMR	UMR	Fertile Mantle			%
SiO_2	48	48	59	50	50	49	35	44	45	44	45	45	SiO_2
MgO	34	34	4	8	8	8	28	41	39	39	37	37	MgO
CaO	3.6	3.1	6.4	11.4	10.7	10	19	2.2	3.2	2.8	3.6	3.7	CaO
Al_2O_3	4.5	3.8	15.8	15.7	16.4	14	1.3	2.3	4.0	3.2	4.6	4.7	Al_2O_3
K_2O	0.03	0.02	1.9	0.1	0.6	0.38	1	0.05	0.01	0.07	0.02	0.02	K_2O
Na_2O	0.4	0.3	3.2	3.1	3.3	2.1	0.8	0.2	0.13	0.44	0.29	0.33	Na_2O
TiO_2	0.2	0.2	0.2	1.5	1.7	2.5	0.5	0.1	0.13	0.19	0.21	0.22	TiO_2
ppm													*ppm*
Ba	6.99	5.22	390	6.8	100	350	2000	33.0	0.563	35.3	5.43	6.08	Ba
Rb	0.64	0.39	58	0.7	12	31	200	1.9	0.05	2.3	0.61	0.71	Rb
U	0.02	0.02	1.42	0.05	0.3	1.0	20	0.1	0.003	0.1	0.02	0.04	U
Th	0.09	0.08	5.60	0.14	1.1	4.0	92	0.7	0.01	0.7	0.07	0.18	Th
Ta	0.04	0.04	1.1	0.1	0.2	2.7	40	0.4	0.01	0.4	0.04	0.08	Ta
Nb	0.71	0.97	12	2	16	48	400	4.8	0.15	5.0	0.65	0.95	Nb
Zr	11	13	123	109	139	280	400	21.0	5.08	28.4	12.31	12.54	Zr
La	0.7	0.6	18	3.5	10	37	200	2.6	0.19	2.9	0.70	0.79	La
Sm	0.4	0.3	3.9	3.6	4	10	30	0.5	0.24	0.7	0.48	0.51	Sm
Yb	0.5	0.3	2	3.4	3	2	1	0.3	0.37	0.5	0.52	0.56	Yb

calculated, by use of partition coefficients for each element. This results in LIL-fractionated and enriched melts and strongly depleted residues. The more likely situation is that melt is continuously extracted as melting proceeds, and some melt is left behind. The extracted melt interacts with the matrix it is percolating through. Incomplete melt extraction on the one hand, and trapping of small-degree melts on the other, result in source rocks that are enriched compared to the theoretical residues of complete melt extraction calculations. The small-degree melts act as enriching or metasomatizing agents for both magmas and mantle xenoliths and are essential components when attempting to do mass-balance calculations. Theoretical attempts to reconstruct the composition of the primordial upper mantle from the compositions of MORB, peridotite and continental crust (CC) rely heavily on data selection, melting models and partition coefficients. Attempts are made to determine the average compositions of the MORB and peridotite endmembers, and to make corrections for *contamination*. An alternative method (Chapter 13) just mixes together materials thought to represent a spectrum of products of mantle melting and metasomatism.

Enriched upper mantle?

The earliest studies of ocean-ridge basalts recognized that some ridge segments display enrichments in highly incompatible elements. The origin of enriched midocean-ridge basalts (EMORB) is controversial. EMORB have been sampled both on- and off-axis at various ridges and fracture zones, and at places not normally considered to be hotspots. EMORB have been attributed to interaction of plumes from the deep mantle with the depleted upper-mantle source of normal midocean-ridge basalts (NMORB). Enriched basalts have also been attributed to preferential melting of veins in the mantle, to small plumes dispersed as small-scale heterogeneities and to melting of enriched eclogitic veins from subducted and stretched oceanic crust recycled into the upper mantle. EMORB along the EPR has been attributed to plume material that was transported 5000 km from Hawaii, through the asthenosphere. Other proposed sources of enrichment include fractionation during melting or metasomatic events. Enriched MORB may occur without any clear relationship to plumes.

Many estimates of the composition of the upper mantle equate the NMORB source with the whole upper mantle. The use of depleted MORB and ultra-depleted residues in the estimates of DUM or DMM gives a lower bound on the LIL content of the upper mantle. Use of BSE as the starting composition for the upper mantle and then removing CC from it leaves a very depleted composition for the subsequent generation of magmas. Calculations in *Theory of the Earth*, abbreviated in Chapter 13 of the present volume, suggest that accretional differentiation pre-enriched the upper mantle by a factor of about 3 compared with primitive abundances, and the CC, and the depleted and enriched components evolved from that starting condition. The whole mantle was processed (mined) during accretion and the continental crust represents only part of the enriched material that was in the starting UM. The NMORB 'reservoir' was depleted by removal of small-degree melts, most of which remained in the shallow mantle and some entered the crust. EMORB and OIB are products of those parts of the mantle that have been enriched by these small-degree melts, as well as those parts of the mantle affected by recycling and delamination of the lower continental crust.

Mantle peridotites, in general, have a wide range of compositions and can be viewed as mixtures of the most depleted ones and an enriched fluid such as kimberlite. Continental peridotites tend to have 1–2 orders of magnitude higher LIL concentrations than theoretical abyssal peridotites from melt extraction calculations. Infertile refractory peridotites can be enriched in trace elements by a low-degree melt, or other metasomatic fluids, without necessarily being very fertile or a suitable source rock for basalts. Other enriched materials, such as recycled eclogite, may also reside at depth, e.g. in the transition region.

The outer 1000 km of the mantle – 40% of the mantle – appears not to be well stirred or homogenous (Meibom and Anderson, 2003). Since the observed compositions of NMORB, EMORB, peridotites, kimberlites and CC can be mixed together to achieve chondritic ratios of the refractory trace elements (Chapter 13) or inferred BSE compositions of LIL, there is no compelling reason to involve the lower mantle in mass balance calculations for the volatile and very incompatible elements, including U, Th and K. The lower mantle, of course, was involved in the original differentiation and is needed to balance the major elements and the chalcophiles and the more compatible elements such as Pb. In fact various geochemical paradoxes may be resolved with a hidden isolated depleted (but not fertile) reservoir, such as parts of the mantle below 1000 km.

Mass balance

In the standard model of mantle geochemistry, the continental crust, CC, is considered to be complementary to the MORB reservoir, MORB extraction leaves behind a complementary depleted residue, and the lower mantle remains primordial (PM);

$$\text{BSE} = \text{CC} + \text{DUM} + \text{PM}$$
$$\text{CC} + \text{DUM} = \text{PM}$$
$$\text{DUM} = \text{MORB} + \text{UMR}$$

In the *RAdial ZOne Refining* model, RAZOR, for Earth accretion and differentiation, the mantle melts, fractionates and differentiates during accretion. The proto-crust and proto-upper-mantle are the buoyant products of this accretional differentiation and a perovskite-rich residue (PV) makes up the deeper mantle.

$$\text{BSE} = \text{CC} + \text{MORB} + \text{UMR} + \text{PV} + \text{Q}$$

There is a large literature on the composition of each of these components and attempts have been made to filter and correct values of natural samples to be representative of uncontaminated material. There are several recent attempts to derive the composition of DUM from literature values of the composition of MORB, UMR and CC (Donnelly *et al.*, 2004; Salters and Stracke, 2004; Workman and Hart, 2005).

Data selection is a problem in this approach. Basalts and peridotites have a large range in composition. Estimates of upper mantle composition depend on which ones are selected to be representative. Even tholeiites have been subdivided into DMORB, TMORB, NMORB, PMORB, EMORB, OIB and so on. Should one attempt to construct averages, or pick endmembers, or pick representative values, or correct observed compositions for contamination?

The MORB sources

Midocean ridge basalts range in composition from depleted (DMORB and normal, N-type MORB) to transitional, enriched and plume-type basalts (TMORB, EMORB, PMORB); NMORB are tholeiites from normal ridge segments; DMORB are particularly depleted basalts that represent endmembers of this class of component. Even MORB from normal ridge segments display a range of compositions, including EMORB; the definitions of N-type MORB and 'normal ridge segments' are arbitrary. From a major element and petrological point of view even OIB tholeiites are similar to MORB and imply similar amounts of melting. Estimates of the composition of the upper mantle, however, focus on the most depleted products.

Salters and Stracke (2004) presented an estimate for the composition of *depleted mantle* (DM), tailored to be a suitable source for midocean-ridge basalts. The depleted mantle reservoir is defined as mantle that can generate 'pure' MORB *uncontaminated by enriched or 'plume components'*; 'pure' MORB is then used to estimate upper-mantle composition. Estimates for some elements are derived from elemental and isotopic compositions of peridotites assumed to be complementary to MORB. The concentrations of some elements are estimated by subtraction of a theoretical low-degree melt from an unfractionated bulk silicate Earth (BSE) composition. The methods used by Salters and Stracke to select and filter the input data are typical of this forward approach to modeling compositions.

A variety of criteria are used to identify 'depleted' (i.e. lacking an enriched component) MORB and to eliminate enriched or potentially enriched samples: the choice of criteria influences the calculated source region – DM – composition. Only those MORB samples 'thought to be representative of DM' are considered.

Attempts are made to avoid accessory phases and hydrothermal alteration. The net result is that estimates of DMM, DM or DUM, and the whole upper mantle in some models, are biased toward very depleted end members. Remaining questions are, what is the fraction of EM and recycled components in the source regions of MORB and OIB, what is the origin of EM, and what is the average composition of the upper mantle? The standard model is that MORB derive from a vigorously convecting well-stirred reservoir and that enriched materials, and hidden heat-producing reservoirs are in the deep mantle.

Although MORB and their presumed residues, abyssal peridotites, have some degree of heterogeneity in radiogenic isotope ratios (Sr–Nd–Pb–Hf), they have a small range of values relative to ocean island basalts and are, with some exceptions, depleted, relative to bulk Earth values, thus requiring a long-term history of low Rb/Sr, Hf/Lu and Nd/Sm (i.e. incompatible element depletion). The small range in composition is partly a result of the data selection and filtering operations just discussed and partly due to the averaging accomplished by natural processes at ridges (Meibom and Anderson, 2003).

Estimates of Donnelly *et al.* (2004), Salters and Stracke (2004) and Workman and Hart (2005) for the NMORB source can be made consistent with chondritic and accretional differentiation constraints, and the compositions in Chapter 13 simply by using different proportions of the components, and by adding enriched components representing small degrees of melt, or recycled material. Mass-balance calculations provide no constraints on the sizes and locations of the various components, reservoirs or blobs. Chondritic constraints, however, can be satisfied even if all the components are in a volume less than the size of the upper mantle.

Table 23.1 contains estimates of BSE, CC, NMORB, EMORB, OIB, KIMB and various ultramafic rocks (UMR). The fertile-mantle values tabulated in Table 23.1 are derived by mixing MORB, EMORB, OIB, KIMB, CC and various estimates of the compositions of spinel and abyssal peridotites in order to retain the LIL and refractory element ratios in BSE (Chapter 13). Peridotites account for 92–94% of the mix and MORB accounts for 4.5–6%. EMORB and OIB can be up to 0.7% of the mix and KIMB can be up to 0.15%. CC is constrained to be 0.55%. Since the upper mantle is 30–40% of the mantle, there is no mass balance reason to suppose that the enriched and fertile components – EMORB, OIB, KIMB – must come from the lower mantle.

Primitive upper mantle (PUM) in the standard model of mantle geochemistry is

$$\mathrm{PUM} = \mathrm{CC} + \mathrm{DUM}$$

In the current model, fertile mantle (FM), prior to removal and stabilization of the continental crust, was

$$\mathrm{FM} = \mathrm{CC} + \mathrm{MORB} + \mathrm{EMORB} + \mathrm{OIB} + \mathrm{KIMB} + \text{residual refractory peridotites (UMR)}$$

The currently accessible outer reaches of the mantle probably contain the MORB-source, the EMORB-source, the OIB-source and the KIMB-source. The non-fertile mantle (BSE minus FM), based on cosmic abundances of the major elements, has more silica and less magnesia than FM and is therefore less olivine-rich.

Enriched blobs

The NMORB-source region, DMM, is just part of the mantle, probably just part of the upper mantle. A representative mass-balance calculation using published estimates of the compositions of DMM (DMORB + abyssal peridotite), continental crust, recycled oceanic crust (now in the mantle as eclogite) and an OIB-source yields, in round numbers, 74% DMM, 5% recyling MORB (eclogite), 21% OIB-source and 0% primitive mantle. The fraction of mafic material in this and similar calculations is less than the mass of oceanic crust generated throughout Earth history, assuming current rates for 4.55 Ga (6–7% the mantle mass).

One point of this exercise is to show that there is no mass-balance contradiction with the hypothesis that *most of the fertile and heat-producing elements are in the upper mantle.* It is only by assumption, in simple box-model calculations, that the entire upper mantle is required to be depleted and low in radioactive elements. This single assumption has created many geochemical paradoxes.

Mass-balance calculations make no statement about the origins, locations or shapes of the enriched domains. They are usually assumed to be large isolated regions of the mantle, such as the lower mantle, or the sizes of grain boundaries and veins. Fertile, or mafic, material gets into the mantle via subduction of oceanic crust and delamination of continental crust, and fertile blobs can have dimensions typical of these tectonic elements. Delaminated lower crust may have dimensions of tens of km. Subducted seamount chains may have dimensions of tens by hundreds of km. If oceanic plateaus can subduct they would generate fertile regions having lateral dimensions of thousands of km.

The present crust is 0.566% of the mass of the mantle and contains more than 50% of the Earth's inventory of some elements (Chapter 13). Therefore, there cannot be more than two crustal masses in the BSE, or one in the mantle. The maximum thickness of the crust, except in actively converging areas, is about the depth to the basalt–eclogite phase change, suggesting that the volume of the present crust may not be controlled by the volume of potential crust but by the depth of a phase change. The potential crust in the mantle may be about equal to the present continental crust. Some of this may appear as enriched components - EM2, say - in midocean-ridge and ocean-island basalts

EMORB is found far from any melting anomalies and bathymetric highs, showing that enriched components are widely available in the shallow mantle; it is chemically similar to OIB. Donnelly *et al.* (2004) estimate that about 2–10% of ridge samples are EMORB in which incompatible elements such as Th and Ba are a factor of 10 more abundant than in typical NMORB sources. The continental crust and kimberlites contain enrichment factors of up to 100 and 1000, respectively - relative to BSE - for the more incompatible elements. End-member kimberlites can account for no more than 0.1% of the mass of BSE, by simple mass balance. This is about one-sixth of the present mass of the continental crust and about one-tenth of the upper bound on the allowable amount of CC in BSE.

Estimates of the degree of melting required to generate MORB and EMORB from a fertile peridotite - ultramafic source - range from 6–20%. Estimates of the degree of melting obviously depend on the source composition. Magmas can also be derived from mafic sources such as recycled ocean crust, delminated lower continental crust and eclogitic cumulates. Eclogite is a type of rock and also a metamorphic facies; it can have an original basaltic extrusive or gabbroic cumulate protolith; or it can be a residue left over from partial melting and melt extraction. These 'eclogites' are not expected to have the same compositions as unprocessed ocean crust unless the metamorphic transformation to eclogite is isochemical. Eclogites are cpx- and ga-rich by definition but may also contain trace to abundant oxide minerals (e.g. rutile) that replace original magmatic ilmenite and magnetite. Phases such as rutile, zircon, amphibole and micas (phlogopite in particular) can dominate some trace-element patterns.

Chapter 24

The nature and cause of mantle heterogeneity

The right to search for truth implies also a duty. One must not conceal any part of what one has discovered to be true.

Albert Einstein

Overview

The lithosphere clearly controls the *location* of volcanism. The *nature* and *volume* of the volcanism and the presence of 'melting anomalies' or 'hotspots,' however, reflect the intrinsic chemical and petrologic heterogeneity of the upper mantle. *Melting anomalies* – shallow regions of ridges, volcanic chains, flood basalts, radial dike swarms – and continental breakup are frequently attributed to the impingement of deep mantle thermal plumes on the base of the lithosphere. The heat required for volcanism in the plume hypothesis is from the core; plumes from the deep mantle create upper mantle heterogeneity. This violates the dictum of good Earth science: *never go for a deep complex explanation if a shallow simple one will do.*

Mantle fertility and melting point variations, ponding, focusing and edge effects, i.e. plate tectonic and near-surface phenomena, may control the volumes and rates of magmatism. The magnitude of magmatism may reflect the fertility and homologous temperature, not the absolute temperature, of the asthenosphere. The chemical and isotopic heterogeneity of the mantle is, in part, due to recycling and, in part, due to igneous processes internal to the mantle. The fertility and fertility heterogeneity of the upper mantle are due to subduction of young plates, aseismic ridges and seamount chains, and to delamination of the lower continental crust, as discussed in Part II. These heterogeneities eventually warm up past the melting point of eclogite and become buoyant low-seismic-velocity diapirs that undergo further adiabatic decompression melting as they encounter thin or spreading regions of the lithosphere.

The heat required for the melting of cold subducted and delaminated material is extracted from the essentially infinite heat reservoir of the mantle, not the core. Melting in the upper mantle does not require an instability of a deep thermal boundary layer or high absolute temperatures. Melts from fertile regions of the mantle, recycled oceanic crust and subducted seamounts, can pond beneath the lithosphere, particularly beneath basins and suture zones, with locally thin, weak or young lithosphere, or they can erupt. The stress state of the lithophere can control whether there is underplating and sill intrusion, or eruption and dike intrusion. Absolute mantle temperature has little to do with this.

The characteristic scale lengths – 150 km to 600 km – of variations in bathymetry and magma chemistry, and the variable productivity of volcanic chains, probably reflect compositional heterogeneity of the asthenosphere, not the scales of mantle convection or the spacing of hot plumes. High-frequency seismic waves, scattering, coda studies and deep reflection profiles are needed to detect the kind of chemical heterogeneity – blobs and small-scale layering – predicted from the recycling and crustal delamination hypotheses.

Mantle homogeneity: the old paradigm

Global tomography and the geoid characterize the large-scale features of the mantle. Higher frequency and higher resolution techniques are required to understand the smaller-scale features, and to integrate geophysics with tectonics and with mantle petrology and geochemistry. The upper mantle is often regarded as being extremely homogenous, based on low-resolution tomographic studies and the chemistry of midocean ridge basalts. Both of these approaches average out the underlying heterogeneity of the mantle. The intrinsic chemical heterogeneity of the shallow mantle, however, is now being recognized. This heterogeneity contributes to the isotope diversity of magmas and the scattering of seismic waves. Melting anomalies themselves – hotspots and swells – reflect lithologic heterogeneity and variations in fertility and melting point of the underlying mantle. The volume of basalt is related more to lithology of the shallow mantle than to absolute temperature. Thus, both the locations of volcanism and the volume of volcanism can be attributed to shallow – lithospheric and asthenospheric – processes, processes that are basically athermal and that are intrinsic to plate tectonics.

Much of mantle geochemistry is based on the assumption of chemical and mineralogical homogeneity of the shallow mantle, with so-called normal or depleted midocean-ridge basalt (NMORB and DMORB) representative of the homogeneity and depletion of the entire upper mantle source (*the convecting upper mantle*). The entire upper mantle is perceived to be a homogenous depleted olivine-rich lithology approximating pyrolite (pyroxene–olivine-rich rock) in composition; all basalts are formed by melting of similar peridotitic lithologies. Venerable concepts such as isolated reservoirs, plumes, temperature–crustal-thickness relations and so on are products of these perceived constraints. Absolute temperature, not lithologic diversity, is the controlling parameter in these models of geochemistry and geodynamics, and in the usual interpretations of seismic images and crustal thickness.

The perception that the mantle is lithologically homogenous is based on two assumptions: (1) the bulk of the upper mantle is roughly isothermal (it has constant potential temperature) and (2) midocean-ridge basalts are so uniform in composition ('the convecting mantle' is geochemical jargon for what is viewed as 'the homogenous well-stirred upper mantle') that departures from the basic average composition of basalts along spreading ridges and within plates must come from somewhere else. The only way thought of to do this is for narrow jets of hot, isotopically distinct, mantle to arrive from great depths and impinge on the plates.

The fact that bathymetry follows the square root of age relation is an argument that the cooling plate is the main source of density variation in the upper mantle. The scatter of ocean depth and heat flow – and many other parameters – as a function of age, however, indicates that something else is going on. Plume influence is the usual, but non-unique, explanation for this scatter, and for depth and chemical anomalies along the ridge. Lithologic (major elements) and isotope homogeneity of the upper mantle are two of the linchpins of the plume hypothesis and of current geochemical reservoir models. Another is that seismic velocities, crustal thicknesses, ocean depths and eruption rates are proxies for `mantle potential temperatures`. The asthenosphere, however, is variable in melting temperature and fertility (ability to produce magma) and this is due to recycling of oceanic crust and delaminated continental crust and lithosphere. In addition, seismic velocities are a function of lithology, phase changes and melting and are not a proxy for temperature alone. Some lithologies melt at low temperature and have low seismic velocities without being hotter than adjacent mantle. These can be responsible for melting and tomographic anomalies.

The isotopic homogeneity of NMORB has strongly influenced thinking about the presumed homogeneity of the upper mantle and the interpretation of 'anomalous' sections of midocean ridges. It is common practice to avoid 'anomalous' sections of the ridge when compiling MORB properties, and to attribute anomalies to 'plume-ridge interactions.' In general, anomalies along the ridge system - elevation, chemistry, physical properties - are part of a continuum and the distinction between 'normal' and 'anomalous' ridge segments is arbitrary and model dependent.

Mantle heterogeneity: toward a new paradigm

It is increasingly clear that the upper mantle is heterogenous in all parameters at all scales. The evidence includes seismic scattering, anisotropy, mineralogy, major- and trace-element chemistry, isotopes, melting point and temperature. An isothermal homogenous upper mantle, however, has been the underlying assumption in much of mantle geochemistry for the past 35 years.

One must distinguish *fertility* from (trace element) *enrichment*, although these properties may be related. *Fertility* implies a high basalt–eclogite or plagioclase–garnet content. *Enrichment* implies high contents of incompatible elements and long-term high Rb/Sr, U/Pb, Nd/Sm etc. ratios. Because of buoyancy considerations, the most refractory products of mantle differentiation - harzburgite and lherzolite - may collect at the top of the mantle and bias our estimates of mantle composition. The volume fractions and the dimensions of the fertile components - basalt, eclogite, pyroxenite, piclogite - of the mantle are unknown. There is no reason to suppose that the upper mantle is equally fertile everywhere or that the fertile patches or veins in hand specimens and outcrops are representative of the scale of heterogeneity in the mantle.

There are three kinds of heterogeneity of interest to petrologists and seismologists, radial, lateral and random, or statistical. Melting and gravitational differentiation stratify the mantle. Given enough time, a petrologically diverse Earth, composed of materials with different intrinsic densities, will tend to stratify itself by density. Plate-tectonic processes introduce heterogeneities into the mantle, some of which can be mapped by geophysical techniques. On the other hand, diffusion, chaotic advection and vigorous unidirectional stirring, are homogenizers. Convection is thought by many geochemists and modelers to homogenize the mantle. Free convection driven by buoyancy is not the same as stirring by an outside agent. Melting of large volumes of the mantle, as at ridges, however, can homogenize the basalts that are erupted, even if they come from a heterogenous mantle.

There are numerous opportunities for generating (and removing) heterogeneities associated with plate tectonics (Figure 24.1). The temperatures and melting temperatures of the mantle depend on plate-tectonic history and processes such as insulation and subduction cooling. Thermal convection requires horizontal temperature gradients; cooling from above and subduction of plates can be the cause of these temperature gradients. The mantle would convect even if it were not heated from below. Radioactive heating from within the mantle, secular cooling, density inhomogeneities and the surface thermal-boundary layer can drive mantle convection. An additional important element is the requirement that ridges and trenches migrate with respect to the underlying mantle. Thus, mantle is fertilized, contaminated and extracted by migrating boundaries - a more energy-efficient process than moving the mantle to and away from stationary plate boundaries, or porous flow of magma over large distances. Lateral return flow of the asthenosphere, and entrained mantle flow, are important elements in plate tectonics. Embedded in these flows can be fertile patches. Even if they are confined to the asthenosphere these patches will move more slowly than plates and plate boundaries, giving the illusion of fixed hotspots.

Creation of mantle heterogeneity

Recycling contributes to chemical and isotopic heterogeneity of the source regions of basalts but it also contributes to the fertility and productivity of the mantle. Temperature

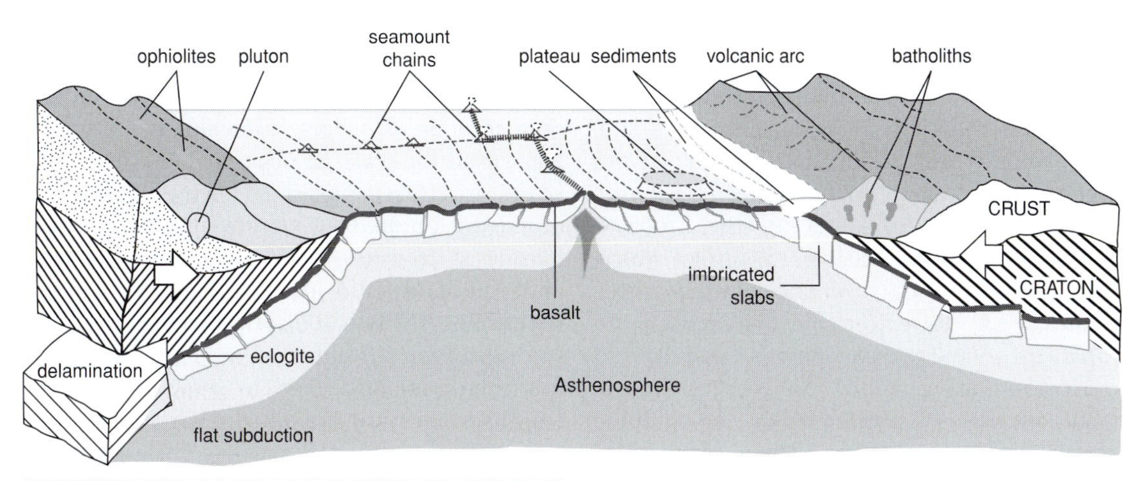

Fig. 24.1 Illustration of one end game of the plate-tectonic cycle – the closure of ocean basins. The other end of the cycle is continental breakup, when this diverse material can be involved in breakup magmatism. The many oceanic plateaus in the newly opened Atlantic and Indian ocean basins may be, in part, this reactivated material.

variations are long wavelength while chemical heterogeneity can be of the scale of slabs and the source regions of volcanoes. Melting anomalies may be primarily due to high homologous, not high absolute, temperature.

Basalts, mafic and ultramafic cumulates and depleted harzburgitic rock are constantly formed along the 60 000-km-long mid-ocean spreading-ridge system. The mantle underlying diverging and converging plate boundaries undergoes partial melting down to depths of order 50–200 km in regions up to several hundred kilometers wide, the processing zone for the formation of magmas - MORB, backarc-basin basalts (BABB) and island-arc basalts (IAB). Midplate volcanoes and off-axis seamounts process a much smaller volume of mantle, and the resulting basalts are therefore - as a consequence of the central limit theorem - much more heterogenous.

Before the oceanic plate is returned to the upper mantle in a subduction zone, it accumulates sediments and the harzburgites become serpentinized; this material enters the mantle (Figure 24.1). Plateaus, aseismic ridges and seamount chains also head toward trenches. About 15% of the current surface area of oceans is composed of young (<20 My) lithosphere and more than 10% of the seafloor area is capped by seamounts and plateaus. The delamination of over-thickened continental crust also introduces fertile material into the asthenosphere; this is warmer and perhaps thicker than subducted oceanic crust, and will equilibrate and melt sooner. These warm delaminates are potential fertile spots and can create melting anomalies. They may account for 5% of all recycled material.

The rate at which young oceanic crust and delaminated lower continental crust enters the mantle is comparable to the global rate of 'hotspot' volcanism, $\sim$2 km^3/yr. *Melting anomalies* may therefore be due to fertile blobs. The fates of older plates with thin crust, and deeper slabs are different.

Fate of recycled material

Large-scale chemical heterogeneity of basalts sampled along midocean-ridge systems occur on length scales of 150 to 1400 km. This heterogeneity exists in the mantle whether a migrating ridge is sampling it or not. Fertile patches, however, are most easily sampled at ridges and may explain the enigmatic relations between physical and chemical properties along ridges. Normal oceanic crust may start to melt after being in the mantle for $\sim$60 million years (Myr), while delaminated crust may melt after only 20–40 Myr because it starts out warmer (Figure 5.2).

Tomographic correlations suggest that middle-aged plates reside mainly in the bottom part of the transition region, near and just below 650 km. Plates that were young (<30 Myr) at the time of subduction (e.g. Farallon slab under western North America) and slabs subducted in the past 30 Myr may still be in the upper mantle. Old, thick slabs appear to collect at 750–900 km, the probable base of the layer accessible to surface volcanoes. The quantitative and statistical methods of determining the depth of subduction and comparison of tomographic and geodynamic models are superior to the visual analysis of selected color tomographic cross-sections – qualitative chromotomography. A chemically stratified mantle will have some deep high-velocity patches and some will appear to correlate with shallower structures.

Scale of mantle heterogeneity

In the plume model isotopic differences are attributed to different large (400–2000 km in extent) reservoirs at different depths. In the marble cake and plum-pudding models the characteristic dimensions of isotopic heterogeneities are centimeters to meters. Chemical differences between ridge and nearby seamount and island basalts may be due, in part, to the nature of the sampling of a common heterogenous region of the upper mantle. In order for this to work there must be substantial chemical differences over dimensions comparable to the volume of mantle processed in order to fuel the volcano in question, e.g. tens to hundreds of km.

Chemical differences along ridges have characteristic scales of 200–400 km. Inter-island differences in volcanic chains, and seamount chemical differences, occur over tens of km e.g. the Loa and Kea trends in Hawaii. If heterogeneities were entirely grain-sized or km-sized, then both OIB and MORB would average out the heterogeneity in the sampling process. If heterogeneities were always thousands of km in extent and separation, then OIB and MORB sampling differences could not erase this. Therefore, there must be an important component of chemical heterogeneity at the tens of km scale, the scales of recycled crust and lithosphere. The hundreds of km scales are comparable to the segmentation of ridges, trenches and fracture zones, and the scales of delaminated crust along island arcs. Chunks of slabs having dimensions of tens by hundreds of km are inserted into the mantle at trenches. They are of variable age, and equilibrate and are sampled over various time scales. The lateral dimensions of plates, and the separation distances of trenches and aseismic ridges are also likely to show up as scale lengths in chemical and physical variations along ridges.

The central limit theorem (CLT) is essential in trying to understand the range and variability of mantle products extracted from a heterogenous mantle. Depending on circumstances, small domains – tens to hundreds of km in extent – can also be isolated for long periods of time until brought to a ridge or across the melting zone. Mineralogy, diffusivity and solubility are issues in determining the size of isolatable domains. When a multi-component mantle warms up to its solidus – not necessarily the same as the surrounding mantle – the erupted magmas can be variable or homogenous; this is controlled by sampling theory, the statistics of large numbers and the CLT. Even under a ridge the melting zone is composed of regions of variable melt content. The deeper portions of the zone, and those regions on the wings, will experience small degrees of melting but these will be blended with high-degree melts under the ridge, prior to eruption. Magma cannot be considered to be uniform degrees of melting from a chemically uniform mantle. Blending of magmas is an alternate to the point of view that convection is the main homogenizing agent of mantle basalts. There are also differences from place to place and with depth, i.e. large-scale heterogeneities.

The isolation time of the upper mantle is related to the time between visits of a trench or a ridge. With current migration rates a domain of the upper mantle can be isolated for as long as 1 to 2 Gyr. These are typical mantle isotopic ages and are usually attributed to a convective-overturn time. Either interpretation is circumstantial.

Composition of OIB sources – eclogite?

Subducted or delaminated basalt converts to eclogite at depths greater than about 50 km. Eclogites are not uniform; they includes extrusives, dikes, sills and diverse cumulates and may be suitable protoliths for compositionally distinct and diverse ocean-island basalts. The possible roles of garnet pyroxenite and eclogite in the mantle sources of flood basalts and ocean islands have recently become a matter of renewed interest. The possibility that the shallow mantle is lithologically variable, containing materials with higher latent basaltic melt fractions than lherzolite, means that the mantle can be more-or-less isothermal on a local and regional scale, yet at given depth closer to the solidi of some of the lithologies than others. Fertile patches can also account for melting anomalies along the global ridge system. Thus, if the upper mantle is sufficiently heterogenous, plumes and high absolute temperatures are not required as an explanation for melting anomalies. The viability of the *thermal* plume hypothesis boils down to the viability of the assumption that the upper mantle is *chemically* homogenous [mantleplumes].

Chapter 25

Crystallization of the mantle

Rocks, like everything else, are subject to change and so also are our views on them.

Franz Y. Loewinson-Lessing

The Earth is cooling and crystallizing. The mantle has evolved considerably from the magma ocean era to the plate tectonic era. Part of the evolution is due to igneous processes, and part is due to plate tectonic processes. The visible rocks are the end, or present, product of mantle evolution. If these were our only source of information, we could come up with a fairly simple scheme of magma genesis, perhaps involving single-stage melting of a homogenous, even primitive, mantle. We could design simple one- and two-layer mantle models, such as were popular in the last century. It now seems unlikely that we will find the 'Rosetta Stone,' a rock or a meteorite fragment that represents 'original Earth' or even the parent or grandparent of other rocks. Rocks and magmas represent products of complex multistage processes, and they are mixtures of components with various melting histories. As we delve deeper into the Earth and further back in time, we depend more and more on isotopes and on modeling of planetary accretion and mantle processes. Melts and rocks are averages of various components and processes. Sampling theory and the manipulation of averages are involved in this modeling. In addition to recycling, and an intrinsically heterogenous mantle, there are igneous processes that cause chemical heterogeneity interior to the mantle. Melt trapping and melt transport can create distinctive chemical and isotopic components. Chemical heterogeneity can be formed internal to the mantle.

Components vs. reservoirs revisited

A variety of studies have lent support to the concept of a `chemically inhomogenous mantle`. The mantle contains a variety of *components* that differ in major elements, intrinsic density, melting point, large-ion-lithophile (LIL)-contents and isotope ratios; they have maintained their separate identities for at least 10^9 years. These *components* have been termed *reservoirs* but only very loose bounds can be placed on the sizes of these so-called *reservoirs*. They could be grain-size, or slab-size, depending on the ability of chemical species to migrate from one to another. A *component* can have dimensions of tens of kilometers, a typical scale for subducted or delaminated assemblages. The melting process gathers together these components, and partial melts therefrom. Some components or reservoirs have high values of Rb/Sr, Nd/Sm, U/Pb and ^{3}He/U. These components are *enriched*, but for historic reasons they are sometimes called *primitive*, *undegassed* or *more-primitive* (than MORB).

Lithology of the upper mantle

Since the seismic velocities and anisotropy of the shallow mantle are consistent with an olivine-rich aggregate, and since most mantle xenoliths are olivine-rich, it has been natural to assume that the shallow mantle, and the source region for various basalts, including OIB and NMORB, are the same and is a peridotite. Since the LIL-depleted components in the mantle have already lost a melt fraction, they should be depleted in garnet – and therefore infertile – unless they are eclogite cumulates, or delaminated lower continental crust. Peridotites depleted in basalt have less Al_2O_3 and garnet than nondepleted or fertile peridotites. But the depletion event may have involved a very small melt fraction, in which case the incompatible trace elements will be affected more than the major elements.

The traditional emphasis on homogenous olivine-rich and peridotite source regions for mantle magmas is based on the following arguments.

(1) Peridotite is consistent with seismic velocities for the shallow mantle; basalts come from the mantle; the upper mantle is therefore mainly peridotite.
(2) Garnet peridotite is stable in the upper mantle.
(3) Garnet peridotites have close compositional relationships to meteorites.
(4) Partial melts of natural samples of garnet peridotite at high pressure have basaltic compositions.
(5) Eclogites are the high-pressure chemical equivalent of basalt and partial melting of eclogite does not recreate the composition of basalt.
(6) Melting of eclogites would have to be very extensive, and melt–crystal segregation would occur before such extensive melting can be achieved.

These arguments are all suggestive rather than definitive. They do not rule out other lithologies for the upper mantle or for the source regions of at least some basalts. There is increasing evidence that large parts of the upper mantle are eclogitic or composed of garnet pyroxenite and are therefore more fertile than most peridotites. Most of the seismic information about the upper mantle is derived from seismic waves with wavelengths from 20 to 300 km. Much of the petrological information comes from midocean ridges and large volcanoes, which sample comparable size regions of the upper mantle. This averaging effect of geophysical and geochemical data can distort views regarding homogeneity of the mantle. But, there is no doubt that peridotites can and do come from the shallow mantle; some regions of the upper mantle and lithosphere are probably mostly peridotite. Even if basalts derive from eclogitic regions of the mantle they traverse and evolve in peridotitic surroundings.

The average composition of the Earth is probably close to ordinary chondrites or enstatite meteorites in major-element chemistry, the mantle therefore contains abundant, although not necessarily predominant, olivine. By the same reasoning the mantle contains even more pyroxene plus garnet. The above arguments do not prove that the source region of the most abundant basalt types, is garnet peridotite or that the regions of the mantle that appear to be peridotitic, on the basis of seismic velocities, are the regions where midocean basalts are generated. Although some of the older ideas about source regions, such as a glassy or basaltic shallow source, can be ruled out, the possibility that basalts involve eclogite, pyroxenite, recycled crust or cumulates, cannot be ruled out. Eclogite, garnet pyroxenite, peridotite–eclogite mixtures, or piclogite, are also candidate source 'rocks'; the 'grains' in such 'rocks' can be tens of km in extent. Hand-specimen-sized rocks are a different scale from what volcanoes and seismic waves see.

The trace-element inhomogeneity of the mantle plus the long-term isolation of the various components suggests that differentiation has been more effective in the long run than mixing. Mixing can be avoided in a chemically inhomogenous – or chemically stratified – mantle if the components are large chunks and/or have large intrinsic density and viscosity contrasts. Garnet has the highest density of any abundant upper-mantle mineral and therefore plays a role in determining the density of various components, and regions, of the mantle. However, the chemical heterogeneity may also be dispersed throughout the upper mantle. Eclogites come in a variety of compositions and densities; they all have low melting points compared to peridotites. Some eclogites have densities similar to some upper mantle peridotites; the density of eclogite is very

Table 25.1 Effect of eclogite and olivine fractionation on primitive magma

Magma	SiO_2	Al_2O_3	FeO	MgO	CaO	TiO_2	Na_2O	K_2O
1. Primitive	46.2	11.1	10.8	20.2	9.4	0.77	1.06	0.08
2. Extract	46.2	13.9	9.3	16.3	11.9	0.81	1.29	0.02
3. Picrite	46.2	8.3	12.3	24.1	6.9	0.74	0.83	0.14
Tholeiites								
4. Model	50.0	13.8	12.4	8.5	11.5	1.23	1.38	0.23
5. Hawaiian	50.0	14.1	11.4	8.6	10.4	2.53	2.16	0.39
6. Continental	50.6	13.6	10.0	8.5	10.0	1.95	2.90	0.54
7. Average oceanic	50.7	15.6	9.9	7.7	11.4	1.49	2.66	0.17

1. Possible primitive magma. The partial melt product of primitive mantle differentiation (O'Hara and others, 1975).
2. Eclogite extract (O'Hara and others, 1975).
3. Residual liquid after 50 percent eclogite (2) removal from primitive magma (1). This is a model picritic primary magma.
4. Residual liquid after a further removal of 40 percent olivine ($Fo_{8.75}$) from liquid (3).
5. Average Hawaiian parental tholeiite.
6. Continental tholeiite (Tasmania) (Frey and others, 1978).
7. Average oceanic tholeiite glass (Elthon, 1979).

dependent on composition and temperature. If the mantle is chemically stratified, mixing will be less vigorous and chemically distinct components can survive.

Part of this gravitational stratification will be irreversible. The coefficient of thermal expansion is high at low-pressure and high-temperature. This means that temperature can overcome intrinsic density differences. However, at high pressure, this is no longer possible and deep dense layers may be trapped. At lower mantle conditions, a chemically distinct layer with an intrinsic density contrast of ~1% can be stable against convective over-turn and mixing. Crystallization of a melt layer or magma ocean leads to a series of cumulate layers, and fractionation of the LIL.

Cumulate layers originally contain interstitial fluids that hold most of the incompatible elements. As crystallization proceeds, these melts may migrate upward. Melts from an eclogite or olivine eclogite cumulate have the characteristics of kimberlites. Removal of late-stage (kimberlite) intercumulus fluids from an eclogite-rich cumulate layer will deplete it and enrich the overlying olivine-rich layer. The enrichment, however, will be selective. It will be uniform in the very incompatible elements, giving apparently primitive ratios of Rb/Sr, Sm/Nd and such, but will impart a pattern of depletion in the HREE, yttrium, sodium, manganese and so on since these are the eclogite-compatible elements. Partial melts from a shallow enriched reservoir will therefore appear to have a garnet-residual pattern, even if this reservoir contains no garnet. This pattern can be transferred to any MORB magmas interacting with this layer.

Access to deep layers

Convection in a chemically stratified system causes lateral variations in temperature, and deformation of the interfaces because of the buoyancy of the uprising currents. If this deformation raises a chemical boundary across the solidus, or if the temperature is perturbed by, for example, continental insulation, then partial melting can generate buoyant diapirs, even in a dense eclogite-rich layer. Subsolidus reactions between garnet and clinopyroxene also occur at high temperature. This results in a temperature-induced density decrease much greater than can be achieved by thermal expansion. Adiabatic ascent of an eclogite blob, or a diapir from a buried eclogite or piclogite layer, can lead to extensive melting. Crystal settling and melt extraction can be avoided in a rapidly rising diapir because of the high temperatures, temperature gradients and stresses, and the surrounding envelope of subsolidus peridotite.

Formation of fertile regions in the mantle

The process of planetary accretion and melting during accretion is akin to a zone-refining process. The surface of the planet, where the kinetic energy of accretion is turned into heat, acts as the furnace; refractory, 'purified' material is fed into the planet. The incompatible elements and melts are preferentially retained near the surface. A deep magma ocean or whole planet melting – at any one time – is not required, or even desirable. It is not desirable since if the surface melt layer is in equilibrium with dense high-pressure phases such as perovskite, there should be anomalies in the trace-element patterns of upper mantle materials that are not observed.

Tables 25.1 and 25.2 and Figure 25.1 illustrate a petrological evolutionary scheme for the mantle. Early melting is likely to be extensive since large amounts of melt, 15–25%, occur in small temperature range just above the solidus. Eclogite and basalt extraction leave a peridotitic residue deficient in the basaltic elements, Ca, Al and Na. Olivine cumulates and enriched residual fluids are complements to the eclogite and basalt regions. Table 25.2 gives a more detailed comparison of the possible products of mantle differentiation.

These tables illustrate the plausibility of eclogite-rich regions in the mantle. An eclogite or basalt layer, or distributed blobs, representing about 10–15% of the mantle reconciles the major-element compositions inferred by cosmochemical, geophysical and petrological techniques, and can account for such elements as Nb, Ti and Zr.

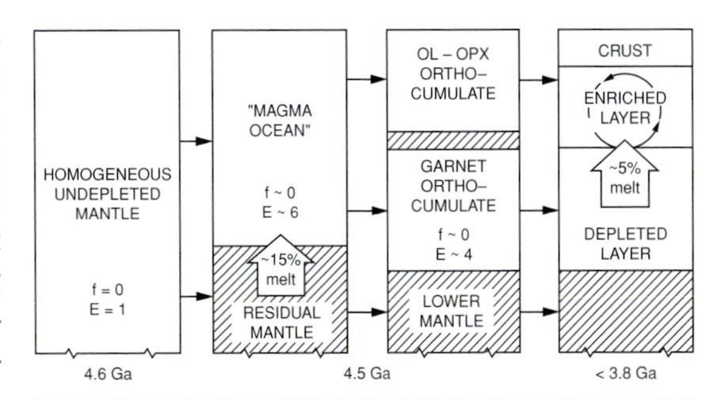

Fig. 25.1 Differentiation of a planet during accretion and early high-temperature evolution. E is the enrichment of incompatible elements, relative to the starting materials. These elements have low crystal-melt partition coefficients and therefore readily enter the melt fraction. *f* is the fractionation factor and gives the ratio of two incompatible elements in the melt, expressed as the difference from the starting material. Very incompatible elements occur in the same ratio in melts as in the original, or primitive, material. Isotopic ratios of these elements will evolve at the same rate as in primitive material. A magma ocean will therefore be enriched but unfractionated. As the magma ocean crystallizes, the fractionating crystals will either float or sink, leaving behind an enriched, fractionated residual liquid layer. This fluid may permeate the shallow mantle, giving an enriched geochemical signature to this region, and to the continental crust. The large difference in crystallization temperature and density of olivine–orthopyroxene (ol–opx), garnet, plagioclase and so on, means that mineralogically distinct regions can form in early Earth history.

The source regions for some basalts may be eclogite-rich cumulates or blobs that have been depleted by removal of a kimberlite-like fluid. Eclogitic layers or blobs become unstable at depth as they warm up. Garnet–clinopyroxene reactions and partial melting contribute to the buoyancy.

Early chemical stratification of the mantle

Chemical stratification resulting from early differentiation of the mantle, upward removal of the melt and fractionation via crystal settling during cooling is one way to explain chemically distinct reservoirs. In the first stage, probably during accretion, the incompatible elements (including Rb, Sr, Nd, Sm and U) are concentrated into melts (zone refined) and the shallow

Table 25.2 | Composition of mantle, upper mantle, picrites and eclogites

Material	SiO_2	Al_2O_3	FeO	MgO	CaO	TiO_2	Na_2O	K_2O
Mantle and Upper Mantle Compositions								
1. Bulk mantle	48.0	5.2	7.9	34.3	4.2	0.27	0.33	
2. Residual mantle	48.3	3.7	7.1	37.7	2.9	0.15	0.15	
3. Pyrolite	45.1	3.3	8.0	38.1	3.1	0.2	0.4	
Possible Picritic Parent Magmas								
4. Eclogite extract	46.2	13.9	9.3	16.3	11.9	0.81	1.29	0.02
5. Oceanic crust	47.8	12.1	9.0	17.8	11.2	0.59	1.31	0.03
6. Tortuga dikes	47.3	13.6	9.8	17.6	9.6	0.79	0.89	0.06
7. High-MgO	46.2	12.6	11.0	16.6	10.5	0.69	1.18	0.02
tholeiites	46.3	13.0	11.3	15.5	10.9	0.71	1.26	0.03
Kimberlite Eclogites								
8. Average	47.2	13.9	11.0	14.3	10.1	0.60	1.55	0.84
9. Roberts Victor	46.5	11.9	10.0	14.5	9.9	0.42	1.55	0.85

1. Bulk mantle composition (Ganapathy and Anders, 1974).
2. Residual after 20 percent extraction of primitive magma (line 1, Table 11.1).
3. This is an estimate of shallow mantle composition (Ringwood, 1975).
4. Possible eclogite extract from primary magma (O'Hara and others, 1975).
5. Average composition of oceanic crust (Elthon, 1979).
6. High magnesia Tortuga dike NT-23 (Elthon, 1979).
7. High magnesia-tholeiites.
8. Average bimineralic eclogite in kimberlite.
9. Eclogite, Roberts Victor 11061 (O'Hara and others, 1975).

mantle is more-or-less uniformly enriched in these elements (Figure 25.1). As the magma layer or magma ocean cools, cumulates, containing intercumulus fluids, form. Peridotitic cumulates at shallower depths. Cumulate layers can have near-primitive ratios of Rb/Sr and Sm/Nd if they contain a moderate amount of interstitial fluid. Transfer of late-stage melts (KREEP or kimberlite) is one mechanism by which parts of the mantle become depleted and other regions enriched. For this type of model, the isotopic ratios will be a function of the crystallization (fractionation) history of the upper mantle and the history of redistribution of LIL-enriched fluids.

A mechanism for creating an LIL-depleted but still fertile reservoir involves an early thick basalt layer in the relatively cold surface thermal-boundary layer. Small degrees of melt can be removed and still leave the basalt fertile. As the Earth cools this basalt layer converts to eclogite and sinks into the mantle, creating a depleted but fertile reservoir.

What is the fate of eclogite in the mantle? Some models assume that it sinks to the core–mantle boundary and is removed from the system; others assume that it is in the transition region or at the base of over-thickened crust. Estimated densities as a function of depth for eclogite and garnet peridotite are shown in Figure 25.2. Some eclogites and garnetites (garnet solid solutions) are denser than peridotite to depths at least as great as 500 km. On the other hand the post-spinel phases of olivine and the perovskite form of orthopyroxene are denser than garnetite or ilmenite eclogite. Eclogite-rich cumulates, or subducted eclogitic lithosphere, are therefore unlikely to sink below 650 km unless they are very garnet-rich and very cold. Whether eclogite can sink below 500 km depends on composition, temperature and the compressibility and thermal expansivity relative to

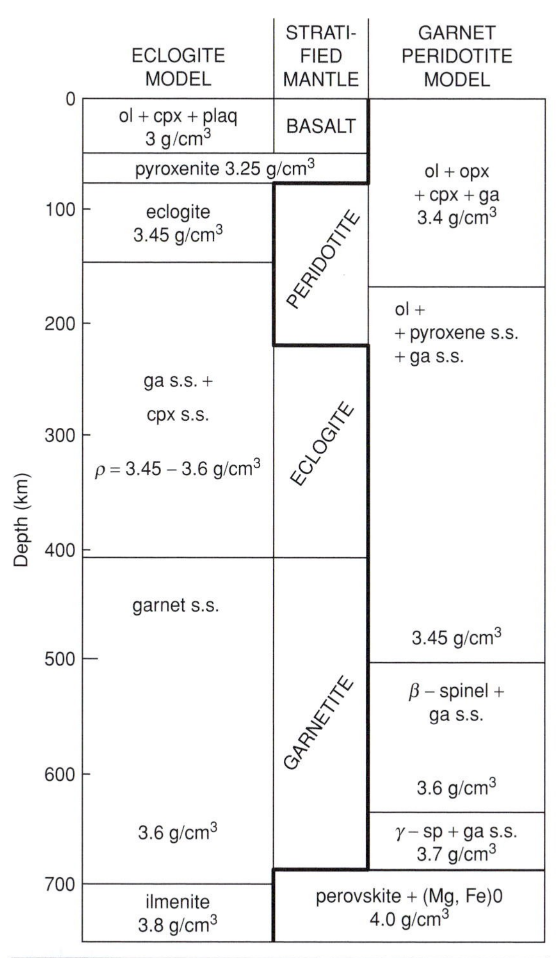

Fig. 25.2 Approximate densities of basalt/eclogite (left) and garnet peridotite (right). Eclogite (subducted oceanic lithosphere or a cumulate in a deep magma ocean) is denser than peridotite until olivine converts to β-spinel. Below some 400 km the garnet and clinopyroxene in eclogite convert to garnet solid solution. This is stable to very high pressure, giving the mineralogical model shown in the center, the gravitationally stable configuration. The heavy line indicates that the bulk chemistry varies with depth (eclogite or peridotite).

peridotite. Seismic tomography and the abrupt termination of earthquakes near 670 km depth suggests that oceanic lithosphere can penetrate to this depth.

Intrinsic density increases in the order basalt, picrite, depleted peridotite, fertile peridotite, garnet-rich eclogite. Some eclogites are less dense than some peridotites. Basalts and picrites crystallizing or recrystallizing below about 50 km can become denser than other upper-mantle assemblages. Garnet-poor and olivine-rich residues or cumulates are likely to remain at the top of the upper mantle since they are less dense than parental peridotites and do not undergo phase changes in the upper 300 km of the mantle. The zero-pressure density of a typical eclogite is about 15% and 3% denser than basalts and fertile peridotites, respectively. With a coefficient of thermal expansion of 3×10^{-5} °C, it would require temperature differences of 1000–5000 °C to generate similar density contrasts by thermal effects alone, as in normal thermal convection, or to overcome the density contrasts in these assemblages. Some eclogites are less dense than some mantle peridotites at depths greater than some 200 km. If the intrinsic density of the deep mantle is only 1% greater than the shallower layers, it could be permanently trapped because of the very low expansivity at high pressure.

Simple Stokes' Law calculations show that inhomogeneities having density contrasts of 0.1 to 0.4 g/cm^3 and dimensions of 10 km will separate from the surrounding mantle at velocities of 0.5 to 2.5 m/yr in a mantle of viscosity 10^{20} poises. This is orders of magnitude faster than average convective velocities. Inhomogeneities of that magnitude are generated by partial melting as material is brought across the solidus in the normal course of mantle convection. The higher mantle temperatures in the past make partial melting in rising convection cells even more likely and the lowered viscosity makes separation even more efficient. It seems unlikely, therefore, that chemical inhomogeneities can survive as blobs entrained in mantle flow for the long periods of time indicated by the isotopic data. Gravitational separation is more likely, and this leads to a chemically stratified mantle like that shown in Figures 25.3 and 25.4. The unlikely alternative is that the reservoirs differ in trace elements but not major elements, intrinsic density or melting point. Small differences in bulk chemistry change the mineralogy and therefore the intrinsic density and melting point; mineralogy is more important than temperature in generating density inhomogeneities.

The density differences among basalt, depleted peridotite, fertile peridotite and eclogite

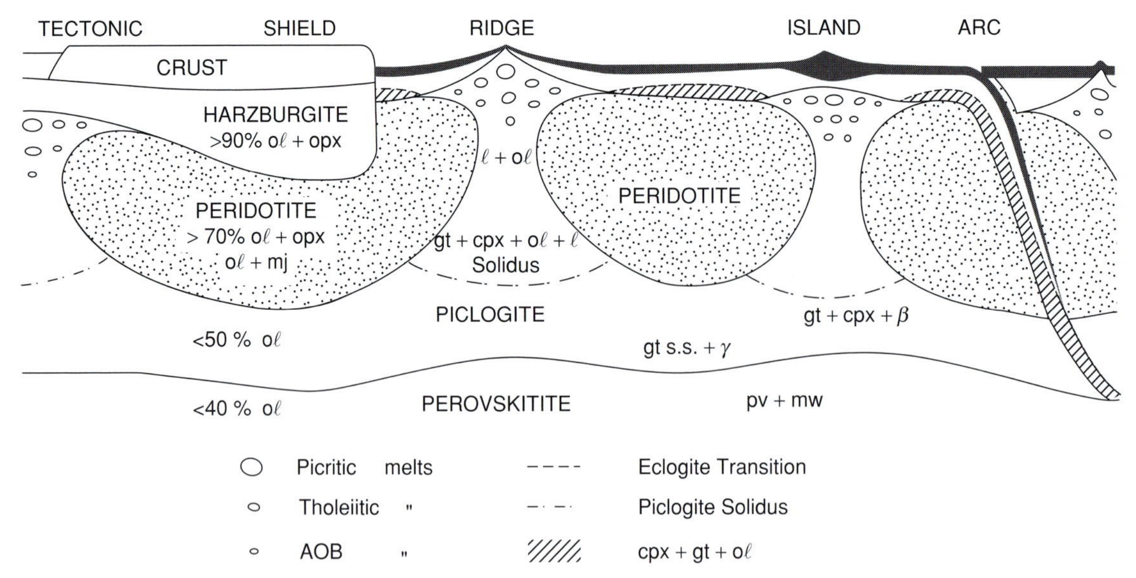

Fig. 25.3 A possible configuration of the major rock types in the mantle. Peridotite accumulates in the shallow mantle because of its low density, particularly residual or infertile peridotite. Fertile peridotite, eclogite, olivine eclogite and piclogite are denser and are, on average, deeper. Partial melting, however, can reverse the density contrast. Upwellings from fertile layers are shown under ridges, islands and tectonic continents; some are passive, some are entrained and some are caused by melt-induced bouyancy. The continental lithosphere is stable and is dominantly infertile harzburgite. It may have been enriched by upward migration of fluids and melts. The shallow mantle is predominantly infertile peridotite (stippled) with blobs of fertile eclogite. Deep fertile layers or blobs can rise into the shallow mantle if they become hot, partially molten or entrained. Melts can underplate the lithosphere, giving a dense, eclogite-rich base to the oceanic lithosphere, which can recycle into the transition region. The top parts of the subducted slab rejuvenate the shallow mantle with LIL. Some slabs are confined to the upper mantle, being too buoyant to sink into the lower mantle. Lower continental crust can also be an eclogite cumulate.

are such that they cannot be reversed by thermal expansion and the kinds of temperature differences normally encountered in mantle convection. However, phase changes such as partial melting and basalt–eclogite involve large density changes. A picritic or pyroxenitic crust or lithosphere, for example, will be less dense than fertile peridotite at depths shallower than about 50 km where it is in the plagioclase or spinel stability field. As the lithosphere or crust thickens and cools, it becomes denser at its base than the underlying mantle and a potential instability develops. If the crust in island arcs, batholiths or compressional mountain belts gets thicker than about 50 km it is prone to delamination. Similarly, if the temperature in a deep garnet-rich layer or blob exceeds the solidus, the density may become less than the overlying layer. The large density changes associated with partial melting, delamination and the basalt–eclogite phase change may be more important in driving mantle convection than thermal expansion. In the deep mantle, chemical variations are more important than temperature variations in controlling density and density contrasts.

Enriching fluids

The composition of a residual fluid in equilibrium with eclogite, as a function of crystallization, is shown in Figures 25.5 and 25.6. The fractionations increase rapidly as the residual melt fraction drops below about 20% (that is, above 80% crystallization).

Kimberlites may represent such late-stage or residual fluids. They appear to have been in equilibrium with eclogite and often contain eclogite xenoliths. They also have LIL patterns that are complementary to MORB. Isotopic evolution in regions enriched by expelled small-degree melts will deviate significantly from the

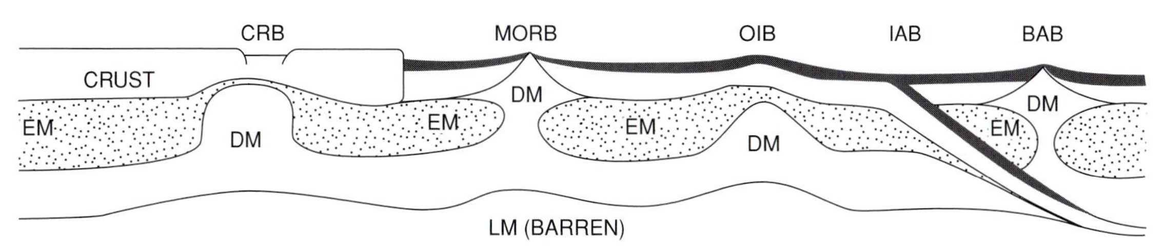

Fig. 25.4 Possible locations of geochemical components (DM, depleted mantle; EM, enriched mantle; LM, lower mantle; CRB, continental rift basalts; MORB, midocean-ridge basalts; OIB, oceanic-island basalts; IAB, island-arc basalts; BAB, back-arc basalts). In this model EM is heterogeneous and probably not continuous. It is isotopically heterogeneous because it has been enriched at various times. LM does not participate in plate tectonic or hotspot volcanism.

primitive-mantle growth curve (Figure 25.6). The U/Pb ratio appears to behave similarly to the Rb/Sr ratio unless sulfides are involved. For a simple crystallization history of the depleting reservoir, the fractionation factor of the melt increases rapidly with time, for example as an exponential or power law of time (Figure 25.5).

The distribution of oceanic ridges and hotspots suggests that a large part of the upper mantle is still above or close to the solidus.

The normalized Rb/Sr ratios of a melt and residual crystals can be written where

$$(\mathrm{Rb/Sr})_{\mathrm{m}} = \frac{F + (1-F)D_{\mathrm{Sr}}}{F + (1-F)D_{\mathrm{Rb}}} = (\mathrm{Rb/Sr})_{\mathrm{res}}\,(D_{\mathrm{Sr}}/D_{\mathrm{Rb}}) = f_{\mathrm{m}} + 1$$

the D are solid–melt partition coefficients and F is the melt fraction. As the cumulate freezes, continuously or episodically losing its fluids to the overlying or underlying layer, it contains less of a more enriched fluid. The net result is a nearly constant f for the cumulate as it evolves. Most of the fractionation that a crystallizing reservoir experiences occurs upon the removal of the first batch of melt.

Depleted reservoir become depleted by the removal of fluids representing late-stage interstitial fluids or small degrees of partial melting (Figure 25.1). The enriched fluid is not necessarily removed to the continental crust; it can also enrich the uppermost mantle and lithosphere. The enriched and depleted layers may differ in major elements and mineralogy, possibly the result of crystallization and gravitational separation. Partial melting of primitive mantle followed by crystallization and gravity separation gives upper-mantle source regions that, at least initially, have LIL ratios, including Rb/Sr and Sm/Nd ratios, similar to primitive mantle. Residual fluids in a cooling Earth, or a cooling cumulate, become more fractionated with time.

Partial melting of the mantle during accretion, melt separation, crystal fractionation and formation of cumulate layers is one model

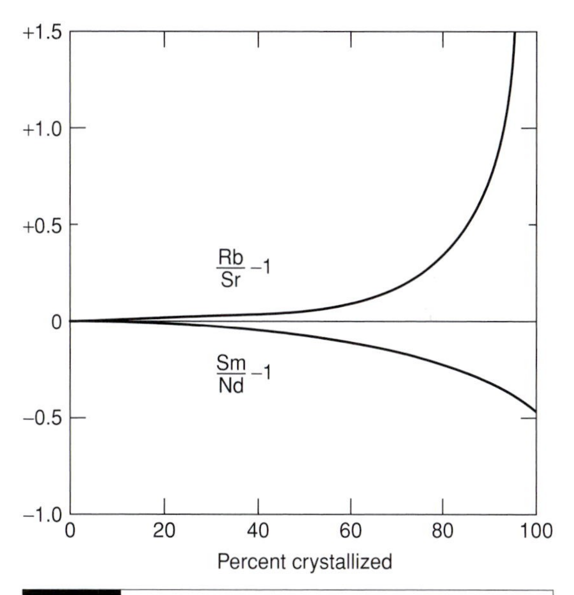

Fig. 25.5 Variation of the normalized Rb/Sr and Sm/Nd ratios in the melt fraction of a crystallizing eclogite cumulate. Equilibrium crystallization is assumed. The fractionation factor in the melt increases as freezing progresses. If melt extracts from this layer are the enriching fluids for upper-mantle metasomatism, then enrichment will increase as crystallization proceeds. Extents of melting and crystallization depend on depth and lateral locations as well as on time. Aggregated melts from a melting regime will have a variety of ratios. The crystalline residue is assumed to be 50:50 garnet: clinopyroxene.

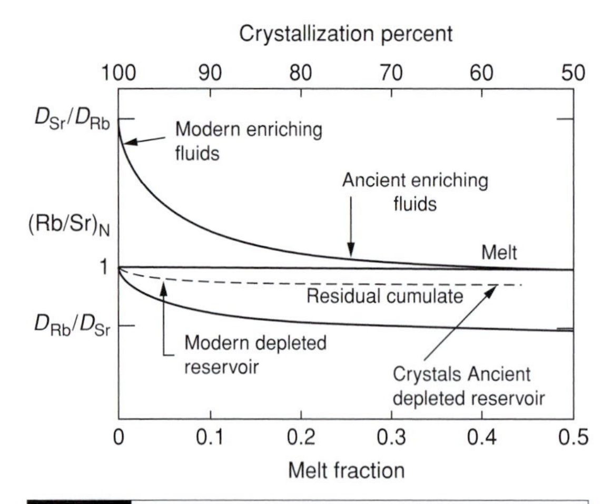

Fig. 25.6 Composition of the melt phase and the residual crystals as a function of melt fraction for a crystallizing eclogite cumulate. A hypothetical residual cumulate is also shown. The residual cumulate is formed by extracting half of the melt fraction. This shows that the composition of the depleted reservoir is nearly constant with time. Therefore, progressive depletion is less important an effect than progressive enrichment. D_{Sr} and D_{Rb} are partition coefficients.

that can explain the geophysical and geochemical observations. The transfer of KREEP-like or kimberlitic material can explain the depletion and enrichment of various lithologies. Similar scenarios have been developed for the Moon.

The role of magma mixing

Melting regions in the mantle involve large degree melts toward the center and low-degree melting on the wings and at greater depth. Melts pond beneath the crust or lithosphere, collect in magma chambers, and mix or blend or hybridize.

When two magmas are mixed, the composition of the mix, or hybrid, is

$$xC_i^1 + (1-x)C_i^2 = C_i^{\text{mix}}$$

where x is the weight fraction of magma i, and C_i^1, C_i^2 and C_i^{mix} are, respectively, the concentration of the ith element in magma 1, magma 2 and the mix. Mixing relations for elements are therefore linear.

The mixing relations for ratios of elements or isotopic ratios are more complicated. For example,

$$(\text{Rb/Sr})_{\text{mix}} = \frac{x(\text{Rb/Sr})_1 + (1-x)(\text{Rb/Sr})_2(\text{Sr}_2/\text{Sr}_1)}{x + (1-x)(\text{Sr}_2/\text{Sr}_1)}$$

$$(^{87}\text{Sr}/^{86}\text{Sr})_{\text{mix}} = \frac{x(^{87}\text{Sr}/^{86}\text{Sr})_1 + (1-x)(^{87}\text{Sr}/^{86}\text{Sr})_2(\text{Sr}_2/\text{Sr}_1)}{x + (1-x)(\text{Sr}_2/\text{Sr}_1)}$$

These are hyperbolas, and the shape or curvature of the hyperbola depends on the enrichment factor E. Depleted magmas when mixed with an enriched magma can appear to be still depleted for some elemental and isotopic ratios, undepleted or 'primitive' for others, and enriched for others, depending on E. This simple observation can explain a variety of geochemical paradoxes. For example, many basalts are clearly enriched, relative to primitive mantle, in some isotope and trace element ratios but enriched in others. Note that trace element and isotope ratios cannot simply be averaged; mixtures are weighted averages. Likewise, the means and variances of ratios have little meaning without information about the concentrations.

Trace-element and isotope data for magmas sometimes appear to be inconsistent. The incompatible elements and strontium and neodymium isotopes show that some abyssal tholeiites (MORB) are from a reservoir that has current and time-integrated depletions of the elements that are fractionated into a melt. MORBs, however, have Pb-isotopic ratios suggesting long-term enrichment in U/Pb. Alkalic basalts and tholeiites from continents and oceanic islands are derived from LIL- and U/Pb-enriched reservoirs. Strontium- and neodymium-isotope ratios, however, appear to indicate that some of these basalts are derived from unfractionated reservoirs and others from reservoirs with time-integrated depletions. These inconsistencies can be reconciled by treating oceanic and continental basalts as mixtures of magmas from depleted and enriched reservoirs, or components, and as blends of small-degree and large-degree melts. MORBs can be thought of as slightly contaminated, depleted magmas, or as large-degree melts of a heterogenous source plus some small-degree melts. Oceanic-island and continental basalts are different, less homogenous, mixtures. The

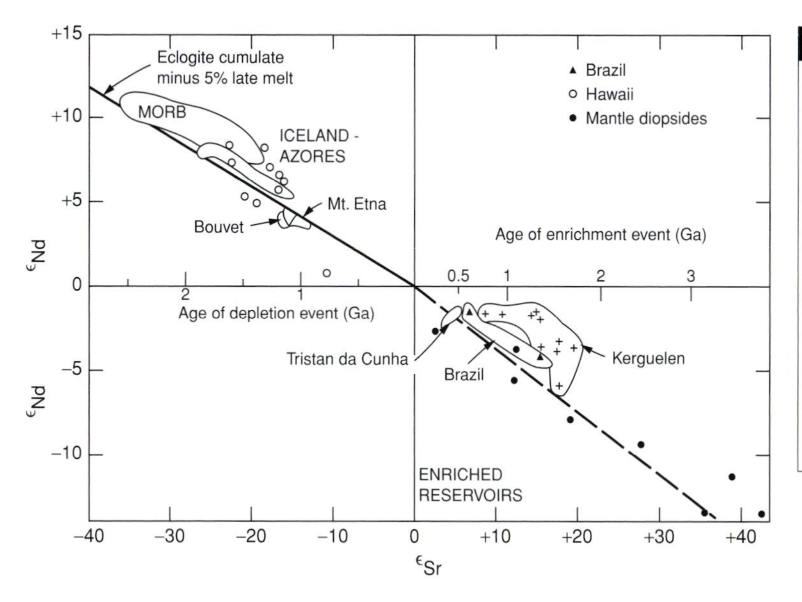

Fig. 25.7 Neodymium and strontium fractionation trends for an eclogite cumulate (solid line) that has been depleted at various times by removal of a melt fraction. The dashed line is the complementary reservoir that has been enriched at various times by the melt extract. Reservoirs are unfractionated until the enrichment/depletion event and uniform thereafter. If present values of enrichment/depletion have been reached gradually over time or if the magmas are mixtures, then ages shown are lower bounds.

mixing relations are such that mixtures can be enriched in U/Pb, Rb/Sr, Nd/Sm or $^{206}Pb/^{204}Pb$ relative to primitive mantle, yet appear to have time-integrated depletions in $^{143}Nd/^{144}Nd$ and $^{87}Sr/^{86}Sr$. A small amount of contamination by an enriched component or material from an enriched reservoir can explain the lead results for MORB (the 'lead paradox'). Depleted basalts are more sensitive to lead than to rubidium or neodymium contamination. Similarly, continental and oceanic-island basalts may represent mixtures of enriched and depleted magmas, or small- and large-degree melts.

Figure 25.7 shows the correlation in differentiation for MORB, oceanic islands, and some continental basalts and mantle diopsides. The depletion and enrichment ages are calculated for a simple two-stage model for the development of the enriched and depleted reservoirs (Figure 25.8). The Rb/Sr and Sm/Nd ratios are assumed to be unfractionated up to the age shown and then fractionated to values appropriate for the depleted and enriched reservoirs. Subsequent isotopic evolution occurs in these fractionated reservoirs.

For this kind of model the MORB reservoirs were apparently depleted and isolated at times ranging from 1.5–2.5 Ga, and the enriched reservoirs (giving magmas in the lower-right quadrant) were enriched between 0.5 and 1.8 Ga. If the enrichment has been progressive, the start of enrichment could have been much earlier. The data shown in Figure 25.7 may be interpreted in terms of mixtures of magmas from depleted and enriched reservoirs. In fact, the compositions of alkali olivine basalts, basanites and continental tholefites are bracketed by MORB and potassium-rich magmas such as nephelinites for most of the major and minor elements as well as for the isotopes. This supports the possibility that many continental and oceanic-island basalt types are mixtures.

The isotopic ratios of end-member MORB are greatly affected by small degrees of contamination, contamination that is probably unavoidable if MORB rises through, or evolves in, lithosphere or enriched upper mantle. Basalts at anomalous ridge segments show clear signs of contamination, as in T- and P-MORB (transitional and plume-type MORB); normal MORB may simply show less obvious signs of contamination. Magmas containing 10–20% contaminant will still appear isotopically depleted for Nd and Sr. Such mixtures will appear to exhibit long-term enrichment in the lead isotopic systems. Clearly, the mixing idea can be extended to multiple components, and to systems that are cooling and fractionating as they mix.

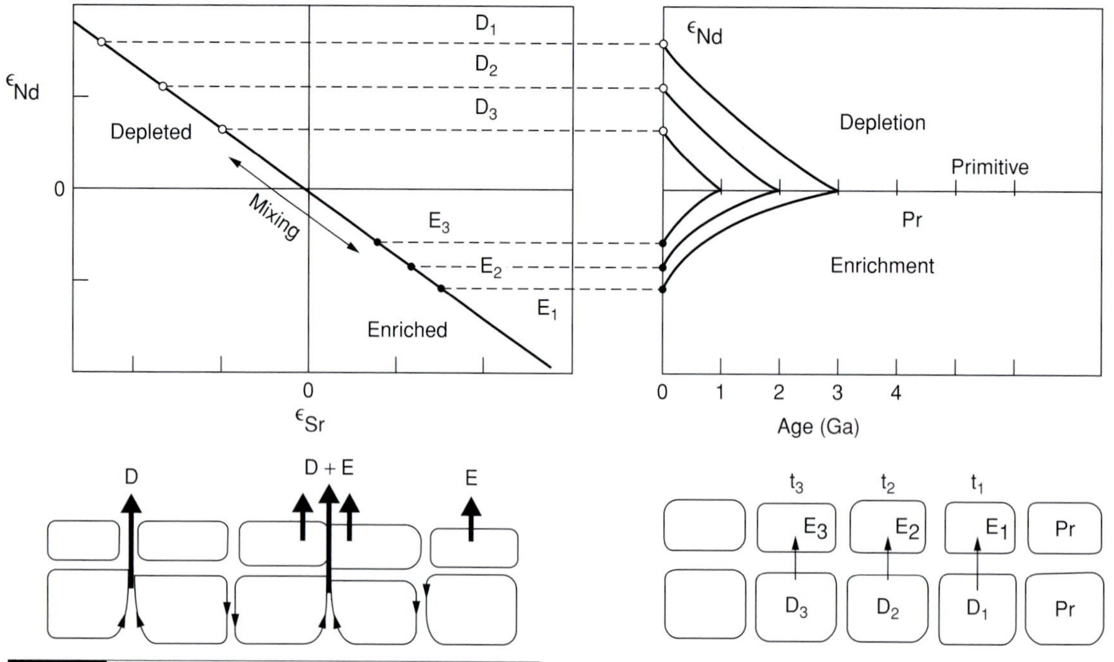

Fig. 25.8 Illustration of isotopic growth in a two-layer mantle. The lower layer is formed of heavy cumulates, perhaps at the base of a magma ocean. As it freezes it expels enriched fluids to the shallow layer (E), thereby becoming depleted (D). As time goes on, and crystallization proceeds, the melts become more enriched and more fractionated. Isotopic growth is more rapid for the parts of the shallow mantle enriched at later times, but the earlier enriched reservoirs (E1) have had more time for isotopic growth. The mantle array can be interpreted as the locus of points representing magmas from different-aged reservoirs or as a mixing array between products of enriched and depleted reservoirs, or some combination. Melts from D may be contaminated by E if they cannot proceed directly to the surface. If E is a trace-element-enriched, but infertile, peridotite, then D may be the main basalt source region, and enriched basalts may simply represent contaminated MORB. The enriched component in E may be kimberlitic. The deeper layer transfers its LIL upward, forming a depleted layer (D) and a complementary enriched layer (E). The growth of ε_{Nd} in the depleted and enriched cells (upper right) combined with similar diagrams for Sr generates the mantle array (upper left). Magma mixing reduces the spread of values, decreasing the apparent ages of the depletion/enrichment events. Layer D may be the transition region. The cumulate layer could also be lower crustal cumulates.

Melts from enriched or heterogenous mantle

Even if the enriched parts of the mantle are homogenous, their partial melts will have variable LIL contents and ratios such as Rb/Sr, Sm/Nd and U/Pb that depend on the extent of partial melting. Magma mixtures, therefore, may appear to require a range of enriched end-members. A plot of an isotopic ratio versus a ratio such as Rb/Sr, Sm/Nd or La/Sm may exhibit considerable scatter about a two-component mixing line even if the end-members are isotopically homogenous. An example is shown in Figure 25.9.

The Rb/Sr of partial melts from this reservoir are also shown. The solid lines are mixing lines between these melts and a melt from the depleted reservoir having properties estimated for 'pure MORB.' The dashed lines are labeled by the fraction of MORB in the mixture. The data points are representative compositions of various basalts; most fall in the field representing 50–95% MORB and an enriched component representing 2–20% melt from the enriched reservoir. The Rb/Sr ratio may also be affected by crystal fractionation and true heterogeneity of one or both of the two source regions.

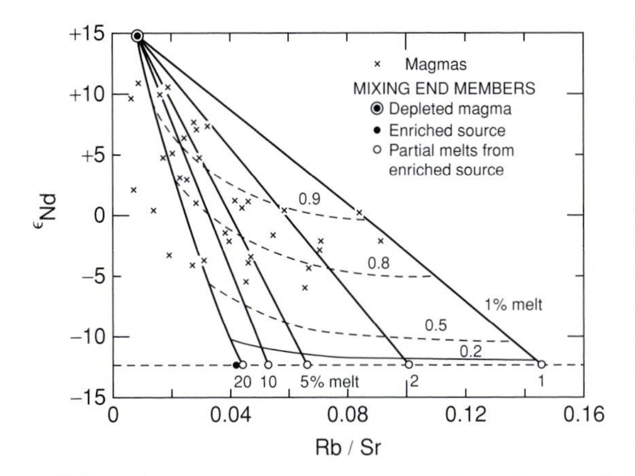

Fig. 25.9 Mixing relations for a depleted magma and partial melts from an enriched peridotite reservoir. The solid lines are mixing lines, and dashed lines give the fraction of the depleted component.

The hypothesis that oceanic and continental magmas represent mixtures of melts from diverse lithologies – enriched and depleted components – appears capable of explaining a variety of geochemical data and resolving some isotopic paradoxes. The hypothesis explains apparently contradictory trace-element and isotopic evidence for enrichment and depletion of the various mantle reservoirs. LIL-enrichment associated with time-integrated isotopic depletion is a simple consequence of mixing relations. Hotspot, oceanic-island and continental flood basalts are mixtures of a variety of components, including recycled oceanic crust, delaminated continental crust, 'contaminated' MORB and so on. If basalts are mixtures, the isotopic ages of their parent components or reservoirs will be underestimated, the sources are more enriched or depleted in their isotopic ratios than the hybrid magmas. The close proximity in time and space of enriched and depleted magmas in all tectonic environments – continental rifts, oceanic islands, fracture zones, midoceanic rifts and arcs – and the diversity of small-scale samples, supports the concept of a heterogenous mantle. Part of the heterogeneity is due to plate-tectonic processes, as discussed in earlier chapters, and part is due to igneous petrology, the transfer around of small degree melts. The final stage of eruption involves commingling and homogenization of melts. Mantle convection and stirring probably has little to do with the homogeneity of midocean-ridge basalts and observed mixing arrays.

Part VII

Energetics

By 2025, a population of 8.2 billion would require an energy use of 55 TW!

WISE News Communique November 27, 1992

Overview

Lord Kelvin assumed that the Earth started as a molten ball and calculated that it cooled to its present condition by thermal conduction. The still molten part was kept uniform by convection and the frozen bits sank to the center. Kelvin knew that Earth's temperature increased downward into deep mines and guessed that the Earth began as molten rock at 7000 °F. By solving Fourier's equation, Kelvin found that it would take a hundred million years for the Earth's temperature gradient to level out to one degree every 50 feet. In numbers haughty for their implied plus or minus nothing, Kelvin's final estimate, in 1897, for the age of the Earth was 24 million years. This calculation established that the Earth had a finite age rather than the prevailing geologic wisdom that there was '*no vestige of a beginning, no prospect of an end*' or that Earth's age was 'incomprehensibly vast'.

On hindsight we know that uncertainties in the assumptions and parameters are such that Kelvin could have concluded that an Earth cooled by conduction was tens to hundreds of times older than he published, even without internal heat sources. A convecting and cooling Earth also can account for the present heat flow, in the time available, even without radioactivity, or, alternatively, with radioactivity but without secular cooling. Clearly, the problem is ill-posed and nonunique without additional constraints. Radioactivity, minor heat sources and secular cooling (fossil heat) contribute to the observed heat flow. Currently, there is controversy about how to interpret the observed heat flow, and how it should be partitioned as to its source. Present conclusions regarding Earth's thermal history are no less dependent on assumptions, mathematics, physics and information from other disciplines, including meteoritics, than was the case in the time of Kelvin, Darwin and Hutton. The initial conditions of Earth's formation are important, even if the Earth has forgotten its initial temperature. Gravitational stratification, differentiation and degassing control the subsequent evolution.

Some geodynamicists claim that their understanding of the thermal evolution of the Earth is in disagreement with geochemical data, such as estimates of radioactive heating. Geochemists believe that the standard model of mantle geochemistry is now stumbling over the difficulty imposed by convection models and their interpretation of mantle tomography. This is similar to the dispute between Lord Kelvin and the geologists regarding the age of the Earth. The geophysical approach uses simplified parameterized models of mantle cooling or approximations of the depth dependence of thermal properties. The calculated rate of cooling of the Earth at the beginning of its history is too rapid to allow a sufficient present-day secular cooling rate to explain the mismatch. Geochemical estimates of U, Th and K in the mantle are too low to explain the observed present mantle heat loss. Cooling rates can be lowered by continental insulation, by chemically layered mantle structures and by having large aspect ratio convection cells, dictated by surface plates rather than the depth of the mantle. The heat transfer rate strongly depends on the wavelength of convection. The length scale of convection in Earth's mantle is that of plate tectonics, implying wide convective cells – large aspect ratio. A large horizontal wavelength of convection can significantly reduce the efficiency of heat transfer. The likely variations of continental insulation and wavelength with time imply important variations of the heat flow on timescale of 100 Ma.

The proposition that the pattern of convection is preferentially the one that maximizes the heat transfer efficiency is not supported by experiments or calculations. Large plates and large convective cells do not favor maximal heat transfer efficiency. The present situation – smallish continents and midocean ridges well distributed over the surface of the Earth – probably leads to an abnormally present high heat loss. Something is probably minimized or maximized in the operation of plate tectonics but efficiency of heat loss is not the control parameter. If the present day heat flow through the surface is higher than average, there may be no heatflow paradox.

Chapter 26

Terrestrial heat flow

During the thirty-five years which have passed since I gave this wide-ranged estimate [of 20–400 million years] experimental investigation has supplied much of the knowledge then wanting regarding the thermal properties of rocks to form a closer estimate of the time which has passed since the consolidation of the earth, we have now good reason for judging that it was more than 20,000,000 and less than 40,000,000 years ago, and probably much nearer 20 than 40.

Lord Kelvin

Heat losses

The nature of the surface boundary condition of the mantle changes with time. Currently, the mantle has a `conduction boundary layer` with a thickness that averages 100–200 km. The boundary layer is assumed to start out at zero thickness at volcanic ridges; it is pierced in places by volcanoes that deliver a small fraction of the Earth's heat to the surface via magma, and it may be invaded at greater depths by sills and dikes that affect the bathymetry and heat flow. Ridges also jump around, migrate or start on a pre-existing TBL. The cooling of the mantle is mainly accomplished by the cooling of the surface plates. In early Earth history a transient magma ocean allowed magmas to transfer their heat directly to the atmosphere. As buoyant material collected at the top, the partially molten interior became isolated from the surface. Magma, however, could break through a possibly thick buoyant layer and create 'heat pipes' to carry magma and heat to the surface. Io, Venus and early Mars are objects that may utilize this mechanism of heat transfer. It is also an alternative to plate tectonics on early Earth. The surface boundary condition in these cases can be viewed as a permeable plate. Intrusion affects the topography and heat flow, making these parameters non-unique functions of age.

The Earth's interior is cooling off by a combination of thermal conduction – and intrusion – through the surface boundary layer and the delivery of cold material to the interior by slabs, a form of advection. An unknown amount of heat is transferred to the surface by hydrothermal circulation. The delamination of the bottom of over-thickened crust also cools off the underlying mantle. The heat generated in the interior of the Earth, integrated over some delay time, is transferred to the surface conduction boundary layer by a combination of solid-state convection, fluid flow, radiation and conduction. The conducted heat through the surface TBL (there may be deeper ones as well) can be decomposed into a steady-state (or declining) background term, a transient term, and a crustal contribution. In

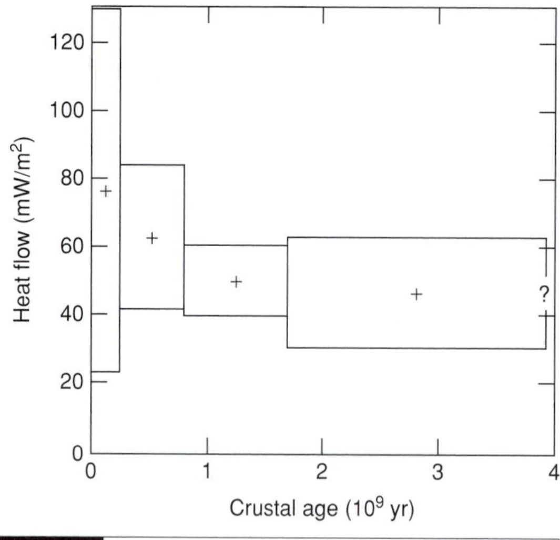

Fig. 26.1 Continental heat flow as a function of crustal age.

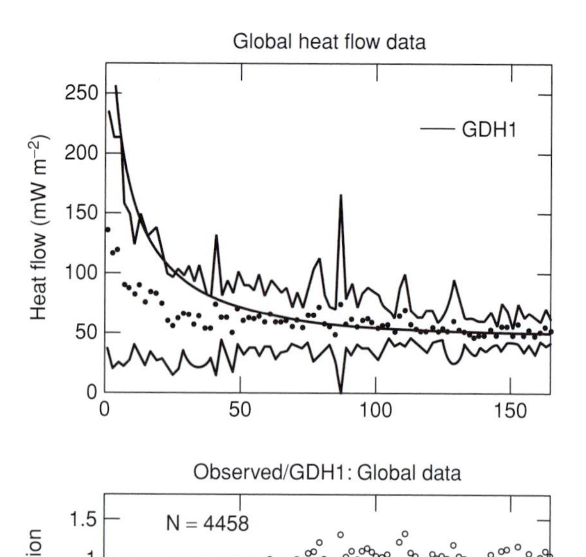

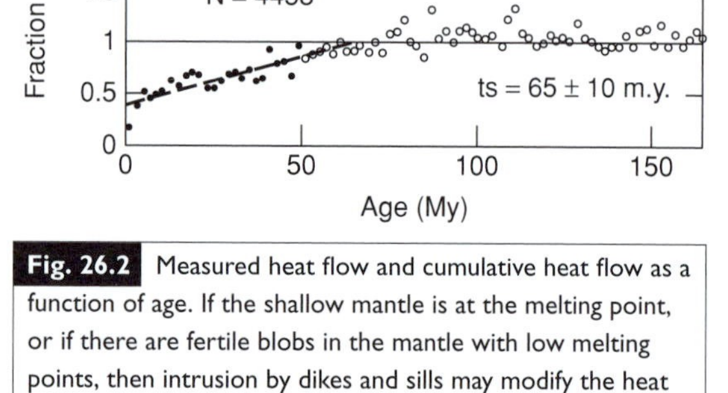

Fig. 26.2 Measured heat flow and cumulative heat flow as a function of age. If the shallow mantle is at the melting point, or if there are fertile blobs in the mantle with low melting points, then intrusion by dikes and sills may modify the heat flow. The top panel shows heat flow vs age with one-standard deviations (Stein and Stein, 1994). [constraints on hydrothermal heat flux]

the continents, heat flow is often plotted as a function of age (Figure 26.1) or the time since the last tectonic or igneous event and the long-term asymptotic value is taken to be the background heat flow. The transient effect has a time constant of more than 200 Myr and this is stretched out further by erosion, which strips off the radioactive-rich outer layers. In some compilations, the transient effect is discounted in estimating continental and global heat flow. Crustal radioactivity is a major contributor to continental heat flow and lateral variations in this heat flow.

In the ocean basins the main contribution to the observed heat flow is the transient effect, the formation of the oceanic crust itself. Theoretically, heat flow should die off as the square-root of age but it is nearly constant after the initial transient (see Figure 26.1). The background oceanic heat flow is nearly the same as under continents, perhaps slightly larger. There is little evidence that hotspots or swells are associated with high heat flow. In contrast to predictions from the plate and cooling half-space models there is little correlation of heat flow with age or depth (Figure 26.2). Measured heat flow is not a unique function of age. This indicates that the mantle is not isothermal (characterized by a single potential temperature) or homogenous (in composition and thermal properties).

The cold outer shell of the Earth is not simply a cooling boundary layer of uniform composition and conductivity, losing heat by conduction alone. Ocean bathymetry is more uniquely a function of square-root age, suggesting that some process affects the near-surface thermal gradient without affecting the integrated density of the outer layers.

Global heat flow

Global heatflow compilations are readily available (www.heatflow.und.edu) (http://www.geo.lsa.umich.edu/IHFC/heatflow.html).

The total heat flow through the surface of the Earth from the interior, based on *measured heat flow*, is about 30 TW. Various corrections and adjustments are made to the data and some workers think the adjusted total heat flow may be closer to 44 TW but this is based on assumptions.

Table 26.1 Summary of heat flow observations

Input	TW
Potential energy contributions	
Mantle differentiation and contraction	3
Heat from core	8
Core differentiation	1.2
Conduction down adiabat	6
Inner core growth	0.5
Earthquakes	2
Tidal friction	1
Current radiogenic (BSE)	28
Delayed radiogenic (1–2 Gyr)	5–15
Secular cooling (50–80 K/Gyr)	9–14
Total	42–57
Radiogenic + other	56–71
Output	
Global heatflow (observed)	30–32
Cooling plate model (theoretical)	44
Regions of excess magmatism	2.4–3.5

About 28 TW are generated by radioactive decay in the interior. There are about 10 TW of non-radiogenic heat sources such as cooling and differentiation of the core, contraction of the mantle, tidal friction and so on. On a convecting planet one expects temporal variations in heat flow of at least 10%. The secular cooling of the Earth contributes somewhere between 30–60% of the measured heat flow. Thus, there is either a good match between heat production and heat flow, or there is a deficit or a surplus of heat. Some workers have declared an energy crisis, or a missing heat-source problem. This crisis is similar to the crisis precipitated by Lord Kelvin and his age of the Earth.

A summary of the energy inputs and outputs of the mantle and core are given in Table 26.1. The total radiogenic and secular cooling amounts to 42–57 TW, while the current radiogenic production is only 28 TW. The observed conducted heat flow loss is 30–32 TW. Of this, about 2.4–3.5 TW is from the vicinity of hotspots. An unknown amount of heat loss is due to hydrothermal circulation. A generous allowance for this brings the heat loss through the surface to 44 TW [mantleplumes heat flow].

The local heat flow from the interior is estimated by drilling holes and measuring temperature gradients and thermal conductivity. Clearly, the surface of the Earth is not densely or uniformly covered by such holes. The most straightforward way of estimating the global terrestrial heat flow is simply to average the data in an appropriate way. A spherical harmonic expansion of heat flow data smooths it, and serves as an interpolation scheme; however, it is not necessarily appropriate for heat flow, tomography, bathymetry or other functions that are not potential functions. Data can be binned (by region, age, tectonomagmatic age and so on) to minimize the uneven spatial distributions of the measurements. In practice, averages are calculated in various tectonic provinces since the global dataset is not uniformly dense. Various 'corrections' are applied to the raw data so that estimates of global power are model dependent. Examples of these corrections are: replacing oceanic measurements with predictions from a theoretical cooling model, adding in an arbitrary or theoretical amount of hydrothermal heat flow – which is well known only near ridges, removing transient effects from tectonic or magmatic events, and eliminating data from areas thought to be affected by hotspots. Some workers argue that it is preferable to base surface heat flow analysis not only on the extensive measurements but also on processes that are thought to bias the measurements. This has become a contentious issue. Global heat flow maps show a strong age dependency that is lacking in the data; this is a result of the correction.

The dramatic effects of hydrothermal circulation on surface heat flux have been extensively documented on young (<20 My) seafloor but theory and data are lacking for old seafloor. The magnitude of the assumed hydrothermal correction to measured values of heat flow is essentially the same as the so-called missing heat source.

Continental and oceanic heat flow data are treated differently. The secular decay of the heat flow in continents is often considered

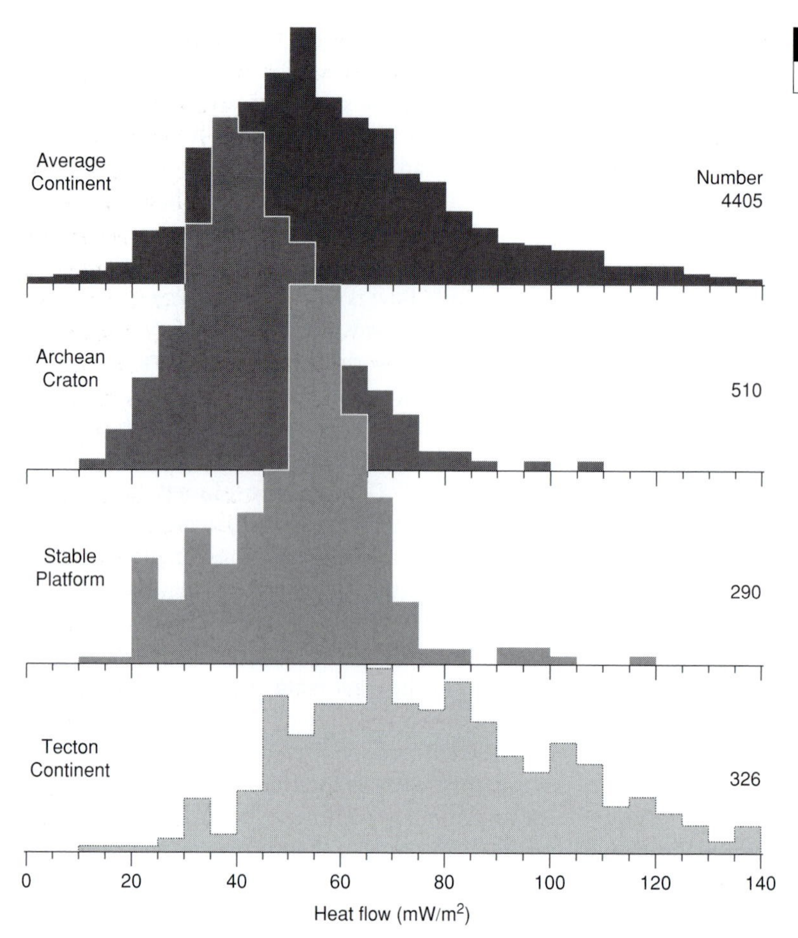

Fig. 26.3 Continental heat flow data.

to be noise (erosion, tectonomagmatic heating) while it is the main signal in oceanic areas (the plate creation process itself). Histograms of heat flow data are far from Gaussian (Figure 26.3) so median values are often tabulated. The uncorrected continental and oceanic heat flow histograms have similar means and medians. Nevertheless, the mean heat flux through the seafloor may be substantially higher than that through continents. It may be just coincidental that the most frequent values and means of the two datasets are so close to each other.

The non-uniform spatial distributions of both continental and oceanic data are partly the result of an emphasis on geothermally active and other anomalous areas, and sedimentary basins. The continental histogram is more peaked than the oceanic one. This is expected if vigorous hydrothermal circulation occurs on the seafloor. Histogram comparisons can be misleading. Data are often edited to eliminate unusual environments, or to force agreement with some theoretical expectation. This is quite common in geochemistry; the impressive chemical uniformity of MORB is partially the result of avoiding *anomalous* areas, and removing data thought to be influenced by plumes. Oceanic heat flow and bathymetry is likewise biased by ignoring or eliminating data thought to be influenced by *hotspots*. Unbiased sampling of a heterogenous population requires that sampling be as uniform as possible in order that statistics can be done. One cannot test a hypothesis if a hypothesis was used to select and prune the data. Means and standard deviations have no meaning if a hypothesis-dependent filtering of the data has taken place prior to the application of the statistics. In spite of data processing and data selection, regions that have been designated as hotspots appear not to have anomalous heat flow or to exhibit anomalous subsidence [mantleplumes].

The various components of the global heat flow budget are given in Table 26.1. [heat flow histograms seismic tomography]

Heat loss through continents

The mean heat flow from continents is about 80 mW/m^2. The median value is closer to 60 mW/m^2. About half the heat flow through continents is from the mantle. Continental crust produces about 0.6–1.2 mW/m^3 of heat and this accounts for 5.8–8 TW of the global heat flow. The continental heat flow that is attributed to the crust itself is 32–40 mW/m^2. Continents affect the style of mantle convection and, in fact, influence the rate at which heat is lost from the mantle. The thermal history of the Earth must take into account the properties of continents and plates and the fact that the surface boundary condition is not uniform or constant.

Heat loss through oceans

The estimated mean oceanic heat flux *including the unmeasured hydrothermal flux* is about 50% larger than the mean continental heat flux. There is a large difference between the average (118 mW/m^2) and the median (65 mW/m^2) value for the heat flux from the ocean floor. In order to get the total heat flux the data must be averaged by age and by area of the seafloor. These weighted estimates give about 62 mW/m^2 for the average. About half of this is a transient effect from the plate-forming process and half is the background flux from the mantle. Measured oceanic heat flow varies from about 300–25 mW/m^2 with 45–55 mW/m^2 being a representative range through old oceanic crust. The theoretical value for half-space cooling gives 100 mW/m^2 but this is sensitive to values adopted for thermal conductivity of the mantle and crust.

Circulating hot water in the dike injection zone of midocean ridges removes heat. Near-axis hydrothermal cooling accounts for about 1 TW of the global heat flux. The extent of hydrothermal cooling due to off-axis circulation of cold water is usually taken as the difference between the predictions of the plate-cooling model and the observed conducted heat flow but there is no theoretical basis for this.

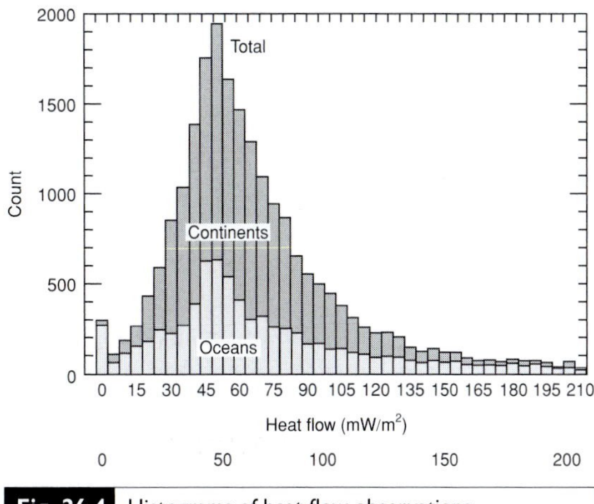

Fig. 26.4 Histograms of heat flow observations. [continental oceanic heat flow]

Deep, wide oceanic basins are the only regions of old seafloor where depth represents thermal isostasy. When the depths of these basins are corrected for sediments and crustal thickness, all the heat-flow data fall at greater depths than predicted by the 'plate' cooling model and there is no relation between heat flow and age. The heat flow data and topography favor discrete or stochastic reheating events. This reheating could be due to intrusions rather than basal heating. Continental and oceanic heat flow values are compared in Figure 26.4.

Expected background variations in heat flow

The boundary layer and plate models attribute all variations in bathymetry and heat flow to conductive cooling as a function of time. However, mantle convection and plate tectonics could not exist and are inconsistent with an isothermal mantle. Lateral temperature variations of the mantle below the plate of at least 200 °C are expected. For a 100 km thick TBL this implies heat-flow variations of about 15% superposed on normal cooling curves. Stochastic intrusions of dikes into the plate also introduce scatter into heat-flow measurements. These effects all change the conducted heat-flow and imply that one should not replace the measurements with a theoretical cooling curve in order to estimate the total heat loss of the mantle. Variations in permeability at the top of the plate cause variations in the hydrothermal component of heat flow. This component of heat flow must be allowed for separately. The important conclusion for present purposes is that there are uncertainties and temporal changes in surface heat flux that must be considered before one declares an 'energy' crisis. Even if there is a mismatch between time averaged heat flow and the energy sources in the interior, the location of the missing energy source, if there is one, cannot be determined from heat-flow data.

Heat sources

Radioactivity

All estimates of terrestrial abundances of the heat-producing elements depend, one way or another, on meteorite compositions. Carbonaceous chondrites are the usual choice of building material but enstatite achondrites and meteorite mixes are also used. In detail, the Earth is unlikely to match any given class of meteorite since it condensed and accreted over a range of temperature from a range of starting materials. The refractory elements are likely to occur in the Earth in cosmic ratios but there is evidence that the volatile elements are depleted. The large metallic core indicates that the Earth, as a whole, is a differentiated and chemically reduced body, although at least the crust and the outer shells of the mantle are oxidized. Enstatite achondrites (EH) match the Earth in the amount of reduced iron (oxidation state) and in oxygen isotopic composition and have been used to estimate terrestrial abundances. Since the terrestrial planets are mainly oxygen, by volume, this is a non-trivial consideration. The U and Th concentrations in various meteorite classes are given in Table 26.3, both for the bulk meteorites and calculated on a volatile and iron-free basic, to approximate mantle or BSE concentrations.

Table 26.2 | Distribution of radioactive heating

	TW
Continental crust	5.8–8.7
Continental lithosphere	1
Oceanic crust	1
Upper 410 km if made of:	
NMORB	13.4
Peridotites	4.6
oceanic	2.1
continental	9.2
Upper 650 km if made of:	
Pacific MORB or eclogite	28
Oceanic peridotites	4
Continental peridotites	18
Picrite	6
Plausible range	10–22
Depleted MORB	1–2
Crust+lithosphere+upper 650-km	18–30
Upper 1000-km if it has all the U, Th, K minus crust	12–24
Upper 1000 km if heat generation is 12.5 fW/g	21
Lower mantle	3–14

Composition of Earth

Estimates of the U, Th and K contents of the bulk silicate earth (BSE) are given in Table 26.2.

The trickiest element to estimate is K since it is not refractory. Estimates of the K content of BSE range from 151–258 ppm (chondrites fall in the 200–550 ppm range, EH are 840 ppm). On a H-, C-, S- and Fe-free basis meteorites range from 490–1315 ppm (Table 26.3). From ^{40}Ar abundances in the atmosphere the minimum K in the Earth is inferred to be 116 ppm. If the degassing efficiency of the mantle is comparable to the fractionation efficiency of LIL into the crust then the K content of BSE may be about 230 to 350 ppm. Most of the K, U and Th may be in the outer shells of Earth – crust, recycling crust, shallow mantle, kimberlites and the MORB-source region. The original processes of accretion and differentiation, and the ongoing processes of recycling and

Table 26.3 | Radioactive elements in meteorites

		Meteorite Class								
		CI	CM	CV	CO	H	L	LL	EH	EL
		Radioactivities in whole meteorite								
K	ppt (‰)	550	370	360	360	780	920	880	840	700
Th	ppb	29	41	58	80	38	42	47	30	38
U	ppb	8	12	17	18	13	15	15	9.2	7
		Radioactivities in volatile-free silicate portion								
K	ppt (‰)	775	511	490	498	1102	1211	1128	1315	971
Th	ppb	41	57	79	111	54	55	60	47	53
U	ppb	11	17	23	25	18	20	19	14	10

slab dehydration, result in strong upward concentrations of the radioactive elements.

Potassium is a minor contributor to present-day heat flow. However, in early Earth history, K and U would have been the major long-lived radioactive heat sources. Estimates of average bulk silicate Earth (BSE) abundances (mantle plus crust) are given in Table 26.2, along with their heat productivities. Estimates for the heating potential of BSE range from 12.7–31 TW. Most analyses give values between 17.6 and 20.4 TW. These are present-day instantaneous values. Heat conducted through the surface was generated some time ago, when the radioactive abundances were higher, so these are lower bounds on the contribution of radioactive elements to the present-day surface heat flow, assuming that the estimates of U, Th and K are realistic. The allowable variation in U and Th contents of the mantle is a large fraction of the so-called discrepancy between production and heatflow. Production of heat can be much larger if potassium contents have been underestimated. Because of the short half-life of ^{40}K, most of the ^{40}Ar in the atmosphere would have been generated in early Earth history. Efficient degassing in early Earth history may explain, in part, the large fraction of the terrestrial ^{40}Ar that is in the atmosphere, compared to the reluctance of ^{4}He to leave the mantle today (the helium-heatflow paradox). Different solubilities of He and Ar in mantle materials may also be involved.

The amount of radioactivity in the crust must be subtracted out in order to obtain mantle abundances and heat productivities. Using 8 TW as the best estimate of crustal productivity gives 9.6–12.4 TW as the mantle heat flow from radioactivity, or 18.8–24 mW/m^2. These can be compared with the basal heat-flow estimates (25–39 mW/m^2). Delayed heat flow and other sources of mantle heating may need to contribute up to about 20 mW/m^2, more than half of the mantle heat flow. Heat from the core (about 9 TW), solid Earth tides (1–2 TW) and thermal contraction (2 TW) are non-radiogenic sources that may add 12 TW to the mantle heat flow, about the same as the current (non-delayed) mantle radiogenic contribution. The radiogenic contribution can be increased by about 25% if it takes 1 Gyr to reach the base of the lithosphere. On top of all this is secular cooling of the mantle. In a chemically stratified mantle, the outer layers cool much faster than the deeper layers. If cooling is confined to the upper 1000 km (Bullen's Region B and C) a temperature drop of 50 °C in a billion years corresponds to a heat flow of 3 TW. Cooling rates of twice this value have been suggested. Thus, there appears to be no need for any exotic heat sources or hidden sources of radioactivity in the mantle. This conclusion is independent of the uncertain contribution of hydrothermal circulation to the surface heat flow. There are implications, however, for the temperatures of Archean – and older – mantle and the style of convection, and the mechanisms of heat removal; the present styles of mantle convection and plate tectonics were unlikely to have been operating.

Distribution of radioactive elements

There is a strong decrease in the concentrations of the radioactive elements as one goes from the upper crust to the lower crust to the upper mantle. Accretion and early differentiation probably swept most of the radioactivities toward the surface. Present recycling involves dehydration at depths of the order of 200–300 km, which removes incompatible elements from the slab and places them in the mantle wedge above the slab. The residual slab carries LIL-poor material to greater depths. A vigorously stirred mantle would tend to be homogenous. In a layered mantle with dehydration, fluid migration and partial melting we expect the processes of fractionation, gravitational separation and differentiation to dominate over processes of convective homogenization, except where melts are co-mingled prior to eruption. Plates, continents, downward increase of viscosity, phase changes, non-Newtonian rheology and chemical stratification all serve to decrease the vigor of convection and to prevent efficient homogenization. The net effect could be a rapid, or exponential, decrease of K, U and Th content with depth, rather than the completely homogeneized and depleted upper mantle and a U-rich lower mantle, as in current geochemical models.

Continental crust

Estimates of the U and Th concentrations of continental crust range from 0.9–1.3 ppm and 3.5–9.0 ppm, respectively (Rudnick and Fountain, 1995). Estimates of the K_2O content show a similar range, 1.1–2.4 wt.%. The inferred heat productions and heatflows range from 0.58–1.31 mW/m^3 and 23–52 mW/m^2. The large range reflects, in part, uncertainty in the composition of the lower crust. More than 30% of the most incompatible elements in BSE are in the continental crust. This is a measure of the differentiation efficiency of the Earth. The continental crust contributes 5.8–8.0 TW to the total energy budget of the Earth.

The estimates of radioactivity in the continental crust are incompatible with geochemical models that attribute its origin to extraction from a chondritic or primitive upper mantle, leaving behind an undifferentiated lower mantle. There is not enough material in the upper mantle portion of undifferentiated mantle to provide the concentrations inferred for the more incompatible elements. One-hundred-percent efficient extraction, and absence of recycling replenishment, is unlikely. It appears that the portion of the mantle from which the crust was extracted was already enriched in the LIL, most likely as a result of a radial zone-refining process concurrent with accretion. A corollary is that crustal extraction did not have to approach 100% efficiency, and that there is still U, Th and K in the upper mantle. The deep mantle may have little heat-production capability; it may be barren or sterile. If indeed most of the radioactive heating in the mantle is shallow then there will be a smaller lag between heat production and heat flow and the bulk of the slab cooling effect will be in the region of main heat production. Some of the arguments against layered convection assume that the lower mantle is U-rich and will therefore overheat. The core would also lose heat less efficiently in this kind of layered model, which may pose problems for the growth of the inner core and maintenance of the dynamo.

Continental lithosphere

Based on xenolith studies, the continental lithosphere (CL) may have radioactivities as high as 10% of the level of average crustal rocks. The CL may contribute half or more of the heat flow in older or low-heatflow continental terrains. Peridotitic xenoliths represent only part of the subcrustal mantle, possibly the most depleted part. Large parts of the Archean cratonic lithosphere appear to be enriched in U and Th. About 15 mW/m^2 of heat is generated in the continental keel (Vitorello and Pollack, 1980). This requires radioactivities about five times higher than measured in depleted peridotites.

The shallow mantle is a sink of subducted sediments, slab-derived fluids, altered ocean crust and serpentinized peridotites and may contain trapped small-degree melts, kimberlites and metasomatic fluids. If so, it is a non-negligible heat source and, because of its large volume, may have an integrated productivity comparable to the continental crust. Even if the continental crust has achieved a steady state between construction and destruction, there is a certain amount of radioactive-rich crustal material circulating in the mantle.

Mantle components

Mantle melts are buoyant compared to residual solids, at least in the shallow mantle, and frozen melts (basalts) are buoyant until they convert to eclogite at about 50 km. These materials strip radioactivity out of the source mantle and deliver it to shallow depths where they pond, underplate, intrude or erupt. The erosion of continents puts radioactive-rich material into seawater and onto the ocean floor where some gets recycled into the mantle. Some of this gets returned quickly to island arc and back-arc volcanoes but some remains in the shallow mantle as an enriched component. Delaminated continental crust also recycles into the mantle. Nevertheless, most estimates of the composition of the shallow mantle adopt the most depleted basalts erupted along the midocean-ridge system and assume that these, along with residual peridotite having little or no radioactivity, are characteristic of the entire mantle above the 650 km phase change. Many mantle peridotites have U and Th contents comparable to depleted basalts and this alone can raise estimates of upper mantle radioactivity by a factor of ten.

Basalts found along the global ridge system have radioactivities that vary by more than an order of magnitude. Kimberlites, carbonatites and alkalic basalts extend the range of upper mantle materials even further. Peridotites also have a large range of compositions. It is conventional to adopt the most depleted of these materials as representative of the upper mantle; this is *ad hoc*. An alternative approach is to mix terrestrial materials together in proportions that satisfy chondritic or cosmic ratios of the refractory elements.

Upper mantle

If the total heat productivity of the Earth is 19–31 TW then the upper mantle share, if uniformly distributed, is 5–10 TW. If the crustal inventory of 8–9 TW was derived only from the upper mantle then the UM would be barren indeed. There is little room for recycled sediments and crust or metasomatic fluids. However, this is only a model. Detailed mass-balance calculations and the amount of ^{40}Ar in the atmosphere suggests that most or all of the mantle must have contributed to the LIL inventory of the crust and UM. This indicates an efficient differentiation and melt extraction process. However, inefficient melt extraction and recycling keep the upper mantle from being completely barren. If the LIL were concentrated into the outer layers of Earth during accretion, then extraction of the crust leaves 10 to 23 TW in the UM. These numbers can be matched by making the UM out of Pacific MORB and peridotitic xenoliths or a mix of enriched and depleted MORB and peridotites and a small fraction of enriched components such as recycling crust or kimberlitic material.

Table 26.2 shows how much heat can be generated in the upper mantle if it has the composition of various mantle samples. The upper 410 km can generate 9 TW if it is entirely made of continental peridotites and 13 TW if made from a representative MORB. The upper 650 km can generate 28 TW if it is made up of Pacific MORB or an average eclogite. A plausible range, using basaltic and peridotitic mantle samples, is from 10–22 TW. This is about an order of magnitude higher than estimates based on depleted MORB.

If most of the U and Th is in the upper mantle then most of the ^{4}He will be generated there. The missing ^{4}He in the integrated volcanic flux implies that He is not readily outgassed. This is consistent with measured U/He ratios and absolute He concentrations in midocean ridge basalts.

The high $^3He/^4He$ ratios found in some basalts probably reflect long-term evolution in a U-poor environment. This need not be a large reservoir but could be single olivine crystals or pieces of depleted lithosphere.

The adiabatic gradient

The horizontally averaged interior temperature of an actively convecting region of a uniform fluid, heated from below, at high Rayleigh number, is adiabatic if one is sufficiently far from thermal boundary layers. Secular cooling and internal heating cause the geothermal gradient to be subadiabatic. The seismic velocity gradient in the deep mantle (~1000 km–2600 km depth) is consistent with an adiabatic gradient, suggesting that this region may satisfy the conditions of being mainly heated from below and not experiencing substantial secular cooling. However, the uncertainties are such that a substantial subadiabatic gradient is also consistent with the data. Although there is little heat entering the lower mantle from the core it may dominate the heating if the lower mantle is depleted in radioactive elements.

The upper mantle is radially and laterally heterogeneous and the seismic properties are affected by phase changes and partial melting. There is no reason to believe that it is adiabatic below the plates. In addition to the surface thermal boundary layer any upward concentration of buoyant material will extend the depth of the conduction layer. Upward migration of melts, underplating and dehydration and melting of the downgoing slab will concentrate U, Th and K into the shallow mantle. The bottoming out of slabs in the upper mantle or transition region contribute to the secular cooling of the shallow mantle, and the maintenance of a subadiabatic gradient. Under these conditions one expects that melting will be confined to the shallow mantle and perhaps to the thermal boundary layers where the thermal gradient is high and positive.

Secular cooling

The heat content of the Earth is immense; about 10^{38} ergs which is equal to a 10 Ga supply of the present flux. Calculations suggest that the upper mantle of the Earth may be cooling by 50–100 °C per Ga. This provides about 15–35% of the global heat flux, but values as high as 50% are not ruled out. The inferred high mantle temperatures in the past probably require a different style of plate tectonics and heat removal, even after the continents were formed and stabilized. Options include the heat-pipe mechanism used to extract heat from the interior of Io, stagnant or buoyant lid convection, multiple short-lived platelets, continuous and widespread volcanism and magma oceans. Heat productivity at 3.5 Ga was at least a factor of three higher than today and a more efficient heat removal mechanism must have prevailed. Tidal friction and energy due to mantle and core differentiation may also have been much higher at that time.

Delayed heat flow

There are several sources of delay in the heat generation vs. heat-flow cycle. The heat flowing through the interior represents heat generated at some time in the past when the heat generation capacity of radioactive elements was higher. One does not expect an instantaneous balance between present heat generation and heat flow. The Earth is cooling, and this contributes to the observed heat flow. These secular effects may contribute at least 10 TW to the heat flow compared to what would be expected from the present level of radioactivity. This is comparable to the shorter term fluctuations due to plate and mantle reorganizations. If the crust is still growing the radioactivity in the mantle, in the past, may have been higher than at the present. If this mechanism is important the U, Th and K in the mantle are decreasing with time both by decay and by removal. These elements are also returned to the mantle by subduction and the balance between removal and return may have changed with time. In the extreme case, all the LIL were stripped out of the bulk of the mantle during accretion and the magma ocean stage and only later returned to the upper mantle when the mantle cooled sufficiently for plate tectonics to operate.

Heat being built up under large plates may be episodically released upon plate reorganizations, stress changes and the formation of new plate boundaries. These transient effects (large igneous provinces, continental flood basalts, seaward dipping reflectors, volcanic chains) may reflect natural transient plate-tectonic events associated with stress changes in the lithosphere. If the asthenosphere is close to the melting point and variable in fertility then the location of volcanism is controlled by plate stress and architecture, and mantle composition, rather than by absolute temperature.

Minor heat sources

Radioactive heating and secular cooling are the major contributors to the energy budget of the Earth's interior. There are many other contributions that are often overlooked. The progressive differentiation of the mantle and core releases heat. As the Earth cools and differentiates the mantle generates about 3 TW of gravitational energy. There is a release of gravitational energy and latent heat by growth of the inner core. This contributes about 1.2 TW. The total power from the core has been estimated to be 8.6 TW. Solid Earth tides contribute slightly to the heating of the mantle. This energy source would have been greater in the past when the Moon was closer to the Earth. Past tidal heating may contribute to present-day heat flow. The Earth, as it cools, releases gravitational energy by contraction. Some fraction of this may be released by earthquakes. The change in gravitational energy associated with earthquakes has been estimated to be as high as 2 TW. A very speculative energy source, $\sim$3 TW, is thermonuclear reactions ('natural nuclear reactors') in regions where radioactivity has been concentrated by natural processes. These minor, and in some cases, speculative, energy sources may account for up to some 17–22 TW of the total heat flow. This is of the order of the 'missing energy source' which has prompted many suggestions for hidden radioactivity in the deep mantle. In addition, heat flow may not be a steadily decreasing function of time if plate reorganizations modulate the heat flow. Current heat flow, for example, may be 5% higher than the mean because of the relatively recent breakup of Pangea. The cumulative contribution of these sources, and the uncertainties in the magnitudes of the major sources, appear to be in the range of the `'missing heat-source paradox'` which has led to a series of speculative papers on hidden radioactive heat sources in the deep mantle or core. The 'missing heat' is also not a problem if the hydrothermal contribution to plate cooling has been over-estimated. The magnitude of the 'missing energy' is comparable to the upward adjustment of the measured heat flow due to its mismatch with theoretical expectations from simple plate-cooling models.

The helium–heat flow paradox

U and Th generate ^{4}He and anti-neutrinos as well as heat. The observed flux of ^{4}He to the oceans from the mantle is about 3.2×10^5 kg/a. The flux predicted from a mantle with 21 ppb U and 95 ppb Th is 3.4×10^6 kg/a. For comparison, the ^{4}He flux predicted from the continental crust is about 10^6 kg/a. The fact that the current flux of ^{4}He from the oceanic mantle is an order of magnitude less than predicted from the mantle U and Th abundance is known as the `helium–heat flow paradox`, just one of many paradoxes associated with the decay chain of uranium. The discrepancy is even larger if there is a substantial delay in the transport of ^{4}He from the source to the surface. On the other hand, He, and CO_2, may be trapped in the shallow mantle. The amount of ^{40}Ar in the atmosphere compared to that released by ^{40}K over the age of the Earth indicates that, on average, the mantle is efficiently outgassed. Because of the short half-life of ^{40}K a large fraction of the Earth's ^{40}Ar was produced in early Earth history. The production of ^{4}He decreased with a longer time constant. Argon does not escape from the atmosphere so the atmospheric inventory of ^{40}Ar can be used to give a lower bound on K in the Earth. Helium escapes so we cannot bound the U and Th in this way.

Helium is degassed, along with its main carrier, CO_2, as basalts rise toward the surface. Helium differs from argon in being highly soluble in magmas. Total degassing requires eruption or intrusion near the surface and even then quickly quenched glasses retain substantial quantities of CO_2 and helium. The missing helium-4 paradox has a parallel in the carbon budget. Carbon is depleted by about an order of magnitude in the exosphere, compared to other volatile elements. On the other hand, the presence of diamonds and carbonatites in the mantle, and CO_2 in magmas, shows that the mantle is a long-time repository for CO_2 and probably helium. The presence of ^3He in mantle magmas shows the same thing. ^3He is a primordial isotope in the sense that it is not created in substantial quantities by reactions in the Earth (although some is brought in or generated by cosmic particles). Some ^3He may have been brought into the Earth by a late veneer but in any case helium has been trapped in the mantle for a long period of time. The most efficient location for bringing magmas to the near-surface and degassing is along the global spreading ridge system. Even in these locations the presence of magma chambers and off-axis volcanoes suggest that gas exsolved from magmas at low-pressure may be trapped in the shallow mantle. However, gas can separate from parent magma, which separates the gas from U and Th. Helium trapped in olivine crystals in cumulates or melt-depleted peridotite will therefore retain its isotopic composition. Ridges migrate readily over the mantle; in fact, ridge migration may be essential to continuous magmatism. It takes about 1 to 2 Gyr for ridges, at their present migration rates, to visit each part of the mantle. At these times, trapped gas has another chance to reach the surface. But the $^3He/^4He$ ratios of some of these gases will be more appropriate for mantle 1 to 2 Gyr older. Gases of various trapped ages may co-mingle, especially in the ridge environment, but in other environments, e.g. seamounts, oceanic islands, a diversity of components of different ages may be evident, including extreme values which would be averaged out at ridges.

It is interesting, and instructive, that products of U and Th decay (heat, lead and helium) have their names attached to so many `geochemical paradoxes and enigmas`. This is a strong signal that current geochemical models are inadequate. Nucleogenic neon is also a product of U and He/Ne ratios are not completely understood. It does not seem possible to use the ^4He budget to usefully constrain the U and Th abundances in the mantle. On the other hand, trapping of He and CO_2 in the upper mantle may explain the low outgassing rates of He and the low crustal abundances of CO_2. It would be interesting to know if the mantle is currently a sink for CO_2 (subduction fluxes exceeding volcanic fluxes) as is occasionally reported. If so, the shallow mantle may well be the repository of the missing CO_2 and ^4He and a storage vessel for ^3He as well. Some midocean ridge basalts have large concentrations of ^3He; the accompanying CO_2 causes the rocks to explode or pop when they are removed from depth by dredging.

Heat from the core

The existence of a geomagnetic field and a solid inner core place constraints on the Earth's thermal history. The solid inner core may be essential for the nature of the current field including reversals. Any model of the Earth's evolution must involve sufficient heat loss from the core to power the dynamo, but not so much as to freeze the core too quickly. These constraints are surprisingly strong. The inner core probably grew with time and may have started to freeze about 2 Ga. The magnetic field existed earlier than this so freezing of the inner core – and compositional stirring of the outer core – may not have been involved in the generation of the early magnetic field. Cooling from a high initial temperature (superheat) may have driven the early dynamo. Whether this could have been maintained for $\sim$2 Gyr is a matter of debate. Convection in the core is highly turbulent and the thermal gradient in the outer core is probably adiabatic. Since the core is an excellent thermal conductor it can conduct heat readily down this thermal gradient. This places a constraint on the minimum amount of heat that enters the mantle from below. This minimum is 8.6 TW. Contraction due to cooling and freezing adds a little to this. If the core is still growing by interactions with the mantle there will be additional gravitational energy terms. Radioactivity, secular cooling and thermal contraction of the core must be minor compared to the equivalent energy sources in the mantle. This minor heat source, however, is the source of heating in the plume hypothesis.

In the fluid outer core, a dynamo process converts thermal and gravitational energy into magnetic energy. The power needed to sustain the dynamo is set by ohmic losses. Estimates of ohmic losses in the core cover a wide range, from 0.1–3.5 TW with more recent estimates in the range of 0.2–0.5 TW. The lower estimates remove the need for radioactive heating in the core and allows the age of the solid inner core to exceed 2.5 billion years. In order to sustain the dynamo, the heat flow from the core must be 5–10 times larger than the ohmic dissipation of 1–5 TW.

The fate of core heat that enters the mantle is controversial. Some workers assume that heat from the core is removed efficiently from the mantle and taken directly to the surface via narrow plumes or heat pipes. Heat from the mantle is removed by large-scale convection and plate tectonics. If this is so, the thermal evolution of the core and the mantle can be treated separately. If mantle convection is strong, instabilities at the core–mantle boundary (CMB) will be swept away and entrained in mantle convection. Core heat then just gets added to mantle heat. Since more heat is generated near the surface of the mantle, and more heat passes through the surface boundary layer than passes through the CMB, and since thermal expansion and viscosity favor rapid turnover at the top, mantle convection must be primarily driven from the top and not the bottom. Narrow instabilities are unlikely to pass unperturbed through the whole mantle, even if they can form in the first place, in the presence of convection driven from above and internal heating.

The most important considerations in the fate of core heat are the thermal properties at lower mantle conditions. Thermal conductivity is much higher than at the top of the mantle because of the combined effects of pressure on the lattice conductivity and the effect of temperature, grain size and composition – including spin-transitions – on radiative conductivity. This promotes the establishment of a thick, and sluggish, conductive TBL. The coefficient of thermal expansion is very low at the CMB, perhaps as much as an order of magnitude lower than at the top of the mantle. This means that an increase in temperature – and heat from the core – does not yield much thermal buoyancy. Only very large features develop enough buoyancy to rise. Pressure increases the viscosity, making it difficult for even very large features to rise. If the deep mantle is chemically denser than the mesosphere it may be trapped since temperature is ineffective in bringing the density down low enough so it can escape. These considerations are neglected in the plume hypothesis for core cooling, which is based on laboratory injection experiments where pressure effects are minimal, and Boussinesq calculations where pressure effects on material properties are ignored. The net effect is that core heat is probably added to mantle heat, and is carried away by conduction or radiation, or by large-scale sluggish upwellings.

Although small in magnitude, core heat may dominate the heat budget at the base of the mantle if most of the radioactive elements are at the top of the mantle, and most of the secular cooling is in the top layers, the case for a chemically stratified mantle. The core mainly cools by rapid conduction into the base of the mantle rather than by short-circuits to the top of the Earth. The temperature gradients at the base of the mantle vary from place to place. The core will lose heat most efficiently through the colder parts of the lower mantle. These are not the places one expects to find hot plumes.

Heat fluxes

Heat is removed from the interior of the mantle by conduction through the surface thermal boundary layer (TBL), intrusion into the TBL, hydrothermal circulation near the surface, and volcanism. In addition, the interior is cooled by subduction of cold slabs and warm delaminated lower crust. The relative importance of these mechanisms changes with time. The conduction layer is, in part, chemical and permanent – although mobile – and, in part, transient. This means that the heat-flux problem is not one-dimensional (1D) or steady-state. Heat can be diverted to regions having thin TBL, and can, to some extent, be temporarily stored in the mantle.

Heat flow shows little correlation to crustal age (Figure 26.2), ocean depth at the sampling site, or the inferred thickness of the plate, it is nearly constant through the Atlantic and Indian ocean crust and Pacific ocean crust older than about 40 million years. Young Pacific crust has a slightly elevated heat flow but much less than predicted from plate tectonic models of cooling with constant thermal conductivity. Heat flow is not significantly elevated over hotspots, swells or superswells, compared with the flux for normal oceanic lithosphere, implying that (1) the underlying mantle is not hotter than average, (2) the associated volcanism does not significantly warm the lithosphere or (3) underplating, intrusion and heating are common and that conditions for extrusion involve stress in the lithosphere rather than excess temperatures or unusual mantle at places that are called hotspots. Even midocean ridges are not always distinguished from 160-Ma lithosphere on the basis of heat flow. A significant problem with interpreting heat-flow data is that hydrothermal circulation suppresses the conductive gradient. Thus, the conduction gradient can be affected by fluid circulation from above or by magma injection from below. These processes are controlled by physical and environmental effects that are not simply functions of age.

The conduction gradient is steep compared to melting curves and the lower part of the TBL can be above the solidus. Melts can be extracted from various depths in the TBL and can therefore yield different potential temperatures. Rapidly ascending melts rise adiabatically and one speaks of the *potential temperature* inferred from these melts. This may not be the same as the potential temperature of the deeper mantle. The mantle beneath the plate is referred to as 'the convecting mantle' with implications about an adiabatic gradient and chemical homogeneity. At high temperatures, the base of the TBL may be weak and fall off, or delaminate, if it is denser than the underlying mantle. It may also deform or flow laterally, if the necessary forces exist, or experience small-scale convection, if the viscosity is low enough.

Departures of bathymetry and heat flow from the theoretical square-root of age relations, are often attributed to thermal perturbations due to deep mantle plumes. These have been used to estimate the buoyancy flux of plumes and as an estimate of core heat. However, the mantle is not homogenous or isothermal, plates are not uniform or impermeable, and thermal properties are not independent of temperature. The upper mantle is close to or above the solidus and the melting point of the mantle is variable. Regions that are shallow for their age do not imply a deep source of heat; there is no reason to believe that the mantle has a single potential temperature or a uniform composition or melting temperature.

Mechanisms for seafloor flattening

Ocean depth and heat flow are usually interpreted in terms of boundary layer and plate theories. In both, the initial condition is an isothermal (or adiabatic) homogenous bath of fluid of zero viscosity. The surface is suddenly dropped to a low temperature. A cold boundary layer grows with time. In the plate model, the thermal boundary layer is taken to be constant thickness and the bottom is at the same temperature as the vertical ridge axis. Heat is conducted through the plate and its average temperature increases with time. After a long time, the heat flow and the temperatures in the plate reach equilibrium values. The constant thickness of the TBL and the constant temperature at its base are not natural boundary conditions, even if the plate is of different material than the underlying mantle. A variant of the plate model is to change the lower boundary condition to one of constant heat flux.

Explanations for seafloor flattening at old age range from the mundane to the exotic; some explanations affect estimates of the global heat budget. Flattening sets in at different times for different sections of the seafloor and sometimes does not occur at all. The most obvious explanation is to accept that the theoretical models are highly artificial and they should not be expected to correspond to reality. The mantle is not isothermal or adiabatic, in the absence of plates, and the plate is subjected to processes other than monotonic conductive cooling. If the plate has a constant heat flow lower boundary condition, rather than a constant temperature one, then one automatically obtains a background heat flux, and a constant heat flux in old ocean basins. The inferred thickness of the outer conduction layer is often much greater than can form by cooling in 200 Myr, consistent with a chemically buoyant region. Plate creation may superpose a new TBL on top of the old one, particularly if ridges migrate and jump. If one can slow down the process of heat conduction for old plates one can slow down the flattening process. Candidates for this include sedimentation, or a decrease in conductivity at depth due to temperature, mineralogy or crystal orientation.

If the mantle at depth is partially molten then dikes may be injected to shallower depths when the lithospheric stress conditions become appropriate. This serves to reheat the plate and partially reset the thermo-magmatic age. This would not uniformly affect all plates or affect them at the same age. The reheating/diking events would be milder versions of the plate creation process itself, at the ridge. The plate does not have to be impacted by a hot plume or over-ride hotter-than-average mantle for this mechanism to be effective.

One explanation for the 'flattening' of the seafloor at old age, and the motivation for the plate model, was that it might be caused by the lower oceanic lithosphere becoming convectively unstable once a critical age is reached. This is a form of delamination. Small-scale convection beneath the plate is assumed to maintain an isothermal boundary at a specific depth, and cause the thermal structure of the cooling lithosphere to resemble that of a finite-thickness plate. The stirring action of small-scale convection alone may act to cool the upper mantle, leading to increased subsidence, not flattening of the depth–age curve. What is needed is a method to reheat the plate or to slow down the cooling. Dike or sill injection, or magma underplating, does this without the involvement of particularly hot mantle.

Hotspots are often held responsible for adding heat to the system and retarding lithospheric subsidence. A low-viscosity plume is very inefficient at thinning the overlying plate, and there is insufficient time for conductive thinning except for slowly moving plates. The excess buoyancy of dike and sill-injected plates could contribute to the anomalous subsidence of the seafloor, without the need for plumes or extra basal heating.

The shallow depths of the seafloor in the western Pacific and the superswell in French Polynesia have been attributed to the residual effect of a hot Cretaceous superplume but evidence used in support of this is largely spurious. The superswell is well explained by a warm or buoyant low-viscosity asthenosphere, not by lithospheric reheating or thinning (McNutt and Judge, 1990); a deep source of heat is not required.

Hypsometric curves predicted by plate cooling models are not a good match to the bathymetry observations. Reheating or intrusion models may be required to explain the heatflow, the hypsometry, the abrupt flattening of median depth with age, and the increased variability of depth at older ages. The cooling half-space model predicts that bathymetry and heat flow will follow a square-root age relation for all time. The plate model assumes that a fixed temperature is maintained at a fixed depth everywhere and so predicts that subsidence curves should flatten at the same age everywhere; they do not. From time to time some areas will probably be extended and intruded; uplift and rejuvenated subsidence may occur anywhere at any time, depending on the stress-state of the plate.

Attempts to 'correct' heat flow and ocean bathymetry by avoiding hotspots or correcting for their effects are misguided if dikes and sills and underplating are ubiquitous in oceanic lithosphere. This mechanism for explaining heat flow and bathymetry is basically a stress mechanism since it does not depend on high temperatures, only stress conditions in the lithosphere. The plate can be treated as semi-permeable to magma and a partially open system, rather than a rigid impermeable LID over the asthenosphere. The stochastic nature of heat sheet penetration is a combination of stress conditions in the plate and the variable fertility and melting point of the asthenosphere. Underplating and intrusion of magmas in the ordinary range of temperatures may be responsible for the background heat flux rather than numerous high temperature deep-mantle plumes.

Flexural rigidity, a measure of elastic plate thickness, appears to decrease and then to increase rapidly after the imposition of a large volcanic load such as at a 'hotspot.' This is usually taken as an indication of *thermal rejuvenation* or heating and weakening of the lithosphere by a plume. However, the load itself causes 'stress-rejuvenation' and an apparent thinning of the plate and does so without a heat flow anomaly. Thus, not only can volcanic chains themselves be caused by stress rather than temperature, but other features such as subsidence and heatflow can also be explained by athermal stress mechanisms. The subsidence of oceanic plateaus and uplift of CFBs are some of the paradoxes of thermal mechanisms. All of these considerations show that, while surface hotspots may represent local rise of normal asthenopspheric mantle to shallow depths, they do not require abnormally hot mantle. In global compilations of heatflow, one should use all the data instead of attempting to mask out hotspots, and one should not use simple theoretical models to correct measured heatflow to the value that it should have for the appropriate age crust.

Temporal changes in heat flux

The question arises: are present estimates of global heatflow representative of the present cycle of continental break-up and seafloor spreading? If the heat flux to the surface has changed significantly in the last 200 Myr it should show up in variations of seafloor spreading rates and sealevel. The creation of new oceanic plates appears to have been constant during the past 180 Myr. Detailed plate reconstructions and sea-level variations preclude large global thermal pulses. There is no evidence for a major increase in plate rates during the Cretaceous. Likewise, there is no evidence for the superplumes, mantle overturns and avalanches that were spawned by the speculation that there are large and rapid variations in plate-creation rates. Long-term (hundreds of millions of years) variations of 5 or 10% in surface heat flow, however, are expected simply from the nature of high-Rayleigh-number convection, even if the boundary conditions are constant. Plate reorganizations may release pent up heat at new plate boundaries since mantle temperatures tend to increase beneath large long-lived plates. Heatflow also depends on the aspect ratio and style of mantle convection and this may change in the order of 50% over a supercontinent cycle of 500 Myr.

Chapter 27

The thermal history of the Earth

Man grows cold faster than the planet he inhabits.

Albert Einstein

Starting with Kelvin there have been many controversies and paradoxes associated with the evolution of the Earth. These are not faith-based controversies, in the ordinary sense; they are based on calculations and assumptions – which are a form of faith. Present-day heat flow is determined by the amount and distribution of radioactive elements, their secular decline with time, the delay between heat production and its appearance at the surface, secular cooling of the interior, and a variety of minor sources of heating which are usually overlooked. Chemical stratification of the mantle slows down the cooling of the Earth; the upward concentration of radioactive elements reduces the time between heat generation and surface heat flow. The initial conditions of the Earth cannot be ignored; they have not been forgotten.

Continents divert mantle heat to the ocean basins and, in addition, tend to move toward cold downwellings, thereby protecting their cold keels. Surface conditions – including the thickness of plates – and mantle viscosity control convective vigor and cooling history of the mantle. A global accounting of the heat lost from the interior by hydrothermal processes at the surface is a lingering issue. A chemically layered mantle with upward concentrations of the radioactive elements, shallow return flow, and a low-viscosity asthenosphere is the most plausible model of mantle dynamics. The migration of ridges and trenches is an important aspect of thermal history and geodynamics.

The heat budget of the Earth cannot be treated as an instantaneous one-dimensional heat-flow problem, or one that involves a homogenous mantle with uniform and static boundary conditions. Both the radial and lateral structure of the Earth must be considered. Viscosity of the mantle is temperature, pressure, composition, location and time-dependent.

Magma oceans, plate tectonics, heat pipes and stagnant-lid convection have transferred heat out of the interior at various times. The current Earth approximates stagnant-lid conditions in one hemisphere and a lid that fully participates in internal convection in the other. On early Earth, the surface may have been covered with thick accumulations of unsubductable basalt, penetrated by pipes of magma, as appears to be the case on Io. In such a scenario the rate of heat loss is regulated by the stress and density of the basalt pile as much as by the viscosity of the interior. The interior cools by depression of the cold surface layer and delamination, which displace hot material upwards. Even on today's planet, eruption and mantle cooling depend on the stress state of the overlying plate.

Kelvin

Lord Kelvin assumed that the Earth started as a molten ball and calculated that it cooled to its present condition by thermal conduction.

Kelvin's assumed initial conditions are probably close to being correct. The energetics of accretion, core formation, giant impacts and formation of the Moon result in a hot, partially molten initial condition; gravitational stratification was unavoidable. Subsequent refinements of Kelvin's thermal history incorporated radioactivity and convection. The acceptance of convection and the rough equality between the heat output of chondritic meteorites and early estimates of the total terrestrial heat flow led to the view that there is essentially a steady state in which heat production and heat loss are balanced. However, this violates the first law of thermodynamics; it also cannot be the case since radioactive heating of necessity decreases with time. The secular cooling of the Earth – the Kelvin effect – is important but has often been neglected or dismissed in recent years. A perceived 'missing energy source,' controversies about 'corrections' to observed surface heat flow, origin of komatiite, and various Archean paradoxes and catastrophes are prominent in current discussions. The fact that there are no thermal or heat flow anomalies associated with hotspots is a paradox for the plume hypothesis [mantleplumes]. Current convection and geochemical models assume uniform radioactivity throughout the mantle, or a depleted upper mantle and a primordial or enriched lower mantle – these also cannot be correct.

Global heat-flow estimates range from 30 to 44 TW. Estimates of the radioactive contribution from the mantle, based on cosmochemical considerations, vary from 19–31 TW. About 5–10 TW enters the mantle from the core. Thus, there is either a good balance between input and output, as once believed, or there is a serious missing heat-source problem, ranging up to a deficit of 25 TW. Attempts to solve the perceived deficit problem by invoking secular cooling, or deeply-hidden heat sources – stealth layers in the mantle – have run into problems, some real, some perceived. Survival of ancient cratonic roots, komatiitic temperatures, an Archean heat flow catastrophe, overheating of the lower mantle, and helium heat flow paradoxes have been cited as potential problems. These concerns have replaced the *chondritic coincidence* that preoccupied geochemists when the first estimates of global heat flow came in, which suggested a steady-state Earth with an instantaneous balance of deep heat production with surface heat flow.

Initial conditions

A type of radial zone refining process accompanies the accretional process. This sweats out the crustal and radioactive elements and keeps them near the surface and drains the dense metallic melts and refractory crystals toward the interior. This chemical stratification, plus radioactive heating, stretches out the subsequent cooling of the mantle. Convection and the subduction part of plate tectonics accelerate the cooling but the surface and internal thermal-boundary layers are still conduction bottlenecks. In the limit of multiple layers the Earth approaches the conduction cooling condition calculated by Lord Kelvin. At the high pressures and temperatures in the deep interior both lattice conductivity and radiative conductivity may be much higher than in the shallow mantle, again making the mantle approach Kelvin's assumptions. Under some conditions the Earth can forget its initial thermal state but chemical stratification and redistribution of radioactive elements cannot be forgotten. A chemically stratified mantle has a low effective Rayleigh number; vigorous convection and chaotic mixing are not expected.

In models of geodynamics and geochemistry that were popular in the last century, and are still in the textbooks, the lower mantle was assumed to have escaped accretional differentiation and to have retained primordial values of radioactivity and noble gases or even to have undetectable radioactively enriched layers. The crust was derived entirely from the upper mantle, making the latter extraordinarily depleted in the radioactive and volatile elements. However, mass balance calculations and the ^{40}Ar content of the atmosphere show that most, if not all, of the mantle must have been processed and degassed in order to explain the concentrations of incompatible and volatile elements in the outer layers of the Earth. The accretional zone-refining process results in an outer shell that contains most of the U, Th and K, at levels about three times chondritic (if the outer shell is equated with the volume of the present upper mantle), from which the proto-crust and basaltic reservoirs were formed. The residual (current) upper mantle retains radioactive abundances greater than chondritic while the bulk of the mantle, including the lower mantle, is essentially barren. Previous chapters explored this possibility. The outer shells of Earth probably also contain, or contained, the bulk of the terrestrial inventory of noble gases. This is the reverse of the classical models and those used in convection simulations.

In more recent models of geodynamics the entire mantle is assumed to have escaped chemical differentiation, except for crust extraction, and to convect as a unit, with material circulating freely from top to bottom. This is `one-layer or whole mantle convection`. Recycled material is quickly stirred back into the whole mantle. In these models, (1) chemical heterogeneities are embedded in a depleted matrix, (2) the whole mantle is uniformly heterogenous and (3) the mantle is relatively cold and the hot thermal-boundary layer above the core plays an essential role in bringing heat to the surface. Convection is assumed to be an effective homogenizer. An amendment to this idea is that there is a radioactive-rich layer deep in the mantle, but this is based on unlikely assumptions about upper-mantle radioactivity. Ancient radioactive-rich regions in the deep mantle (if they survive accretional differentiation) will overheat, overturn and deliver their heat and heat-producing elements to the shallow mantle.

Style of mantle convection

The deep mantle convects sluggishly because of the effects of pressure on thermal expansion, thermal conductivity and viscosity. Large grain size, high temperature and partitioning of iron may increase the ability of the lower mantle to transmit heat by conduction and radiation. The buoyancy flux of hotspots is often equated with the heat flux from the core but this heat may also be conducted readily into the base of the mantle and become part of the general background flux of the mantle. In any case the heat flow from the core and the heat generation of a depleted lower mantle are minor compared to the heat generation, secular cooling and plate insulation effects, and lateral temperature gradients in the upper mantle. Plate tectonics, recycling and magmatism are forms of convection but the active layer may not extend deep into the mantle where the effects of pressure, low-heat sources and more uniform thermal gradients suppress the importance of convection. The plate tectonic style of convection and heat removal may not extend too far back in time.

At upper-mantle conditions, viscosity, conductivity and thermal expansion are strongly dependent on temperature. When this is so, mantle convective vigor adjusts itself to remove more heat as the temperature rises, unless the plates at the surface do not allow this. At high pressure, temperature is less effective in changing thermal and physical properties that depend on volume and there is less negative feedback. Viscosity also depends on water content. Recycling serves to reintroduce water into the upper mantle, which promotes melting and a lowering of viscosity.

One might think that a hotter Archean Earth convected more vigorously than the current mantle. But if melt and volatile removal are more important than recycling then a hot dry early mantle may have convected more slowly than the current mantle. If so, present-day secular cooling can contribute to the observed heat flow with no `Archean thermal catastrophe`. The ability of continents to divert heat to the ocean basins, and to move toward cold mantle, means that the stability of continental crust and mantle does not imply the absence of secular cooling for the mantle as a whole. The mantle may have cooled more slowly during the Archean than now because of the combined roles of subduction depth, recycling and mantle viscosity on mantle heat loss. The fact that the core is still mostly molten and the upper mantle is still near or above the melting point implies a long drawn-out cooling process. This favors a chemically stratified mantle but also one with most of the radioactive elements toward the surface. The drawn-out cooling also implies that mantle viscosity is not just a function of temperature or that surface boundary conditions may also regulate the vigor and style of mantle convection. A chemically stratified mantle, even with the radioactivity toward the top, will cool slowly. On the other hand, if the deep layers are undepleted or enriched, they will warm up with time, and the system may overturn.

Regional Earth models

The mantle loses two to three times as much heat per unit area in oceanic areas as in continental areas. One cannot discuss heat flow through continents, at present or in the past, without taking into account this partitioning. A corollary is that geological evidence from continents provides little constraint on average mantle temperatures or secular cooling rates. Continents affect the style of mantle convection, the rate of heat loss and the locations of downwellings. The growth rate of continents and the partitioning of mantle heat loss below oceans and continents may serve to maximize mantle heat loss. The motions of continents, plate reorganizations and the creation of new plate boundaries permit pent-up heat to escape and these introduce temporal variability into global heat flow.

The petrogenesis and inferred temperatures of Archean komatiites and the survival of cratonic roots may argue against an extremely hot mantle in the Archean and, thus, against substantial secular cooling. It has been thought that such cooling could be ruled out since the very high temperatures and convective vigor predicted for the mantle do not show up in continental geology; this is the Archean paradox. These arguments are based on one-dimensional and static continent considerations. Continents both affect the distribution of heat flux and move around. A large fraction of the heat flow, and the evidence for possibly elevated mantle temperatures in the past, is therefore in the missing ocean basins and does not show up in the thermal history of continents. The Americas and Australia are currently over-riding mantle cooled by recent subduction. The African crust and lithosphere have little knowledge of temperatures and convective vigor of the Pacific mantle. Realistic thermal history calculations include the coexistence of regional forms of convection and the multiple branches available to mantle evolution and cooling. Continents also introduce a long-wavelength lateral temperature gradient at the surface of the mantle. This alone can drive mantle flow. These lessons show us that secular cooling – the Kelvin effect – may not be lightly dismissed as an important contributor to surface heat flow, that the heat loss in oceans is affected by the presence of continents, and that mantle convection can be organized and driven by conditions at the surface. The plates may control the cooling of the mantle.

Thermal history

There can be a long delay between the generation of heat in Earth's interior and its arrival at the surface. A tabulation of current heat sources and heat losses is therefore only part of the story. To understand the Earth's thermal budget requires information about thermal history. Some estimates of global heat flow appear to greatly exceed that available from current radioactive heating in the mantle. The ratio of heat production to total heat flux is called the Urey ratio.

The viscosity of the mantle is an important issue. Viscosity is temperature dependent and high temperatures result in low viscosity and increased convective vigor. This negative feedback serves to buffer mantle temperature. However, water content also affects viscosity, and the melting point. Magmatism extracts water from the mantle causing the viscosity to increase, even if the temperature is high. Therefore, high convective vigor and extensive magmatism in the past could actually lead to decreased convective vigor and lower heat flow. Magmatism plays a minor role in current estimates of heat flow but it may have been a more important mechanism for removing heat in the past. The current plate-tectonic mode removes heat by plate tectonics (plate cooling at the surface and slab cooling in the interior). Io, a moon of Jupiter, may be an example where continuous volcanism, and the subsidence of the surface layer – the heat-pipe mechanism – rather than plate tectonics, is the mechanism for coping with the prodigious tidal heating. The current style of plate tectonics also involves massive recycling of water into the interior. This means that the melting point of the mantle is low and the viscosity is low, in spite of relatively low temperatures compared to the Archean. Magmatism is primarily restricted to plate boundaries and regions of tensile stress; the latent heats of crystallization contribute much less than the heat associated with cooling from high temperatures. This may not always have been the case. The cooling rate and cooling mechanism of the mantle are unlikely to have been the same throughout time.

The vigor of mantle convection does not depend only on mantle temperature and viscosity. Plates, slabs and continents affect the style of mantle convection. Plates dissipate energy by colliding, bending, breaking and moving past each other and the underlying mantle. The buoyancy that drives convection can be dissipated in the plate system as well as in the mantle. Plates may actually regulate mantle convection and over-ride the buffering effect of mantle viscosity. Thick plates, and a jammed plate system, can allow the mantle to over-heat, compared to the viscosity-regulated effect. Thus, the arguments against high mantle temperatures in the past, the viscosities associated with such temperatures, the intuitive relationships between temperature and heat flow and the self-regulation of mantle temperature must be reconsidered [fate of fossil heat].

Chemically stratified mantle?

We return once again to the possibility of a chemically stratified mantle. This possibility has apparently been ruled out by some workers based on visual inspection of a few color tomographic cross-sections. One-layer or whole mantle convection is the current reigning paradigm in mantle geochemistry and geodynamics. The various paradoxes that plague studies of mantle dynamics and geochemistry can be traced to the assumptions that underlie the models. In one class of models the mantle is divided into large reservoirs that can interchange, and exchange or entrain material. The deeper layers are highly radioactive and usually gas-rich. Slabs and plumes traverse the whole mantle, carrying the various components that are thought to characterize hotspot magmas. Ocean-island basalts (OIB) are assumed to come from the lower mantle, or the undegassed mantle. The upper-mantle reservoir is called 'the convecting mantle,' implying that it is well stirred by vigorous convection and is therefore homogenous.

Chemical stratification is the natural result of a planet growing in its own gravity field. The effect of pressure on thermal properties makes this stratification irreversible on a large planet. The standard models of mantle geochemistry and geodynamics, however, are quite different. The reasons are not hard to find; unphysical assumptions have been used to rule out the possibility of chemical stratification and irreversible differentiation. These include the Boussinesq approximation, assumption of a radioactive-rich lower mantle, assumed flat chemical interfaces between layers, and the assumption that the only plausible chemical boundary is the 650-km phase change. The textbook model of a depleted upper mantle, an undepleted, undegassed or primordial lower mantle, and a major isotopic boundary at 650 km, is unsatisfactory on many counts but this has been used to argue against all stratified models.

The Kelvin effect, revisited

The production of heat by decay of radioactive elements and the secular cooling of the mantle are the two main contributors to the present observed mean heat flux Q at the surface of the Earth. The first term is traditionally written as the Urey ratio, defined as the ratio between the heat production by radioactive elements and the total heat loss of the mantle. Attempts to reconstruct the thermal history of the Earth from a geophysical point of view have, since the time of Lord Kelvin, been in apparent disagreement with geochemical and geological observations. The geophysical approach uses simplified parameterized models of mantle cooling, usually involving whole-mantle convection. Effects of time-dependence, sphericity, pressure, plates and continents are generally ignored. The rate of cooling in early Earth history obtained in these models is generally too rapid to allow a sufficient present-day secular cooling rate. Geochemical estimates of radioactive element concentrations in the mantle appear to be too low to explain the observed present mantle heat loss. With present estimates of radioactive heat production in the mantle, simple parameterized models of whole mantle convection lead to a cooling of the Earth at the beginning of its history that is too fast and they are unable to explain the present heat loss of the Earth. This is one of the heat flow paradoxes of geophysics and it is related to the age of the Earth paradox of Lord Kelvin's time. There are several paradoxes associated with U, Th and K and their daughters such as heat, helium, lead and argon.

The balance between internal heat production and efficiency of heat transfer for a mantle with a temperature-dependent viscosity is such that the system is not sensitive to the initial conditions and self-regulates at each time step. A local increase in heat production or temperature results in a lower viscosity and more vigorous convection that carries away the excess heat.

Mantle convection may not be self-regulating. Cooling may be regulated by plate tectonics – involving the sizes and stiffnesses of the plates and the distribution of continents – leading to a weak dependence of the heat flow on the mantle viscosity and temperature. A hotter mantle results in a deeper onset of melting, more extensive melting, and a thicker buoyant crust and a dryer, stiffer lithosphere, which can reduce the total heat loss. Mantle viscosity and plate rheology depend on water content as well as temperature. These effects may be more important than the lowering of mantle viscosity by high temperature; a hotter mantle can actually result in lower heat flow.

It is common to assume the existence of a hidden mantle reservoir – a stealth or phantom reservoir – enriched in radiogenic elements and sequestered from global mantle circulation, to explain the present-day global heat budget. A deep, invisible undegassed mantle reservoir enriched in heat-producing elements has been the traditional explanation for the various heat flow and U–Pb–He paradoxes. But it is difficult to store heat-producing elements in the deep mantle. It is more likely that heat flow is variable with time, that plate tectonics controls the cooling rate of the mantle, and that the depleted MORB-reservoir does not occupy the whole of the upper mantle. There are various missing element paradoxes in geochemical mass-balance calculations and there is good evidence that the deeper mantle may be irreversibly stratified. But the inaccessible regions are more likely to be refractory and depleted. It is the assumption that depleted mid-ocean ridge basalts and their depleted residues represent the entire upper mantle that is responsible for the idea that there are missing heat-producing elements. Recycled and delaminated crust, not currently at the Earth's surface, and ultra-enriched magmas, such as kimberlites, can readily make up the perceived He, Th and K deficits.

Wrap up

One-dimensional and homogenous mantle models, or models with a downward increase in radioactive heating have dominated the attention of convection modelers. Paradoxes such as the Archean catastrophe, overheating of the lower mantle, persistence of cold continental keels and the missing heat-source 'problem' can be traced to these non-realistic assumptions and initial and boundary conditions.

A variety of evidence indicates that high temperatures and efficient gravitational differentiation determined the initial conditions of the Earth. On an Earth-sized body the effects of accretional energy and high pressure in the interior make chemical segregation essentially irreversible. The outer parts of Earth contain most of the heat-producing elements and supply much of the secular-cooling part of the present heat flow.

The major outstanding problems in the Earth's thermal budget and thermal history involve the role of hydrothermal circulation near the top, and the role of radiative transfer of heat near the bottom of the mantle. Convection modeling has not yet covered the parameter range that seems most pertinent from physical considerations and geophysical data. The largest failings in this regard are the neglect of pressure effects on material properties, and the use of parameterized convection. What is needed is a thermodynamically self-consistent approach that includes the temperature, pressure and volume – dependence of physical properties, realistic initial and boundary conditions, and the ability to handle melting. Modeling has focused on models that are do-able, and which are perceived by modelers to represent constraints from other fields. These include whole-mantle convection or layered convection models with mass transfer between layers, the persistence of large isolated mantle reservoirs, and the need for vigorous chaotic stirring.

The perceived mismatch between heat sources and surface heat flow and the assumption of a homogenous upper mantle and undepleted lower mantle have led to a series of complex proposals regarding mantle overturns and depths of recycling. Continents affect the form of mantle heat loss; continents drift so as to be over cold downwellings or near trenches. The stability of cratonic roots and the temperatures of komatiites do not constrain the thermal evolution of ocean basins or the rate of mantle convection. Higher mantle temperatures may have led to more vigorous convection (the Tozer effect) but increased melting also removes volatiles and increases the viscosity and strength of the upper mantle-plate system. The Hadean and Archean mantle may have been capped by a thick buoyant unsubductable layer and may have been drier and higher viscosity than the current mantle. There appears to be no mismatch between observed heat flow and plausible sources of heating. Current radioactive heating is just part of the equation. There are many open questions and opportunities for new approaches and ideas are enormous. The uncertainties are large and we must not make the same mistake as Lord Kelvin, who was confident in his estimate of the age of the Earth.

References and notes

There are three kinds of references in this monograph: (1) Key words and phases, and search strings, which can be inserted into a search engine such as Google. These are set in `special type`, and are called `Googlets`; (2) web addresses; and (3) normal author–date references. Supplementary material can be found on:

www.gps.caltech.edu/~dla/
www.caltechbook.library.caltech.edu/14/
www.mantleplumes.org

This material can also be found with the following `Googlets: don l anderson, theory of the earth` and `mantleplumes`. The entire first edition of *Theory of the Earth*, abbreviated TOE, can be found on www.caltechbook.library.caltech.edu/14/, and can be downloaded in its entirety or by specific chapter. The book *Plates, Plumes and Paradigms* and the site `mantleplumes` have many figures, tables and color maps that complement some of the material published here.

The following search strings are also useful for finding material that supplements the text:
`treatise on geochemistry`
`physics of the earth's interior`
`convection in the earth and planets`

Notes on Googlets

The Web has completely changed the way researchers and students do research, teach and learn. Search engines can be used to supplement textbooks and monographs. Conventional references are included in this book, but occasionally a `Googlet` is inserted with key search words for a given topic. These `Googlets` when used with a search engine can find pertinent recent references, color picture, movies and further background on the subject of interest. For example, if one wants to investigate the relationship between the *Deccan traps* and *Reunion* one can insert `[Reunion Deccan mantleplumes]` into Google. If one wants to know more about shear-wave splitting or the Love Rayleigh discrepancy one just types into *Google* `[shear-wave splitting]` or `[Love Rayleigh discrepancy]`. Often the author and a keyword can replace a list of references e.g. `[Anderson tomography]`. These `Googlets` are sprinkled throughout the text. The use of these is optional and the book can be used without interrogating the web. But if this resource is available and, if used, it can cut down the time required to find references and supplementary material. For the ordinary reader, these `Googlets` should be no more distracting than *italics* or **boldface** and much less distracting that the usual form of referencing and footnoting. They can be treated as keywords, useful but not essential. The key phrases have been designed so that, when used in a search, the top hits will contain relevant information. There may be, of course, some un-useful and redundant hits in the top five. Supplementary and current material can be found with keywords `Don L Anderson` and `mantleplumes`.

The *General References*, *Supplementary Reading* and *Recommended Reading* sections that are common in many monographs of this type have been largely replaced by `Googlets` that come before the main reference section of each chapter.

References

General references

There are many fads, false starts and dead ends in all scientific endeavors, but there are also some rock-solid foundation and milestone papers. The following is a highly selective compilation of significant publications in geodynamics, Earth structure and mantle geochemistry, most of which has stood the test of time and have set the stage for either our present understanding or demonstrate that we have wandered from a fruitful beginning. If the net and Google disappeared a good start on the subject matter of this book could be made from a subset of the publications in the following list. In many respects, we have lost ground!

Classical references

Birch, F. (1952) Elasticity and constitution of the Earth's interior. *J. Geophys. Res.*, **57**, 227–86.
Bowen, N. L. (1928) *The Evolution of Igneous Rocks*. Princeton, NJ, Princeton University Press.
Bowie, W. (1927) *Isostasy*. New York, Dutton.
Bullen, K. E. (1975) *The Earth's Density*. London, Chapman and Hall.
Bullen, K. (1947) *An Introduction to the Theory of Seismology*. Cambridge, Cambridge University Press.
Chandrasekhar, S. (1961) *Hydrodynamic and Hydromagnetic Stability*. Oxford, Clarendon Press.

Daly, R. A. (1933) *Igneous Rocks and the Depths of the Earth*. New York, McGraw Hill.

Daly, R. A. (1940) *Strength and Structure of the Earth*. Englewood Cliffs, NJ, Prentice-Hall.

Darwin, G. H. (1879) On the bodily tides of viscous and semi-elastic spheroids and on the ocean tides upon a yielding nucleus. *Phil. Trans. Roy. Soc. London A*, **1970**.

Dietz, R. S. (1961) Continent and ocean basin evolution by spreading of the sea floor. *Nature*, **190**, 854–7.

Du Toit, A. L. (1937) *Our Wandering Continents*. Edinburgh, Oliver and Boyd.

Elder, J. (1976) *The Bowels of the Earth*. Oxford, Oxford University Press.

Elsasser, W. M. (1969) Convection and stress propagation in the upper mantle. In *The Application of Modern Physics to the Earth and Planetary Interiors*, ed. Runcorn, S. K. New York, John Wiley & Sons, pp. 223–49.

Gast, P. W. (1969) The isotopic compositon of lead from St. Helena and Ascension Islands. *Earth Planet. Sci. Lett.*, **5**, 353–9.

Gutenberg, Beno (1939) *Internal Constitution of the Earth*, ed. Gutenberg, B. New York, McGraw-Hill. (2nd Edition, New York, Dover Publications, 1951.)

Haskell, N. A. (1937) The viscosity of the asthenosphere. *Am. J. Sci.*, **33**, 22–30.

Hess, H. H. (1962) History of ocean basins. In *Petrologic Studies: A Volume in Honor of A. F. Buddington*, eds. Engel, A. E. J., James, H. L. and Leonard, B. F. Boulder, CO, Geological Society of America, pp. 599–620.

Hirth, J. P. and Lothe, J. (1982) *Theory of Dislocations*, New York, John Wiley & Sons.

Holmes, A. (1928) Radioactivity and continental drift. *Geol. Mag.*, **65**, 236–8.

Holmes, A. (1944) *Principles of Physical Geology*. Edinburgh, Thomas and Sons Ltd.

Holmes, A. (1946) An estimate of the age of the Earth. *Nature*, **57**, 680–4.

Howard, L. N. (1963) Heat transport by turbulent convection. *J. Fluid Mech.*, **17**, 405–32.

Jeffreys, H. (1926) The stability of a layer of fluid heated from below. *Phil. Mag.*, **2**, 833–44.

Jeffreys, H. (1976) *The Earth*, Sixth Edition. Cambridge, Cambridge University Press.

Jeffreys, H. (1939) *Theory of Probability*. Oxford, Oxford University Press; with new editions in 1948 and in 1961 (also in the Oxford Classic Texts in the Physical Sciences series).

Jeffreys, H. and Swirles, B. (eds.) (1971–77) *Collected Papers of Sir Harold Jeffreys on Geophysics and other Sciences* (in six volumes). London, Gordon & Breach.

Kaula, W. M. (1968) *An Introduction to Planetary Physics, the Terrestrial Planets*, New York, John Wiley & Sons.

Patterson, C. (1956) Age of meteorites and the Earth. *Geochim. Cosmochim. Acta*, **10**, 230–7.

Pekeris, C. L. (1935) Thermal convection in the interior of the Earth. *Mon. Not. R. Astron. Soc., Geophys. Suppl.*, **3**, 343–67.

Rayleigh, Lord (1916) On convection currents on a horizontal layer of fluid when the higher temperature is on the under side. *Phil. Mag.*, **32**, 529–46. (See also Pearson, J. R. A. (1958) On convection cells induced by surface tension. *J. Fluid Mech.*, **4**, 489–500.)

Rutherford, E. (1907) Some cosmical aspects of radioactivity. *J. Roy. Astr. Soc. Canada*, **May–June**, 145–65.

Thompson, D'Arcy (1917) *On Growth and Form*. Cambridge, Cambridge University Press.

Wegener, A. (1924) *The Origin of Continents and Oceans*. New York, Dutton.

CHAPTER REFERENCES

PART I. PLANETARY PERSPECTIVE

Chapter 1. Origin and early history

Solar system abundances
Origin of Moon

References

Anders, E. (1968) Chemical processes in the early solar system, as inferred from meteorites. *Acct. Chem. Res.*, **1**, 289–98.

Fuchs, L. H., Olsen, E. and Jensen, K. (1973) Mineralogy, mineral-chemistry, and composition of the Murchison (C2) meteorite. *Smithsonian Contrib. Earth Sci.*, **10**, 39.

Grossman, L. (1972) Condensation in the primitive solar nebula. *Geochim. Cosmochim. Acta*, **36**, 597–619.

Grossman, L. and Larimer, J. (1974) Early chemical history of the solar system. *Rev. Geophys. Space Phys.*, **12**, 71–101.

Morgan, J. W. and Anders, E. (1980) Chemical composition of the Earth, Venus, and Mercury. *Proc. Natl. Acad. Sci.*, **77**, 6973.

Safronov, V. S. (1972) Accumulation of the planets. In *On the Origin of the Solar System*, ed. Reeves, H. Paris,

Centre Nationale de Recherche Scientifique, pp. 89–113.

Selected reading

Abe, Y. (1997) Thermal and chemical evolution of the terrestrial magma ocean. *Phys. Earth Planet. Inter.*, **100**, 27–39.

Agee, C. B. (1990) A new look at differentiation of the Earth from melting experiments on the Allende meteorite. *Nature*, **346**, 834–7.

Cameron, A. G. W. (1997) The origin of the moon and the single impact hypothesis. *Icarus*, **126**, 126–37.

Canup, R. M. and Asphaug, E. (2001) The Moon-forming impact. *Nature*, **412**, 708–12.

Carrigan, C. R. (1983) A heat pipe model for vertical, magma-filled conduits. *J. Volc. Geotherm. Res.*, **16**, 279–88.

Grossman, L. (1972) Condensation in the primitive solar nebula. *Geochim. Cosmochim. Acta*, **36**, 579–619.

Murthy, V. R. (1991) Early differentiation of the earth and the problems of mantle siderophile elements: a new approach. *Science*, **253**, 303–6.

Ohtani, E. (1985) The primordial terrestrial magma ocean and its implications for the stratification of the mantle. *Phys. Earth Planet. Inter.*, **38**, 70–80.

Ringwood, A. E. (1979) *Origin of the Earth and Moon*. New York, Springer-Verlag.

Chapter 2. Comparative planetology

origin of the Moon and the single impact hypothesis
origin of the Earth and Moon
timing of metal -silicate fractionation
Elementary and nuclidic abundances in the solar system
timing of metal-silicate fractionation W Hf

References

Hartman, W., Phillips, R. and Taylor, G. (1986) *Origin of the Moon*. Houston, Lunar and Planetary Institute.

Taylor, S. R. (1982) *Planetary Science, A Lunar Perspective*. Houston, Lunar and Planetary Institute.

Taylor, S. R. and McLennan, S. (1985) *The Continental Crust: Its Composition and Evolution*. London, Blackwell.

Weaver, B. L. and Tarney, J. (1984) Major and trace element composition of the continental lithosphere. *Phys. Chem Earth*, **15**, 39–68.

Further reading

Anderson, D. L. (1972) The internal constitution of Mars. *J. Geophys. Res.*, **77**, 789–95.

Ganapathy, R. and Anders, E. (1974) Bulk compositions of the Moon and Earth estimated from meteorites. *Proc. Lunar Sci. Conf.*, **5**, 1181–206.

Taylor, S. R. and McLennan, S. M. (1981) The composition and evolution of the Earth's crust; rare earth element evidence from sedimentary rocks. *Phil. Trans. Roy. Soc. Lond. A*, **301**, 381–99.

Chapter 3. The building blocks of planets

composition of the earth
enstatite model earth oxygen

References

Anders, E. and Ebihara, M. (1982) Solar system abundances of the Elements. *Geochim. Cosmochim. Acta*, **46**, 2363–80.

Breneman, H. H. and Stone, E. C. (1985) Solar coronal and photospheric abundances from solar energetic particle measurements. *Astrophys. J. Lett.* **294**, L57–62.

BVP, Basaltic Volcanism Study Project (1980) *Basaltic Volcanism on the Terrestrial Planets*. New York, Pergamon.

Drake, M. J. and Righter, K. (2002) Determining the composition of the Earth. *Nature*, **416**, 39–44.

Ganapathy, R. and Anders, E. (1974) Bulk compositions of the Moon and Earth estimated from meteorites. *Proc. Lunar Sci.Conf.*, **5**, 1181–206.

Grossman, L. (1972) Condensation in the primitive solar nebula. *Geochim. Cosmochim. Acta*, **36**, 597–619.

Javoy, M. (1995) The integral enstatite chondrite model of the Earth. *Geophys. Res. Lett.*, **22**, 2219–22.

Mason, B. (1962) *Meteorites*. New York, John Wiley & Sons.

Morgan, J. W. and Anders, E. (1980) Chemical composition of the Earth, Venus, and Mercury. *Proc. Natl. Acad. Sci.*, **77**, 6973.

Ringwood, A. E. (1977) *Composition and Origin of the Earth*. Publication No. 1299. Canberra, Research School of Earth Sciences, Australian National University.

Wood, J. A. (1962) Chondrules and the origin of the terrestrial planets. *Nature*, **194**, 127–30.

Further reading

Anders, E. and Owen, T. (1977) Mars and Earth: origin and abundance of volatiles. *Science*, **198**, 453–65.

Cameron, A. G. W. (1982) Elementary and nuclidic abundances in the solar system. In *Essays in Nuclear*

Astrophysics, eds. Barnes, C. A. *et al.* Cambridge, Cambridge University Press.
Duffy, T. S. and Anderson, D. L. (1989) Seismic velocities in mantle minerals and the mineralogy of the upper mantle. *J. Geophys. Res.*, **94**, 1895–912.
Grossman, L. and Larimer, J. (1974) Early chemical history of the solar system. *Rev. Geophys. Space Phys.*, **12**, 71–101.
Mazor, E., Heymann, D. and Anders, E. (1970) Noble gases in carbonaceous chondrites. *Geochim. Cosmochim. Acta*, **34**, 781–824.
von Zahn, V., Kumar, S., Niemann, H. and Prim, R. (1983) Composition of the Venus atmosphere. In *Venus*, eds. Hunten, D. M., Colin, L., Donahue, T. and Moroz, V. Tucson, University of Arizona Press, pp. 299–430.
Wacker, J. and Marti, K. (1983) Noble gas components of Albee Meteorite. *Earth Planet. Sci. Lett.*, **62**, 147–58.
Wanke, H., Baddenhausen, H., Blum, K., Cendales, M., Dreibus, G., Hofmeister, H., Kruse, H., Jagoutz, E., Palme, C.,Spettel, B., Thacker R. and Vilcsek, E. (1977) On chemistry of lunar samples and achondrites; Primary matter in the lunar highlands; A re-evaluation. *Proc. Lunar Sci. Conf. 8th*, 2191–13.
Weidenschilling, S. J. (1976) Accretion of the terrestrial planets. *Icarus*, **27**, 161–70.

PART II. EARTH: THE DYNAMIC PLANET

Chapter 4. The outer shells of Earth

shallow flat subduction
far-from-equilibrium self-organized system
non-adiabaticity in mantle convection
tomographic geodynamic mantle models
fracture Tessellation
plate Tectonics and Platonic Solids
generation of plate tectonics from mantle convection

References and notes

Clare, B. W. and Kepert, D. L. (1991). The optimal packing of circles on a sphere. *J. Math. Chem.*, **6**, 325–49.
Foulger, G. L., Natland, J. H., Presnall, D. C. and Anderson, D. L. (eds.) (2005) *Plates, Plumes and Paradigms*. Boulder, CO, Geological Society of America, Special Paper 388.
Morgan, W. J. (1971) Convective plumes in the lower mantle. *Nature*, **230**, 42–3.
Rowley, D. B. (2002) Rate of plate creation and destruction: 180 Ma to present. *Geolog. Soc. Am. Bull.*, **114**, 927–33.
Van Hunen, J., van den Berg, A. P. and Vlaar, N. (2002) On the role of subducting oceanic plateaus in the development of shallow flat subduction, *Tectonophysics*, **352**, 317–33.
Wilson, J. T. (1973) Mantle plumes and plate motions, *Tectonophysics*, **19**, 149–64.

Plate Driving Forces; Shallow Return Flow

geodynamics plate motion counterflow Chappell Harper Elsasser Chase Parmentier Oliver Hager O'Connell

It is often assumed that mantle convection drives the plates but complex computer simulations of mantle convection have not yielded anything resembling plate tectonics. By about 1980 the driving mechanisms of plate tectonics were fairly well understood; the plates and slabs drove themselves and were responsible for density variations in the underlying mantle. These plate models of mantle convection assumed shallow return flow and upper mantle convection. The advent of supercomputers and high Rayleigh number convective simulations of whole mantle convection have not given us much more additional insight. The upper mantle counterflow and top-down models of plate tectonics are mostly overlooked in modern treatises on mantle convection and geodynamics, which emphasize whole mantle convection. The milestone papers and books of mantle geodynamics and plate tectonic theory are listed below.

Further reading

Chappell, W. M. and Tullis, T. E. (1977) Evaluation of the forces that drive plates. *J. Geophys. Res.*, **82**, 1967–84.
Chase, C. G. (1979) Asthenospheric counterflow: a kinematic model. *Geophys. J. R. Astron. Soc.*, **56**, 1–18.
Chase, C. G. (1979) Subduction, the geoid, and lower mantle convection. *Nature*, **282**, 464–8.
Elder, J. W. (1967) Convective self-propulsion of continents. *Nature*, **214**, 657–60.
Elsasser, W. M. (1969) Convection and stress propagation in the upper mantle. In *The Application of Modern Physics to the Earth and Planetary Interiors*, ed. Runcorn, S. K. New York, John Wiley & Sons, pp. 223–49.

Forsyth, D. and Uyeda, S. (1975) On the relative importance of the driving forces of plate motion. *Geophys. J. R. Astr. Soc.*, **43**, 163–200.
Hager, B. H. (1983) Global isostatic geoid anomalies for plate and boundary layer models of the lithosphere. *Earth Planet. Sci. Lett.*, **63**, 97–109.
Hager, B. H. and O'Connell, R. J. (1979) Kinematic models of large-scale flow in the Earth's mantle. *J. Geophys. Res.*, **84**, 1031–48.
Hager, B. H. and O'Connell, R. J. (1981) A simple global model of plate dynamics and mantle convection. *J. Geophys. Res.*, **86**, 4843–67.
Harper, J. F. (1978) Asthenosphere flow and plate motions. *Geophys. J. R. Astr. Soc.*, **55**, 87–110.
Jacoby, W. R. (1970) Instability in the upper mantle and global plate movements. *J. Geophys. Res.*, **75**, 5671–80.
Kaula, W. M. (1972) Global gravity and tectonics. In *The Nature of the Solid Earth*, ed. Robertson, E. C. New York, McGraw-Hill, pp. 386–405.
Kaula, W. M. (1980) Material properties for mantle convection consistent with observed surface fields. *J. Geophys. Res.*, **85**, 7031–44.
Parmentier, E. M. and Oliver, J. E. (1979) A study of shallow global mantle flow due to the accretion and subduction of lithospheric plates. *Geophys. J. R. Astr. Soc.*, **57**, 1–21.
Ramberg, H. (1967) *Gravity, Deformation and the Earth's Crust*. London, Academic Press.

Plate Driving Forces; Whole Mantle Convection

It is commonly assumed that plate motions are the surface expression of mantle convection rather than vice versa. How the dynamical system should be broken down in order to identify and analyze the driving forces has been debated since plate tectonics was recognized. The approaches can be classified into a force balance on the plates and slabs, in order to achieve a force equilibrium, or a full convection calculation where the sources of buoyancy in the mantle are approximated, usually by density–velocity scalings from tompographic images. In one case, the forces on the plates drive themselves *and* mantle convection, a top-down system. In the other extreme, plate motions are viewed as the result of mantle motions, plumes and density anomalies.

A variety of mantle circulation models can explain the motions of the tectonic plates. The anisotropy of the mantle, the gravity field and the stresses in the plates might be able to distinguish between the various flow models and the style of mantle convection. Modern attempts to understand plate tectonics start with density anomalies in the mantle, rather than the forces on the plates. Plates and plate tectonics are often regarded as the results of whole mantle convection driven by internal density variations and significant bottom heating.

Becker, T. W. and O'Connell, R. J. (2001) Predicting plate motions with mantle circulation models. *Geochemistry, Geophysics, Geosystems* **2**, 2001GC000171.
Bercovici, D. (1995) A source-sink model of the generation of plate tectonics from non-Newtonian mantle flow. *J. Geophys. Res.*, **100**, 2013–30.
Tackley, P. (2000) The quest for self-consistent generation of plate tectonics in mantle convection models. In *History and Dynamics of Global Plate Motions, Geophys. Monogr. Ser.*, eds. Richards, M. A., Gordon, R. and van der Hilst, R. Washington, DC, American Geophysical Union, pp. 47–72.
Trompert, R. & Hansen, U. (1998) Mantle convection simulations with rheologies that generate plate-like behavior. *Nature*, **395**, 686–9.
Lithgow-Bertelloni, C. & Richards, M. A. (1998) The dynamics of Cenozoic and Mesozoic plate motions. *Rev. Geophys.* **36**, 27–78.

Chapter 5. The eclogite engine

Crustal delamination
sierran xenoliths
lower crustal convective instability
mantleplumes
eclogite engine
mantleplumes lithospheric removal

Plates vs. plumes

plates vs plumes Jacoby Elsasser
Plate tectonics represents the behavior of the outer fragmented shell of the Earth. Plumes are believed to be responsible for features that have traditionally been treated outside the framework of rigid-plate tectonics and are thought to originate in a lower thermal boundary layer. There has recently been much interest in testing the plume hypothesis and comparing it with alternate mechanisms for forming volcanic chains and melting anomalies. Melting anomalies, so-called hotspots, may be due to fertility anomalies as well as thermal anomalies. The following publications discuss and compare the plate and plume theories, addenda to plate tectonics, plate forces and recycling mechanisms, and test the deep mantle plume hypothesis.

Further reading

Allen, R. & Tromp, J. (2005) Resolution of regional seismic models: Squeezing the Iceland anomaly. *Geophys. J. Inter.*, **161**, 373–86.

Anderson, D. L. (2005) Scoring hotspots: The plume and plate paradigms. In *Plates, Plumes, and Paradigms*, eds. Foulger, G. R., Natland, J. H., Presnall, D. C. and Anderson, D. L. Boulder, CO, Geological Society of America, Special Paper 388, pp. 31–54.

Anderson, D. L. (2002) How many plates? *Geology*, **30**, 411–14.

Anderson, D. L. and Natland, J. H. (2005) A brief history of the plume hypothesis and its competitors: Concept and controversy. In *Plates, Plumes and Paradigms*, eds. Foulger, G. R., Natland, J. H., Presnall, D. C. and Anderson, D. L. Boulder, CO, Geological Society of America, Special Paper 388, pp. 119–46.

Anderson, D. L. and Schramm, K. A. (2005). Hotspot catalogs. In *Plates, Plumes and Paradigms*, eds. Foulger, G. R., Natland, J. H., Presnall, D. C. and Anderson, D. L. Boulder, CO, Geological Society of America, Special Paper 388, pp. 19–30.

Becker, T. W. and Boschi, L. (2002) A comparison of tomographic and geodynamic mantle models. *Geochem. Geophys. Geosyst.*, **3**, 2001GC000168.

Cates, M. E., Wittmer, J. P., Bouchaud, J.-P. & Claudin, P. (1998) Jamming, force chains and fragile matter. *Phys. Rev. Lett.*, **81**, 1841–4.

Conrad, C. P. & Hager, B. H. (2001) Mantle convection with strong subduction zones. *Geophys. J. Inter.*, **144**, 271–88.

Davies, G. F. (2000) *Dynamic Earth: Plates, Plumes and Mantle Convection*. Cambridge, Cambridge University Press.

Chapter 6. The shape of the Earth

Dynamic geoid tomography

References

Hager, B. H. (1984) Subducted slabs and the geoid; constraints on mantle theology and flow. *J. Geophys. Res.*, **89**, 6003–15.

Hager, B. H., Clayton, R., Richards, M., Comer, R. and Dziewonski, A. (1985) Lower mantle heterogeneity, dynamic topography and the geoid. *Nature*, **313**, 541–5.

Rapp, R. H. (1981) The Earth's gravity field to degree and order 180 using Seaset altimeter data, terrestrial gravity data, and other data. *Report 322, Dept. Geodetic. Sci. and Surv.* Columbus, OH, Ohio State University.

Richards, M. A. and Hager, B. H. (1984) Geoid anomalies in a dynamic Earth. *J. Geophys. Res.*, **89**, 5987–6002.

Selected reading and notes

Apparent polar wander was a key element in recognizing the mobility of continents. True polar wander, or the shift of the Earth around the rotation axis, was recognized later, although the physics is better understood. The shape of the Earth was not really understood until seismic tomography came along (Hager, 1984; Hager *et al.*, 1985). There were some false starts involving fossil rotational bulges and a rigid or ultrahigh viscosity lower mantle. The fixed plume hypothesis was initially based on the idea that the lower mantle was not convecting. Key papers are listed below.

Darwin, G. (1877) On the influence of geological changes on the earth's axis of rotation. *Phil. Trans. R. Soc. Lond. A*, **167**, 271–312.

Goldreich, P. and Toomre, A. (1968) Some remarks on polar wandering. *J. Geophys. Res.*, **74**, 2555–67.

Hager, B. H. and Richards, M. (1989) Long-wavelength variations in Earth's geoid: physical models and dynamical implications. *Phil. Trans. R. Soc. Lond. A*, **328**, 309–27.

Kaula, W. M. (1972) Global gravity and tectonics. In *The Nature of the Solid Earth*, ed. Robertson, E. C. New York, McGraw-Hill, pp. 386–405.

Kaula, W. M. (1980) Material properties for mantle convection consistent with observed surface fields. *J. Geophys. Res.*, **85**, 7031–44.

Further reading

Parsons, B. and Sclater, J. G. (1977) An analysis of the variation of ocean floor bathymetry and heat flow with age. *J. Geophys. Res.*, **82**, 803–27.

Chapter 7. Convection and complexity

Far-from-equilibrium self-organized system

Top down tectonics

Layered mantle convection

Selected reading and notes

Layered mantle convection

```
chemical stratification layered
  convection
layered mantle convection eclogite
thermal coupling slab-like anomalies
```

It is very difficult to simulate, in a thermodynamically self-consistent way, the convection in a chemically layered system. Most attempts to simulate layered convection use viscosity or phase changes to stratify the mantle, and use flat interfaces and uniform internal heating. The effects of pressure on thermal properties are generally ignored. Therefore, most of the published convection models for the mantle assume whole-mantle convection and the Boussinesq 'approximation.' There are a few calculations of simplified two-layer models, and none at all for multiple layers. There is much to be done in this area of research.

Anderson, D. L. (1979) Chemical stratification of the mantle. *J. Geophys. Res.*, **84**, 6297–8.

Anderson, D. L. (2002) The case for irreversible chemical stratification of the mantle. *Int. Geol. Rev.*, **44**, 97–116.

Ciskova, H., Cadek, O., van den Berg, A. P. & Vlaar, N. (1999) Can lower mantle slab-like seismic anomalies be explained by thermal coupling between the upper and lower mantles? *Geophys. Res. Lett.*, **26**, 1501–4.

Ciskova, H. & Cadek, O. (1997) Effect of a viscosity interface at 1000 km depth on mantle convection. *Studia geoph. geod.*, **41**, 297–306.

Glatzmaier, G. A. & Schubert, G. (1993) Three-dimensional spherical models of layered and whole mantle convection. *J. Geophys. Res.*, **98**, 969–76.

Gu, Y., Dziewonski, A. M. & Agee, C. (1998) Global de-correlation of the topography of transition zone discontinuities. *Earth Planet. Sci. Lett.*, **157**, 57–67.

Honda, S. (1984) A preliminary analysis of convection in a mantle with a heterogeneous distribution of heat-producing elements. *Phys. Earth Planet. Inter.*, **34**, 68–76.

Honda, S. (1986) Strong anisotropic flow in a finely layered asthenosphere. *Geophys. Res. Lett.*, **13**, 1454–7.

Nataf, H-C., Moreno, S. and Cardin, Ph. (1988) What is responsible for thermal coupling in layered convection? *J. Phys. France*, **49**, 1707–14.

Phillips, B. R. and Bunge, H.-P. (2005) Heterogeneity and time dependence in 3D spherical mantle convection models with continental drift. *Earth Planet. Sci. Lett.*, **233**, 121–35.

Schubert, G., Turcotte, D. and Olson, P. (2001) *Mantle Convection in the Earth and Planets*. Cambridge, Cambridge University Press.

Silver, P. G., Carlson, R. W. and Olson, P. (1988) Deep slabs, geochemical heterogeneity and the large-scale structure of mantle convection: Investigation of an enduring paradox. *Ann. Rev. Earth Planet. Sci.*, **16**, 477–541.

Todesco, M. & Spera, F. (1992) Stability of a chemically layered upper mantle. *Phys. Earth Planet. Inter.*, **71**, 85–99.

van Keken, P. E. & Ballantine, C. (1998) Whole-mantle versus layered mantle convection and the role of a high-viscosity lower mantle in terrestrial volatile evolution. *Earth Planet. Sci. Lett.*, **156**, 19–32.

Wen, L. & Anderson, D. L. (1997) Layered mantle convection: a model for geoid and topography. *Earth Planet. Sci. Lett.*, **146**, 367–77.

PART III. RADIAL AND LATERAL STRUCTURE

Chapter 8. Let's take it from the top: the crust and upper mantle

```
PREM model
Mantle reflections
P'P' precursors mantle
www.quake.wr.usgs.gov/research/structure/
  CrustalStructure/crust/
```

References

Babuska, V. (1972) Elasticity and anisotropy of dunite and bronzitite. *J. Geophys. Res.*, **77**, 6955–65.

Christensen, N. I. and Lundquist, J. N. (1982) Pyroxene orientation within the upper mantle. *Geol. Soc. Amer. Bull.*, **93**, 279–88.

Christensen, N. I. and Smewing, J. D. (1981) Geology and seismic structure of the northern section of the Oman ophiolite. *J. Geophys. Res.*, **86**, 2545–55.

Clark, S. P., Jr. (1966) *Handbook of Physical Constants. Geol. Soc. Amer. Mem.* **97**.

Condie, K. L. (1982) *Plate Tectonics and Crustal Evolution*, Second edition. New York, Pergamon.

Duffy, T. S. and Anderson, D. L. (1988). Seismic velocities in mantle minerals and the mineralogy of the upper mantle. *J. Geophys. Res.*, **94**, 1895–912.

Dziewonski, A. M. and Anderson, D. L. (1981). Preliminary reference Earth model. *Phys. Earth Planet. Inter.*, **25**, 297–356.

Elthon, D. (1979) High magnesia liquids as the parental magma for ocean floor basalts. *Nature*, **278**, 514–18.

Given, J. and Helmberger, D. (1981) Upper mantle structure of northwestern Eurasia. *J. Geophys. Res.*, **85**, 7183–94.

Grand, S. P. and Helmberger, D. (1984a) Upper mantle shear structure of North America. *Geophys. J. Roy. Astr. Soc.*, **76**, 399–438.

Grand, S. P. and Helmherger, D. (1984b) Upper mantle shear structure beneath the Northwest Atlantic Ocean. *J. Geophys. Res.*, **89**, 11 465–75.

Jordan, T. H. (1979) Mineralogies, densities and seismic velocities of garnet lherzolites and their geophysical implications. In *The Mantle Sample*, eds. Boyd, F. R. and Meyer, H. O. A. Washington DC, American Geophysical Union, pp. 1–14.

Lehmann, I. (1961) S and the structure of the upper mantle, *Geophys. J. R. Astron. Soc.*, **4**, 124–38.

Manghnani, M. H. and Ramananotoandro, C. S. P. (1974) Compressional and shear wave velocities in granulite facies rocks and eclogites to 10 kbar. *J. Geophys. Res.*, **79**, 5427–46.

Mooney, W. D., Laske, G. and Masters, G. (1998) A new global crustal model at 5 × 5 degrees: CRUST5.1. *J. Geophys. Res.*, **103**, 727–47.

Regan, J. and Anderson, D. L. (1984) Anisotropic models of the upper mantle. *Phys. Earth Planet. Inter.*, **35**, 227–63.

Salisbury, M. and Christensen, N. L. (1978) The seismic velocity structure of a traverse through the Bay of Islands ophiolite complex, Newfoundland, an exposure of oceanic crust and upper mantle. *J. Geophys. Res.*, **83**, 805–17.

Sumino, Y. and Anderson, O. L. (1984) Elastic constants of minerals. In *Handbook of Physical Properties of Rocks* 3, ed. Carmichael, R. S. Boca Raton, FL, CRC Press, pp. 39–138.

Taylor, S. R. and McLennan, S. (1985) *The Continental Crust: Its Composition and Evolution*. London, Blackwell.

Walck, M. C. (1984) The P-wave upper mantle structure beneath an active spreading center: The Gulf of California. *Geophys. J. R. Astr. Soc.*, **76**, 697–723.

Further reading

Anderson, D. L. and Bass, J. D. (1984) Mineralogy and composition of the upper mantle. *Geophys. Res. Lett.*, **11**, 637–40.

Bullen, K. (1947) *An Introduction to the Theory of Seismology*. Cambridge, Cambridge University Press.

Deuss, A. and Woodhouse, J. H. (2002) A systematic search for mantle discontinuities using SS-precursors. *Geophys. Res. Lett.*, **29**, 8, doi: 10.1029/2002GL014768.

Grand, S. P. (1994). Mantle shear structure beneath the Americas and surrounding oceans. *J. Geophys. Res.*, **99**, 591–621.

Montelli, R., Nolet, G., Dahlen, F. A., Masters, G., Engdahl, E. R. and Hung, S. H. (2004) Finite-frequency tomography reveals a variety of plumes in the mantle. *Science*, **303**, 338–43.

Shimamura, H., Asada, T. and Kumazawa, M. (1977) High shear velocity layer in the upper mantle of the Western Pacific. *Nature*, **269**, 680–2.

Weidner, D. J. (1986) Mantle models based on measured physical properties of minerals. In *Chemistry and Physics of Terrestrial Planets*, ed. Saxena, S. K. New York, Springer-Verlag, pp. 251–74.

Weidner, D. J., Sawamoto, H., Sasaki, S. and Kumazawa, M. (1984) Single-crystal elastic properties of the spinel phase of Mg_2SiO_4. *J. Geophys. Res.*, **89**, 7852–60.

Whitcomb, J. H. and Anderson, D. L. (1970) Reflection of P'P' seismic waves from discontinuities in the mantle. *J. Geophys. Res.*, **75**, 5713–28.

Chapter 9. A laminated lumpy mantle

`crustal delamination`
`eclogite sinkers`
`marble cake mantle`
`seismic scattering upper mantle coda`
`seismic scattering phonon`

References

Deuss, A. and Woodhouse, J. H. (2002) A systematic search for mantle discontinuities using SS-precursors. *Geophys. Res. Lett.*, **29**, 8, doi: 10.1029/2002GL014768.

Revenaugh, J. & Sipkin, S. A. (1994) Seismic evidence for silicate melt atop the 410-km mantle discontinuity. *Nature*, **369**, 474–6.

Nolet, G. & Zielhuis, A. (1994) Low S velocities under the Tornquist–Teisseyre zone: evidence for water injection into the transition zone by subduction. *J. Geophys. Res.*, **99**, 15813–20.

Song, T., Helmberger, D. & Grand, S. (2004) Low velocity zone atop the 410 seismic discontinuity in the northwestern U. S. *Nature*, **427**, 530–3.

Vinnik, L., Kumar, M. R., Kind, R. & Farra, V. (2003) Super-deep low-velocity layer beneath the Arabian plate. *Geophys. Res. Lett.*, **30** (1415), doi:10.1029/2002GL016590.

Whitcomb, J. H. & Anderson, D. L. (1970) Reflection of P′P′ seismic waves from discontinuities in the mantle. *J. Geophys. Res.*, **75**, 5713–28.

Chapter 10. The bowels of the earth

```
First principles thermoelasticity
stratification of the earth's mantle
low spin iron mantle
PKIKP
PKJKP
```

References

Birch, F. (1952) Elasticity and constitution of the Earth's interior. *J. Geophys. Res.*, **57**, 227–86.

Lay, T. & Helmberger, D. V. (1983) Body-wave amplitude and travel-time correlations across North America. *Bull. Seism. Soc. Am.*, **73**, 17–30.

Lehmann, I. (1936) *Bur. Centr. Seism. Inst. A* 14, 3–31.

Julian, B., Davies, D. & Sheppard, R. (1972) PKJKP. *Nature*, **235**, 317–18.

Further reading

Bullen, K. (1947) *An Introduction to the Theory of Seismology*. Cambridge, Cambridge University Press.

Wen, L. & Helmberger, D. V. (1998) Seismic evidence for an inner core transition zone. *Science*, **279**, 1701–3.

Ishii, M. & Dziewonski, A. M. (2002) The innermost inner core of the earth: evidence for a change in anisotropic behavior at the radius of about 300 km. *PNAS*, **99**, 14026–30.

Chapter 11. Geotomography: heterogeneity of the mantle

```
Steve Grand's mantle models
Upper mantle shear structure
Probabilistic tomography maps
mantleplumes
Robust inversion of IASP91
seismic inversion resolution mantle
superplumes from the core-mantle
large-scale mantle heterogeneity
```

References

Anderson, D. L. (2005) Scoring hotspots: the plume and plate paradigms, in *Plates, Plumes, and Paradigms*, pp. 31–54, ed. Foulger, G. R., Natland, J. H., Presnall, D. C., and Anderson, D. L., Boulder, C. O., Geological Society of America Special Paper 388.

Baig, A. M. and Dahlen, F. A. (2004) Travel time biases in random media and the S-wave discrepancy. *Geophys. J. Inter.*, **158**, 922–38.

Becker, T. W. and Boschi, L. (2002) A comparison of tomographic and geodynamic mantle models. *Geochem. Geophys. Geosyst.*, **3**, 2001GC000168.

Bijwaard, H., Spakman, W. and Engdahl, E. (1998) Closing the gap between regional and global travel time tomography. *J. Geophys. Res.* **103**, 30 055–78.

Dziewonski, A. M. (2005) The robust aspects of global seismic tomography. In *Plates, Plumes and Paradigms*, eds. Foulger, G. R., Natland, J. H., Presnall, D. C. and Anderson, D. L. Boulder, CO, Geological Society of America, pp. 147–54.

Forsyth, D. and Uyeda, S. (1975) On the relative importance of the driving forces of plate motion. *Geophys. J. R. Astr. Soc.*, **43**, 163–200.

Foulger, G. L., Natland, J. H., Presnall, D. C. and Anderson, D. L., eds. (2005) *Plates, Plumes and Paradigms*. Boulder, CO, Geological Society of America, Special Paper 388.

Grand, S. P. (1986) Shear velocity structure of the mantle beneath the North American plate, Ph.D. Thesis, California Institute of Technology.

Grand, S. P. (1994) Mantle shear structure beneath the Americas and surrounding oceans. *J. Geophys. Res.*, **99**, 591–621.

Grand, S. P. and Helmberger, D. (1984a) Upper mantle shear structure of North America. *Geophys. J. Roy. Astron. Soc.*, **76**, 399–438.

Grand, S. P. and Helmberger, D. (1984b) Upper mantle shear structure beneath the Northwest Atlantic Ocean. *J. Geophys. Res.*, **89**, 11,465–75.

Grand, S. P., van der Hilst, R. and Widiyantoro, S. (1997) Global seismic tomography: a snapshot of convection in the Earth. *GSA Today*, **7**, 1–7.

Gu, Y. J., Dziewonski, A. and Ekström, G. (2001) Preferential detection of the Lehmann discontinuity beneath continents. *Geophys. Res. Lett.*, **28**, 4655–8.

Hager, B. H. (1983) Global isostatic geoid anomalies for plate and boundary layer models of the lithosphere. *Earth Planet. Sci. Lett.*, **63**, 97–109.

Hager, B. H. and Clayton, R. W. (1986) Constraints on the structure of mantle convection using seismic observations, flow models and the geoid. In *Mantle Convection*, ed. Peltier, W. R. New York, Gordon and Breach Science Publishers, pp. 657–763.

Ishii, M. and Tromp, J. (2004) Constraining large-scale mantle heterogeneity using mantle and inner-core sensitive normal modes. *Physics of the Earth and Planetary Interiors*, **146**, 113–24.

Nakanishi, I. and Anderson, D. L. (1983) Measurements of mantle wave velocities and inversion for lateral heterogeneity and anisotropy: Part I, Analysis of great circle phase velocities. *J. Geophys. Res.*, **88**, 10 267–83.

Nakanishi, I. and Anderson, D. L. (1984a) Aspherical heterogeneity of the mantle from phase velocities of mantle waves. *Nature*, **307**, 117–21.

Nakanishi, I. and Anderson, D. L. (1984b). Measurements of mantle wave velocities and inversion for lateral heterogeneity and anisotropy: Part II, Analysis by single-station method. *Geophys. J. Roy. Astron. Soc.*, **78**, 573–617.

Polet, J. and Anderson, D. L. (1995) Depth extent of cratons as inferred from tomographic studies. *Geology*, **23**, 205–8.

Ray, T. W. and Anderson, D. L. (1994) Spherical disharmonics in the Earth sciences and the spatial solution; ridges; ridges, hotspots, slabs, geochemistry and tomography correlations. *J. Geophys. Res.*, **99**, 9605–14.

Scrivner, C. and Anderson, D. L. (1992) The effect of post Pangea subduction on global mantle tomography and convection. *Geophys. Res. Lett.*, **19**, 1053–6.

Shearer, P. M. and Earle, P. S. (2004) The global short-period wavefield modelled with a Monte Carlo seismic phonon method. *Geophys. J. Int.*, **158**, 1103–17.

Shapiro, N. M. and Ritzwoller, M. H. (2004) Thermodynamic constraints on seismic inversions. *Geophys. J. Int.* **157**, 1175–88, doi:10.1111/j.1365-246X.2004.02254.x, 2004.

Spakman, W. and Nolet, G. (1988) Imaging algorithms, accuracy and resolution in delay time tomography. In *Mathematical Geophysics*, eds. Reidel.

Spakman, W., Stein, S., Van der Hilst, R. and Wortel. R. (1989). Resolution experiments for NW Pacific Subduction Zone Tomography. *Geophys. Res. Lett.*, **16**, 1097–100.

Su, W.-J. and Dziewonski, A. M. (1991) Predominance of long-wavelength heterogeneity in the mantle. *Nature*, **352**, 121–6.

Su, W.-J. and Dziewonski, A. M. (1992) On the scale of mantle heterogeneity. *Phys. Earth Planet. Inter.* **74**, 29–54.

Tanimoto, T. (1991) Predominance of large-scale heterogeneity and the shift of velocity anomalies between the upper and lower mantle. *J. Phys. Earth*, **38**, 493–509.

Tanimoto, T. and Anderson, D. L. (1984) Mapping convection in the mantle. *Geopkys. Res. Lett.*, **11**, 287–90.

Tanimoto, T. and Anderson, D. L. (1985) Lateral heterogeneity and azimuthal anisotropy of the upper mantle: Love and Rayleigh waves 100–250 sec. *J. Geophys. Res.*, **90**, 1842–58.

Thoraval, C., Machetel, Ph. and Cazanave, A. (1995) Locally layered convection inferred from dynamic models of the Earth's mantle. *Nature*, **375**, 777–80.

Trampert, J., Deschamps, F., Resovsky, J. and Yuen, D. (2004) Probabilistic tomography maps chemical heterogeneities throughout the lower mantle. *Science*, **306**, 853–6.

Vasco, D. W., Johnson, L. R. and Pulliam, J. (1995) Lateral variations in mantle velocity structure and discontinuities determined from P, PP, S, SS, and SS-S$_{-}$pS travel time residuals. *J. Geophys. Res.*, **100**, 24 037–59.

Walck, M. C. (1984) The P-wave upper mantle structure beneath an. active spreading center: The Gulf of California. *Geophys. J. R. Astron. Soc.*, **76**, 697–723.

Wen, L. and Anderson, D. L. (1995) The fate of slabs inferred from seismic tomography and 130 million years of subduction. *Earth Planet. Sci. Lett.*, **133**, 185–98.

Wen, L. and Anderson, D. L. (1997) Slabs, hotspots, cratons and mantle convection revealed from residual seismic tomography in the upper mantle. *Phys. Earth Planet. Inter.*, **99**, 131–43.

Whitcomb, J. H. and Anderson, D. L. (1970) Reflection of P′P′ seismic waves from discontinuities in the mantle. *J. Geophys. Res.*, **75**, 5713–28.

Further reading

Cizkova, H., Cadek, O., van den Berg, A. P. and Vlaar, N. (1999) Can lower mantle slab-like seismic anomalies be explained by thermal coupling between the upper and lower mantles? *Geophys. Res. Lett.*, **26**, 1501–4.

Gu, Y., Dziewonski, A. M. and Agee, C. B. (1998) Global de-correlation of the topography of transition zone discontinuities. *Earth Planet. Sci. Lett.*, **157**, 57–67.

Ritsema, J. (2005) Global tomography. In *Plates, Plumes and Paradigms*, eds. Foulger, G. L., Natland, J. H., Presnall, D. C. and Anderson, D. L. Boulder, CO, Geological Society of America, Special Paper 388, pp. 11–18. www.mantleplumes.org/TopPages/TheP3Book.html, Web Supplement

Vasco, D. W. and Johnson, L. R. (1998) Whole Earth structure estimated from seismic arrival times. *J. Geophys. Res.*, **103**, 2633–71.

PART IV. SAMPLING THE EARTH

Chapter 12. Statistics and other damned lies

`statistical upper mantle`
`probabilistic tomography`
`spectral density mantle tomography`

Further reading

Anderson, D. L. (1989) `www.caltechbook.library.caltech.edu/14/`

Meibom, A. and Anderson, D. L. (2003) The statistical upper mantle assemblage. *Earth Planet. Sci. Lett.*, **217**, 123–39.

Trampert, J., Deschamps, F., Resovsky, J. and Yuen, D. (2004) Probabilistic tomography maps chemical heterogeneities throughout the lower mantle. *Science*, **306**, 853–6.

Chapter 13. Making an Earth

References

Anderson, D. L. (1983) Kimberlites and the evolution of the mantle. In *Kimberlites and Related Rocks*, ed, J. Kornprobst, pp. 395–403.

Jacobsen, S. B., Quick, J. and Wasserburg, G. (1984) A Nd and Sr isotopic study of the Trinity Peridotite; implications for mantle evolution, *Earth Planet. Sci. Lett.*, **68**, 361–78.

Maaloe, S. and Steel, R. (1980) Mantle composition derived from the composition of lherzolites. *Nature*, **285**, 321–2.

Morgan, J. W. and Anders, E. (1980) Chemical composition of the Earth, Venus, and Mercury. *Proc. Natl. Acad. Sci.*, **77**, 6973–80.

Ringwood, A. E. and S. Kesson (1976) A dynamic model for mare basalt petrogenesis, *Proc. Lunar Sci. Conf.*, **7**, 1697–722.

Further reading

Anderson, D. L. (1999) A theory of the Earth: Hutton and Humpty Dumpty and Holmes. In *James Hutton – Present and Future*, eds. Craig, G. and Hull, J. London, Geological Society of London, Special Publication 150, pp. 13–35.

Chapter 14. Magmas: windows into the mantle

`kimberlites`
`komatiites`
`ocean island basalts`

References

Condie, K. L. (1982) *Plate Tectonics and Crustal Evolution*, Second edition. New York, Pergamon.

Crawford, A. J., Falloon, T. J. and Green, D. H. (1989) Classification, petrogenesis and tectonic setting of boninites. In *Boninites*, ed. Crawford, A. London, Unwin Hyman, pp. 1–49.

Dawson, J. B. (1980) *Kimberlites and their Xenoliths*, Springer-Verlag, Berlin, 252 pp.

Gill, J. B. (1976) Composition and age of Lau basin and ridge volcanic rocks; implications for evolution of an interarc basin and remnant arc. *Geol. Soc. Am. Bull.*, **87**, 1384–95.

Hawkins, J. W. (1977) Petrology and geochemical characteristics of marginal basin basalt. In *Island Arcs, Deep Sea Trenches, and Back-Arc Basins*, eds. Talwani, M. and Pittman, W. C. Washington, DC, American Geophysical Union, pp. 355–77.

Parman S. W., Grove T. L. and Dann J. C. (2001) The production of Barberton komatiites in an Archean subduction zone. *Geophys. Res. Lett.*, **28**, 2513–16.

Wedepohl, K. H. and Muramatsu, Y. (1979) The chemical composition of kimberlites compared with the average composition of three basaltic magma types. In *Kimberlite, Diatremes and Diamonds*, eds. Boyd, F. R. and Meyer, H. Washington, DC, American Geophysical Union, pp. 300–12.

Further reading

Anderson, D. L. (1983a) Kimberlite and the evolution of the mantle. In *Kimberlites and Related Rocks*, ed. J. Kornprobst, pp. 395–403.

Anderson, D. L. (1983b) Chemical composition of the mantle. *J. Geophys. Res.*, **88** suppl., B41–52.

Jacobsen, S. B., Quick, J. and Wasserburg, G. (1984) A Nd and Sr isotopic study of the Trinity Peridotite; implications for mantle evolution. *Earth Planet. Sci. Lett.*, **68**, 361–78.

Ringwood, A. E. (1966) Mineralogy of the mantle. In *Advances in Earth Science*. Cambridge, MA, MIT Press, pp. 357–99.

Ringwood, A. E. and Kesson, S. (1976) A dynamic model for mare basalt petrogenesis. *PLC*, **7**, 1697–722.
Taylor, S. (1982) Lunar and terrestrial crusts. *Phys. Earth Planet. Inter.*, **29**, 233.

Chapter 15. The hard rock cafe

Pyrolite rock
Mantle eclogite
piclogite

References

Basu, A. R. and Tatsumoto, M. (1982) Nd isotopes in kimberlites and mantle evolution. *Terra Cog.*, **2**, 2–14.
Beus, A. A. (1976) *Geochemistry of the Lithosphere*, Moscow, MIR Publications.
Boyd, F. R. (1986) High- and low-temperature garnet peridotite xenoliths and their possible relation to the lithosphere–asthenosphere boundary beneath southern Africa. In *Mantle Xenoliths*, ed. Nixon, P. New York, John Wiley & Sons, pp. 403–12.
Boyd, F. R. and Mertzman, S. A. (1987) Composition and structure of the Kaapvaal lithosphere, southern Africa. In *Magmatic Processes: Physicochemical Principles*, ed. Mysen, B. O. University Park, Pennsylvania, The Geochemical Society, Special Publication 1, pp. 13–24.
Clarke, D. B. (1970) Tertiary basalts of Baffin Bay; possible primary magma from the mantle. *Contrib. Mineral. Petrol.*, Special Publication 1, **25**, 203–24.
Echeverria, L. M. (1980) Tertiary komatiites of Gorgona Island. *Carnegie Instn. Wash. Ybk.*, **79**, 340–4.
Elthon, D. (1979) High magnesia liquids as the parental magma for ocean floor basalts. *Nature*, **278**, 514–18.
Frey, F. A. (1980) The origin of pyroxenites and garnet pyroxenites from Salt Crater, Oahu, Hawaii: trace element evidence. *Am. J. Sci.*, **280-A**, 427–49.
Green, D. H. and Ringwood. A. (1963) Mineral assemblages in a model mantle composition. *J. Geophys. Res.*, **68**, 937–45.
Green, D. H., Hibberson, W. and Jaques, A. (1979) Petrogenesis of mid-ocean ridge basalts. In *The Earth: Its Origin, Structure and Evolution*, ed. McElhinny, M. W. New York, Academic Press, pp. 265–95.
Jahn, B.-M., Auvray, B., Blais, S. *et al.* (1980) Trace element geochemistry and petrogenesis of Finnish greenstone belts. *J. Petrol.*, **21**, 201–44.
Maaløe, S. and Aoki, K. (1977) The major element composition of the upper mantle estimated from the composition of lherzolites. *Contrib. Mineral. Petrol.*, **63**, 161–73.
Ringwood, A. E. (1975) *Composition and Petrology of the Earth's Mantle*. New York, McGraw-Hill.
Smyth, J. R. and Caporuscio, F. (1984) Petrology of a suite of eclogite inclusions from the Bobbejaan Kimberlite; II, Primary phase compositions and origin. In *Kimberlites*, ed. Kornprobst, J. Amsterdam, Elsevier, pp. 121–31.
Wedepohl, K. H. and Muramatsu, Y. (1979) The chemical composition of kimberlites compared with the average composition of three basaltic magma types. In *Kimberlites, Diatremes, and Diamonds*, eds. Boyd, F. R. and Meyer, H. O. Washington, DC, American Geophysical Union, pp. 300–12.

Further reading

Bowen, N. L. (1928) *The Evolution of the Igneous Rocks*. Princeton, NJ, Princeton University Press.
Carmichael, I. S. E., Turner, F. and Verhoogen, J. (1974) *Igneous Petrology*. New York, McGraw-Hill.
Chen, C. and Frey, F. (1983) Origin of Hawaiian theiite and alkali basalt. *Nature*, **302**, 785.
Frey, F. A., Green, D. and Roy, S. (1978) Integrated models of basalts petrogenesis: A study of quartz tholeiites to olivine melilitites from southeastern Australia utilizing geochemical and experimental petrological data. *J. Petrol.*, **19**, 463–513.
Green, D. H. (1971) Composition of basaltic magmas as indicators of conditions of origin; application to oceanic volcanism. *Phil. Trans. R. Soc.*, **A268**, 707–25.
Green, D. H. and Ringwood, A. (1967) The genesis of basaltic magmas. *Contrib. Mineral. Petrol.*, **15**, 103–90.
Hiyagon, H. and Ozima, M. (1986) Partition of noble gases between olivine and basalt melt. *Geochim. Cosmochim. Acta*, **50**, 2045–57.
Jaques, A. and Green, D. (1980) Anhydrous melting of peridotite at 0–15 Kb pressure and the genesis of tholeitic basalts. *Contrib. Mineral. Petrol.*, **73**, 287–310.
Menzies, M., Rogers, N., Zindle, A. and Hawkesworth, C. (1987) In *Mantle Metasomatism*, ed. Menzies, M. A. New York, Academic Press.
Nataf, H.-C., Nakanishi, I. and Anderson, D. L. (1986) Measurements of mantle wave velocities and inversion for lateral heterogeneities and anisotropy. Part III, Inversion. *J. Geophys. Res.*, **91**, 7261–307.
O'Hara, M. J. (1968) The bearing of phase equilibria studies in synthetic and natural systems on the

origin and evolution of basic and ultrabasic rock. *Earth Sci. Rev.*, **4**, 69–133.

Rigden, S. S., Ahrens, T. J. and Stolper, E. M. (1984) Densities of liquid silicates at high pressures. *Science*, **226**, 1071–4.

Ringwood, A. E. (1962) Mineralogical constitution of the deep mantle. *J. Geophys. Res.*, **67**, 4005–10.

Ringwood, A. E. (1966) Mineralogy of the mantle. In *Advances in Earth Science*. Cambridge, MA, MIT Press, pp. 357–99.

Ringwood, A. E. (1979) *Origin of the Earth and Moon.* New York, Springer-Verlag.

Smyth, J. R., McCorrnick, T. and Caporuscio, F. (1984) Petrology of a suite of eclogite inclusions from the Bobbejaan Kimberlite; I, Two unusual corundum-bearing kyanite eclogites. In *Kimberlites*, ed. Kornprobst, J. Amsterdam, Elsevier, pp. 109–19.

Walck, M. C. (1984) The P-wave upper mantle structure beneath an active spreading center; the Gulf of California. *Geophys. J. Roy. Astron. Soc.*, **76**, 697–723.

Chapter 16. Noble gas isotopes

MORB helium statistics
Helium paradoxes
Popping rock helium
high 3He/4He ratios

References

Anderson, D. L. (1998a) The helium paradoxes. *Proc. Nat. Acad. Sci.*, **95**, 4822–7.

Anderson, D. L. (1998b) A model to explain the various paradoxes associated with mantle noble gas geochemistry. *Proc. Nat. Acad. Sci.*, **95**, 9087–92.

Anderson, D. L. (2000a) The statistics of helium isotopes along the global spreading ridge system and the central limit theorem. *Geophys. Res. Lett.*, **27**, 2401–4.

Anderson, D. L. (2000b) The statistics and distribution of helium in the mantle. *Int. Geology Rev.*, **42**, 289–311.

Javoy, M. and Pineau, F. (1991) The volatiles record of a "popping" rock from the mid-Atlantic ridge at 14°N: Chemical and isotopic composition of gas trapped in the vesicles. *Earth Planet. Sci. Lett.*, **107**, 598–611.

Sarda, P. and Graham, D. (1990) Mid-ocean ridge popping rocks: implications for degassing at ridge crests. *Earth Planet. Sci. Lett.*, **97**, 268–89.

Seta, A., Matsumoto, T. and Matsuda, J.-I. (2001) Concurrent evolution of $^3He/^4He$ ratio in the Earth's mantle reservoirs for the first 2 Ga. *Earth Planet. Sci. Lett.*, **188**, 211–19.

Staudacher, T., Sarda, P., Richardson, S. H., Allegre, C. J., Sagna, I. and Dmitriev, L. V. (1989) Noble gases in basalt glasses from a Mid-Atlantic Ridge topographic high at 14°N: geodynamic consequences. *Earth Planet. Sci. Lett.*, **96**, 119–33.

Further reading

Meibom, A., Anderson, D. L., Sleep, N., Frei, R., Chamberlain, C., Hren, M. and Wooden, J. (2003) Are high $^3He/^4He$ ratios in oceanic basalts an indicator of deep-mantle plume components? *Earth Planet. Sci. Lett.*, **208**, 197–204.

Moreira, M., and Sarda, P. (2000) Noble gas constraints on degassing processes. *Earth Planet. Sci. Lett.*, **176**, 375–86.

Chapter 17. The other isotopes

References

Chase, C. G. (1981) Oceanic island Pb; two-stage histories and mantle evolution. *Earth Planet. Sci. Lett.*, **52**, 277–84.

Dalrymple, G. B. (2001) The age of the Earth in the twentieth century – a problem (mostly) solved. In *The Age of the Earth – from 4004 BC to AD 2002*, eds. Lewis, C. L. E. and Knell, S. J. London, The Geological Society, Special Publication 190, pp. 205–21.

Eiler, J. M. (2001) Oxygen isotope variations of basaltic lavas and upper mantle rocks. In *Stable Isotope Geochemistry*, eds. Valley, J. W. and Cole, D. R. *Rev. Mineral.*, **43**, 319–64.

Eiler, J. M., Farley, K. A., Valley, J. W., Hauri, E., Craig, H., Hart, S. R. and Stolper, E. M. (1997) Oxygen isotope variations in ocean island basalt phenocrysts. *Geochim. Cosmochim. Acta*, **61**, 2281–93.

Eiler, J. M., Valley, J. and Stolper, E. (1996a) Oxygen isotope ratios in olivine from the Hawaiian Scientific Drilling Project. *J. Geophys. Res.*, **101**, 11 807–13.

Eiler, J. M., Farley, K., Valley, J., Hofmann, A. and Stolper, E. (1996b) Oxygen isotope constraints on the sources of Hawaiian volcanism. *Earth Planet. Sci. Lett.*, **144**, 453–68.

Meibom, A., Sleep, N. H., Chamberlain, C. P., Coleman, R. G., Frei, R., Hren, M. T., and Wooden, J. L. (2002) Re–Os isotopic evidence for long-lived heterogeneity and euilibration processes in the Earth's upper mantle. *Nature*, **418**, 705–8.

Patterson, C. (1956) Age of meteorites and the Earth. *Geochim. Cosmochim. Acta*, **10**, 230–7.

Roy-Barman, M. and Allegre, C. J. (1994) $^{187}Os/^{186}Os$ ratios of midocean ridge basalts and abyssal peridotites. *Geochim. Cosmochim. Acta*, **58**, 5043–54.

Shirey, S. B. and Walker, R. J. (1998) The Re–Os isotope system in cosmochemistry and high-temperature geochemistry. *Ann. Rev. Earth Planet. Sci.*, **26**, 423–500.

Smith, A. D. (2003) Critical evaluation of Re–Os and Pt–Os isotopic evidence on the origin of intraplate volcanism. *J. Geodyn.*, **36**, 469–84.

Selected reading and notes

Mantle geochemistry

Isotope geochemistry introduced a new chapter into studies of the Earth's composition and evolution. Lead isotopes were among the earliest to be used and the first of many geochemical paradoxes were identified. Isotopic components were initially attributed to isolated reservoirs, and assigned location such as the 'lower mantle,' and 'the convecting upper mantle.' Many of these components are now recognized as recycled materials. Ironically, lead-isotopes and the daughter isotopes of uranium and thorium decay were not used in what is now considered the `Standard Model of mantle geochemistry`, and they contradict it in many ways. Modern data favor early differentiation of the Earth and do not favor primordial or undegassed reservoirs. Key papers are listed below.

Armstrong, R. L. (1981) Radiogenic isotopes: the case for crustal recycling on a near-steady-state no-continental-growth Earth. *Phil. Trans. R. Soc. Lond. A*, **301**, 443–72.

Chase, C. G. (1981) Oceanic island Pb: two-stage histories and mantle evolution. *Earth Planet. Sci. Lett.*, **52**, 277–84.

Clarke, W. B., Beg, M. and Craig, H. (1969) Excess 3He in sea: evidence for terrestrial primordial helium. *Earth Planet. Sci. Lett.*, **6**, 213–20.

Craig, H. and Lupton, J. (1976) Primordial neon, helium, and hydrogen in oceanic basalts. *Earth Planet. Sci. Lett.*, **31**, 369–85.

Gast, P. W. (1968) Trace element fractionation and the origin of tholeiitic and alkaline magma types. *Geochim. Cosmochim. Acta*, **32**, 1057–86.

Further reading

Anderson, D. L. (1981) Hotspots, basalts, and the evolution of the mantle. *Science*, **213**, 82–9.

Garlick, G., MacGregor, I. and Vogel, D. (1971) Oxygen isotope ratios in eclogites from kimberlites. *Science*, **171**, 1025–7.

Gregory, R. T. and Taylor, H. P. Jr. (1981) An oxygen isotope profile in a section of Cretaceous oceanic crust, Samail ophiolite, Oman. *J. Geophys. Res.*, **86**, 2737–55.

Dalrymple, G. B. (1991) *The Age of the Earth*. Stanford, CA, Stanford University Press.

Hofmeister A. M. (1999). Mantle values of thermal conductivity and the geotherm from phonon lifetimes. *Science*, **283**, 1699–706.

Ishii, M. and Tromp, J. (1999) Normal mode and free-air gravity constraints on lateral variations in velocity and density of Earth's mantle. *Science*, **285**, 1231–6.

Meibom, A. and Anderson, D. L. (2003) The Statistical Upper Mantle Assemblage. *Earth Planet. Sci. Lett.*, **217**, 123–39.

Meibom, A., Sleep, N. H., Zahnle, K. and Anderson, D. L. (2005) Models for noble gases in mantle geochemistry: Some observations and alternatives. In *Plates, Plumes, and Paradigms*, eds. Foulger, G. R., Natland, J. H., Presnall, D. C. and Anderson, D. L. Boulder, CO, Geological Society of America, Special Paper 388, pp. 347–63.

Reisberg, L. and Zindler, A. (1986) Extreme isotopic variations in the upper mantle; evidence from Ronda. *Earth and Planetary Science Letters*, **81**, 29–45.

PART V. MINERAL PHYSICS

Chapter 18. Elasticity and solid state physics

Further reading

Duffy, T. S. and Anderson, D. L. (1989) Seismic velocities in mantle minerals and the mineralogy of the upper mantle. *J. Geophys. Res.*, **94**, 1895–912.

Dziewonski, A. M. and Anderson, D. L. (1981) Preliminary reference Earth model. *Phys. Earth Planet. Inter.*, **25**, 297–356.

Ishii, M. and Tromp, J. (2004). Constraining large-scale mantle heterogeneity using mantle and inner-core sensitive modes. *Phys. Earth Planet. Inter.*, **146**, 113–24.

Karki, B. B., Stixrude, L. and Wentzcovitch, R. (2001) High-pressure elastic properties of major materials of Earth's mantle from first principles. *Rev. Geophys.*, **39**, 507–34.

Nakanishi, I. and Anderson, D. L. (1982) World-wide distribution of group velocity of mantle Rayleigh waves as determined by spherical harmonic inversion. *Bull. Seis. Soc. Am.*, **72**, 1185–94.

Nataf, H.-C., Nakanishi, I. and Anderson, D. L. (1984) Anisotropy and shear-velocity heterogeneities in the upper mantle. *Geophys. Res. Lett.*, **11**, 109–12.

Nataf, H.-C., Nakanishi, I. and Anderson, D. L. (1986) Measurements of mantle wave velocities and inversion for lateral heterogeneities and anisotropy. Part 111, Inversion. *J. Geophys. Res.*, **91**, 7261–307.

Trampert, J., Deschamps, F., Resovsky, J. and Yuen, D. (2004) Probabilistic tomography maps chemical heterogeneities throughout the lower mantle. *Science*, **306**, 853–6.

Chapter 19. Dissipation

References

Anderson, D. L. and Given, J. (1982) Absorption band Q model for the Earth. *J. Geophys. Res.*, **87**, 3893–904.

Kanamori, H. and Anderson, D. L. (1977) Importance of physical dispersion in surface wave and free oscillation problems. *Rev. Geophys. Planet. Sci.*, **15**, 105–12.

Minster, J. B. and Anderson, D. L. (1981) A model of dislocation controlled rheology for the mantle. *Phil. Trans. Roy. Soc. London*, **299**, 319–56.

Spetzler, H. and Anderson, D. L. (1968) The effect of temperature and partial melting on velocity and attenuation in a simple binary system. *J. Geophys. Res.*, **73**, 6051–60.

Selected reading

Anderson, D. L., Ben-Menahem, A. and Archambeau, C. B. (1965) Attenuation of seismic energy in the upper mantle. *J. Geophys. Res.*, **70**, 1441–8.

O'Connell, R. J. and Budiansky, B. (1978) Measures of dissipation in viscoelastic media. *Geophys. Res. Lett.*, **5**, 5–8.

Chapter 20. Fabric of the mantle

References

Anderson, D. L. and Dziewonski, A. M. (1982) Upper mantle anisotropy; evidence from free oscillations. *Geophys. J. Royal Astr. Soc.*, **69**, 383–404.

Anderson, D. L., Minster, J. B. and Cole, D. (1974) The effect of oriented cracks on seismic velocities. *J. Geophys. Res.*, **79**, 4011–15.

Ando, M. Y., Ishikawa and Yamazaki, F. (1983) Shear wave polarization anisotropy in the upper mantle beneath Honshu, Japan. *J. Geophys. Res.*, **10**, 5850–64.

Babuska, V. (1981) Anisotropy of Vp and Vs in rock-forming minerals, *J. Geophys.*, **50**, 1–6.

Backus, G. E. (1962). Long-wave elastic anisotropy produced by horizontal layering. *J. Geophys. Res.*, **67**, 4427–40.

Christensen, N. I. and Lundquist, S. (1982) Pyroxene orientation within the upper mantle. *Bull. Geol. Soc. Am.*, **93**, 279–88.

Christensen, N. I. and Salisbury, M. (1979) Seismic anisotropy in the upper mantle: Evidence from the Bay of Islands ophiolite complex. *J. Geophys. Res.*, **84**, B9, 4601–10.

Dziewonski, A. M. and Anderson, D. L. (1981) Preliminary reference Earth model. *Phys. Earth Planet. Inter.*, **25**, 297–356.

Fukao, Y. (1984) Evidence from core-reflected shear waves for anisotropy in the Earth's mantle. *Nature*, **309**, 695–8.

Hager, B. H. and O'Connell, R. J. (1979) Kinematic models of large-scale flow in the Earth's mantle. *J. Geophys. Res.*, **84**, 1031–48

Montagner, J.-P. (2002) Upper mantle low anisotropy channels below the Pacific plate. *Earth Planet. Sci. Lett.*, **202**, 263–74.

Montagner, J.-P. and Nataf, H.-C. (1986) A simple method for inverting the azimuthal anisotropy of surface waves, *J. Geophys. Res.*, **91**, 511–20.

Nataf, H.-C., Nakanishi, I. and Anderson, D. L. (1986) Measurements of mantle wave velocities and inversion for lateral heterogeneities and anisotropy, Part III. Inversion, *J. Geophys. Res.*, **91**, 72161–3070.

Nicolas, A. and Christensen, N. I. (1987) Formation of anisotropy in upper mantle peridotites. In *Composition, Structure and Dynamics of the Lithosphere/Asthenosphere System*, eds. Fuchs, K. and Froidevaux, C. Washington, DC, American Geophysical Union, pp. 111–23.

Regan, J. and Anderson, D. L. (1984) Anisotropic models of the upper mantle. *Phys. Earth Planet. Inter.*, **35**, 227–63.

Sawamoto, H., D. J. Weidner, S. Sasaki and M. Kumazawa (1984) Single-crystal elastic properties of the modified spinel phase of magnesium orthosilicate, *Science*, **224**, 749–51.

Tanimoto, T. and Anderson, D. L. (1984) Mapping convection in the mantle. *Geophys. Res. Lett.*, **11**, 287–90.

Further reading

Christensen, N. I. and Crosson, R. (1968) Seismic anisotropy in the upper mantle. *Tectonophysics*, **6**, 93–107.

Gilvarry, J. J. (1956) The Lindemann and Grüneisen Laws. *Phys. Rev.*, **102**, 308–16.

Morris, E. M., Raitt, R. and Shor, G. (1969) Velocity anisotropy and delay times of the mantle near Hawaii. *J. Geophys. Res.*, **74**, 4300–16.

Raitt, R. W., Shor, G., Francis, T. and Morris, G. (1969) Anisotropy of the Pacific upper mantle. *J. Geophys. Res.*, **74**, 3095–109.

Chapter 21. Nonelastic and transport properties

```
thermal conductivity geotherm phonon lifetimes
Thermal conductivity of rock forming minerals
Thermal conductivity of mantle
```

References

Freer, R. (1981) Diffusion in silicate minerals: a data digest and guide to the literature. *Contrib. Mineral. Petrol.*, **76**, 440–54.

Horai, K. (1971) Thermal conductivity of rock-forming minerals. *J. Geophys. Res.*, **76**, 1278–308.

Horai, K. and Simmons, G. (1970) An empirical relationship between thermal conductivity and Debye temperature for silicates. *J. Geophys. Res.*, **75**, 678–82.

Gilvarry, J. J. (1956) The Lindemann and Grüneisen Laws. *Phys. Rev.*, **102**, 308–16.

Keyes, R. (1963) Continuum models of the effect of pressure on activated processes. In *Solid under Pressure*, eds. Paul, W. and Warschauer, D. M. New York, McGraw-Hill, pp. 71–99.

Kobayzshigy, A. (1974). Anisotropy of thermal diffusivity in olivine, pyroxene and dunite. *J. Phys. Earth*, **22**, 359–73.

Further reading

Minster, J. B. and Anderson, D. L. (1981) A model of dislocation controlled rheology for the mantle. *Phil. Trans. R. Soc. London A*, **299**, 319–56.

Ohtani, E. (1983) Melting temperature distribution and fractionation in the lower mantle. *Phys. Earth Planet. Int.*, **33**, 12–25.

Schatz, J. F. and Simmons, G. (1972) Thermal conductivities of Earth materials at high temperatures. *J. Geophys. Res.*, **77**, 6966–83.

Chapter 22. Squeezing: phase changes and mantle mineralogy

```
fate of subducted basaltic crust
mantle phase diagrams
Generation mid-ocean ridge basalts pressures
upper mantle transition region: Eclogite?
Phase transformations and seismic structure
density of subducted basaltic crust
```

Links to phase diagrams are listed below.

www.uni-wuerzburg.de/mineralogie/links/teach/diagramteach.html
www.agu.org/reference/minphys.html
www.agu.org/reference/minphys/16_presnell.pdf
www.mpi.stonybrook.edu/ResearchResults/PhaseRelationsGasparik/figures.htm
www.ruf.rice.edu/~ctlee/Chapter3.pdf
www.springeronline.com/sgw/cda/frontpage/0,11855,1-10010-22-2155856-0,00.html
www.mantleplumes.org/TransitionZone.html

References

Presnall, D. C. (1995) Phase diagrams of Earth-forming minerals. In *Handbook of Physical Constants, Mineral Physics and Crystallography*, ed. Ahrens, T. J. Washington, DC, American Geophysical Union, AGU Reference Shelf 2. www.agu.org/reference/minphys/16_presnell.pdf

Further reading

Akaogi, M. and Akimoto, S. (1977) Pyroxene-garnet solid solution equilibrium. *Phys. Earth Planet. Inter.* **15**, 90–106.

Akimoto, S. (1972) The system MgO-FeO-SiO2 at high pressure and temperature. *Tectonophysics*, **13**, 161–87.

Ito, E. and Yamada, H. (1982) Stability relations of silicate spinels, ilmenite and perovskites. In *High-Pressure Research in Geophysics*, eds. Akimoto, S. and Manghnam, M. Dordrecht, Reidel, pp. 405–19.

Kato, T. and Kumazawa, M. (1985) Garnet phase of MgSiO, filling the pyroxene-ilmenite gap at very high temperature. *Nature*, **316**, 803–5.

Kuskov, O. L. and Galimzyanov, R. (1986) Thermodynamics of stable mineral assemblages of the mantle transition zone. In *Chemistry and Physics of the Terrestrial Planets,* ed. Saxena, S. K. New York, Springer-Verlag, pp. 310–61.

Litasov, K., Ohtani, E., Suzuki, A., Kawazoe, T. and Funakoshi, K. (2004) Absence of density crossover between basalt and peridotite in the cold slabs passing through 660 km discontinuity. *Geophys. Res. Lett.*, **31** (24) doi:10.1029/2004GL021306.

Ohtani, E. (1983) Melting temperature distribution and fractionation in the lower mantle. *Phys. Earth Planet. Inter.*, **33**, 12–25.

Ono, S., Ohishi, Y., Isshiki, M. and Watanuki, T. (2005) In situ X-ray observations of phase assemblages in peridotite and basalt compositions at lower mantle conditions: Implications for density of subducted oceanic plate. *J. Geophys. Res.*, **110**, B02208, doi: 10.1029/2004JB003196.

PART VI. ORIGIN AND EVOLUTION OF LAYERS AND BLOBS

Chapter 23. The upper mantle

`SUMA mantle`
`depleted upper mantle`
`olivine free mantle`
`delamination and mantle heterogeneity`
`thickness and composition of continental lithosphere`

References

Anderson, D. L. (1967) Phase changes in the upper mantle. *Science*, **157**, 1165–73.

Anderson, D. L. (1989) www.caltechbook.library.caltech.edu/14/

Donnelly, K. E., Goldsteina, S. L., Langmuir, C. H. and Spiegelman, M. (2004) Origin of enriched ocean ridge basalts and implications for mantle dynamics. *Earth Planet. Sci. Lett.*, **226**, 347–66.

Escrig, S, Capmas, F, Dupré, B. and Allègre, C. J. (2004) Osmium isotopic constraints on the nature of the DUPAL anomaly from Indian mid-ocean-ridge basalts. *Nature*, **431**, 59–63.

Escrig, S., Doucelance, R., Moreira, M. and Allegre, C. J. (2005) Os isotope systematics in Fogo Island: Evidence for lower continental crust fragments under the Cape Verde Southern Islands. *Chem. Geol.*, **219**, 93–113.

Gao, S., Rudnick, R. L., Yuan, H.-L., Liu, X.-M., Liu, Y.-S., Ling, W.-L., Ayers, J. and Wang, X.-C. (2004) Recycling lower continental crust in the North China craton. *Nature*, **432**, 892–7.

Meibom, A. and Anderson, D. L. (2003) The statistical upper mantle assemblage. *Earth Planet. Sci. Lett.*, **217**, 123–39.

Salters, V. J. M. and Stracke, A. (2004) Composition of the depleted mantle. *Geochem. Geophys. Geosyst.*, **5**, Q05004, doi:10.1029/2003GC000597.#

Sobolev, A. V., Hofmann, A. W., Sobolev, S. V. and Nikogosian, I. K. (2005) An olivine-free mantle source of Hawaiian shield basalts. *Nature*, **434**, 590–7. doi:10.1038/nature03411.

Workman, R. K. & Hart, S. R. (2005) Major and trace element composition of the depleted MORB mantle (DMM). *Earth Planet. Sci. Lett.*, **231**, 53–63.

Selected reading and notes

Chemical heterogeneity of the mantle

Modern geochemical and geodynamic models tend to favor a homogenous peridotite mantle. But the mantle is clearly lithologically heterogenous and chemical variations can explain some of the phenomena that have been attributed to thermal anomalies. Eclogite or garnet pyroxenite is an important constituent of the mantle and of the source regions of some basalts.

Birch, F. (1958) Differentiation of the mantle. *Bull. Geol. Soc. Am.*, **69**, 483–6.

Gerlach, D. C. (1990) Eruption rates and isotopic systematics of ocean islands: further evidence for small-scale heterogeneity in the upper mantle. *Tectonophysics*, **172**, 273–89.

Ito, K. and Kennedy, G. C. (1971) An experimental study of the basalt-garnet granulite–ecologite transition. In *The Structure and Physical Properties of the Earth's Crust*, ed. Heacock, J. G. Washington, DC, American Geophysical Union. *Geophys. Monogr.*, **14**, 303–14.

Kay, R. W. and Kay, S. (1993) Delamination and delamination magmatism. *Tectonophysics*, **219**, 177–89.

Rudnick, R. L. and Gao, S. (2003) The composition of the continental crust. In *The Crust, Vol. 3, Treatise on Geochemistry*, vol. eds. Holland, H. D. and Turekian, K. K. Oxford, Elsevier, pp. 1–64.

Rudnick, R. L. and Fountain, D. M. (1995) Nature and composition of the continental crust: a lower crustal perspective. *Rev. Geophysics*, **33**, 267–309.

Chapter 24. The nature and cause of mantle heterogeneity

`top-down tectonics`
`irreversible chemical stratification mantle`
`far-from-equilibrium self-organized system`
`statistical upper mantle assemblage`
`scale of mantle heterogeneity`

comparison of tomographic and geodynamic mantle models
stagnant slabs
fate of slabs
model for the layered upper mantle
plate tectonics hotspots third dimension
post pangea subduction tomography
putative mantle plume source

Further reading

Hofmann A. W. (1997) Mantle geochemistry: the message from oceanic volcanism. *Nature*, **385**, 219–29.

Hofmann A. W. (2003) Sampling mantle heterogeneity through oceanic basalts: isotopes and trace elements. In *Treatise on Geochemistry*, Vol. 2, eds. Carlson, R. W., Holland, H. D. and Turekian, K. K. pp. 61–101.

Chapter 25. Crystallization of the mantle

References

Elthon, D. (1979) High magnesia liquids as the parental magma for ocean ridge basalts. *Nature*, **278**, 514–18.

Frey, F. A., Green, D. and Roy, S. (1978) Integrated models of basalts petrogenesis: a study of quartz tholeiites to olivine melilitities from southeastern Australia utilizing geochemical and experimental petrological data. *J. Petrol.*, **19**, 463–513.

O'Hara, M. J., Saunders, A. and Mercy, E. (1975) Garnet peridotite, primary ultrabasic magma and eclogite; interpretation of upper mantle processes in kimberlite, *Phys. Chem. Earth.*, **9**, 571–604.

Ringwood, A. E. (1975) *Composition and Petrology of the Earth's Mantle*. New York, McGraw Hill.

Further reading

Anderson, D. L. (1985) Hotspot magmas can form by fractionation and contamination of MORB. *Nature*, **318**, 145–9.

DePaolo, D. J. and Wasserburg, G. (1979) Neodymium isotopes in flood basalts from the Siberian Platform and inferences about their mantle sources. *Proc. Natl. Acad. Sci.*, **76**, 3056.

O'Hara, M. J. (1968) The bearing of phase equilibria studies in synthetic and natural systems on the origin and evolution of basic and ultrabasic rocks. *Earth Sci. Rev.*, **4**, 69–133.

PART VII. ENERGETICS

Chapter 26. Terrestrial heat flow

heat flow age
heat flow age images
bathymetric depths superswell
Earth's heat flux revised:
Comments on Earth's heat fluxes
www.mantleplumes.org/Superswell.html
www.heatflow.und.edu.

References

McNutt, M. K. and Judge, A. (1990) The superswell and mantle dynamics beneath the South Pacific. *Science*, **248**, 969–75.

Rudnick, R. L. and Fountain, D. M. (1995) Nature and composition of the continental crust: A lower crustal perspective. *Rev. Geophysics*, **33**, 267–309.

Stein, C. and Stein, S. (1994) Comparison of plate and asthenospheric flow models for the evolution of oceanic lithosphere, *Geophys. Res. Lett.*, **21**, 709–12.

Vitorello, I. and Pollack, H. N. (1980) On the variation of continental heat flow with age and the thermal evolution of continents. *J. Geophys. Res.*, **85**, 983–95.

Selected reading

DeLaughter, J., Stein, S. and Stein, C. (1999) Extraction of a lithospheric cooling signal from oceanwide geoid data. *Earth Planet. Sci. Lett.*, **174**, 173–81.

Gubbins, D. (1977) Energetics of the Earth's core. *J. Geophys.*, **43**, 453.

Pollack, H. N., Hurter, S. and Johnson, J. (1993) Heat flow from the earth's interior: analysis of the global data set. *Rev. Geophysics*, **31**, 267–80.

Rudnick, R. L. and Nyblade, A. A. (1999) The composition and thickness of Archean continental roots: constraints from xenolith thermobarometry. In *Mantle Petrology: Field Observations and High-Pressure Experimentation: A Tribute to Francis R. (Joe) Boyd*, eds. Fei, Y.-W., Bertka, C. M. and Mysen, B. O. Geochemical Society Special Publication 6, pp. 3–12.

Sclater, J., Parsons, B. and Jaupart, C. (1981) Oceans and continents: similarities and differences in the mechanism of heat transport. *J. Geophys. Res.*, **86**, 11 535–52.

Stacey, F. D. & Stacey, C. H. B. (1999) Gravitational energy of core evolution: implications for thermal history and geodynamo power. *Phys. Earth Planet. Inter.*, **110**, 83–93.

Stein, C. A. and Stein, S. A. (1994) Constraints on hydrothermal heat flux through the oceanic lithosphere from global heat flow. *J. Geophys. Res.*, **99**, 3081–95.

Stein, C. A. and Stein, S. A. (1992) A model for the global variation in oceanic depth and heat flow with lithospheric age. *Nature*, **359**, 123–8.

Van Schmus, W. R. (1995) Natural radioactivity of the crust and mantle. In *Global Earth Physics, A Handbook of Physical Constants*, ed. Ahrens, T. J. Washington, DC, American Geophysical Union, pp. 283–91.

Von Herzen, R., Davis, E. E., Fisher, A., Stein, C. A. and Pollack, H. N. (2005) Comments on Earth's heat fluxes. *Tectonophysics*, doi:10.1016/j.tecto.2005.08.003. www.es.ucsc.edu/~afisher/Research/Appen/H_Cresponse2005.pdf www.es.ucsc.edu/~afisher/Research/Appen/H_C_comment.htm

Further reading

Pollack, H., Hurter, S. and Johnson, J. (1993) Heat flow from the Earth's interior: analysis of the global data set. *Rev. Geophys.*, **31**, 267–80.

Chapter 27. The thermal history of the Earth

Fate of fossil heat

archean thermal catastrophe

heat flow U Pb paradoxes

W. Thompson (Lord Kelvin) secular cooling

radiogenic heat source contents of the Earth and Moon

On the secular cooling of the Earth

Appendix

Table A.1 | Earth model PREM and its functionals evaluated at a reference period of 1 s. Above 220 km the mantle is transversely isotropic (see Table A.2); the parameters given are 'equivalent' isotropic moduli and velocities

Level	Radius (km)	Depth (km)	Density (g/cm^3)	V_p (km/s)	V_s (km/s)	Q_μ	Q_K	Q_P	Φ (km^2/s^2)	K_S (kbar)	μ (kbar)	σ	Pressure (kbar)	dK/dP	B.P.	Gravity (cm/s^2)
1	0.	6371.0	13.08	11.26	3.66	85	1328	431	108.90	14253	1761	0.440	3638.5	2.33	0.99	0.
2	100.0	6271.0	13.08	11.26	3.66	85	1328	431	108.88	14248	1759	0.440	3636.1	2.33	0.99	36.5
3	200.0	6171.0	13.07	11.25	3.66	85	1328	431	108.80	14231	1755	0.440	3628.9	2.33	0.99	73.1
4	300.0	6071.0	13.06	11.24	3.65	85	1328	432	108.68	14203	1749	0.440	3617.0	2.33	0.99	109.6
5	400.0	5971.0	13.05	11.23	3.65	85	1328	432	108.51	14164	1739	0.441	3600.3	2.33	0.99	146.0
6	500.0	5871.0	13.03	11.22	3.64	85	1328	433	108.29	14114	1727	0.441	3578.8	2.33	0.99	182.3
7	600.0	5771.0	13.01	11.20	3.62	85	1328	434	108.02	14053	1713	0.441	3552.7	2.33	0.99	218.6
8	700.0	5671.0	12.98	11.18	3.61	85	1328	436	107.70	13981	1696	0.441	3522.0	2.34	0.99	254.7
9	800.0	5571.0	12.94	11.16	3.59	85	1328	437	107.33	13898	1676	0.442	3486.6	2.34	0.99	290.6
10	900.0	5471.0	12.91	11.13	3.57	85	1328	439	106.91	13805	1654	0.442	3446.7	2.34	0.99	326.4
11	1000.0	5371.0	12.87	11.10	3.55	85	1328	440	106.45	13701	1630	0.442	3402.3	2.34	0.99	362.0
12	1100.0	5271.0	12.82	11.07	3.53	85	1328	443	105.94	13586	1603	0.443	3353.5	2.34	1.00	397.3
13	1200.0	5171.0	12.77	11.03	3.51	85	1328	445	105.38	13462	1574	0.443	3300.4	2.34	1.00	432.5
14	1221.5	5149.5	12.76	11.02	3.50	85	1328	445	105.25	13434	1567	0.443	3288.5	2.34	1.00	440.0
15	1221.5	5149.5	12.16	10.35	0.	0	57822	57822	107.24	13047	0	0.500	3288.5	3.75	1.03	440.0
16	1300.0	5071.0	12.12	10.30	0.	0	57822	57822	106.29	12888	0	0.500	3245.4	3.65	1.02	463.6
17	1400.0	4971.0	12.06	10.24	0.	0	57822	57822	105.05	12679	0	0.500	3187.4	3.54	1.01	494.1
18	1500.0	4871.0	12.00	10.18	0.	0	57822	57822	103.78	12464	0	0.500	3126.1	3.46	1.01	524.7
19	1600.0	4771.0	11.94	10.12	0.	0	57822	57822	102.47	12242	0	0.500	3061.4	3.40	1.00	555.4
20	1700.0	4671.0	11.87	10.05	0.	0	57822	57822	101.12	12013	0	0.500	2993.4	3.35	1.00	586.1
21	1800.0	4571.0	11.80	9.98	0.	0	57822	57822	99.71	11775	0	0.500	2922.2	3.32	1.00	616.6
22	1900.0	4471.0	11.73	9.91	0.	0	57822	57822	98.25	11529	0	0.500	2847.8	3.30	1.00	647.0
23	2000.0	4371.0	11.65	9.83	0.	0	57822	57822	96.73	11273	0	0.500	2770.4	3.29	1.00	677.1
24	2100.0	4271.0	11.57	9.75	0.	0	57822	57822	95.14	11009	0	0.500	2690.0	3.29	1.00	706.9
25	2200.0	4171.0	11.48	9.66	0.	0	57822	57822	93.48	10735	0	0.500	2606.8	3.29	1.00	736.4
26	2300.0	4071.0	11.39	9.57	0.	0	57822	57822	91.75	10451	0	0.500	2520.9	3.30	1.00	765.5
27	2400.0	3971.0	11.29	9.48	0.	0	57822	57822	89.95	10158	0	0.500	2432.4	3.32	1.00	794.2
28	2500.0	3871.0	11.19	9.38	0.	0	57822	57822	88.06	9855	0	0.500	2341.6	3.34	1.00	822.4
29	2600.0	3771.0	11.08	9.27	0.	0	57822	57822	86.10	9542	0	0.500	2248.4	3.36	1.00	850.2
30	2700.0	3671.0	10.97	9.16	0.	0	57822	57822	84.04	9220	0	0.500	2153.1	3.39	1.00	877.4

(cont.)

Table A.1 (*cont.*)

Level	Radius (km)	Depth (km)	Density (g/cm^3)	V_p (km/s)	V_s (km/s)	Q_μ	Q_K	Q_P	Φ (km^2/s^2)	K_S (kbar)	μ (kbar)	σ	Pressure (kbar)	dK/dP	B.P.	Gravity (cm/s^2)
31	2800.0	3571.0	10.85	9.05	0.	0	57822	57822	81.91	8889	0	0.500	2055.9	3.41	1.00	904.1
32	2900.0	3471.0	10.73	8.92	0.	0	57822	57822	79.68	8550	0	0.500	1956.9	3.44	1.00	930.2
33	3000.0	3371.0	10.60	8.79	0.	0	57822	57822	77.36	8202	0	0.500	1856.4	3.47	1.00	955.7
34	3100.0	3271.0	10.46	8.65	0.	0	57822	57822	74.96	7846	0	0.500	1754.4	3.49	1.00	980.5
35	3200.0	3171.0	10.32	8.51	0.	0	57822	57822	72.47	7484	0	0.500	1651.2	3.52	1.00	1004.6
36	3300.0	3071.0	10.18	8.36	0.	0	57822	57822	69.89	7116	0	0.500	1546.9	3.54	0.99	1028.0
37	3400.0	2971.0	10.02	8.19	0.	0	57822	57822	67.23	6743	0	0.500	1441.9	3.56	0.99	1050.6
38	3480.0	2891.0	9.90	8.06	0.	0	57822	57822	65.04	6441	0	0.500	1357.5	3.57	0.98	1068.2
39	3480.0	2891.0	5.56	13.71	7.26	312	57822	826	117.78	6556	2938	0.305	1357.5	1.64	0.99	1068.2
40	3500.0	2871.0	5.55	13.71	7.26	312	57822	826	117.64	6537	2933	0.304	1345.6	1.64	1.00	1065.3
41	3600.0	2771.0	5.50	13.68	7.26	312	57822	823	116.96	6440	2907	0.303	1287.0	1.64	1.01	1052.0
42	3630.0	2741.0	5.49	13.68	7.26	312	57822	822	116.76	6412	2899	0.303	1269.7	1.64	1.01	1048.4
43	3630.0	2741.0	5.49	13.68	7.26	312	57822	822	116.76	6412	2899	0.303	1269.7	3.33	1.01	1048.4
44	3700.0	2671.0	5.45	13.59	7.23	312	57822	819	115.08	6279	2855	0.302	1229.7	3.29	1.01	1040.6
45	3800.0	2571.0	5.40	13.47	7.18	312	57822	815	112.73	6095	2794	0.301	1173.4	3.24	1.01	1030.9
46	3900.0	2471.0	5.35	13.36	7.14	312	57822	811	110.46	5917	2734	0.299	1118.2	3.20	1.00	1022.7
47	4000.0	2371.0	5.30	13.24	7.09	312	57822	807	108.23	5744	2675	0.298	1063.8	3.17	1.00	1015.8
48	4100.0	2271.0	5.25	13.13	7.05	312	57822	803	106.04	5575	2617	0.297	1010.3	3.15	1.00	1010.0
49	4200.0	2171.0	5.20	13.01	7.01	312	57822	799	103.88	5409	2559	0.295	957.6	3.13	1.00	1005.3
50	4300.0	2071.0	5.15	12.90	6.96	312	57822	795	101.73	5246	2502	0.294	905.6	3.13	1.00	1001.5
51	4400.0	1971.0	5.10	12.78	6.91	312	57822	792	99.59	5085	2445	0.292	854.3	3.14	1.00	998.5
52	4500.0	1871.0	5.05	12.66	6.87	312	57822	788	97.43	4925	2388	0.291	803.6	3.16	1.00	996.3
53	4600.0	1771.0	5.00	12.54	6.82	312	57822	784	95.26	4766	2331	0.289	753.5	3.19	0.99	994.7
54	4700.0	1671.0	4.95	12.42	6.77	312	57822	779	93.06	4607	2273	0.288	704.1	3.23	0.99	993.6
55	4800.0	1571.0	4.89	12.29	6.72	312	57822	775	90.81	4448	2215	0.286	655.2	3.27	0.99	993.1
56	4900.0	1471.0	4.84	12.16	6.67	312	57822	770	88.52	4288	2157	0.284	606.8	3.32	0.99	993.0
57	5000.0	1371.0	4.78	12.02	6.61	312	57822	766	86.17	4128	2098	0.282	558.9	3.38	0.99	993.2
58	5100.0	1271.0	4.73	11.88	6.56	312	57822	761	83.76	3966	2039	0.280	511.6	3.45	0.99	993.8
59	5200.0	1171.0	4.67	11.73	6.50	312	57822	755	81.28	3803	1979	0.278	464.8	3.52	0.99	994.6
60	5300.0	1071.0	4.62	11.57	6.44	312	57822	750	78.72	3638	1918	0.275	418.6	3.59	0.99	995.7

(cont.)

Table A.1 *(cont.)*

Level	Radius (km)	Depth (km)	Density (g/cm^3)	V_p (km/s)	V_s (km/s)	Q_μ	Q_K	Q_P	Φ (km^2/s^2)	K_S (kbar)	μ (kbar)	σ	Pressure (kbar)	dK/dP	B.P.	Gravity (cm/s^2)
61	5400.0	971.0	4.56	11.41	6.37	312	57822	743	76.08	3471	1856	0.273	372.8	3.67	0.98	996.9
62	5500.0	871.0	4.50	11.24	6.31	312	57822	737	73.34	3303	1794	0.270	327.6	3.75	0.98	998.3
63	5600.0	771.0	4.44	11.06	6.24	312	57822	730	70.52	3133	1730	0.266	282.9	3.84	0.97	999.8
64	5600.0	771.0	4.44	11.06	6.24	312	57822	730	70.52	3133	1730	0.266	282.9	2.98	0.97	999.8
65	5650.0	721.0	4.41	10.91	6.09	312	57822	744	69.51	3067	1639	0.273	260.7	3.00	0.97	1000.6
66	5701.0	670.0	4.38	10.75	5.94	312	57822	759	68.47	2999	1548	0.279	238.3	3.03	0.98	1001.4
67	5701.0	670.0	3.99	10.26	5.57	143	57822	362	64.03	2556	1239	0.291	238.3	2.40	0.37	1001.4
68	5736.0	635.0	3.98	10.21	5.54	143	57822	362	63.32	2523	1224	0.291	224.3	2.38	0.37	1000.8
69	5771.0	600.0	3.97	10.15	5.51	143	57822	362	62.61	2489	1210	0.290	210.4	2.37	0.37	1000.3
70	5771.0	600.0	3.97	10.15	5.51	143	57822	362	62.61	2489	1210	0.290	210.4	8.09	1.98	1000.3
71	5821.0	550.0	3.91	9.90	5.37	143	57822	363	59.60	2332	1128	0.291	190.7	7.88	1.92	999.6
72	5871.0	500.0	3.84	9.64	5.22	143	57822	364	56.65	2181	1051	0.292	171.3	7.67	1.86	998.8
73	5921.0	450.0	3.78	9.38	5.07	143	57822	365	53.78	2037	977	0.293	152.2	7.46	1.79	997.9
74	5971.0	400.0	3.72	9.13	4.93	143	57822	366	50.99	1899	906	0.294	133.5	7.26	1.73	996.8
75	5971.0	400.0	3.54	8.90	4.76	143	57822	372	48.97	1735	806	0.298	133.5	3.37	0.83	996.8
76	6016.0	355.0	3.51	8.81	4.73	143	57822	370	47.83	1682	790	0.297	117.7	3.33	0.82	995.2
77	6061.0	310.0	3.48	8.73	4.70	143	57822	367	46.71	1630	773	0.295	102.0	3.30	0.80	993.6
78	6106.0	265.0	3.46	8.64	4.67	143	57822	365	45.60	1579	757	0.293	86.4	3.26	0.79	992.0
79	6151.0	220.0	3.43	8.55	4.64	143	57822	362	44.50	1529	741	0.291	71.1	3.23	0.78	990.4
80	6151.0	220.0	3.35	7.98	4.41	80	57822	195	37.80	1270	656	0.279	71.1	−0.73	−0.12	990.4
81	6186.0	185.0	3.36	8.01	4.43	80	57822	195	38.01	1278	660	0.279	59.4	−0.72	−0.12	989.1
82	6221.0	150.0	3.36	8.03	4.44	80	57822	195	38.21	1287	665	0.279	47.8	−0.70	−0.12	987.8
83	6256.0	115.0	3.37	8.05	4.45	80	57822	195	38.41	1295	669	0.279	36.1	−0.68	−0.13	986.6
84	6291.0	80.0	3.37	8.07	4.46	80	57822	195	38.60	1303	674	0.279	24.5	−0.67	−0.13	985.5
85	6291.0	80.0	3.37	8.07	4.46	600	57822	1447	38.60	1303	674	0.279	24.5	−0.67	−0.13	985.5
86	6311.0	60.0	3.37	8.08	4.47	600	57822	1447	38.71	1307	677	0.279	17.8	−0.66	−0.13	984.9
87	6331.0	40.0	3.37	8.10	4.48	600	57822	1446	38.81	1311	680	0.279	11.2	−0.65	−0.13	984.3
88	6346.6	24.4	3.38	8.11	4.49	600	57822	1446	38.89	1315	682	0.279	6.0	−0.64	−0.13	983.9
89	6346.6	24.4	2.90	6.80	3.90	600	57822	1350	25.96	753	441	0.254	6.0	−0.00	−0.00	983.9
90	6356.6	15.0	2.90	6.80	3.90	600	57822	1350	25.96	753	441	0.254	3.3	0.00	0.00	983.3
91	6356.0	15.0	2.60	5.80	3.20	600	57822	1456	19.99	520	266	0.281	3.3	0.00	0.00	983.3
92	6368.0	3.0	2.60	5.80	3.20	600	57822	1456	19.99	520	266	0.281	0.3	−0.00	−0.00	982.2
93	6368.0	3.0	1.02	1.45	0.	0	57822	57822	2.10	21	0	0.500	0.2	−0.00	−0.00	982.2
94	6371.0	0.	1.02	1.45	0.	0	57822	57822	2.10	21	0	0.500	0.0	0.00	0.00	981.5

Dziewonski, A. M. and D. L. Anderson (1981) Preliminary reference Earth model, *Phys. Earth Planet. Inter.*, *25*, 297–356

Table A.2 Crust and upper mantle of PREM including directional velocities, anisotropic elastic constants and 'equivalent' isotropic velocities. Evaluated at reference periods of 1 s (top) and 200 s (bottom)

Radius (km)	Depth (km)	Density (g/cm^3)	V_{PV} (km/s)	V_{PH} (km/s)	V_{SV} (km/s)	V_{SH}	η	Q_μ	Q_K	A (kbar)	C (kbar)	L (kbar)	N (kbar)	F (kbar)	V_p (km/s)	V_s (km/s)
6151.0	220.0	3.359 50	7.800 50	8.048 62	4.441 10	4.436 29	0.976 54	80	578 22	2176	2044	663	661	831	7.989 70	4.418 85
6171.0	200.0	3.361 67	7.823 15	8.063 10	4.436 49	4.454 23	0.968 77	80	578 22	2186	2057	662	667	835	8.002 35	4.425 80
6191.0	180.0	3.363 84	7.845 81	8.077 60	4.431 89	4.472 18	0.960 99	80	578 22	2195	2071	661	673	839	8.014 94	4.432 85
6211.0	160.0	3.366 02	7.868 47	8.092 09	4.427 28	4.490 13	0.953 21	80	578 22	2204	2084	660	679	843	8.027 47	4.440 00
6231.0	140.0	3.368 19	7.891 13	8.106 59	4.422 67	4.508 07	0.945 43	80	578 22	2213	2097	659	685	847	8.039 92	4.447 24
6251.0	120.0	3.370 36	7.913 78	8.121 08	4.418 06	4.526 02	0.937 65	80	578 22	2223	2111	658	690	851	8.052 31	4.454 58
6271.0	100.0	3.372 54	7.936 44	8.135 58	4.413 45	4.543 97	0.929 87	80	578 22	2232	2124	657	696	854	8.064 63	4.462 01
6291.0	80.0	3.374 71	7.959 09	8.150 06	4.408 85	4.561 91	0.922 10	80	578 22	2242	2138	656	702	857	8.076 88	4.469 53
6291.0	80.0	3.374 71	7.959 11	8.150 08	4.408 84	4.561 93	0.922 09	600	578 22	2242	2138	656	702	857	8.076 89	4.469 54
6311.0	60.0	3.376 88	7.981 76	8.164 57	4.404 24	4.579 87	0.914 32	600	578 22	2251	2151	655	708	860	8.089 07	4.477 15
6331.0	40.0	3.379 06	8.004 42	8.179 06	4.399 63	4.597 82	0.906 54	600	578 22	2260	2165	654	714	863	8.101 19	4.484 86
6346.6	24.4	3.380 76	8.022 12	8.190 38	4.396 03	4.611 84	0.900 47	600	578 22	2268	2176	653	719	866	8.110 61	4.490 94
6346.6	24.4	2.900 00	6.800 00	6.800 00	3.900 00	3.900 00	1.000 00	600	578 22	1341	1341	441	441	459	6.800 00	3.900 00
6356.0	15.0	2.900 00	6.800 00	6.800 00	3.900 00	3.900 00	1.000 00	600	578 22	1341	1341	441	441	459	6.800 00	3.900 00
6356.0	15.0	2.600 00	5.800 00	5.800 00	3.200 00	3.200 00	1.000 00	600	578 22	875	875	266	266	342	5.800 00	3.200 00
6368.0	3.0	2.600 00	5.800 00	5.800 00	3.200 00	3.200 00	1.000 00	600	578 22	875	875	266	266	342	5.800 00	3.200 00
6368.0	3.0	1.020 00	1.450 00	1.450 00	0.	0.	1.000 00	0	578 22	21	21	0	0	21	1.450 00	0.
6371.0	0.	1.020 00	1.450 00	1.450 00	0.	0.	1.000 00	0	578 22	21	21	0	0	21	1.450 00	0.

Radius (km)	Depth (km)	Density (g/cm^3)	V_{PV} (km/s)	V_{PH} (km/s)	V_{SV} (km/s)	V_{SH}	η	Q_μ	Q_K	A (kbar)	C (kbar)	L (kbar)	N (kbar)	F (kbar)	V_p (km/s)	V_s (km/s)
6151.0	220.0	3.359 50	7.729 30	7.979 75	4.347 48	4.342 77	0.976 54	80	578 22	2139	2007	635	634	849	7.920 08	4.324 95
6171.0	200.0	3.361 67	7.752 30	7.993 80	4.342 97	4.360 33	0.968 77	80	578 22	2148	2020	634	639	853	7.932 36	4.332 12
6191.0	180.0	3.363 84	7.775 31	8.007 86	4.338 46	4.377 90	0.960 99	80	578 22	2157	2034	633	645	856	7.944 57	4.339 34
6211.0	160.0	3.366 02	7.798 31	8.021 92	4.333 95	4.395 47	0.953 21	80	578 22	2166	2047	632	650	859	7.956 73	4.346 65
6231.0	140.0	3.368 19	7.821 32	8.035 98	4.329 43	4.413 04	0.945 43	80	578 22	2175	2060	631	656	863	7.968 82	4.354 06
6251.0	120.0	3.370 36	7.844 32	8.050 04	4.324 92	4.430 61	0.937 65	80	578 22	2184	2074	630	662	866	7.980 84	4.361 55
6271.0	100.0	3.372 54	7.867 32	8.064 10	4.320 41	4.448 18	0.929 87	80	578 22	2193	2087	630	667	869	7.992 79	4.369 14
6291.0	80.0	3.374 71	7.890 31	8.078 15	4.315 90	4.465 74	0.922 10	80	578 22	2202	2101	629	673	871	8.004 68	4.376 82
6291.0	80.0	3.374 71	7.949 82	8.140 37	4.396 45	4.549 11	0.922 09	600	578 22	2236	2133	652	698	859	8.067 15	4.457 17
6311.0	60.0	3.376 88	7.972 51	8.154 80	4.391 86	4.567 00	0.914 32	600	578 22	2246	2146	651	704	862	8.079 28	4.464 80
6331.0	40.0	3.379 06	7.995 22	8.169 24	4.387 26	4.584 90	0.906 54	600	578 22	2255	2160	650	710	865	8.091 35	4.472 53
6346.6	24.4	3.380 76	8.012 95	8.180 51	4.383 68	4.598 87	0.900 47	600	578 22	2262	2171	650	715	867	8.100 74	4.478 63
6346.6	24.4	2.900 00	6.791 51	6.791 51	3.889 04	3.889 04	1.000 00	600	578 22	1338	1338	439	439	460	6.791 51	3.889 04
6356.0	15.0	2.900 00	6.791 51	6.791 51	3.889 04	3.889 04	1.000 00	600	578 22	1338	1338	439	439	460	6.791 51	3.889 04
6356.0	15.0	2.600 00	5.793 28	5.793 28	3.191 01	3.191 01	1.000 00	600	578 22	873	873	265	265	343	5.793 28	3.191 01
6368.0	3.0	2.600 00	5.793 28	5.793 28	3.191 01	3.191 01	1.000 00	600	578 22	873	873	265	265	343	5.793 28	3.191 01
6368.0	3.0	1.020 00	1.449 96	1.449 96	0.	0.	1.000 00	0	578 22	21	21	0	0	21	1.449 96	0.
6371.0	0.	1.020 00	1.449 96	1.449 96	0.	0.	1.000 00	0	578 22	21	21	0	0	21	1.449 96	0.

Dziewonski, A. M. and D. L. Anderson (1981) Preliminary references Earth model, *Phys. Earth Planet. Inter.*, *25*, 297–356

Table A.3 | Conversion Factors

To Convert	To	Multiply by
angstrom, Å	cm	10^{-8}
	nm	10
bar	atm	0.987
	dyne cm^{-2}	10^{-6}
	MPa	10^{-1}
calorie (g), cal	joule	4.184
dyne	g cm s^{-2}	1
	newton	10^{-5}
erg	cal	2.39×10^{-8}
	watt-second	1
	joule	10^{-7}
gamma	gauss	10^{-5}
	tesla	10^{-9}
gauss	tesla	10^{-4}
heat-flow unit (H.F.U.)	μcal/cm^2/s	1
	m W / m^2	41.84
micrometer, μm	cm	10^{-4}
poise	g cm^{-1} s^{-1}	1
	Pa-s	0.1
stoke	cm^2/s	1
watt	J/s	1
year	s	3.156×10^7

Table A.4 | Physical Constants

Speed of light	c	2.998×10^8 m/s
Electronic charge	e	-1.602×10^{19} C
Permeability of vacuum	μ_0	$4\pi \times 10^{-7}$ N/A^2
Permittivity of vacuum	ϵ_0	8.854×10^{-12} F/m
Planck constant	h	6.626×10^{-34} J
Boltzmann constant	k	1.381×10^{-23} J/K
Stefan–Boltzmann constant	σ	5.67×10^{-8} W/m^2/K^4
Gravitational constant	G	6.673×10^{-11} m^3/kg/s^2
Electron rest mass	m_e	0.911×10^{-10} kg
Avogadro's number	N_A	6.022×10^{23}/mol
Gas constant	R	8.314 J/mol/K

Table A.5 | Earth Parameters

Equatorial radius, a	6378.137 km
Polar radius, c	6356.752 km
Equivolume sphere radius	6371.000 km
Surface area	5.1×10^8 km^2
Geometric flattening, $(a - c)/a$	1298.257
Ellipticity, $(a^2 - c^2)/(a^2 + c^2)$	1/297.75
GM	3.986005×10^{14} m^3 s^{-2}
Mass, M	5.97369×10^{24} kg
Mean density	5.5148×10^3 kg m^{-3}
Moments of inertia	
Polar, C	8.0378×10^{37} kg m^2
Equatorial, A	8.0115×10^{37} kg m^2
C/Ma^2	0.33068
Dynamic ellipticity (precession constant), $H = \frac{C - A}{C} = 1/305.51$	
Ellipticity coefficient, $J_2 = \frac{C - A}{Ma^2}$	10826.3×10^{-7}
Angular velocity, Ω	7.2921×10^{-5} rad s^{-1}
Angular momentum, $C\Omega$	5.8604×10^{33} kg m^2 s^{-1}
Normal gravity at equator	9.7803267 m s^{-2}
Normal gravity at poles	9.832186 m s^{-2}

Index